FRAGMENTATION OF MOLECULAR CLOUDS
AND STAR FORMATION

INTERNATIONAL ASTRONOMICAL UNION

UNION ASTRONOMIQUE INTERNATIONALE

FRAGMENTATION OF MOLECULAR CLOUDS AND STAR FORMATION

PROCEEDINGS OF THE 147TH SYMPOSIUM OF THE
INTERNATIONAL ASTRONOMICAL UNION,
HELD IN GRENOBLE, FRANCE, JUNE 12–16, 1990

EDITED BY

E. FALGARONE
Laboratoire de Radioastronomie Millimétrique,
Ecole Normale Supérieure, Paris, France

F. BOULANGER
Laboratoire de Radioastronomie Millimétrique,
Ecole Normale Supérieure, Paris, France

and

G. DUVERT
Observatoire de Grenoble, France

SPRINGER-SCIENCE+BUSINESS MEDIA, B.V.

Library of Congress Cataloging-in-Publication Data

International Astronomical Union. Symposium (147th : 1990 : Grenoble,
France)
    Fragmentation of molecular clouds and star formation : proceedings
of the 147th Symposium of the International Astronomical Union, held
in Grenoble, France, June 12-16, 1990 / edited by E. Falgarone, F.
Boulanger, and G. Duvert.
        p.   cm.
    Includes index.
    ISBN 978-0-7923-1159-1   ISBN 978-94-011-3384-5 (eBook)
    DOI 10.1007/978-94-011-3384-5
    1. Molecular clouds--Congresses.   2. Stars--Formation--Congresses.
I. Falgarone, E. (Edith)   II. Boulanger, F.   III. Duvert, G.
IV. Title.
QB791.4.I57   1990
523.8--dc20   91-7265
                                                      CIP
ISBN 978-0-7923-1159-1

Contents

Poster Sessions

Preface

A few years ago, a motivation for organizing one more IAU Symposium on star formation in Grenoble, was the anticipated completion of the IRAM interferometer on the Plateau de Bures, close to Grenoble. This choice was also a sort of late celebration of the genius of Joseph Fourier, born in Grenoble, whose work is the very fondation of interferometry. At the time when we finally announced the advent of this conference, the first reactions we got from the community were expressions of saturation and even reject, the Symposium being unfortunately scheduled almost simultaneously as two other major meetings on closely related topics, and sponsored by different organizations. A wave of disappointment then reached the organizers. Some of us were enthusiastic enough to help the others overcome their discouragement. Let them be thanked here.

There was, indeed, a deeper motivation for organizing this conference. It was to trigger the meeting and communication of physicists and astrophysicists since many of the difficulties met now in understanding the physics of the interstellar medium and its evolution toward star formation are common to several, if not most, other fields of physics. They are assigned to one origin: complexity. In particular, and at the opposite of what was thought one decade ago, the process of star formation is no longer understood as a local process but depends on more global parameters in the sense that the non linearity of the mechanisms which regulate the physics of the interstellar medium makes all the scales coupled. Another major difficulty, also common to other fields, is the fact that the systems under study are far out of equilibrium. At last, the observed quasi equipartition of the various forms of energy (thermal, radiative, magnetic, turbulent, cosmic rays...) is an additional source of complexity. Several physicists were invited, they expressed their interest in the meeting and agreed to give talks. In the end, only few of them actually came, perhaps because of the foreseen difficulty in communicating, perhaps also because the complexity of the interstellar medium is closer to that of biological systems than that met in pure physics and the approach we follow in astrophysics is still too phenomenological.

In spite of that, we believe that this conference helped most of us, getting some grounds in fields of physics we were not familiar with. We heard: "In all the conferences where I have been in the past, each time I could not understand what was being said, I believed that I was the only one to be in that state. At this conference, it was clear at once that everyone was like me" (Don Cox). We also heard from an young hydrodynamicist: "A length for me has always been 2π. I am glad to have discovered what a real length is." And if there was as always in that sort of meeting a lot of healthy controversy, there may have been, according to many, much more modesty in everyone's mind than usual. We were all ready to accept the failure of some of the models we had been working on for years and were discovering, paper after paper, the unavoidable coexistence of the remarkable uniformity and order of the interstellar medium when looked at on large scale which singularly contrasts with the ever increasing (and somewhat disheartening) level of diversity and complexity that small scale observations steadily reveal to us.

We made the choice to record in this volume all the contributions to the Symposium. The only exception is what was said in the panel discussion, chaired by Alex Dalgarno and devoted to "Observational tests of fragmentation processes" because most of the ideas and arguments which emerged during the discussion are indeed included in the contributions. The review and invited talks are gathered in the first half, together with the few oral contributions. We found that the large body of information presented in the poster sessions deserved publication and we present them in the second half of the book.

Edith Falgarone

Scientific Organizing Committee

A.P. Boss (USA), E. Falgarone (chairperson, France), M. Fujimoto (Japan), R. Genzel (Germany), M. Guélin (France), A. Hjalmarson (Sweden), G. Morfill (Germany), T. Nakano (Japan), J. Scalo (USA), P. Thaddeus (USA).

Local Organizing Committee

F. Boulanger, A. Castets, D. Downes, G. Duvert, P. de Jonge, C. Kahane, A. Omont (chairperson).

Acknowledgements

This Symposium would not have been possible without the generous financial support of a myriad of institutions which allowed us to invite speakers and participants from a large number of countries: the International Astronomical Union, the National Center for Scientific Research (CNRS), the European Space Agency (ESA), the National Center for Space Studies (CNES), the Centers for Atomic Energy (CEA) of Saclay and Bruyères-le-Châtel, the Ministery of National Education (MEN), the Institut for Millimeter Radio Astronomy (IRAM), the Joseph Fourier University in Grenoble which hosted the conference, the city of Grenoble, and the General Council of Isère. We also acknowledge the support of two companies, Sofradir and Air Liquide.

But it is the hard work of a handful of efficient and enthusiastic persons who definitely ensured the success of the conference. They were Carrah Wright and Mike Rogers of the California Institute of Technology who helped me during the last year of my stay there. Later, Elizabeth Palleau of the Grenoble Observatory took the entire responsability of the detailed organization of the conference in Grenoble. I am particularly grateful to her because of her initiative, her energy, her skills, and also her good mood. She solved the innumerable problems raised day after day, allowing me, in the various places where I was in 1989 and 1990, to concentrate on the scientific organization. Françoise Bouillet of the Grenoble Observatory, Gisela Matoso and Claude Vianey-Liaud of IRAM shared this task with her and did it in such a way that, in the end, most participants were feeling welcomed. I sincerely thank all of them.

I also acknowledge the indispensable support of a few of my colleagues who never doubted of the interest of such a conference and kept encouraging my efforts to make this meeting shed a different light on the physics of the interstellar medium.

Edith Falgarone.

List of Participants

Dr. P. Abrahám
Konkoly Observatory
P.O. Box 67
Budapest H-1525
HUNGARY

Dr. A. Baudry
Observatoire de Bordeaux
33270 Floirac
FRANCE

Dr. F. Bertoldi
Princeton University Observatory
Peyton Hall
Princeton, N.J. 08544-1001
USA

Dr. P. Boissé
Ecole Normale Supérieure
24 rue Lhomond
75231 Paris Cedex 05
FRANCE

Dr. L. Bronfman
Departamento de Astronomia
Universitad de Chile, Casilla 36-D
Santiago
CHILE

Dr. A. Castets
Observatoire de Grenoble
CERMO BP53X
38041 Grenoble Cedex
FRANCE

Dr. J.-P. Chièze
CEA-BIII Service PTN
BP12
91680 Bruyères-le-Châtel
FRANCE

Dr. N. Anderson
University of Illinois
Department of Physics
Urbana, IL 61801
USA

Dr. J.J. Benayoun
Observatoire de Grenoble
CERMO BP 53X
38041 Grenoble Cedex
FRANCE

Dr. C. Bertout
Observatoire de Grenoble
CERMO BP 53X
38041 Grenoble Cedex
FRANCE

Dr. W. Boland
NFRA P.O. Box 2
7990 AA Dwingeloo
THE NETHERLANDS

Dr. V.V. Burdyuzha
Lebedev's Physical Institute,
Academy of Sciences of USSR
Profsoyuznaya, 84/32
117810 Moscow
USSR

Dr. E. Caux
CESR
9 Av. du Colonel Roche
B.P. 4346
31029 Toulouse Cedex
FRANCE

Dr. F.O. Clark
AFGL/OPC
Hanscom AFB, MA 01731
USA

xviii

Dr. R.M. Crutcher
Department of Astronomy
University of Illinois
1011 W. Springfield Avenue
Urbana, IL 61801
USA

Dr. D. Despois
Observatoire de Bordeaux
BP 89
33270 Floirac
FRANCE

Dr. B.T. Draine
Peyton Hall
Princeton University
Princeton, NJ 08544
USA

Dr. G. Duvert
Observatoire de Grenoble
CERMO BP 53X
38041 Grenoble Cedex
FRANCE

Dr. E. Falgarone
Ecole Normale Supérieure
24 rue Lhomond
75231 Paris Cedex 05
FRANCE

Dr. R.C. Fleck
Department Math. Phys. Sci.
Embry-Riddle Aeronautical University
Daytona Beach, FL 32114
USA

Dr. M. Fujimoto
Department of Physics
Nagoya University
Nagoya 464-01
JAPAN

Dr. C. de Boisanger
CEA-BIII Service PTN
BP 12
91680 Bruyères-Le-Châtel
FRANCE

Dr. A. Di Fazio
Osservatorio Astronomico di Roma
Viale del Parco Mellini 84
00136 Roma,
ITALY

Dr. A. Dudorov
Astronomy Department
Moscow State University
Universitetskij prospect, 13
119899 Moscow V-234
USSR

Dr. B.G. Elmegreen
IBM T.J. Watson Research Ctr.
P.O. Box 218
Yorktown Heights, NY 10598
USA

Dr. C. Feldt
Hamburger Sternwarte
Gojenbergsweg 112
2050 Hamburg 80
GERMANY

Dr. T. Forveille
Observatoire de Grenoble
CERMO BP 53X
38041 Grenoble Cedex
FRANCE

Dr. Y. Fukui
Department of Physics
Nagoya University
Furocho Chikusaku
Nagoya 464
JAPAN

Dr. J. Garcia-Barreto
IRAM
300 rue de la Piscine
38406 St. Martin d'Hères Cedex
FRANCE

Dr. S. Garcia-Burillo
IRAM
300 rue de la Piscine
38406 St. Martin d'Hères Cedex
FRANCE

Dr. K.M. Gierens
I. Physikalisches Institut
Universität zu Köln
Zülpicher Str. 77
5000 Köln
GERMANY

Dr. D. Gillet
Observatoire de Haute Provence
04870 Saint Michel l'Observatoire
FRANCE

Dr. A.E. Glassgold
New York University
Department of Physics
2 Washington Pl.
New York, NY 10003
USA

Dr. P. Goldsmith
Department of Physics and Astronomy
University of Massachusetts
619 Lederle Graduate Research Center
Amherst, MA 01002
USA

Dr. A.A. Goodman
Astronomy Department
University of California
Berkeley CA 94720
USA

Dr. U.U. Graf
Max Planck Institut
für extraterrestrische Physik
8046 Garching bei München
GERMANY

Dr. M. Grewing
IRAM
300 rue de la Piscine
38406 St. Martin d'Hères Cedex
FRANCE

Dr. M. Guélin
IRAM
300 rue de la Piscine
38406 St. Martin d'Hères Cedex
FRANCE

Dr. R. Güsten
Max Planck Institut für Radioastronomie
Auf dem Hügel 69
5300 Bonn 1
GERMANY

Dr. J. Harju
Observatory and Astrophysics Laboratory
University of Helsinki
Tähtitorninmäki
00130 Helsinki 13,
FINLAND

Dr. R.N. Henriksen
Department of Physics
Queen's University
Kingston, Ontario
CANADA K7L 3N6

Dr. A. Hjalmarson
Onsala Space Observatory
43900 Onsala
SWEDEN

Dr. K.W. Hodapp
Institute for Astronomy
2680 Woodlawn Drive,
Honolulu, HI 96822
USA

Dr. A.I. Issa
Academy of Scientific Research
and Technology
National Research Institute
Astronomy and Geophysics
Helwan, Cairo
EGYPT

Dr. C. Kahane
Observatoire de Grenoble
CERMO BP 53X
38041 Grenoble Cedex
FRANCE

Dr. E.R. Keto
Institute of Geophysics and Planetary Sciences
LLNL
POB 808-L413
Livermore CA 94550
USA

Dr. C. Koempe
IRAM
Avd. Divina Pastora 7
Núcleo Central
Granada 18012
SPAIN

Dr. W. Kundt
Astronomisches Institut der
Universität Bonn
Auf dem Hügel 71
5300 Bonn 1
GERMANY

Dr. R.B. Larson
Yale Astronomy Dept
Box 6666
New Haven, CT 06511
USA

Dr. S. Ishizuki
Nobeyama Radio Observatory
Minamimaki, Minamisaku,
Nagano 384-13
JAPAN

Dr. T. Jacq
Observatoire de Bordeaux
33270 Floirac
FRANCE

Dr. R. Kawabe
Nobeyama Radio Observatory
Minamimaki, Minamisaku,
Nagano 384-13
JAPAN

Dr. V. Khersonskij
Leningrad Department of
Special Astrophysical Observatory
196140 Leningrad Pulkovo
USSR

Dr. I. Kolesnik
Main Astronomical Observatory
Ukrainian Academy of Sciences
252127 Kiev
USSR

Dr. E. Lada
Center for Astrophysics
60 Garden Street
Cambridge, MA 02138
USA

Dr. J. Lattanzio
L-413, LLNL, Box 808
Livermore, CA 94550
USA

Dr. P. Léna
Observatoire de Meudon
92195 Meudon Principal Cedex
FRANCE

Dr. T. Liljeström
Helsinki University Observatory
Tähtitorninmäki
00130 Helsinki
FINLAND

Dr. C. Loup
Observatoire de Grenoble
CERMO BP 53X
38041 Grenoble Cedex
FRANCE

Dr. A.W. Masheder
Department of Physics
University of Bristol
Tyndall Avenue
Bristol BS8 1TL
ENGLAND

Dr. V. Migenes
University of Manchester
Nuffield Radio Astronomy Labs
Jodrell Bank, Macclesfield
Cheshire SK11 9DL
ENGLAND

Dr. J.L. Monin
Observatoire de Grenoble
CERMO BP 53X
38041 Grenoble Cedex
FRANCE

Dr. P. Myers
Center for Astrophysics
60 Garden Street
Cambridge, MA 02138
USA

Dr. J. Lepine
Universidade de Sao Paulo
Instituto Astronomico e Geofisico
C.P. 30.627
Sao Paulo 01051 SP,
BRAZIL

Dr. A. Lioure
CEA Centre BIII
Service PTN
BP 12
91680 Bruyères le Châtel
FRANCE

Dr. R. Lucas
IRAM
300 rue de la Piscine
38406 St. Martin d'Hères Cedex
FRANCE

Dr. R. Mauersberger
Max-Planck Institut für
Radioastronomie
Auf dem Hügel 69
5300 Bonn 1
GERMANY

Dr. A. Mizuno
Department of Astrophysics
Nagoya 464
Nagoya University
JAPAN

Dr. S. Moy
University of Kent
Electronics Laboratory
Canterbury
ENGLAND

Dr. T. Nakano
Nobeyama Radio Observatory
Nobeyama, Minamisaku,
Nagano 384-13
JAPAN

Dr. N. Ohashi
Nobeyama Radio Observatory
Nobeyama, Minamisaku,
Nagano 384-13
JAPAN

Dr. S. Okumura
University of Tokyo
Department of Earth Science & Astron(
Komaba Meguro Ku
153 Tokyo
JAPAN

Dr. L. Pagani
DEMIRM
Observatoire de Meudon
92195 Meudon Principal Cedex
FRANCE

Dr. F. Palla
Osservatorio Astrofisico de Arcetri
Universita di Firenze
50125 Firenze,
ITALY

Dr. N.D. Parker
Department of Physics
Queen Mary College
Mile End Road
London E1 4NS
ENGLAND

Dr. M. Pashchenko
Sternberg State Astronomical Institute
Universitatskij Prospect 13
119899 Moscow
USSR

Dr. S. Pinnock
Physics Department
Queen Mary College
Mile End Road
London E1 4NS
ENGLAND

Dr. H. Pongracic
Department of Physics
University of Wales
P.O. Box 913
Cardiff CF1 3TH
WALES

Dr. S. Prasad
University of California
Physics and Astronomy, SHS 274
University Park MC1341
Los Angeles CA 90089-1341
USA

Dr. T. Prusti
Laboratory for Space Research
P.O. Box 800
9700 AV Groningen
THE NETHERLANDS

Dr. J.L. Puget
Institut d'Astronomie Spatiale
Université Paris-Sud
Bâtiment 120
91406 Orsay Cedex
FRANCE

Dr. V. Radhakrishnan
Raman Research Institute
Iiebal, PO Box 6
Bangalore 560 080
INDIA

Dr. E. Roueff
Observatoire de Meudon
92195 Meudon Principal Cedex
FRANCE

Dr. G.M. Rudnitskij
Radio Astronomy Department
Sternberg Astronomical Institute
119899 Moscow V-234
USSR

Dr. P. Schilke
Max-Planck-Institut
für Radioastronomie
Auf dem Hügel 69
5300 Bonn 1
GERMANY

Dr. E. Serabyn
Caltech 320-47
Pasadena, CA 91125
USA

Dr. M. Smith
Department of Physics
University of Durham
South Road
Durham DH1 3LE
ENGLAND

Dr. R. Stark
Sterrewacht Leiden
P.O. Box 9513
2300 RA Leiden,
THE NETHERLANDS

Dr. K. Sugitani
I. Physikalisches Institut
Universität zu Köln
Zülpicher Str. 77
5000 Köln
GERMANY

Dr. J.A. Tauber
Radio Astronomy Lab
601 Campbell Hall
University of California
Berkeley CA 94720
USA

Dr. J. Torrelles
Instituto de Astrofisica
de Andalucia
Apdo Correos 2144
18000 Granada
SPAIN

Dr. J. Schmid-Burgk
Max-Planck-Institut für
Radioastronomie
Auf dem Hügel 69
5300 Bonn 1
GERMANY

Dr. C. Shukre
Raman Research Institute
Sadashivnagar
Bangalore 560 080
INDIA

Dr. Y. Sofue
Institute of Astronomy
The University of Tokyo
Mitaka, Tokyo 181
JAPAN

Dr. L. Stenholm
Stockholm Observatory
13300 Saltsjobaden
SWEDEN

Dr. K. Sunada
Nobeyama Radio Observatory
Nobeyama, Minamisaku,
Nagano 384-13
JAPAN

Dr. S. Terebey
IPAC
Caltech 100-22
Pasadena CA 91125
USA

Dr. T. Tosaki
Tohoku University
Aoba, Aramaki, Aoba
Sendai 980
JAPAN

Dr. T.H. Troland,
Phys. Astron. Department
University of Kentucky,
Lexington, KY 40506
USA

Dr. E. Vazquez
University of Texas
Department of Astronomy
Austin TX 78712
USA

Dr. C.M. Walmsley
Max Planck Institut
für Radioastronomie
Auf dem Hügel 69
5300 Bonn 1
GERMANY

Dr. A. Wootten
NRAO
Edgemont Road
Charlottesville VA 22903-2475
USA

Dr. I. Val'tts
Lebedev's Physical Institute,
Academy of Sciences of USSR
Profsoyuznaya, 84/32
117810 Moscow
USSR

Dr. L. Verstraete
Max Planck Institut
für extraterrestrische Physik
8046 Garching bei München
GERMANY

Dr. H. Weidkard
Observatoire de Grenoble
CERMO BP 53X
38041 Grenoble Cedex
FRANCE

Dr. D. Ward-Thompson
Lancashire Polytechnic
Fylde Rd.
Preston PR1 2TQ
ENGLAND

I - LARGE SCALE STRUCTURE

CHARACTERISTICS OF THE DIFFUSE INTERSTELLAR MEDIUM

D. P. COX
Department of Physics
University of Wisconsin-Madison
1150 University Ave.
Madison, Wisconsin 53706
U.S.A.

ABSTRACT. There have been several recent changes in perspective on the diffuse interstellar environment, including recognition of a thick disk of warm gas, cosmic rays, and magnetic field. In addition, evidence for a pervasive hot phase driven by supernova disruption has weakened to the point that a quasihomogeneous warm intercloud gas may occupy most of the interstellar volume at midplane, with individual bubbles created by supernovae and OB associations occupying perhaps 10 and 20 per cent respectively. The bubble population is sufficient to explain the high stage ions (O VI, N V, C IV, perhaps Si IV) found in the disk, though possibly not those found at higher z. The estimated midplane pressure has increased, leaving the thermal pressure inside clouds almost negligible. The reduced porosity of the medium, its greater thickness, and its larger pressure all act to suppress fountain activity, either arising from the disk generally, or from the blowout of superbubbles. Finally, there appears to be a peculiar coincidence between the cloud heating mechanism and the activity determining the interstellar pressure.

1. INTRODUCTION

I would like to draw the attention of students of the denser parts of the interstellar medium (ISM) to changes that have taken place in the picture of the surrounding medium. I expect that those changes will affect how one describes the formation of molecular clouds, as well as the boundary conditions relevant to their stability.

Perhaps the most significant results in this context are: that the total midplane pressure corresponds roughly to p/k $\approx$ 25,000 cm^{-3} K, with roughly 1/3 each in cosmic ray, magnetic, and kinetic forms; and that the most probable density in the ISM could well be that in warm gas (WIM or WNM), about 0.2 cm^{-3}.

2. THE THICK DISK

Evidence for a thick disk extending to $|z| \sim 2$ kpc is found in radio synchrotron studies, dispersion and rotation measures to pulsars (and rotation measures of extragalactic objects), measures of cosmic ray pathlength and trapping timescale, observations of HI and Ti II which traces HI, observations of H$^+$ and Al III which traces H$^+$ (c.f., Boulares and Cox, 1990). The weight of the large amount of low density HI and H$^+$ high off the plane is substantial, causing the increase in the estimated midplane pressure to the value quoted above, even in the presence of lowered estimates for the gravitational acceleration.

This increase in pressure is in line with recent reevaluations of the magnetic field strength, cosmic ray pressure, and kinetic pressure associated with the broad wings of the 21 cm line. The dominance of magnetic over thermal pressure (perhaps 8000 cm^{-3} K magnetic versus 3000 cm^{-3} K thermal in clouds) is also consistent with the nearly density independent magnitude of B. These points are discussed further in, for example, Cox (1988, 1989, 1990a, 1990b), Spitzer (1990), Boulares and Cox (1990), and Cox and Slavin (1990), with references to the large body of relevant observational work.

3. THE PROBABLE DEMISE OF THE INTERCONNECTED HOT PHASE

Several lines of evidence once led us to think that a relatively large fraction of the interstellar volume was occupied by very low density gas (n $\lesssim 10^{-2}$ cm^{-3}) at high temperature (T $\gtrsim 3 \times 10^5$ K). None of these, however, has proven to be a certain indicator of such gas. The soft X-ray background appears to arise from a single Local Bubble of 10^6 K gas surrounding the Sun. There are other such bubbles, generally too large to be created by individual supernovae and normally associated with OB associations. The origin of the Local Bubble remains mysterious, but its identification and bounded character remove its interior conditions from being an example of interstellar material everywhere. This is discussed further in Cox and Reynolds (1987), Snowdon et al. (1990), Cox and Slavin (1990).

A rather convincing argument for the widespread presence of hot gas in the ISM was made by McKee and Ostriker (1977). They demonstrated that at the anticipated rate, supernovae (SNe) would violently disrupt a warm intercloud medium into a froth of hot gas and dense included bits of shells, or clouds. Because the warm intercloud gas, with typical density ~ 0.2 cm^{-3}, had been the only candidate other than hot gas for filling the volume, it appeared certain that only hot gas was actually left as a possibility.

This argument has been critically reviewed by Cox and Slavin (1990). They find that a combination of factors (lower SN rate, improved remnant model, higher interstellar pressure) lowers the estimated porosity generated by SNe in the intercloud medium by at least a factor of 30 from the previous results. The best current estimate appears to be q ~ 0.1.

As a consequence we can no longer be certain that supernovae are capable of the disruption that would guarantee the existence of a pervasive hot phase. Their influence appears instead to be the production of individual localized bubbles of hot gas occupying about 10% of the interstellar volume.

This re-analysis made use of a new model for SNR evolution by Slavin and Cox (1990). The model included significant magnetic pressure and followed the remnant evolution until its hot bubble completely disappeared. With an explosion energy of 5×10^{50} ergs, an ambient density of 0.2 cm^{-3}, and a magnetic field of 5 μG, the bubble achieved a maximum radius of 56 pc and disappeared after 5.5×10^6 years.

Furthermore, the evolution-averaged contents of the ions O VI, N V, and C IV in the bubble had the correct ratios and magnitudes to indicate that the observed average interstellar densities of these ions would be just those expected from the population of bubbles. Thus the other chief indicator of interstellar hot gas, the high stage ions, is better understood as being due to individual SNR bubbles than from any previous description.

4. THE PROBABLE DEMISE OF GALACTIC FOUNTAINS

Without a hot phase pervading the ISM, models with galactic fountains that rise diffusely out of the general disk have no source term (except for a few SNe at very high z).

With a thick disk including warm interstellar gas (with scale height of perhaps 500 pc), as well as the pressure of cosmic rays and magnetic field (with even greater scale height and continued existence to at least 2 kpc), the breakout of OB association driven bubbles from the disk of the Galaxy is made much more difficult. Models of fountains using such breakouts as their source function will now find that far fewer, if any, of the Galactic OB associations will be sufficiently vigorous to serve that function. Small bubbles should be smaller than previously estimated, large ones avoiding breakout will grow even larger in the plane. The magnetic pressure will make the shells thicker, and cause them to rebound more quickly to eliminate both bubble and shell.

5. OTHER DYNAMICAL ASPECTS OF THE THICK DISK

I think there are several other dynamical features that will be interesting to explore, but are now only vague thoughts. One is connected to the fact that the Alfven speed is roughly 30 km s^{-1} at midplane and probably increases with z to perhaps 50 km s^{-1} at 1 kpc.

My suspicion is that waves generated by SN, OB associations, molecular cloud motions, and density waves will increase in energy density until saturated (i.e. $\delta B/B \sim 1$). Then nonlinear effects provide dissipation. If this is true at all z, it implies $\rho v^2 \sim B^2/8\pi$ and a rough equipartition between diffuse gas kinetic pressure and the field pressure. It also implies that at high z there will be dissipative shocks produced with characteristic velocities of order 50 km s^{-1}, possibly even higher at higher z. These high velocity components are a potential source of high stage ions far off the plane, though one would tend to expect highly variable amounts of the very highest stage ions, O VI and N V relative to the lower stages like C IV and Si IV in individual features.

In addition, the outer layers of the disk should be very prone to large excursions from equilibrium, like the upper end of an exponential atmosphere. In situations with large relative velocities between interarm and arm material, hydraulic jumps could be an interesting aspect of spiral structure.

6. THE DIFFUSE CLOUD COINCIDENCE

Diffuse clouds, or at least their denser parts, occupy no more than a few percent of the interstellar volume at midplane, and only a rather small fraction of one percent of the total thick disk. I think of them as a condensate in the more diffuse environment, low in the gravitational potential.

One tends to think of diffuse clouds gathering up material and magnetic flux from the intercloud phase as they are formed, and therefore having higher magnetic field as well as higher density. Except for transients during formation, however, that is impossible. The clouds have no surrounding pressure capable of keeping their magnetic field elevated. Even ram pressure cannot be very significant for a population whose dispersion velocity is significantly lower than the Alfven speed in the surrounding gas. Thus it cannot be too surprising that the magnetic fields in diffuse clouds are no larger than those in the surroundings. The field pressure is too large for it to be otherwise.

As a consequence, the diffuse cloud phase equilibrium should not be thought of as taking place at a constant thermal pressure imposed by the surroundings, except on perhaps irrelevantly long time scales.

On short timescales, I think this means that any thermal pressure is allowed within the clouds, so long as it is not so large as to blow them apart, i.e., it should not exceed the roughly 8000 cm^{-3} K available from the external field pressure. It also means that higher thermal pressure should be accompanied by lower field pressure inside the clouds.

So the minimum requirement for the existence of diffuse clouds is that they not be heated so much that their minimum thermal pressures exceed the available magnetic pressure. As it happens, this requirement is met only marginally (see Cox 1988 for diagram and discussion), suggesting that there is probably some link between these two pressures, though in existing models, there isn't.

There are at least two possible forms for such a link.

The first, which was explored somewhat by Field, Goldsmith, and Habing (1969), works in analogy with a fluid in a closed container, or the Earth's ocean and atmosphere. If there is too little fluid in the vapor phase to provide the needed pressure to confine the liquid, then more of the liquid will vaporize. The equilibrium vapor pressure is approached automatically. In this picture, the clouds accept only excess material from the environment above that needed to establish the pressure needed to confine them. Since we usually think of the heating and cooling mechanisms of diffuse clouds being tied to the stellar radiation field, the elemental abundances, and the grain population, we find ourselves looking for mechanisms by which the clouds can decide on the value of the surrounding pressure, particularly the contribution of the magnetic field.

In the opposite perspective, one can inquire instead about what mechanisms operate directly to set the intercloud pressure. The value must equal the weight of the intercloud gas, thus the total intercloud column density and scale height are involved. Certainly the approach above acts directly on the column density, but another factor decides the scale height, and thus the average gravity experienced by the gas.

If one supposed that supernova stirring somehow creates the motions required to establish the magnetic field, that cosmic rays and wave fields equilibrate with that, then the supernovae could be the active agent deciding on the scale height, establishing the pressure versus column density relationship, completing a picture which involves supernovae plus cloud heating and cooling to set the equilibrium.

This has got to be too simplistic. There will almost certainly be a model like this that works, but it cannot be complete. It relies on a coincidence. When all is said and done, out of all the orders of magnitude available, the required scale height for the gas turns out to be very similar to that of the stars themselves.

Another way of viewing this coincidence is to assume at the outset that the scale height of the diffuse gas will be comparable to that of the stirring agent, the diffusely located supernovae. Then the explosions are not needed to specify the pressure/column density relationship, but the picture doesn't work unless the supernovae can stir the system to run the dynamo.

In short, either the cloud heating or supernova rate is accidentally just right, or one of them is controlled in some way by the other.

I have given some thought to how such control might come about. My favorite idea at the moment (c.f. Cox 1990a) is that dust grains are expelled from dense clouds into the intercloud medium with properties that will not allow that material to condense immediately into clouds. But grain processing in that environment eventually leads the grains to photoelectrically absorb less of the starlight, until finally they have the right properties to allow stable clouds to condense. It may not be the right idea, but I am told at least (Savage, private communication) that there is observational evidence that the UV extinction of dust is larger in low density regions.

ACKNOWLEDGEMENTS

This work was supported in part by the National Aeronautic and Space Administration under grant number NAG5-629.

REFERENCES

Boulares, A. and Cox, D. P. (1990) Ap. J. in press (Dec. 20)
Cox, D. P. (1988) in Supernova Remnants and the Interstellar Medium, eds. R. S. Roger and T. L. Landecker, (Cambridge Univ. Press) p. 73.
Cox, D. P. (1989) in Structure and Dynamics of the Interstellar Medium, eds. G. Tenorio-Tagle, M. Moles, and J. Melnick (Springer-Verlag), p. 500.
Cox, D. P. (1990a) in The Interstellar Medium in Galaxies, eds. H. A. Thronson and J. M. Shull (Kluwer) p. 181.
Cox, D. P. (1990b) in Proceedings of IAU Symposium 144, The Interstellar Disk-Halo Connection in Galaxies, ed. H. Bloemen (Kluwer) in press.
Cox, D. P. and Reynolds, R. J. (1987) Ann. Rev. Astron. and Astrophys., 25, 303.
Cox, D. P. and Slavin, J. D. (1990) Ap. J., submitted.
Field, G. B., Goldsmith, D. W., and Habing, H. J. (1969), Ap. J. (letters), 155, L149.
McKee, C. F. and Ostriker, J. P. (1977) Ap. J. 218, 148.
Slavin, J. D. and Cox, D. P. (1990) Ap. J. submitted.
Snowden, S. L., Cox, D. P., McCammon, D., and Sanders, W. T. (1990) Ap. J. 354, 211.
Spitzer, L., Jr. (1990) Ann. Rev. Astron. and Astrophys., 28, in press.

ON THE STRUCTURE AND KINEMATICS OF MOLECULAR CLOUDS FROM LARGE SCALE MAPPING OF MM-LINES

J. Bally, W.D. Langer, R.W. Wilson, A.A. Stark, and M.W. Pound
AT&T Bell Laboratories
HOH-L245, Holmdel, NJ 07733

ABSTRACT

Molecular gas in the interior of the Orion superbubble consists of sheets, filaments, and partial shells in which the active star forming dense cloud cores are embedded. The main body of the Orion A and B clouds and at least 14 smaller clouds in Orion region are cometary in appearance suggesting strong interaction with massive stars in the Orion OB association. While the small scale (< 1 pc) structure of the clouds may be determined primarily by internal magnetic fields, gravity, and the effects of outflows from young stellar objects, the large scale morphology and kinematics is affected by the energy injected by massive stars. Supernovae, stellar winds, and radiation have compressed, accelerated, ablated, and dispersed molecular gas over the last 10^7 years. Most GMC/OB star complexes in the Solar neighborhood exhibit morphological and kinematic properties similar to the Orion region. We argue that energy injection by massive stars plays a vital role in the evolution of the ISM and may be responsible for much of the observed large-scale structure and kinematics of molecular clouds.

INTRODUCTION

Over the last 6 years, the Bell Laboratories 7-meter antenna has been used to observe over 700,000 individual positions in 24 regions of the sky containing molecular clouds. The purpose of this survey is to search for morphological and kinematic clues to the evolution and life cycle of molecular clouds. Large scale mapping of the J = 1-0 lines of ^{12}CO and ^{13}CO, and to a more limited extent the J = 2-1 line of CS, allow investigation of the structure and kinematics of molecular gas on the scale of an entire Giant Molecular Cloud (GMC). In this paper, we analyze the Orion region (based on about 300,000 spectra obtained on a 1' grid with a 100" diameter beam) and apply the results to a discussion of the evolution and star formation history of clouds throughout the Galaxy.

We describe the physical and velocity structure of the Orion region and relate this structure to available sources of energy, exploring the possibility that kinetic energy injection from recently formed stars, combined with the influence of gravity and magnetic fields, determines the present state of the Orion clouds. In the last section, we discuss application of these considerations to models of the origins and evolution of molecular clouds on a galactic scale.

MOLECULAR GAS IN THE ORION REGION

Figure 1 is a color image of the ^{12}CO, ^{13}CO, and CS emission from the major concentrations of molecular gas in the Orion region. The images show that most of the $10^5 M_\odot$ of molecular gas is confined to three major clouds: the Orion A and L1641 cloud in the south, the Orion B cloud containing NGC 2024 and NGC 2023 in the middle, and the NGC 2071 cloud containing NGC 2068 in the north. The large-scale morphology and kinematics of the gas is similar in both ^{12}CO and ^{13}CO. However, the filamentary internal structure of the cloud is easier to see in the more optically thin species. Most of the structure in the CO images shows the gas to consist of filaments or sheets and partial shells, suggesting that the gas is responding to hydrodynamic forces or magnetic fields.

Fragmentation is evident in CO or ^{13}CO only on the largest scales, where the gas in the Orion complex can be subdivided into the 3 major clouds and several dozen smaller clouds or cores. In the L1641 region of Orion, only about 5% of the ^{13}CO emission can be assigned to well defined clumps (Bally et al. 1987). The intensity of ^{13}CO emission ranges from about 3 to 60 K km s^{-1}, indicating that the column density of molecular gas towards the major clouds in Orion is about 3 to 5 times greater than towards the Taurus dark clouds (Ungerechts and Thaddeus 1987; Myers and Benson 1983). The higher gas column density and line opacity in Orion renders cloud cores more difficult to see in ^{12}CO or ^{13}CO compared to lower column density cloud complexes such as Taurus or the Rosette Molecular Cloud (Blitz and Stark 1986).

Many individual cloud cores closely associated with star formation can, however, be seen in the high density tracer CS (Lada 1989; Lada, Bally, and Stark 1990). Most clouds exhibit highly filamentary structure in the CO lines. Cloud cores and fragmentation are best observed in the CS lines or other high-dipole-moment molecules which probe clumping on a scale closer to that on which star formation occurs.

On the smallest scales resolved by our survey (0.2 to 3 pc), we see many small cavities, bubbles, and regions having large ^{13}CO linewidths which may have been produced by molecular outflows or stellar winds from young stellar objects. Examples of these structures include small cavities surrounding the HH 1 and 2 region near NGC 1999, the broad lines seen 20′ north of this region (near the cluster L1641N), and the nearly complete bubble located about 1° to the southwest of NGC 2068.

On linear scales of a few to tens of parsecs, the cloud consists of a network of filaments and partial shells which have a variety of characteristics. Some of the filaments extend for many degrees but are only a few arc minutes wide — aspect ratios (length / width) in excess of 30:1 are common. Examples of long straight filaments are the ones marking the western edge of L1641, and the southern boundary of the Orion B cloud. The images in Bally et al. (1987) show over a dozen filaments in the Orion A (L1641) cloud which have velocity jumps and reversals along their length, yet remain spatially continuous. The straight filaments may contain magnetic field lines along which gas collects to form "magnetized ropes". Magnetic turbulence in the form of Alfvén waves may prevent (or regulate) gravitational collapse along the field lines while magnetic pressure resists flow orthogonal the field lines.

Curved filaments or partial shells are seen in other parts of the cloud. Examples include several overlapping arcs or filaments located 1° east of NGC 2024, the 1°-long tongue of gas projecting towards the east from between the NGC 2023 and NGC 2024 cloud cores, the "bubble" projecting to the southeast of NGC 2068, several nearly concentric partial shells about 20′ in radius surrounding the NGC 1999 reflection nebula, and "arches" of gas lying about 20′ north of NGC 2071. These features may be limb-brightened partial shells or sheets seen edge-on and may indicate the presence of shocks. Energy release from massive stars or groups drive shock waves into the molecular cloud, altering its structure. These structures suggest that magnetic fields and large scale shocks waves play an important role in determining cloud structure, dynamics, and evolution in the Orion region.

The Orion molecular clouds show organization on the scale of a hundred parsecs. The cloud shapes and velocity gradients have a preferred orientation pointing toward the center of the Orion OB association, evidence that energy release from these massive stars must have played an important role in the evolution of these clouds.

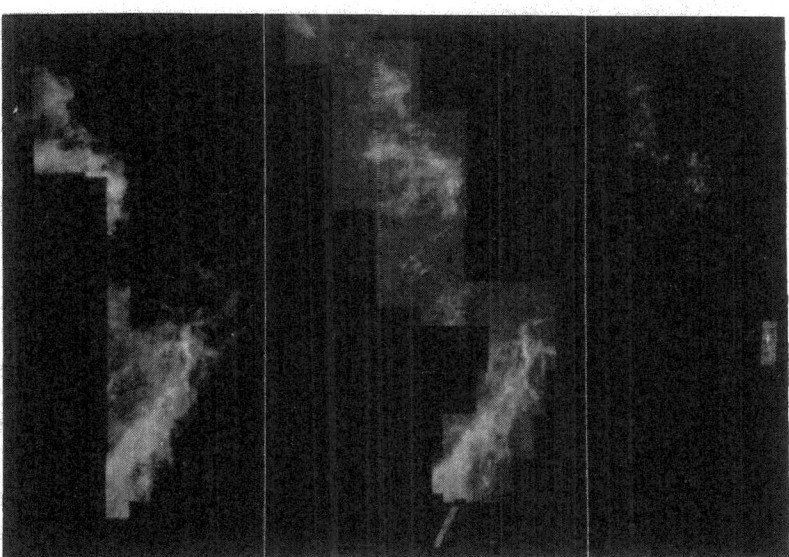

Figure 1: Three color images of the Orion region showing the major molecular clouds in ^{12}CO (left), ^{13}CO (middle), and CS J=2-1 (right). Each image shows the emission integrated from $V_{LSR} = -4$ km s^{-1} to $V_{LSR} = 16$ km s^{-1}, color coded with velocity so that blue corresponds to the velocity interval (-4,5), green (5,9) and red (9,16). The reflection nebulae NGC 2071 and NGC 2068 are located in the main concentration near the top of the figure, NGC 2023 and NGC 2024 is in the center, and Orion A (M42) is the region near the top of the comet-shaped L1641 region at the bottom. *(See Color-plates section)*

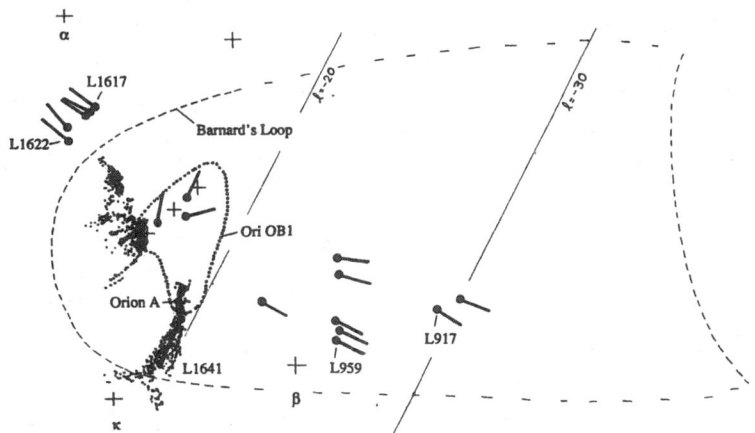

Figure 2: A large-scale view of the Orion region showing the location of Barnard's Loop on the left and the Eridanus Loop on the right which mark the optical edge of the sphere of influence of the Orion OB association. A fragmented shell of 21 cm HI emission lies outside the dashed line which marks the position of faint Hα nebulosity. The location of the Orion OB1 association stars is indicated by the dotted region. The comet-shaped objects mark the location and orientation of 16 cometary molecular clouds. All of these clouds point toward the OB stars located near the center.

CLOUD SYSTEMATICS IN ORION

The extent of the sphere of influence of the Orion OB association can be estimated from the size and location of the bubble of gas marked on the east by Barnard's Loop and by the Eridanus Loop 30° to the west (Sivan 1974). A 21 cm HI bubble surrounds this region, indicating that at least a 100pc × 300pc region containing all of the molecular gas discussed above has been influenced by the energy released by the OB stars in the Orion region. Ultraviolet observations of the near side of this expanding shell (Cowie, Songaila, and York 1979) show a typical expansion velocity of order 30 to 50 km s^{-1} in the low density gas seen along the line-of-sight towards the Orion OB stars.

At least 16 molecular clouds within the boundaries of the Orion superbubble exhibit cometary morphology. These clouds have dense heads with tails pointing away from the center of the Orion OB association as shown in Figure 2. Even the large clouds in Orion exhibit edges and filaments pointing radially away from the OB stars. These structures can be recognized in both the CO and IRAS infrared images of the Orion region. As shown by Bally et al. (1987), the entire southern part of the Orion A cloud (L1641) has a cometary morphology.

Figure 3 shows the detailed morphology of eight cometary clouds in the Orion region. Five clouds which lie in the northeastern portion of Barnard's Loop are shown in the two panels on the left. These clouds point towards the southwest, towards the center of the OB association. The cometary cloud L1622 is surrounded by a "boomerang"-shaped ridge of strong 12 μm emission which coincides with the optically bright southwestern edge of the cometary cloud. This feature indicates that strong radiation fields illuminate the cloud from the direction of the Orion OB1 association, which is in the opposite direction from the Galactic plane. A small cluster of low luminosity stars has formed in the head of this cometary cloud. The overall shape, kinematics, and morphology resembles the models of cometary clouds produced by ionizing radiation discussed by Bertoldi (1989a,b) and Bertoldi and McKee (1990).

These clouds are located slightly outside Barnard's Loop, which marks the present location of the ionization front generated by the OB association in the low density gas located between the major molecular clouds. Since the gas near L1622 and L1617 has already been extensively altered by interaction with the OB stars, it seems likely that in the past, radiation from an OB association has extended beyond the present location of Barnard's Loop. As new OB stars are born and die, the Lyman continuum luminosity of the association fluctuates, producing ionization fronts which repeatedly expand and retreat. Also, the cometary globules in this portion of the Loop may have originated well inside Barnard's Loop and been accelerated radially outward by the rocket effect.

The upper and lower right panels of Figure 3 show two cometary clouds in Eridanus which again point toward the Orion OB association. L917 and L906 are located nearly 30° below the plane, which at the distance of Orion corresponds to about 300 pc or over 4 molecular cloud scale heights from the galactic plane. These clouds must have been either ejected from near the main Orion clouds or condensed from the expanding HI shell surrounding Orion.

There are several systematic patterns in the morphology and kinematics of the gas in Orion: (1) **Large scale velocity gradients point towards the center of the OB association.** In the Orion A, Orion B, and in the NGC 2071 clouds, the portion of the cloud closest to the Orion OB association (whose oldest sub-group is centered about 3 degrees west of NGC 2024) have the most positive velocity (about $V_{lsr} \approx 10$km s^{-1}) while most of the gas farther from the OB stars has a velocity in the range $V_{lsr} \sim 1$ to 8 km s^{-1}. (2) **The densest parts of the molecular clouds face the OB association.** Eight degrees south of the OB association, the southern part of the Orion A cloud has a width of about 1.5° and a mean density of about 10^3cm^{-3}. Four degrees

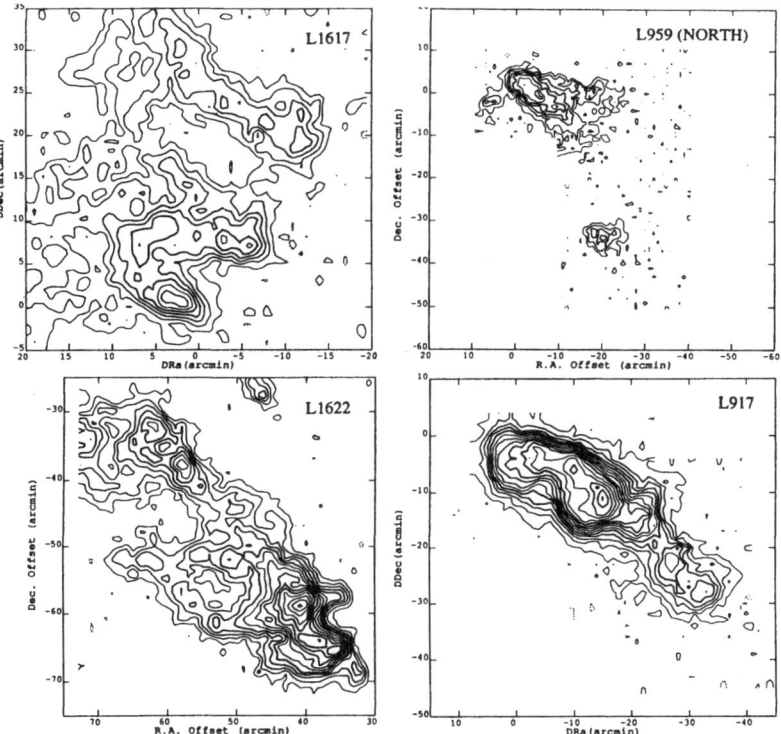

Figure 3: Four panels showing eight different cometary clouds in the Orion region. Note that the objects shown on the left, which lie to the northeast of the Orion OB association point to the southwest while the objects shown on the right, which lie on the other side of the OB stars point in the opposite direction.

(Upper Left) A group of 3 cometary clouds in Lynds 1617, which contain the Herbig-Haro Objects HH-110, HH-111, and HH-113 (Reipurth 1989). Map shows ^{13}CO integrated from V_{LSR} = 8 to 10 km s^{-1}. Contour interval is 1 K km s^{-1} and the (0,0) position is $\alpha(1950) = 05^h49^m09^s$, $\delta(1950) = 02°47'50''$, the source of the HH-111 jet.

(Lower Left) The cometary cloud L1622. At least 6 low mass stars have formed near the head. A smaller cometary globule is seen near the top of the figure. Map shows ^{13}CO integrated from V_{LSR} = 0 to 3 km s^{-1}. Contour interval is 1 K km s^{-1} and the (0,0) position is the same as in the upper left panel.

(Upper Right) Two cometary clouds located to the west of M42 in L959. Map shows ^{13}CO integrated from V_{LSR} = -4 to 0 km s^{-1}. Contour interval is 1 K km s^{-1} and the (0,0) position is $\alpha(1950) = 05^h05^m00^s$, $\delta(1950) = -6°18'00''$.

(Lower Right) A large cometary cloud, L917, located 29 degrees below the Galactic plane in Eridanus. Assuming a distance of 500 pc, this cloud lies 300 pc outside of the Galactic plane. Map shows ^{12}CO integrated from V_{LSR} = −6 to −4 km s^{-1}. Contour interval is 1 K km s^{-1} and the (0,0) position is $\alpha(1950) = 04^h46^m20^s$, $\delta(1950) = -5°52'00''$.

south of the association, near M42, the emission is concentrated into a narrow (5′ wide) ∫-shaped filament which is bright in the J = 2-1 CS line, implying an average density n > $10^4 cm^{-3}$. In the Orion B region, the western edge of the cloud contains a CS-bright compressed ridge facing the HII region IC434 (which lies due west of the Horsehead Nebula) while several degrees to the east, farther away from the OB association, the cloud consists of low-density filaments. Further to the north, the NGC 2071 cloud contains dense cores facing toward the southwestern edge of the cloud. Here too, the average density drops with increasing distance from the OB association. (3) **The most active star formation has taken place in the dense gas nearest the OB association.** Near infrared surveys conducted with 2μm cameras demonstrate the existence of four dense concentrations of embedded young stars in Orion B. Dense clusters are also known to exist in M42, NGC 1977, and in the northern part of L1641, in the Orion A cloud. These clusters contain between 50 and 1000 individual members with core densities approaching 1000 stars per cubic parsec. (See Lada 1989 for the NGC 2023, NGC 2024, NGC 2068, and NGC 2071 clusters). Inspection of co-added IRAS images at 12μm shows large concentrations of stars throughout most of the dense CS emitting regions of the Orion clouds.

ORIGIN OF KINETIC ENERGY IN MOLECULAR CLOUDS

The above observations indicate that there may be a connection between the structure, morphology, and kinematics of the Orion clouds and identifiable stellar energy sources. On small scales (~ 0.1 to 1 pc), outflows from young stellar objects produce shocks, sweep-up gas into compressed shells, and inject kinetic energy which may contribute to the regulation of cloud collapse. On ~ 100 pc scales, the Orion clouds are shaped, accelerated, and compressed by the collective effect of shocks and ionizing radiation produced by stars in the OB association. On the largest scales (> 100 pc) important effects are: gravitation, tidal shear manifested in the differential rotation of the Galaxy, mutual magnetic and gravitational interaction between giant molecular clouds, spiral density waves, and expanding supershells. The kinetic energy content of clouds is ultimately derived from the cumulative effects of all of these influences.

In active star forming regions like Orion, the combined effects of individual stars on the gas and the cumulative influence of the entire OB association over its lifetime result in the injection of kinetic energy on all scales, ranging from several hundred parsecs to under a parsec. Therefore, the spatial and velocity structure on the small scales can not be viewed as the result of a turbulent cascade which transports energy from the largest scale to the smallest. As can be seen from the cometary clouds located over 15° away from the center of the OB association, stellar energy injection can operate over large distances. The dependence of the observed linewidth on the size of the region sampled can be described as a power law, $<\Delta V> \propto <r>^\alpha$ where $\alpha = 0.3\pm0.1$. Although this power law often characterizes turbulent cascades, it may be produced by other mechanisms. For example, this size-linewidth relationship naturally occurs in an ensemble of virialized clouds having mean density scaling as a power law of the size, $<\rho> \propto <r>^{2(\alpha-1)}$.

(a) Energy Injection in Orion

Stars can inject kinetic energy or momentum by several mechanisms: (1) Shock waves generated by embedded outflows from young stellar objects, (2) stellar winds and supernovae produced by older stars located in the vicinity of the cloud, and (3) ionization by hard UV radiation.

On small scales, winds and outflows from young stellar objects can produce 0.01 to 1 pc cavities inside molecular clouds. There are several dozen optical and molecular outflows in the Orion region, injecting momentum into the cloud at a rate

$$\dot{P}_{flow} = 0.05 \ M_{\odot} km \ s^{-1} yr^{-1} \left[\frac{N_{flow}}{20} \right] \left[\frac{M_{flow}}{2.5 M_{\odot}} \right] \left[\frac{V_{flow}}{10 km \ s^{-1}} \right] \left[\frac{10^4 years}{\tau_{dyn}} \right]$$

where N_{flow} is the number of flows, M_{flow} is their average molecular mass, and V_{flow} is their average characteristic flow velocity. The Orion clouds contain about $10^5 M_{\odot}$ of molecular gas with a typical line FWHM of order 3 km s^{-1}. The time required to supply the momentum contained in the gas is $P/\dot{P} \approx 6 \times 10^6$ years. Fukui (1989) discuss the role outflows play in regulating cloud collapse.

On large scales, shocks driven by supernovae or main sequence winds of massive stars sweep up low-density intercloud gas lying in front of the shock into a high density snow-plowed layer, while the tenuous plasma behind the shock is heated, raising the post shock pressure. The Orion region has produced several dozen O stars which have exploded as supernovae in the last 10^7 years (Blaauw 1964), releasing roughly 10^{51} ergs of energy about every 5×10^5 years. During their main sequence phase these stars have already injected a comparable amount of energy in the form of stellar winds. Thus the mechanical luminosity of the Orion OB association is at least $2 \times 10^{51} ergs/5 \times 10^5 years \approx 3 \times 10^4$ L$_{\odot}$. Since the O stars lie close to the molecular clouds from which they were born, the clouds subtend about 10% of the sky as seen from the star. Thus the kinetic energy coupled into the cloud is $\eta \cdot 3 \times 10^3$ L$_{\odot}$, where η is the efficiency with which the SN or wind energy is converted to kinetic energy of motion in the cloud. An efficiency of $\eta = 0.003$ implies a kinetic energy injection rate of 10 L$_{\odot}$ into the clouds which would generate the internal turbulent energy of the cloud (~ 10^{49} ergs) in about 8 Myr.

Since the supershell shocks propagate rapidly into the low-density ISM and move slowly into denser clouds, these clouds are surrounded and encorporated into the interior of the bubble. Initially, the interior of the bubble is heated to a temperature of about 10^6 K. Cooling by radiation, electron conduction, and the evaporation of cold clouds within the cavity lowers the internal temperature over an expansion time scale. Since reheating occurs periodically because of the explosion of new supernovae and the steady input of energy from the winds of massive stars, the interior of the large bubbles surrounding OB associations will have a higher pressure that the surrounding ISM. This effect helps to compress clouds (cloud crushing — Bedogni and Woodward 1990), possibly triggering gravitational collapse and star formation.

Ionizing radiation heats gas to 10^4 K in the process of HII region formation. Most HII regions develop in highly non-uniform environments, which result in the the development of highly anisotropic motions in the ionized medium (the "champagne flow" of Bodenheimer, Tenorio-Tagle, and York 1979) with velocities close to the sound speed in the ionized gas (~ 10 km s^{-1}). As the average density of the HII regions decreases due to expansion, an ever-increasing fraction of the neutral cloud is ionized. A reaction force on the neutral cloud results which can compress and accelerate the neutral cloud to a velocity $V_{HII} \approx c_{HII} M_{HII}/M_{neutral}$ (the "rocket" effect). Photons with a wavelength longer than 912 Å, which can penetrate beyond the HII region, are generated in greater abundance than shorter wavelength ionizing photons and can heat cloud surfaces in photo-dissociation regions (PDRs) to about 10^3 K, producing low-velocity ablation flows and a mild rocket effect.

As shown by Bally and Scoville (1980), clouds can be accelerated to about 5 km s^{-1} if 1/4 of their mass is ablated by ionization, and near the sound speed in ionized gas if 1/2 of the mass is ablated. The dense gas in Orion tends to be seen at higher velocities than the low density gas. If both the Orion A and Orion B clouds are located behind the OB association, then a radial acceleration away from its center produces a positive radial velocity shift for the accelerated gas. This geometry is consistent with the pattern of illumination on the surface of the L1641 region, which suggests that it lies behind the illuminating stars. Furthermore, M42 is known to lie in front of its associated cloud. This geometry is also possible for Orion B, although in this case the evidence is less clear.

Ionizing radiation typically converts photon energy into kinetic energy with an efficiency greater than 0.001 (Tielens and Hollenbach 1985). The Lyman continuum luminosity of an O7 star is about 10^{49} photons sec^{-1}, so that an estimate of the total present Lyman luminosity of the Orion region is about 2×10^{49} photons sec^{-1} (Goudis 1982). Assuming that the clouds subtend 10% of the entire sky as seen from the ionizing stars, the kinetic energy injection rate due to ionization is more than 10 $L_{\odot}$, as much as the value for supernova/stellar wind power injection. Note that the magnitude of both effects depend on inaccurate estimates of small efficiency factors.

OB associations can disrupt giant molecular cloud complexes. Evidence for disruption has been presented by Bally and Scoville (1980) and Leisawitz (1989). Winds, supernovae, and ionization can disperse much of the molecular cloud into smaller accelerated cloud fragments and expanding HI clouds formed by the recombined gas ablated from the molecular clouds. The cometary shapes of clouds which point away from the center of the OB association and the large-scale velocity gradients indicate that cloud disruption is taking place in the Orion region.

(b) OB Associations and the Evolution of Molecular Clouds

Within 500 pc of the Sun, there are two other large OB associations besides Orion — Per OB2 (d=350 pc) and the Sco-Cen association (d=125 to 200 pc; Blaauw 1964; deGeuss 1989). Like Orion, these OB associations are surrounded by massive expanding shells of atomic hydrogen, and are associated with GMC's. All clouds within 700 pc of the Sun having M > $10^4 M_{\odot}$ are associated with OB associations and large HI shells. For example, the ρ-Oph cloud (associated with the upper-Sco sub-group of the Sco-Cen OB association) exhibits wind-blown and cometary structure with the densest gas and active site of star formation facing the OB stars. The Sco-Cen association is surrounded by a network of nested HI shells which subtend over a steradian as seen from the Sun. We suggest that cloud crushing, ablation, and acceleration produced by OB associations is responsible for many of the structural and kinematic features of the molecular gas within the Solar neighborhood.

In the Solar neighborhood, the mean separation between GMCs, OB associations, and supershells having ages less than 10^7 years is about 300 pc. The associated shells expand with a velocity of order 10 to 40 km s^{-1} into the low-density medium. The distribution of O, B, and A stars and molecular clouds lying within 800 pc of the Sun appears to define an expanding, tilted system, the Gould Belt, whose center of expansion and age seems to coincide with the 25 Myr old Taurus-Cas group of B and A stars (Blaauw 1964; Olano and Poppel 1987). The nearby HI shells expand with a typical velocity of 10 km s^{-1}. If their motion is approximated as a momentum-conserving flow driven by a quasi-steady wind (a reasonable approximation for shells driven by multiple supernovae — McCray and Kafatos, 1987), the shells intersect on a time scale of order 30 Myr. These conclusions are only slightly altered by the inclusion of the shear produced by the differential rotation of the Galaxy (Tenorio-Tagle and Palous 1987). Thus no point within the Solar neighborhood remains unaffected by OB-association-driven blast waves for a time longer than this. Supershells must therefore play an important role in the evolution of molecular clouds and the ISM in the Solar neighborhood.

The rocket effect produced by OB stars may be responsible for much of the cloud-to-cloud velocity dispersion observed to in the Galaxy. Stark (1979) and Stark and Brand (1988) noted that the molecular cloud population is characterized by a 1-d rms cloud-to-cloud velocity dispersion of 6 to 8 km s^{-1}. A feature of the velocity dispersion is that it is approximately constant for clouds ranging in mass from under $10^2 M_\odot$ to over $10^5 M_\odot$, implying that the cloud population does not reach collisional equipartition of energy, except for the most massive clouds (Stark 1979). The ISM of the Galaxy contains about $2 \times 10^9 M_\odot$ of HI and a similar amount of H_2. Assuming that most of the observed HI has been ablated from molecular clouds in champagne flows during a molecular cloud dissipation time, the rocket effect should have accelerated the molecular cloud population to a velocity of order the sound speed in ionized gas since the masses of the two phases are similar. This is, in fact, what seems to occur. The total Lyman continuum production rate in our Galaxy is about $Q_{tot} \approx 5 \times 10^{52}$ photons per second (Smith et al. 1978), each with over 13.6 eV of energy. The total kinetic energy content of GMCs is $1.5 \, M_{tot} \sigma_{1D}^2 \approx 3 \times 10^{54}$ ergs. If on average the conversion efficiency of UV photon energy into kinetic energy is 0.001, then the Lyman continuum luminosity of the Galaxy can sustain the observed cloud-to-cloud dispersion if the dissipation timescale for random cloud motions is over 10^8 years. Although at any one time only a very small fraction of the ISM is ionized (< 1%), photo-ionization is the dominant mechanism for converting H_2 into HI, with an overall Galactic rate of ~ 20 to 100 $M_\odot yr^{-1}$. The cloud destruction time scale for the $2 \times 10^9 M_\odot$ of H_2 in the Galaxy due to interactions with OB stars is of order 10^7 to 10^8 years. In order to maintain the ISM in a quasi-steady state, molecular clouds must reform at a similar rate.

Expanding superbubbles may provide an efficient mechanism which sweeps up low density HI from the ISM into dense shells which become gravitationally unstable at a characteristic mass scale of order $10^5 M_\odot$ (McCray and Kafatos 1987; Tenorio-Tagle and Palous 1987; MacLow, McCray, and Norman 1989). All-sky surveys of the distribution of HI have shown that much of the HI consists of shells or partial shells (the supershells and "worms" of Heiles, 1984). These features of the 21 cm sky may correspond to the layers of low density gas swept up by the expanding shocks associated with superbubbles surrounding OB associations. McCray and Kafatos (1987) demonstrate that as supershells sweep-up more gas and decelerate, they become unstable to self-gravitation. For conditions appropriate to the Solar vicinity, the onset of this instability takes place at a characteristic mass scale of order $5 \times 10^4 M_\odot$, essentially the mass of nearby GMCs.

Cooling of the swept-up layers of the expanding supershells should lead to the conversion of HI into H_2. It is thus possible that the youngest molecular clouds may form from gravitational instabilities of old expanding supershells. A prediction of this theory of cloud formation is that the densest parts of old HI shells should contain molecular cores. There is some observational evidence that this is the case — many of the small high latitude molecular clouds discovered by Blitz, Magnani, and Mundy (1984) are located along the HI shells produced by the nearest OB associations in Sco-Cen, Perseus, and Orion.

Clouds formed by the gravitational fragmentation of old shells may be further compressed by self-gravity or the influence of younger OB groups. The mutual gravitational interaction of the clouds produced by shell fragmentation may be enhanced by swing amplification, streaming motion, and magnetic effects, which are more important in spiral arms. In a spiral arm, the resulting cloud build-up leads to the formation of much more massive cloud complexes than those in the Solar neighborhood. Spiral arm GMCs may produce much more energetic OB associations than those currently near the Sun. Such massive OB groups must have been responsible for the OB association which has set into motion the Gould Belt we see surrounding us. The current generation of OB associations near the Sun may be the result of secondary star formation resulting from the fragmentation of the expanding network of cloud which makes up the Gould Belt.

Our observations of molecular clouds imply that massive stars play an important role in determining cloud structure and kinematics. Recently-formed stars still embedded inside clouds produce shocks which contribute to cloud support and regulate the cloud collapse rate. On large scales, ionization ablates and accelerates clouds by the rocket effect. The expanding ionization fronts drive shocks into the gas which may trigger gravitational collapse and induced star formation. The stellar winds and supernovae produced by massive stars in OB associations generate large-diameter superbubbles and supershells which sweep up the low density ISM. Cooling and compression in the shells may lead to molecule formation — deceleration and gravitational instability eventually leads to the formation of $10^5 M_\odot$ clouds. These GMCs form new OB associations, continuing the evolutionary cycle. Galactic scale processes such as tidal shear, streaming and orbit crowding in spiral arms, mutual gravitational and magnetic interactions between GMCs and swing amplification modulate this cycle to produce the observed structures in the Galaxy.

REFERENCES

Bally, J., Langer, W.D, Wilson, R.W. and Stark, A.A. 1987 *Ap.J.(Letters)*, **312**, L45.

Bally, J., and Scoville, N.Z. 1980, *Ap.J.*, **239**, 121.

Bertoldi, F. 1989a, Ph.D. Thesis, University of California, Berkeley.

Bertoldi, F. 1989b, *Ap.J.*, **346**, 735.

Bertoldi, F., and McKee, C.F. 1990, *Ap.J.*, **354**, 529.

Blaauw, A. 1964, *Ann.Rev.Astr.Ap.*, **2**, 213.

Blitz, L., and Stark, A.A. 1986, *Ap.J. (Letters)*, **300**, L89.

Blitz, L., Magnani, L., and Mundy, L. 1984, *Ap.J. (Letters)*, **282**, L9.

Bedogni, R., and Woodward, P.R. 1990, *Astron. Astrophys.*, **231**, 481.

Bodenheimer, P., Tenorio-Tagle, G., and York, H.W. 1979, *Astron. Astrophys.*, **233**, 85.

Cowie, L.L., Songaila, A., and York,D.G. 1979, *Ap.J.*, **230**, 469.

Fukui, Y. 1989, in *ESO Workshop on Low Mass Star Formation and Pre-Main Sequence Objects*, ed: B. Reipurth (ESO Garching).

deGeuss, E. 1988 Ph.D. Thesis, Leiden.

Goudis, C. 1982, *The Orion Complex: A Case Study of Interstellar Matter*, (Dordrecht: D. Reidel).

Heiles, C. 1984, *Ap.J.Suppl.*, **55**, 585.

Lada, E.A. 1989, Ph.D. Thesis, University of Texas, Austin.

Lada, E.A., Bally, J., and Stark, A.A. 1990, *Ap.J.*, (in press).

Leisawitz, D., Bash, F.N., and Thaddeus, P. 1989, *Ap.J.Suppl.*, **70**, 731.

MacLow, M., and McCray, R., and Norman, C. 1989, *Ap.J.*, **337**, 141.

McCray, R., and Kafatos, M. 1987, *Ap.J.*, **317**, 190.

Myers, P.C., and Benson, P.J. 1983, *Ap.J.*, **266**, 309.

Olano, C.A., and Poeppel 1987, *Astron. Astrophys.*, **179**, 202.

Reipurth, B. 1989, *Nature*, **340**, 42.

Sivan, J.P. 1974, *Astron. Astrophys. (Suppl.)*, **16**, 163.

Smith, L.F. ,Biermann, P., and Mezger, P.G. 1978,*Astron. Astrophys.*, **66**, 65.

Stark, A.A. 1979, Ph.D. Thesis, Princeton University.

Stark, A.A., and Brand, J. 1989, *Ap.J.*, **339**, 763.

Tenorio-Tagle, G., and Palous, J. 1989, *Astron. Astrophys.*, **186**, 287.

Tielens, A.G.G.M., and Hollenbach, D.J. 1985, *Ap.J.*, **291**, 722.

Ungerechts, H., and Thaddeus, P. 1987, *Ap.J.Suppl.*, **63**, 645.

OBSERVATIONAL MANIFESTATIONS OF SEQUENTIAL STAR FORMATION IN GIANT STAR-GAS COMPLEXES OF THE GALAXY

T. G. SITNIK
Sternberg Astronomical Institute
13 Universitetskij prospect
Moscow V-234,119899
USSR

ABSTRACT. The age distribution of stars and stellar gro-upings was studied in the galactic large-scale star-gas complexes (SGCs).

1. GALACTIC STAR - GAS COMPLEXES

Seventeen giant (170-700 pc) SGCs apart from the Local Sys-tem have been detected within 3 kpc from the Sun (Fig.1) [1]. These SGCs include about 90% of stellar groupings (OB - associations and open o-b2 clusters) younger than 2-3 10^7 yrs and molecular clouds with masses 10^5-$10^6 M_0$. The largest SGCs with the dimensions of 300-700 pc are connected with HI superclouds' remnants, whose initial mass could be about 10^7-10^8 M_0. The large-scale complexes contain ten or more stellar groupings.

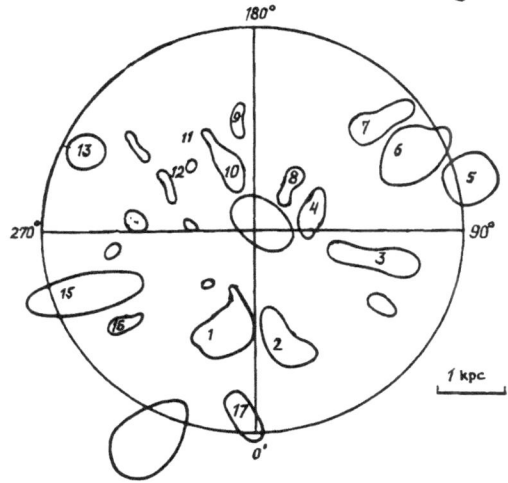

Figure 1. SGCs projec-ted onto the galactic plane. The Sun is at the center

22

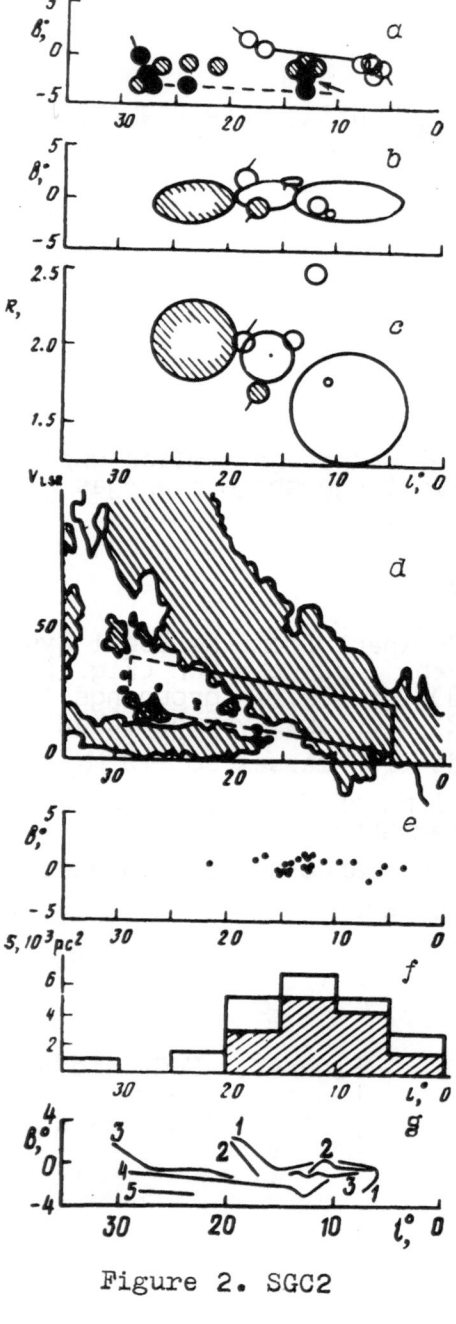

Figure 2. SGC2

2. THE AGE GRADIENT OF STAR GROUPINGS

We have held the detailed investigation of all eight large-scale SGCs and of the nearby aged clusters whose spectral types were later than b2 [2]. It was found that in the seven SGCs there is a gradient of the OB-associations' ages equal to $(0.3-1.2) \cdot 10^7$ yrs for the distances 0.3-0.5 kpc. Moreover the age of the stellar groups is changing sequentially from one edge of the SGC (HI supercloud) to the other one. In the interstellar medium the existence of the regions with the different ages manifests itself in the following way. The giant molecular clouds with masses $M \geqslant 10^5 M_\odot$, young H_2O masers and HII regions are associated mostly with stellar associations which are not older than $6.5 \cdot 10^6$ yrs. The discussed situation is illustrated by the example of the SGC2 (Fig.2). SGC2 (Fig.1) is located in the Sgr arm at the distances 1.2-2.0 kpc. Clusters of early spectral types (open circles on Fig.2a), young OB-associations with $t \leqslant 6.5 \cdot 10^6$ yrs (they are not shaded on Fig.2b, c), molecular clouds (inside the marked parallelogram on Fig.2d), HII regions (Fig.2f) and young H_2O-masers (Fig.2e) are localized near the inner border of the arm, whereas the old groups (dark and shaded circles on Fig.2a,b,c) are closer to the outer one.

3. THE STRATIFICATION OF THE INDIVIDUAL STARS OF VARIOUS AGES

We have also explored the distribution of the youngest and the evolved stars in the SGCs [3] . The youngest objects are OV stars with the initial masses greater than $30M_{\odot}$. The evolved objects are less massive supergiants with $M_{in} \ll$ $15M_{\odot}$. WR stars and Cepheids. We included the σ-aO open clusters and H_2O masers, connected with the sites of star formation into our examination. The age of the stars was determined by means of the evolutionary tracks with mass loss [4]. The Cepheids' age was determined from the period age relationship [5]. The age of the open clusters was found from the absolute stellar magnitude of the brightest main-sequence star [6]. All these objects were used for the construction of the isochrones. The obtained results are shown again by the example of the SGC2. As it is seen from Fig. 2g, the spatial stratification of the stars of different ages is observed in the SGC2 even in (l,b) plots without using the less accurate coordinate - the distance.

The isochrone 1 corresponds to $t \ll 6.5\ 10^6$ yrs, $2 - t \ll 8\ 10^6$ yrs, $3 - (1.8-3.7)\ 10^7$ yrs, $4 - (4-5.4)\ 10^7$ yrs, $5 - t \gg 8\ 10^7$ yrs. Only about 9% of the 76 objects investigated lies out of the obtained isochrones. Thus in the SGC2 the relative displacement of the individual stars of different ages is observed, similar to that found earlier for the stellar groupings. The age change occurs across the S_{gr} arm at some angle to the plane $b = 0^{\circ}$: the older regions are mainly localized below the plane $b = 0^{\circ}$ and at greater l's , young regions - above the older ones and at smaller l's.

4. CONCLUSIONS

The obtained picture of the distribution of different components inside the SGCs - star groupings and individual stars of various ages, interstellar clouds and young H_2O masers-is typical for seven out of eight largescale SGCs. A natural explanation is the following. The shock wave is propagating across the HI supercloud with molecular clouds inside. The generations of stars with different ages are born sequentially behind the shock front. The observed age gradient across the Car-Sgr arm gives evidence that star formation in all the three SGCs (1, 2,15 on Fig.1) of this arm is connected with a spiral density wave. Perhaps the Cygnus arm is lying near the corotation radius, since there is no age gradient across this arm [2]. The direction of the age's changing is different in all the three Cas-Per arm's SGCs. It cannot be excluded that in SGC6 a "reverse" age gradient is observed.

The existence of the described above SGCs confirms the hypothesis about large-scale star formation concentrated in supergiant HI clouds with embedded giant molecular clouds [7,8]. The components of stellar population and of the interstellar medium earlier united into SGCs on the basis of their close spatial location are indeed physically interconnected.

REFERENCES

1. Efremov Yu.N., Sitnik (1988). Soviet. Astron. Letters 14, 347.
2. Sitnik T.G. (1989). Pisma v Astron. Zh. 15, 897.
3. Sitnik T.G. (1991). Pisma v Astron. Zh. to be published.
4. Maeder A. (1981). Astron. Astrophys. 102, 401; (1983), 120, 113.
5. Efremov Yu.N. (1989). Stellar complexes. Astrophys. Space Sci. Rev. 7.
6. Mermilliod J.C. (1981). Astron. Astrophys. 97, 235.
7. Efremov Yu.N. (1978). Soviet Astron. Letters 4,66; (1979), 5, 12.
8. Ellmegreen B.G., Elmegreen D.M. (1983). Month. Notic. Roy. Astron. Soc. 203, 31.

A CS SURVEY OF MASSIVE STARS EMBEDDED IN MOLECULAR CLOUDS

L. BRONFMAN and J. MAY, Universidad de Chile
L.A. NYMAN, Onsala Space Observatory
P. THADDEUS, Center for Astrophysics

ABSTRACT. The CS J=2→1 molecular line at 98 GHz, a normally optically thin line requiring high densities to be excited , has been detected with SEST (Swedish ESO Submillimeter Telescope) toward 294 IRAS point-like sources having the characteristic FIR colors of embedded stellar objects and apparently associated with the largest molecular cloud complexes in the southern Milky Way. We present here their Galactocentric radial distribution and a correlation between their FIR and CS luminosities.

1. FAR INFRARED EMISSION FROM GALACTIC MOLECULAR CLOUDS

Large scale CO surveys can be used together with the IRAS database to study the properties of FIR emission from large molecular clouds in the Galactic plane. Comparison of velocity-integrated CO intensity contour maps and corresponding IRAS surface-brightness images show substantial infrared counterparts for almost every CO concentration. Because the infrared data provides no information about the distance of the source and, at least in the inner Galaxy, there are several molecular clouds at different distances in most lines of sight, it is not easy to tell which fraction of the infrared emission is being produced by a particular cloud. Different methods developed to compare the diffuse FIR emission from IRAS with CO data (Mooney and Solomon 1988, Scoville and Good 1989) apply mostly to isolated clouds. Here we take a different approach, based on the observation of molecular lines toward selected IRAS point-like sources.

Molecular clouds, often associated with extended HII regions, are also associated with infrared point-like sources, very likely to be compact HII regions heated by embedded massive stars. Wood and Churchwell (1988; WC), using a two-color selection criterion for the IRAS point source catalog based on the FIR spectral characteristics of known ultracompact HII regions, identified 1,646 candidates to be massive stars embedded in molecular clouds in the Galactic plane. These stars provide a large fraction of the internal heating of molecular clouds and may play an important role in their fragmentation. We have observed now the CS J-2→1 line, normally optically thin and requiring high densities to be excited, toward embedded star candidates in the third and fourth Galactic quadrants and used the kinematic information from the CS line, together with our CO survey of the southern Galaxy, to calculate their distances and estimate the FIR luminosity of their surrounding dust.

A fully sampled CO survey of the fourth Galactic quadrant, covering from b=-2° to 2° at an angular resolution of 0°.125, allowed us to derive the gross distribution of molecular gas and to identify the largest molecular cloud complexes within the solar circle (Bronfman et al. 1988a, 1988b, 1989). Molecular clouds outside the solar circle have been described by Grabelsky et al. (1988) and May et al. (1988).

2. CS OBSERVATIONS

Sources in the fourth Galactic quadrant, within 2° of the plane, were selected from the IRAS Point Source Catalog using WC criterion. From l=307°, near the tangent of Centaurus Spiral arm, to l=348°, the limit of the region presently analyzed of our CO survey, every source was observed in the CS line. Between l=280 and 308, candidates were observed first in the CO 1-0 line, and those with non-local velocities observed in CS. For the third quadrant, we observed in CS all IRAS point sources with non-local velocities from the CO catalog of Wouterloot and Brand (1989), plus those associated with molecular clouds recently discovered in a 0°.125 resolution survey of the third Galactic quadrant (May et al. 1990).

Observations of the CS 2→1 line at 98 GHz, at 50" resolution, were made in frequency switching mode during 3 runs of 30 hours each in October 1989 and May 1990 with the SEST Telescope at the European Southern Observatory, in La Silla. Spectra were corrected for atmospheric attenuation using standard calibration methods; no correction for beam efficiency was applied. Up to third order baselines were subtracted from the spectra; single gaussian fits were used to compute the mean velocity, full-width at half-maximum, and integrated intensity of the lines. Single velocity components, corresponding to one of the CO molecular clouds in the line of sight, were observed in practically every case. The CS line is normally excited at gas densities above $10^4 cm^{-3}$, a condition that appears to be fulfilled near the Galactic center and in the dense star-forming regions in the Galactic plane, but seldom elsewhere. Thus, the observed CS line can be associated with fair certainty with the dense gas surrounding compact HII regions detected by IRAS. Some profiles depart from a normal distribution, but a detailed study of profile characteristics is posponed to further consideration and hopefully more observations.

3. DISTRIBUTION AND FIR LUMINOSITY OF EMBEDDED MASSIVE STARS

Because the CO survey data were analized so as to enhance the contrast for molecular clouds with CO masses above $5 \times 10^5 M_\odot$, the striking correspondance between CS detections and CO contour maps (Fig. 1) strongly supports the generally accepted view that massive stars form in large molecular clouds, like the ones identified in our CO surveys. Such association is used below to help determine the distances for the

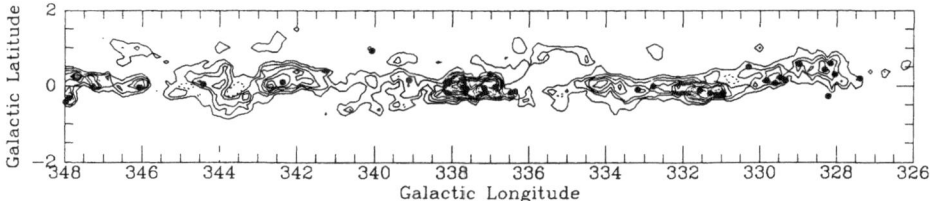

FIGURE 1. Contour-plot of CO emission integrated in velocity over a region corresponding to the Norma Spiral Arm. Contour interval is 8K km/s. Filled circles are CS detections in the same velocity range.

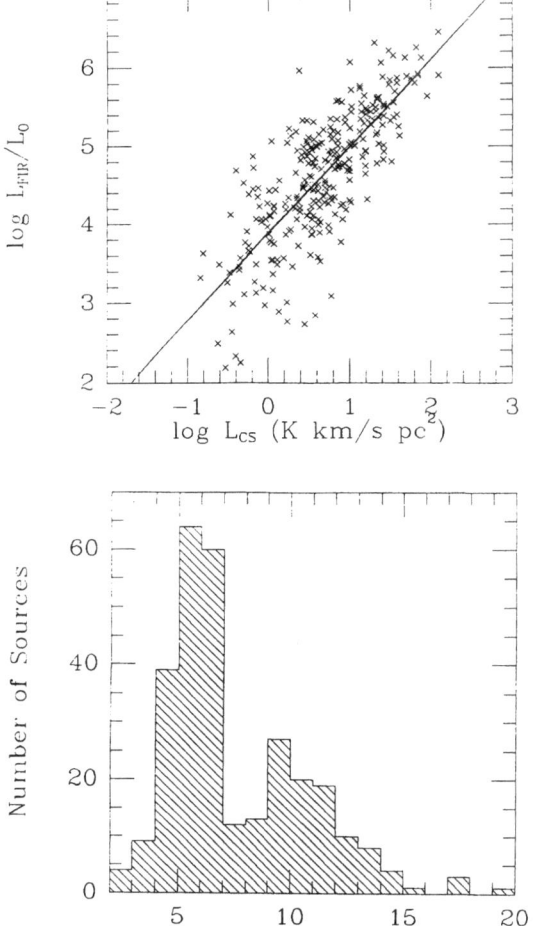

FIGURE 2. Radial distribution of embedded massive stars in the third and fourth Galactic quadrants. Galactocentric radii were derived from CS line velocities using a CO rotation curve within the solar circle and a constant velocity $V(R)=220$ km/s outside.

FIGURE 3. FIR luminosity from IRAS against CS luminosity measured with SEST for dust and dense molecular gas heated by embedded massive stars. The best linear fit is $\log L_{FIR}=(1.109\pm0.051) \log L_{CS}+(3.884\pm0.045)$.

embedded stars within the solar circle.

Assuming purely circular motion about the Galactic center and through the use of a standard rotation curve, the CS line velocities were used to derive Galactocentric radii for the embedded stars. In the inner Galaxy massive stars concentrate in a ring about 3 kpc wide (FWHM) at a mean Galactocentric radius of 5 kpc (Fig. 2), coincident with but somewhat narrower than the independently derived distribution of molecular clouds (Bronfman et al. 1988a). For sources within the solar circle a two-fold distance ambiguity must be solved; we have used here the distance ambiguity resolution for their parent molecular clouds, obtained through standard techniques like OH and H_2CO absorption, distance to the Galactic plane, and size to linewidth ratio. The secondary peak in Fig. 2, outside the solar circle, corresponds to the Carina and outer spiral arms. The distribution there is slightly shifted toward negative latitudes following closely the warp of the Galactic plane.

Having determined the distances of the embedded stars, it is possible to compute the FIR luminosities of their dust coccoons; the luminosity from an IRAS pixel of 2'x2' is computed here as $L=4\pi D^2 \Sigma \nu F_\nu$ (Boulanger and Perault 1988), where D is the kinematic distance, F_ν is the flux at frequency ν, and the sum goes over all IRAS bands. For the same FIR colors CS is detected more often in point sources within the solar circle than outside, where molecular clouds seem to be underluminous also in CO (Digel et al. 1990). There is a fair correlation between CS and FIR luminosities (Fig. 3) possibly meaning that sources undetected in CS correspond to stars of relatively low luminosity, unimportant for our sampling of massive stars embedded in large molecular clouds in the Galactic plane.

L.B. acknowledges partial support by FONDECYT (República de Chile) through grant 90-1079.

REFERENCES

Boulanger, F., and Perault, M. 1988, Ap.J. 330, 964.
Bronfman, L., Cohen, R., Alvarez, H., May, J., and Thaddeus, P. 1988a, Ap. J. 324, 248.
Bronfman, L., Nyman., and Thaddeus, P. 1988b, Lecture Notes in Physics 331, 139.
Bronfman, L., Cohen, R., Alvarez, H., and Thaddeus, P. 1989, AP.J.Supp. 71, 481.
Grabelsky, D., Cohen, R., Bronfman, L., and Thaddeus, P. 1988, Ap.J.331, 181.
Digel, S., Bally, J., and Thaddeus, P. 1990, Ap.J.(Letters) 357, L29.
May, J., Murphy, D.C., and Thaddeus, P. 1988, Astron. Astrophys. Suppl. Ser, 73, 51-83.
May, J., Alvarez, H., and Bronfman, L. 1990, in preparation.
Mooney, T.J., and Solomon, P.M. 1988, Ap.J.(Letters) 334, L51-L54.
Scoville, N.Z., and Good, J.C. 1989, Ap.J. 339, 149.
Wood, D. and Churchwell, E. 1988, Ap.J. 340, 265.
Wouterloot, J., and Brand, J. 1989, Astron.Astrophys.Suppl.Ser. 80, 149.

DIFFUSE MOLECULAR CLOUDS
AT HIGH GALACTIC LATITUDE

RONALD STARK

Sterewacht Leiden,
P.O. Box 9513, NL-2300 RA Leiden, The Netherlands
E-mail STARK@HLERUL51.BITNET

Abstract. The IRAS 60 and 100 μm flux from cirrus clouds are commonly explained by dust continuum emission. But this explanation in some cases requires unexpectedly high dust temperatures.

We argue that the contribution of fine structure emission of neutral oxygen, $O°(63$ μm), can be significant in the IRAS 60 μm band. The $O°(63$ μm) line emission together with the dust continuum emission offers a plausible explanation of the observed flux ratio $I(60)/I(100)$, as well as of its variation across individual clouds. We also discuss the clumpy/filamentary structure of these clouds.

Keywords : Atomic processes; Infrared radiation; Interstellar medium : clouds : cirrus

1. Introduction

The IRAS mission revealed a wide spread emission component in the interstellar medium, the now so-called cirrus clouds (Low *et al.*[1]). Most of the cirrus clouds appear as optically thin clouds : A_V <1 mag (e.g. de Vries and le Poole[2]). Because their distance from the Sun is of the order 100 pc (Magnani and de Vries[3]), cirrus clouds can be studied in great detail. The best way to study their nature is to look for isolated cirrus clouds at high galactic latitude, where : (1) the geometry of the dominant radiation field is known (i.e., the general galactic radiation field), (2) there is no blending with possible background features.

The IRAS 100 μm brightness of cirrus is very well correlated with the visual extinction seen at optical wavelengths[2,3]. It is generally assumed that both phenomena are caused by one population of big (radii of order of 0.01-0.1 μm) dust particles, which are in thermodynamic equilibrium with the radiation field (Mathis *et al.*[4]). These dust particles should have an equilibrium temperature below 20 K (Spencer and Leung[5], Black[6]; Chlewicki[7]).

Figure 1 and 2 show respectively the IRAS 100 μm emission and the visual scattering at 0.5 μm of the cirrus cloud G230-28N.

The 60 μm emission of cirrus is also found to be correlated with the 100 μm emission (Harwit *et al.*[8], Laureijs *et al.*[9]). If the 60 μm emission were due to the same population of dust particles that radiate at 100 μm, we expect a brightness ratio $I(60)/I(100) = 0.11$[4]. However the observed 60/100 ratio of these clouds varies from region to region and is on average higher by a factor of two[8,9], requiring grain temperatures of the order of 24 K.

Figure 3 shows the temperature map of the cloud G230-28N derived from the 60 and 100 μm emission, under the assumption that the emission is caused by one population of dust particles radiating as a blackbody with an emissivity proportional to λ^{-1}. Note the nearly constant temperature in the outer parts of the cloud corresponding to 60/100=0.2 and the steep rise towards the centre of the cloud (60/100=0.24).

Recently the dust temperatures of cirrus clouds have been estimated through mm measurements by Andreani *et al.*[10], who find dust temperatures well below 20 K.

We therefore conclude that the IRAS 60/100 μm brightness ratio cannot be used to derive the cirrus dust temperature.

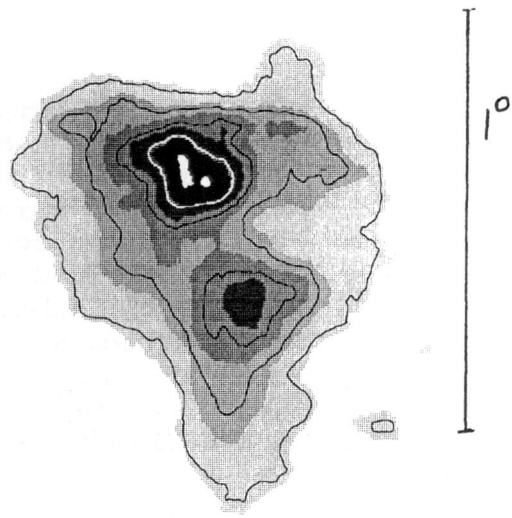

Fig. 1.   IRAS 100 μm co-add map of the cirrus cloud G230-28N. The angular resolution is : 3.8′ × 6.4′. Grey scales range from 0-5 MJy ster^{-1}, with darker grey scales corresponding to higher brightnesses. Contours are at 1...(1)...5 MJy ster^{-1}.

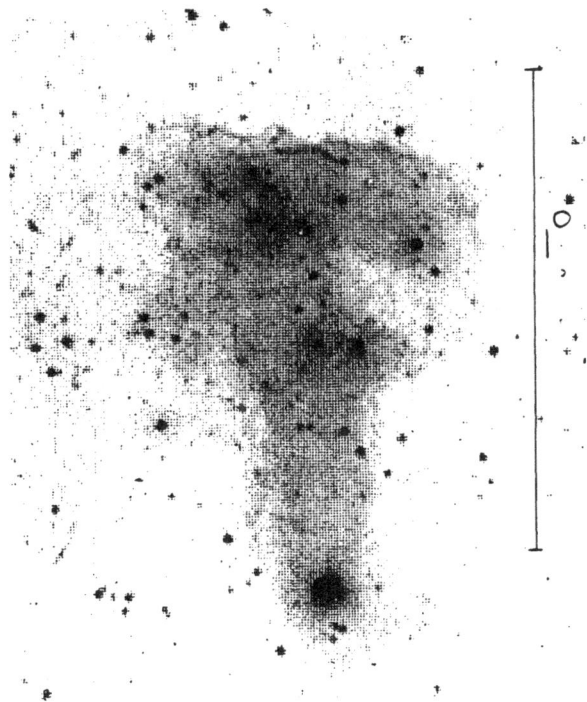

Fig. 2. Digitized (SERC IIIaJ) 0.5 μm surface brightness map of G230-28N, all but the brightest ($m_B < 10$ mag) stars have been removed. Darker grey scales correspond to higher brightnesses. Angular resolution : $22'' \times 22''$.

2. THE IRAS 60 μm EMISSION

What emission component can we expect to be present in the 60 μm band apart from dust continuum emission?

A population of very small grains (Draine and Anderson[11]) or large aromatic molecules (Puget et al.[12]) has been postulated to explain the IRAS infrared emission at 12 and 25 μm. This population of dust particles could also have an emission contribution to the 60 μm band. However the 12 and 25 μm emission have a different spatial distibution than the 60 μm emission[9,13].

Laureijs et al.[9] postulate a population of small iron grains with such a size distribution that the peak of the emission spectrum has a maximum in the IRAS 60 μm band. These grains would also contribute to the IRAS 25 and 100 μm bands.

A more plausible explanation is to assume a contribution of the neutral oxygen fine structure line, $O°(63 \mu m)$, to the IRAS 60 μm band. This idea was first expressed by Harwit et al.[8], who suggested that the IRAS FIR radiation could

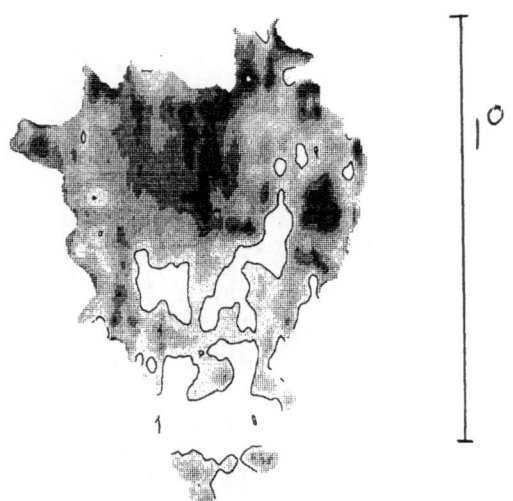

Fig. 3.   Dust temperature map of G230-28N, assuming both 60 and 100 μm are caused by continuum emission of the same dust, modelled with a modified Planck function. Grey scales range from 22-28 K, darker grey scales correspond to higher temperatures. The contour is at 25 K.

be largely fine structure radiation of $O^{\circ}(63~\mu m)$ and $O^{2+}(88~\mu m)$. This explanation did not stick in the literature, probably because of the good correlation that was generally found of the 100 μm radiation with visual scattering and hydrogen column density, ruling out a contribution of O^{2+} emission.

Stark[14] has argued in a plausible way that a contribution of the neutral oxygen line to the dust continuum emission at 60 μm can account for the observed 60/100 brightness ratio and its variation in individual clouds. The observed 60 μm radiation can then be written as

$$I_{obs}(60~\mu m) = I_{O^{\circ}}(63~\mu m) + I_{dust}(60~\mu m) \qquad (1)$$

If we assume

$$I_{obs}(100~\mu m) = I_{dust}(100~\mu m) \quad \text{and} \quad \frac{I_{dust}(60~\mu m)}{I_{dust}(100~\mu m)} = 0.1$$

we can write

$$I_{O^\circ}(63\ \mu m) = I_{obs}(60\ \mu m) - 0.1 I_{obs}(100\ \mu m) \tag{2}$$

Figure 4 shows the thus obtained "oxygen-map" for G230-28N. The intensities range from 0.1 MJy ster^{-1} at the outerparts of the cloud to 0.7 MJy ster^{-1} at the cloud centre.

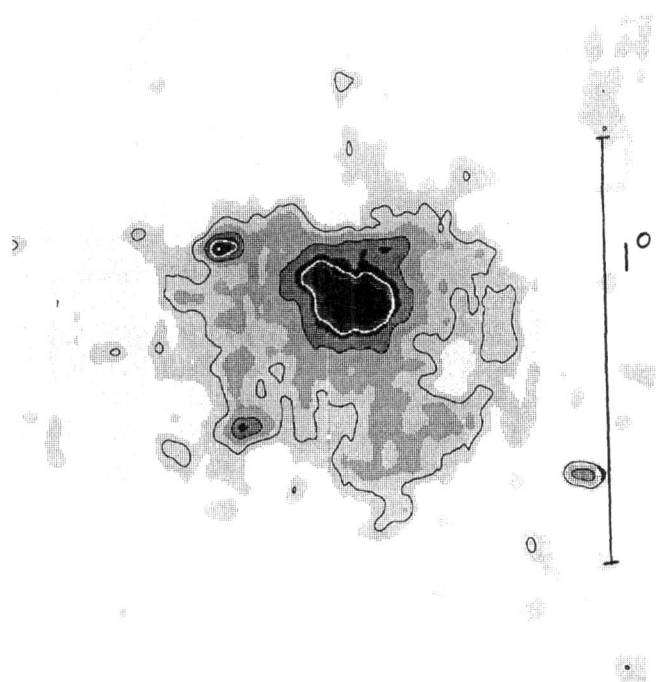

Fig. 4. "Oxygen-map"of G230-28N, derived from the IRAS 60 and 100 μm maps. Grey scales range from 0...0.7 MJy ster^{-1}. Contours are at 0.2, 0.4 and 0.6 MJy ster^{-1}.

3. DISCUSSION

We will discuss here the intensity of the oxygen line.

The intensity of an optically thin oxygen line can be written as :

$$I = \frac{1}{4\pi} n_H n_O \{ L_H(O,T) + x_e L_e(O,T) \} d \tag{3}$$

where L_H and L_e are the cooling efficiencies for collisions between oxygen atoms with respectively hydrogen and electrons; n_H and n_O are the volume densities of hydrogen and O° respectively; x_e is the fractional ionization and d is the thickness of the cloud.

Stark[14] showed that a brightness of 0.1 MJy ster^{-1}($\equiv$2.6 10^{-6} erg cm^{-2} s^{-1} ster^{-1}) is obtained for $T = 100$ K n_H=270 cm^{-3} , $n_O = [O]/[H]n_H = 8.3$ 10$^{-4}n_H$ (van Dishoeck and Black[15]) and $d = 1$ pc.

An increase of the oxygen line intensity towards the cloud centre (Fig. 4) cannot be explained from homogeneous cloud models[15,16]. However, if the cloud has a clumpy/filamentary structure, see Fig. 2, the radiation field (and thus the gas temperature) may be more or less constant through the whole cloud. As a result the line intensity goes as $n_H^2 d$. The increase of the oxygen line by a factor of 7 can then be explained as follows : $n_H d$ increases a factor of 3 and likewise does the density. The increase of $n_H d$ is in agreement with the observed increase of A_V from the edge (0.4) to the centre (1.1) of the cloud[13].

Additional evidence for a clumpy/filamentary structure comes from radio observations. Deul and Burton[20] find from H° measurements that cirrus clouds consist of superpositions of kinematically distinct components. Sometimes cirrus clouds contain significant molecular material : N(CO)$\sim 10^{16}$ cm^{-2} (Magnani et al.[21]). It is remarkable that clouds with such a low A_V and thus little shielding against photodissociation, show a significant CO column density. However, the CO can survive if small dense clumps exist. Falgarone and Pérault[22] find evidence for clumpiness from CO measurements on a very small scale (0.02 pc) : discrete clumps of ^{13}CO coherently moving with the low density ^{12}CO gas. They adress these density contrasts to turbulent motions within the clouds.

The oxygen line could also be collisional excited through shocks. Shocks are possible if the shock speed exceeds the Alfvén speed, which is of the order of a few km s^{-1} under typical interstellar conditions. Indications for shocks come from digitized optical plates (e.g Fig. 2), which show sometimes sharp transitions from the cloud to the background sky at one side of the cloud, whereas the other side of the cloud shows a very smooth transition. This so called head-tail like structure is observed in a number of cirrus clouds (Odenwald and Rickard[17]).

However, shock layers are expected to be very narrow (of the order 0.03 pc[18,19]) and an increase of the oxygen line from the edge of the cloud towards its centre over a distance of 0.4 pc (see Fig. 4 and assuming a distance from the cloud to the Sun of 100 pc) is therefore hard to explain, unless the shock is coming in from behind the cloud.

4. CONCLUSIONS

We have found several indications that the oxygen line might contribute significantly to the IRAS 60 μm band. The brightnesses we find correspond to a line intensity of the order of $10^{-6} - 10^{-5}$ erg cm^{-2} s^{-1} ster^{-1}, from the cloud edge to its centre respectively. A direct measurement of the O°(63 μm) line is therefore badly needed. In addition, measurements of both the C^+(158 μm) and the

O°(63 μm) FIR lines in combination with radio H° and CO line studies would give important insights in the gas temperature and the energy budget of these diffuse clouds.

Using the values of n_H and T from Sect. 3 and $n_{C^+} = 0.4[C]/[H]n_H = 1.9\ 10^{-4}n_H$ (van Dishoeck and Black[15]), we predict (analogous to Eq.(3)) the C^+ line intensity to be a factor 10 stronger than the O° line.

Today, only the NASA Kuiper Airborne Observatory is well suited to perform such a FIR study. The ESA Infrared Space Observatory, to be launched in 1993, will be capable to study many of the important FIR lines in a large number of cirrus clouds.

Acknowledgements

I would like to thank Ewine van Dishoeck and Harm Habing for their valuable comments on an earlier version of this manuscript.

References

1. Low, F.J. et al. : 1984 Astrophys. J. 278, L19
2. de Vries, C.P., le Poole, R.S. : 1985 Astron. Astrophys. 145, L7
3. Magnani, L., de Vries, C.P. : 1986 Astron. Astrophys. 168, 271
4. Mathis, J.S., Mezger, P. Panagia, N. : 1983 Astron. Astrophys. 128, 212
5. Spencer, R.G., Leung, C.M. : 1978 Astrophys. J. 222, 140
6. Black, J.H. : 1987 in Interstellar Processes, 731, eds,. D.J. Hollenbach, H.A. Thronson, Jr. (Reidel)
7. Chlewicki, G. : 1987 Astron. Astrophys. 181, 127
8. Harwit, M., Houck, J.R., Stacey, G.J. : 1986 Nature, 319, 646
9. Laureijs, R.J., Chlewicki, G., Clark, F.O. : 1988 Astron. Astrophys. 192, L13
10. Andreani, P. et al : 1990 preprint
11. Draine, B.T., Anderson, N. : 1985 Astrophys. J. 292, 494
12. Puget, J.L., Léger, A., Boulanger, F. : 1985 Astron. Astrophys. 142. L19
13. Stark, R. et al. : 1990 in preparation
14. Stark, R. : 1990 Astron. Astrophys. 230, L25
15. van Dishoeck, E.F., Black, J.H. : 1988 Astrophys. J. 334, 771
16. van Dishoeck, E.F., Black, J.H. : 1986 Astrophys. J. Suppl. Ser. 62, 109
17. Odenwald, S.F., Rickard, L.J. : 1987, Astrophys. J. 318, 702
18. Pineau des Forêts, G., Flower, D.R., Hartquist, T.W., Dalgarno, A. : 1986 M.N.R.A.S. 220, 801
19. Draine, B.T. : 1986, Astrophys. J. 310, 408
20. Deul, E.R., Burton, W.B. : 1990, Astron. Astrophys. 230, 153
21. Magnani, L., Blitz, L., Mundy, L. : 1985, Astrophys. J. 295, 402
22. Falgarone, E., Pérault, M. : 1988, Astron. Astrophys. 205. L1

The atomic hydrogen/molecular cloud association : an unavoidable relationship

G. JONCAS
Dept. de Physique
Université Laval
and Observatoire Astronomique du mont Mégantic
Québec, Québec
Canada G1K 7P4

1. INTRODUCTION

The presence of HI in the interstellar medium is ubiquitous. HI is the principal actor in the majority of the physical processes at work in our Galaxy. Restricting ourselves to the topics of this symposium, atomic hydrogen is involved with the formation of molecular clouds and is one of the byproducts of their destruction by young stars. HI has different roles during a molecular cloud's life. I will discuss here a case of coexisting HI and H_2 at large scale and the origin of HI in star forming regions. For completeness' sake, it should be mentionned that there are at least three other aspects of HI involvement : HI envelopes around molecular clouds, the impact of SNRs (see work on IC 443), and the role of HI in quiescent dark clouds (see van der Werf's work).

2. HIGH RESOLUTION OBSERVATIONS OF A HIGH LATITUDE GAS COMPLEX

This study was prompted by the IRAS discovery of large filamentary structures baptized cirrus. These have been correlated with local interstellar clouds, either atomic or molecular. Joncas, Boulanger and Dewdney (1990, in preparation) undertook the HI line study of such a cloud in order to understand their kinematical behaviour and density structure. The observations were obtained with the aperture synthesis radio telescope of the Dominion Radio Astrophysical Observatory (DRAO). The data set was completed by filling the inner portion of the uv planes with single dish observations. The resolution (FWHM) was 1'.0 x 1'.06 x 0.66 km s^{-1}. The results will be used to make comparisons with the structure of molecular clouds to get a better understanding of the transformation from one kind of cloud to the other.

The area we chose is part of a 21° x 12° (l x b) loop of atomic gas called the North Celestial Pole loop. This object is a complex of molecular and atomic clouds as demonstrated by de Vries et al. (1987) in a 9' (HPBW) survey of an 8° x 8° section of the loop. Our observations cover the section having the brightest HI emission. We resolved the HI feature into several filaments and clumps having different radial velocities. Emission was detected in the 14.21 to -3.1 km s^{-1} velocity range. Figure 1 is a gray scale representation of one of the velocity maps. The feature's morphology is truly amazing, showing criss-crossing filaments and clumps of all sizes. Examination of the HI spectra reveals complicated multi-component profiles with FWHM varying between 1 and 3 km s^{-1}. Assuming optically thin conditions and uncertain geometries, densities smaller than 100 cm^{-3} are derived. Using these quantities and the strength of the local magnetic field (Heiles 1989), it appears that the magnetic pressure dominates the forces at work on this HI cloud. The magnetic field may thus secure the coherence of the cloud's substructures. We know from de Vries et al.'s work that the HI and CO clouds are closely linked. One may now ask wether the HI feature originates from the dissociation of the nearby molecular cloud or if a molecular cloud is now forming after the crossing of the shock front that created the loop. The quiescence of the molecular cloud and the location of the features in the loop argue against a shock front presently crossing the cloud. So far our HI observations and CO observations of others indicate that both clouds are not in equilibrium. Since they are small entities (7.5 pc. if located at 100 pc. as proposed by de Vries 1987), they may represent a very early stage of molecular cloud formation.

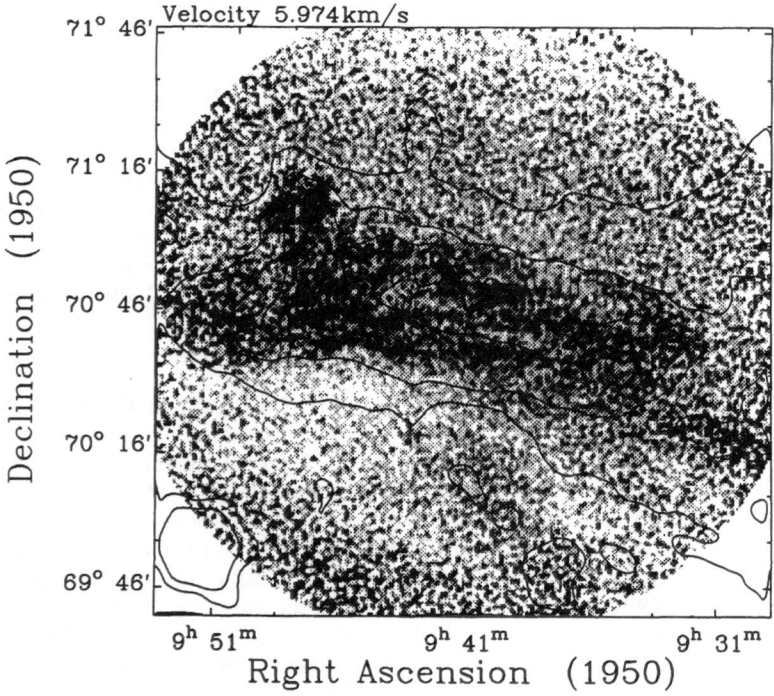

Figure 1. Map of HI line emission over a 2° field of view. There are 11 gray scale levels starting at 5 K with steps of 2.5 K. The overlaid contours are 100 μm emission at 2, 4, and 6 MJy/sr.

3. DISSOCIATED HYDROGEN

This section is a short summary of an ongoing project consisting on the study of HI associated with HII region/molecular cloud complexes. Such an association can be expected since newly formed massive stars produce substantial amounts of UV photons longward of the Lyman limit. Their absorption by the hydrogen molecule will produce its dissociation. Hill and Hollenbach (1978) proposed such a scenario where an H_2 dissociating front precedes the ionization front (IF) into the molecular cloud producing a layer of atomic gas. The HII region could dynamically affect this interface. Indeed the IF may have a shock front preceding it, advancing through the atomic layer. This interface, also called a photodissociation region, is important to a molecular cloud specialist, it is a source of information about the physics, chemistry and structure of molecular clouds not forgetting its impact on star formation. Read (1981, and references therein), using aperture synthesis, studied the kinematical behaviour of the HI component of 3 star forming complexes. His results brought the first strong qualitative evidence for the presence of dissociated hydrogen. This kind of work was quantitatively continued by Dewdney and Roger at DRAO (Roger and Dewdney 1986 and references therein).

One of the proofs for the existence of dissociated hydrogen came from a study of the S142 complex (Joncas et al. 1985). On the eastern side of the nebula is an HI feature (≈ 1200 M_o) whose west side contours follow the slope and curvature of those on the radio continuum quite accurately. The LSR velocity of the feature is identical to that of the CO cloud. Figure 2 is a plot showing a radial profile of continuum and HI emission, both averages over a 90° sector centered on the exciting star. Noteworthy is the local minimum of HI emission near the star and the fall off of the continuum emission as the HI emission rises up. The characteristics (mass and thickness) of the HI ridge are reproduced by the Dewdney-Roger (DR) model.

Another good example is the S187 complex (Joncas et al. 1990 in preparation). S187 is a textbook example of a young photodissociation zone. The HII region is surrounded by a clumpy HI shell. Its mean thickness is 1.1 pc and its total mass is 60 M_o. Applying the DR model such an amount of HI is produced in $\approx 10^5$ years using an observed density of 10^4 cm^{-3} for the molecular cloud. Theoretically the dissociation front is now 0.6 pc away from the exciting star. Close to the observed value of 0.9 pc (half intensity of the shell) for the eastern side. The mean velocity of the shell (-16.2 km s^{-1}) is somewhat blueshifted with respect to the velocity of the molecular cloud (-15 km s^{-1}). To better understand the behaviour of the HI shell, multi-molecule observations (^{12}CO, ^{13}CO, C^{18}O, CS) of the shell area were secured using the FCRAO 14 m antenna. Figure 3 is a ^{13}CO antenna temperature contour map of a 16' field centered on the HII region. The full line circle represents the extent of the ionized gas while the dashed line depicts the outer boundary of the HI shell. Both seem to sit in a molecular hole, confirming the eroding action of the exciting star. Also noticeable is the presence of the CO core depriving the HI shell of its sphericity. In fact the shell is twice as thin on its eastern side evidently because the dissociating photons have a harder time penetrating the denser material (more dust absorption).

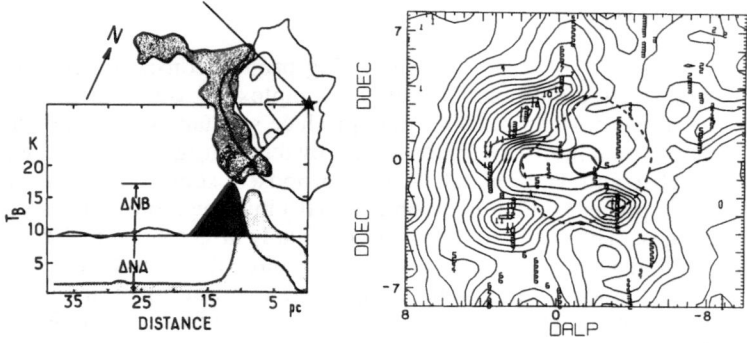

Figure 2 (left) Radial profiles of radio continuum brightness temperature (lower) and HI column density (upper) from S142. They are shown relative to the continuum emission (contours) and adjacent HI emission (shaded contour).

Figure 3 (right) ^{13}CO antenna temperature (T_A*) contour map of a section of the S187 molecular cloud. The full circle and dashed line circumscribe the HII region and the HI shell respectively.

Also apparent in Figure 3 is the presence of 3 major condensations. The northern one (the core) does not appear in our CS map but does so in $C^{18}O$. The southeastern one has a strong CS and $C^{18}O$ counterpart. This can be explained if the core is heated by the star forming activity (the HI shell seems in contact with it). The latter component is denser but not colder and does not seem in contact with the HI shell, it contains however a bipolar outflow. The data have not been completely analyzed yet. One of the goals is the comparison of the kinematical activity of the shell against the molecular cloud's. As mentioned previously the HI shell should be made up, at least in part, of shocked gas. However shock fronts of this type have never been observed. One possible reason for this is that the pressure transition usually thought to exist at an IF which would give rise to a shock is not present because of substantial heating of the atomic gas in advance of the front. If this is the case the HI gas will be expanding away from the molecular cloud.

To conclude I will emphasize further the importance of such studies by mentioning the discovery of a dissociating star by Dewdney et al. : a B4 star having a very small HII region (6.6 x 10^{-4} M_o) but a large amount of atomic gas surrounding it (1.4 M_o). HI observations could thus become tracers of B stars in molecular clouds.

de Vries, H. W., Heithausen, A., and Thaddeus, P. 1987, Ap. J. 319, 723.
Heiles, C. 1989, Ap. J.336, 808.
Hill, J. K. and Hollenbach, D. J. 1978, Ap. J. 225, 390.
Joncas, G., Dewdney, P.E., Higgs, L. A., and Roy, J.-R. 1985, Ap. J. 298, 596.
Read, P. L. 1981, M.N.R.A.S., 195, 371.
Roger, R. S. and Dewdney, P. E. 1986, in IAU symposium 115, Star Forming Regions, ed. M. Peimbert and N. Kaifu, (Dordrecht:Reidel), P. 203.

II - VELOCITY FIELD
AND
MAGNETIC FIELD

MAGNETIC FIELDS IN DENSE REGIONS

CARL HEILES
1989-1990 Visiting Fellow, Joint Institute for Laboratory Astrophysics
University of Colorado, Boulder CO 80309-0440 USA

ALYSSA A. GOODMAN
University of California, Berkeley, CA 94720 USA

CHRISTOPHER F. MCKEE
University of California, Berkeley, CA 94720 USA

ELLEN G. ZWEIBEL
University of Colorado, Boulder CO 80309-0440 USA

This paper concentrates on reviewing magnetic fields in dense regions; see Heiles (1988, 90) for a review of fields in diffuse regions. The past few years have increased our observational knowledge of magnetic fields in dense regions by an enormous factor—not because there are many measurements, but because we started from zero. The observable is polarization, which is small and subject to systematic errors. The fact that the advances have occurred only recently is a result of several factors: technological development; the interest and commitment of experimentally-minded astronomers; and the maturing of molecular and infrared astronomy to the point that really new results require either new insights or more difficult techniques.

This paper is a severe condensation of a more complete observational review (Heiles *et al.* 1990), which has an accompanying theoretical review denoted herein as Paper II (McKee *et al.* 1990). Previous observational reviews (Crutcher 1988, Troland 1990) summarize all existing observational results; here we cover results for only a few objects, but discuss them extensively. A word on notation. Equation (II 4.2) means equation (4.2) in Paper II. We use $\vec{B}$ to designate both the direction and magnitude of the magnetic field, $\hat{B}$ to designate the direction alone (within the two-fold ambiguity of 'which way the vector points'), and B to designate the magnitude alone. The subscripts $\parallel$ and $\perp$ designate the line-of-sight and the plane-of-the-sky components, respectively.

1. MANIFESTATIONS OF MAGNETIC FIELDS IN DENSE CLOUDS

1.1. Zeeman splitting.

The Zeeman effect arises from the coupling of an atom's or a molecule's magnetic moment with an external magnetic field. For any species with an unpaired electron spin the splitting $\Delta\nu_Z \sim 2$ Hz μG^{-1}, while for the much more common case of no unpaired electron spin but non-zero nuclear magnetic moment, such as H_2O, the splitting is smaller by roughly the ratio of the proton mass to the electron mass.

Under essentially all interstellar conditions except in OH masers, $\Delta\nu_Z \ll \Delta\nu$, where $\Delta\nu$ is the line width. This makes Zeeman splitting very difficult to detect. Success is favored by a species with both a large magnetic moment and a low line frequency (which increases the ratio $\Delta\nu_Z/\Delta\nu$), and also (of course!) by a high field strength. In this usual case, only the line-of-sight component $B_\parallel$ is obtained and all measured splittings must be adjusted for the projection factor to derive the full field strength. The median correction factor is $B = 2B_\parallel$ (Heiles and Troland 1982).

To date, the Zeeman effect has been detected from only four species in the interstellar medium: H I, OH, C_2S (Güsten and Fiebig 1990), and H_2O (Fiebig and Güsten 1989). B ranges from a few μG in H I to over 40 mG in H_2O masers.

The unpaired electron and low transition frequencies make OH the best tracer of magnetic fields in molecular regions. But just how well does OH trace H_2? The answer is not clear. In clouds that are not too dense, both observations and theory imply that OH is a good tracer of H_2. However, in cold dense clouds there are no observational indications, and theory implies that OH is *not* a good tracer of H_2.

Observationally, for moderate extinctions of < 7 mag the fractional abundance of OH with respect to H (total H, H I $+ 2H_2$) $X_{OH} \approx 4 \times 10^{-8}$ (Crutcher 1979). The major observational work that extends the range of *column* density as high as 7 mag is the detailed study of the ρ Oph dark cloud by Myers *et al.* (1978); this study finds the corresponding *volume* density to be $n_H \approx 2500$ cm^{-3}. Crutcher (1988) argues that OH traces n_H up to $\sim 3 \times 10^4$ cm^{-3} in TMC-1, but in our opinion this argument needs to be better substantiated and to be made quantitative. Theoretically, in clouds that are not too dense starlight dominates the ionization and is involved both in the production and destruction of many molecules. Under these conditions X_{OH} is predicted to be roughly independent of n_H for the range $250 < n_H < 1000$ cm^{-3}, although X_{OH} does depend on temperature, particularly for $T \gtrsim 50$ K (van Dishoeck and Black 1986). The way in which X_{OH} varies with n_H should be reliably predicted by the theory, but the actual theoretical value of X_{OH} depends on the exact process responsible for its formation (see below), which is uncertain; thus the theory cannot be expected to accurately reproduce the absolute abundance of OH. It seems reasonable to use the observations as a guide and extend the upper limit of the range of n_H from 1000 to at least 2500 cm^{-3}.

At high densities we must rely on theoretical astrochemical calculations. The question of what constitutes a 'high' density depends on the degree to which starlight is excluded, which in turn depends on the degree of nonuniformity of a cloud, and in particular its porosity. The absence of starlight makes a big difference because the sequence of reactions that form OH involves ion-molecule reactions, whose rate is limited by the ionization rate. Without starlight, ionization results from cosmic rays, which to first order makes n_{OH} independent of n_{H_2} (Herbst and Klemperer 1973). Calculations that include a more complete set of reactions (e.g. Leung *et al.* 1984) indicate a weak dependence, roughly $n_{OH} \propto n_{H_2}^{0.2}$. Recent uncertainties in the exact process by which OH is actually produced (Lepp, Dalgarno, and Sternberg 1987; Gredel *et al.* 1989) affect the absolute abundance of OH but should not affect this approximate result that n_{OH} is almost independent of n_{H_2}.

The independence of n_{OH} with respect to n_{H_2} is definitely not universally true at very high densities. Theoretically, high-temperature regions should produce copious amounts of OH. High temperatures can be produced by shocks, either with magnetic fields (Draine and Katz 1986) or without (Mitchell and Watt 1985); X_{OH} can be much higher than usual with shock velocities of somewhat less than 10 km s^{-1}. In OH masers, X_{OH} is high and $n_H \sim 10^7$ cm^{-3}; the high OH abundance can be understood only if these regions have been subjected to high temperatures.

1.2. Polarization of radio spectral lines arising from radiative transfer effects.

1.2.1. Emission lines: linear polarization. Linear polarization provides information on $\widehat{B}_\perp$, the projection of $\widehat{B}$ on the plane of the sky. The physical mechanism that produces linear polarization rests on the fact that π and σ components, which are orthogonally polarized, have different directional dependencies on their interaction with radiation when an external magnetic field orients the quantization axis. To obtain a significant population difference the molecule must be subject to a sufficiently intense anisotropic radiation field and the collisional rate must not be too high. In addition, the Zeeman splitting must be larger than the collisional and the spontaneous emission rates, but small compared to the line width. These conditions are easily satisfied for many transitions.

To discuss the predicted observables we consider two illustrative examples introduced by Kylafis (1983a) in which the anisotropy is provided by the velocity field. The velocity field is symmetric with respect to the z axis. In the 1-d example, the cloud expands in the z direction with uniform velocity gradient. In the 2-d example, the cloud expands axisymmetrically with no motion in the z direction. The magnetic field is parallel to the axis of symmetry and is perpendicular to the line of sight.

a. Intensity: Kylafis (1983a) provided solutions over the full range of physical parameters for his two examples, but restricted the treatment to a two-level system. Deguchi and Watson (1984) extended the treatment to the full rotational ladder for CO and CS, and found that the polarization was smaller than for the two-level case by a factor ~ 2; we use their results here.

The *fractional polarization*, denoted as $p_l = (U^2 + V^2)^{1/2}/I$, depends sensitively on C/A, the ratio of collisional rate to Einstein A, and also on the optical depth. As $C/A \to 0$, $p_l \overset{\sim}{\to} 0.10$; as C/A increases, p_l decreases significantly. The linearly *polarized intensity* $P_l = (U^2 + V^2)^{1/2}$ peaks at $C/A \sim 1$, at which point $p_l \sim 0.01$ for the 2-d and 0.03-0.12 (depending on the viewing angle) for the 1-d case.

These two simple examples show that p_l depends extremely sensitively on the velocity field. Simply going from 1-d to 2-d reduces the polarization to the point of being observable only with difficulty. It seems reasonable to conclude that for more realistic velocity fields the polarization will be very small indeed.

b. Direction: The polarization is either parallel or perpendicular to $\widehat{B}_\perp$, but it is not so easy to determine which. Again we consider the two examples of Kylafis (1983a). Consider the behavior as a function of α, the direction of the magnetic field. In the 1-d case, the polarization is perpendicular to the magnetic field for $54.7° < \alpha < 125.3°$, and parallel otherwise. In contrast, the 2-d case is *precisely opposite.*

In the absence of calculations for other, more realistic situations, we can only speculate. The direction of polarization is a sensitive function of the details of the velocity field. It seems likely that in more complicated cases the polarization direction might change direction rapidly with position, both on the sky and along the line of sight, and this should lead to much smaller p_l than is predicted for the simple situations. In this spirit, the negative results of attempts to observe polarization by Wannier, Scoville, and Barvainis (1983) and by Lis *et al.* (1988) come as no surprise.

1.2.2. Absorption lines: linear and circular polarization.   *a. Linear polarization:* Absorption lines seen against a background source are polarized by the same mechanism as emission lines. But there is an additional possibility. If the source is nearby, so that it subtends an appreciable solid angle as seen by the gas, then the source itself provides a non-isotropic source of radiation. This radiation can be either at the frequency of the line or at other relevant frequencies for excitation of the line such as far-IR for the OH molecule (Burdyuzha and Varshalovich 1972). This anisotropic radiation can affect the level populations to a greater degree than other local physical conditions if the source is strong. An extreme example is circumstellar material, in particular SiO masers.

Two idealized examples by Kylafis (1983b) produced detectably large polarization, and it seems worth making a serious observational effort to detect the linear polarization of absorption lines. Interpretation of results will not be without ambiguity, because as in the case of polarized emission lines the polarization is either parallel or perpendicular to $\widehat{B}_\perp$, depending on geometry. The requirement that the background source subtend a large solid angle is crucial to the mechanism, and observers should select sources for which other evidence favors this geometry.

b. Circular polarization: Circular polarization can exist whenever linear polarization can be produced by the mechanisms discussed above, which rely on differing opacities for the polarizations that are parallel and perpendicular to $\widehat{B}_\perp$ (the 'optical axes'). These opacities are the imaginary part of the index of refraction. According to the Kramers-Kronig dispersion relation, this inevitably leads to different phase velocities for orthogonal linear polarizations that are aligned along the optical axes. This turns linear polarization into circular polarization by a process that is similar to Faraday rotation, which occurs when the phase velocities are different for the orthogonal circular polarizations.

First, consider a linearly polarized background source. If the polarization direction is *not* parallel to either optical axis, then the wave can be decomposed into two waves in both axes. The differing phase velocities then produce elliptical polarization; the intensity of the circular component varies sinusoidally with distance along the direction of propagation. The polarization always changes sign across the line center, and looks similar to signature of the Zeeman effect. This is known as 'linear birefringence' (Kylafis and Shapiro 1983). V can be comparable to the change in linear polarization.

Next, consider an unpolarized background source. Here, the differing opacities produce linear polarization within the cloud. This polarization lies along one of the optical axes, so there is no possibility for circular polarization. However, if the direction of $\widehat{B}_\perp$ changes along the line of sight—the field 'twists' along the line of sight—then the optical axes follow the twist and linear birefringence can occur. The circular polarization depends on the amount of twist. For typical cases we expect $p_c \lesssim p_l^2$, because the circular polarization is a second-order effect.

c. Fake Zeeman splitting: In the latter 'unpolarized background source' case, p_c should be typically so small that it would be very difficult to detect, and we would thereby recommend that observers concentrate on other, less difficult observations. Nevertheless, the effect may loom large in importance because it mimics—and can easily be mistaken for—the Zeeman effect. The Zeeman effect produces small p_c, about $1.5\Delta\nu_Z/\Delta\nu$. This amounts to a percent or less in many cases, which is small enough to be produced by linear birefringence. If so, linear polarization should also be present.

1.2.3. Polarization in masers.   *a. Zeeman splitting:* Fiebig and Güsten (1989) have detected weak circular polarization in H_2O masers for one maser in each of four H II regions and derive typical $B_\parallel \sim 35$ mG; H_2O masers sample densities of 10^8 to 10^{10} cm^{-3} (Elitzur, Hollenbach, and McKee 1989; EHM). The circular polarization $p_c \sim 0.001$, which is extremely small. If linear polarization is present at the level of a few percent, the circular polarization could be a result of linear birefringence (section 1.2.2b, c). Future observations should include all Stokes parameters, which makes the task more difficult.

For OH masers associated with H II regions, Reid and Silverstein (1990) have compiled and assessed many measurements. Field strengths range up to 7 mG and average 3.6 mG. OH masers sample densities of 10^6 to 10^8 cm^{-3} (Reid and Moran 1981). If we ignore the possibility that the H_2O and OH field strengths are biased towards high values because of observational selection effects, then we conclude, very roughly, that in this regime $B \propto n^{1/2}$.

Reid and Silverstein come to a remarkable conclusion: the *directions* of the fields in OH masers are systematically aligned over large segments of the Galaxy. This indicates that the field direction is largely preserved during the contraction of clouds to high volume densities. This is surprising, because during this contraction process one would expect the competing effects of gravitation, angular momentum, and shocks to randomize the field in the dense cloud with respect to the surroundings. We discuss a particular example, Orion, below in section 2, and conclude that the coincidence of the field directions in Orion and the Galaxy may be accidental; this makes the conclusion of Reid and Silverstein either suspect or even more remarkable.

b. Linear polarization: Linear polarization is produced by the same basic mechanism as discussed above in section 1.2.1. The production of linear polarization in J=1–0 transitions requires that several constraints on ratios of pumping rates, decay rates, and Zeeman splitting be satisfied (Goldreich, Keeley, and Kwan 1973). However, extremely recent calculations by Deguchi and Watson (1990) and Nedoluha and Watson (1990) show that for higher-J transitions, such as in H_2O masers, the constraints are more severe. The existence of detectable linear polarization in H_2O masers implies field strengths of ~ 30 mG, which is commensurate with the few results obtained directly from Zeeman splitting of H_2O masers.

1.3. Linear polarization caused by aligned grains.

Optical starlight is almost universally linearly polarized, to a degree that increases with extinction. The polarization arises from absorption by systematically oriented dust grains. Only magnetic orientation is sufficiently general and powerful to provide the universality, although in individual circumstances other agents, such as gas streaming or photons, may dominate (e.g. section 2.2). An excellent contemporary review of these matters is given by Hildebrand (1988).

In magnetic alignment, a needle-like grain spins primarily end-over-end around an axis that is parallel to $\widehat{B}$. The polarization that is most strongly absorbed by this field of systematically aligned grains lies along the longest projected grain area, which is perpendicular to $\widehat{B}_\perp$. Therefore, the observed linear polarization of absorbed light is parallel to $\widehat{B}_\perp$. The alignment also produces linear polarization of the grains' thermal emission at far-infrared and sub-mm wavelengths, which is perpendicular to $\widehat{B}_\perp$.

2. OBSERVATIONAL EXAMPLES: ORION

The Orion region occupies a place of central importance because it is the nearest region of massive star formation. Orion A is a 'blister' H II region, protruding from the near side of a dense molecular cloud within which the current star formation activity is occurring.

2.1. $B_{\parallel}$ from Zeeman splitting. Volume densities sampled by the different observations range from $\gtrsim 400$ to $\sim 10^9$ cm^{-3}, a range of about 6 orders of magnitude, and the total magnetic field strengths B range from ~ 0.05 to ~ 40 mG, a range of about 3 orders of magnitude. Thus, very roughly, $B \propto n^{1/2}$.

a. Ambient gas: The ambient gas on the near side of the H II region is sampled by absorption lines. The H I absorption line comprises two velocity components. In our opinion, this H I is not in the photodissociation region (PDR) and does not directly abut the H II region. We support our opinion by two facts. One, the H I velocities of ~ 0 and 6 km s^{-1} are 3 and 9 km s^{-1} more *positive* than the H II velocity, while the absorption gas is in front of the H II region and should be moving at *negative* velocity with respect to the H II region. Two, the H I is cold, but given the exciting stars for Orion the H I in the PDR should be warm (Tielens and Hollenbach 1985).

In the H I absorption lines, Troland, Heiles, and Goss (1989) mapped the field strength with 25" resolution over most of the H II region, which occupies an area about 5' in diameter. They found $B_{\parallel}$ to range from -43 to -107 μG. In both OH and H I absorption lines, Troland, Crutcher, and Kazès (1986) measured Zeeman splitting for a single spot on the eastern edge of the H II region; they obtained -125 and -49 μG, respectively. We adopt $B = -100$ μG, and if this uniformly fills a 5'-diameter circle, the magnetic flux is -2.0 mG arcmin^2.

b. The dense molecular gas: This gas lies on the far side of the H II region. The molecular 'doughnut' (Plambeck *et al.* 1982) is a disk of dense (up to $\sim 10^7$ cm^3) gas with outer diameter $\sim 44"$ and a hole in the middle. It surrounds IRc2 and expands at ~ 20 km s^{-1}. Its total mass is ~ 15 M$_{\odot}$, which seems to be too massive to have been ejected directly from IRc2. The expansion time scale (radius/velocity) ~ 3000 yr and the kinetic energy $\sim 5 \times 10^{46}$ erg, which amounts to $\sim 0.5\%$ of the luminosity of IRc2 in 3000 yr. Thus it is reasonable to consider the doughnut as being composed of ambient material that has been swept up—i.e., shocked—by less massive, faster-moving or high-temperature ejecta from IRc2.

There are two classes of H$_2$O maser, the 18 km s^{-1} ('low-velocity') flow and the 30-100 km s^{-1} ('high-velocity') flow (Genzel *et al.* 1981). The low-velocity masers lie in the plane of the doughnut and trace its outside with diameter $\sim 44"$; they have $n_H \sim 10^9$ cm^{-3} and $B \sim 30$ mG, corresponding to an Alfvén velocity $V_A \simeq 1.8$ km s^{-1}. The high-velocity masers lie on the axis of the doughnut and are probably the equivalent of Herbig-Haro objects.

Most of the OH masers lie within 8" of IRc2 and have velocities within 18 km s^{-1} of IRc2 (Johnston, Migenes, and Norris 1989). Thus it seems that the OH masers trace the *inside* of the doughnut, because both the velocities and gas densities match. The OH masers have $n_H \sim 10^7$ cm^{-3} and $B \sim 1.4$ mG, which gives $V_A \simeq 0.8$ km s^{-1}. Comparing the H$_2$O and OH masers implies that the Alfvén velocity is constant to within a factor ~ 2, or *very* roughly, $B \propto n_H^{1/2}$.

How are the masers excited? EHM have suggested that H$_2$O masers arise in shocks. Their calculations were for the case of fast, dissociative shocks, but the conditions in slower, non-dissociative shocks can be similar, so it is likely that the low-velocity H_2O masers arise in shocks. Much of the emission in non-dissociative shocks occurs near the velocity of the *unshocked* gas, where the heating due to ambipolar diffusion is greatest. However, in order to get saturated maser emission in masers of the observed size (which is comparable to the shock thickness), the density must be of order 10^9 cm^{-3} (EHM); this suggests that the maser emission occurs at the density, and hence the velocity, of the *shocked* gas. The location of the H_2O masers at the outer perimeter of the doughnut is consistent with the idea that they are in recently shocked gas.

The OH maser is a ground state transition, so exciting it does not require the temperature to be above several hundred degrees as does the water maser. (Production of OH requires either high temperatures [section 1.1] or photodissociation of H_2O, though.) The location of the OH masers at the inner edge of the doughnut is consistent with either radiative pumping or with weak shocks due to variations in the wind luminosity of IRc2. However, we are left with the puzzle as to why the pressure in the H_2O masers is so much greater than that in the OH masers, even though their radial velocities are the same.

c. Connection of the ambient to the dense gas: The OH masers see the magnetic field near the inside of the doughnut, ~ -1.4 mG, and the H_2O masers at the outside, ~ -30 mG. We don't know how the field strength varies within the doughnut, but if the average field strength is -5.4 mG then the doughnut, which we take as a circle $44''$ in diameter with a $16''$ diameter hole, contains the same magnetic flux as the ambient gas, ~ -2.0 mG arcmin^2.

There is, in fact, no reason to assume that the fluxes are equal because the ambient-medium flux was obtained from the H I absorption observations and thereby depends on the size of the H II region, which is determined by completely different considerations. However, if the fluxes are, in fact, equal then the field lines can connect the dense gas to the ambient gas, while satisfying the constraint $\nabla \cdot \vec{B} = 0$, without bending unduly sharply. This is because the H II region is roughly spherical (Balick, Gammon, and Hjellming 1974), so that as the field lines diverge from the doughnut, which is located behind the H II region, to the ambient gas, which is located in front, they need not bend by more than $30°$.

The *sign* of $B_{\parallel}$ is negative for *all size scales*. This would seem to be a single example of the general trend for OH masers that the maser field directions reflect the Galactic or ambient field directions (section 1.2.3a.). However, the situation is not quite so clear-cut: Zeeman observations of OH in absorption against Orion B (NGC2024), only 4 degrees away from Orion A, show a *positive* field (Heiles and Stevens 1986). Thus, Orion A and Orion B cannot both reflect the ambient Galactic field. We suspect that the last word has not yet been written concerning the general trend for OH maser fields to reflect the ambient Galactic field.

2.2. $\widehat{B}_{\perp}$ from linear polarization. Next, we consider linear polarization, which indicates the direction of $\widehat{B}_{\perp}$. Polarization of the far-IR continuum emission from dust grains has been measured with $40''$ angular resolution by Novak *et al.* (1989) and Gonatas *et al.* (1990), and that of mm-wave emission with about $22''$ resolution by Barvainis, Clemens, and Leach (1988) and Novak, Predmore, and Goldsmith (1990). Position angles for the far-IR and mm wavelengths are similar, which means that the polarization is produced by aligned grains. If the grains are oriented by the magnetic field, then the position angle of the field as derived from dust, $\theta_{B,dust}$, is about 110°.

Aitken et al. (1985) made spectropolarimetric observations of the diffuse mid-IR (8-13 μm) continuum with 4″ angular resolution on six IR sources within $\sim$ 12″ of IRc2. For all positions except IRc2, the spectral dependence of the polarization follows that of the overall grain opacity, which means that the polarization is produced by absorption of aligned grains; polarization directions differ from those at shorter wavelengths, where polarization results from scattering (Werner, Dinerstein, and Capps 1983). Importantly, the position angle varies considerably among the six positions: the polarization vectors all tend to point towards IRc2. Aitken et al. use this fact to argue that the grains are oriented not by the magnetic field, but by the photons from IRc2.

Whatever the mechanism of grain alignment, the observational fact is that 4″ resolution reveals structure in the mid-IR polarization. Thus the alignment direction of the grains changes on scales of a few arcsec, and the coarser angular resolution that has been used at far-IR and mm wavelengths is inadequate. For the BN position, the values of $\theta_{B,dust}$ derived from the mid-IR and far-IR agree, possibly because the average of the mid-IR Stokes parameters over the 40″ far-IR beam yields a position angle of 109° (Novak et al. 1989).

$\theta_{B,dust} \approx$ 110° is in very good agreement with the position angle of $\sim$ 100° for optical polarization of background stars over an area $\sim$ 1° in extent. Optical polarization indicates the field direction in the ambient matter surrounding the dense molecular cloud. Thus we obtain a nice picture, one that is consistent with the Zeeman results: the field direction in the dense cloud is similar to that in the immediately-adjacent medium.

Position angles describe several other phenomena in the innards of Orion, and we might expect them to be related to the local direction of the field. These phenomena include H_2O maser polarization and the orientations of various dynamical or structural features. We exhibit them in Figure 1 and discuss them in the following paragraphs:

(1) The low-velocity H_2O masers tend to lie on a line having position angle $\theta_{line} \sim$ 30° (Knowles and Batchelor 1978). EHM predict that the maser luminosity $\propto B_{0\perp}^4$, where $B_{0\perp}$ is the component of the preshock magnetic field perpendicular to the shock velocity. The shock moves radially outward from IRc2. $B_{0\perp}$ is largest, and the masers should be strongest, where the shock moves orthogonally to $\widehat{B}_\perp$. Thus θ_{line} should lie 90° away from $\theta_{B,dust}$; the actual value is only $\sim$ 10° different.

(2) The group of most intense OH masers, populations 1 and 2 of Norris (1984), lie on a line having position angle $\sim$ 60°. This differs by 40° from the value expected from the EHM theory, which may not apply to OH masers. The OH masers lie toward the inside of the doughnut, while the H_2O masers lie on the outside; the field directions may differ in these regions.

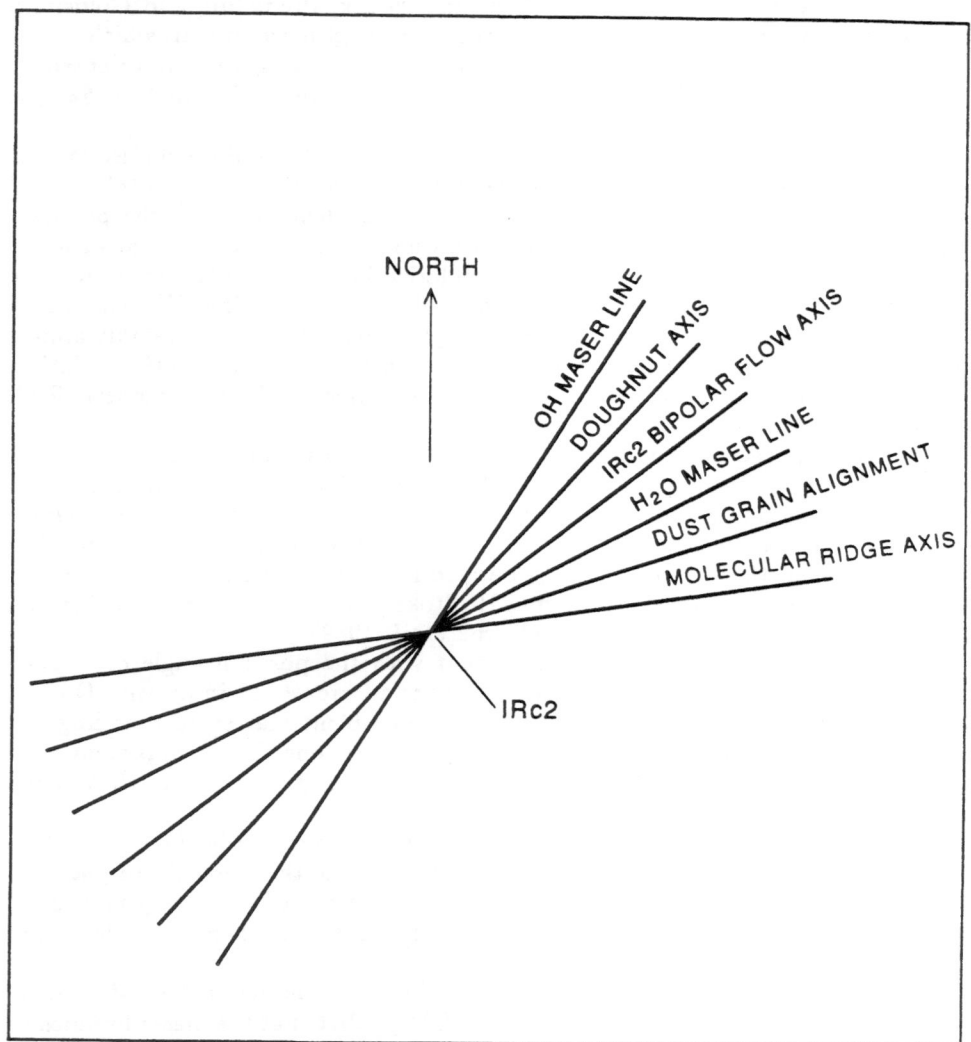

Figure 1. Directions of $\widehat{B}_\perp$ near IRc2 in Orion derived from the possible indicators discussed in the text. Aside from observational uncertainties, differences might be caused by changes in $\widehat{B}_\perp$ with position.

(3) On the very smallest scale, the position angle of the bipolar flow from IRc2 is ∼ 130°; this is our estimate using the SO map of Erickson *et al.* (1982) and the SiO map of Wright *et al.* (1983). This is only about 20° from $\theta_{B,dust}$.

(4) Judging from the map of Vogel *et al.* (1984), the axis of the molecular doughnut has position angle 140°, ∼ 30° from $\theta_{B,dust}$.

(5) On the much larger scale of $\sim 0.5°$ (~ 4 pc), the 'molecular ridge' in which the Orion complex is imbedded has a position angle of $\sim 10°$, and it contains filaments running along its length (Bally *et al.* 1987). This is nicely aligned (perpendicularly) to $\theta_{B,dust}$.

Figure 1 shows that the inferred $\theta_B = 130° \pm 20°$. It is quite well defined given the uncertainties in measurement and interpretation. This is particularly true because the projection from 3-d to the 2-d plane-of-the-sky tends to amplify angular differences.

We conclude that position angles of various structural features may provide significant information on the magnetic field. However, we need much more than a statistical sample of one H II region! There is much to be done in the way of further observations, both on Orion itself and on other H II regions. We again remark that the mid-IR results show that high angular resolution is required, and that alignment of dust grains may occur by nonmagnetic processes.

3. OBSERVATIONAL EXAMPLES: DERIVING $\vec{B}$

Myers and Goodman (1990; MG90) have devised a method to almost completely specify the uniform component of $\vec{B}$. Consider the total magnetic field $\vec{B}$ at any point in space to be the vector sum of a straight 'uniform' field, $\vec{B_0}$, and a spatially varying, nonuniform field, $\vec{B_r}$. Individual Zeeman observations sample $B_{\|}$ and provide estimates of both $B_{\|,0}$ and the dispersion of the line-of-sight component of B_r, $\sigma_{B_{\|}}$. Similarly, optical polarization maps provide the mean position angle of $\widehat{B}_{\perp}$, $\langle \theta_B \rangle$, and its dispersion σ_{θ_B}.

$B_{\perp,0}$ cannot be derived directly from the measured polarization because too little is known about the grain alignment process. Nevertheless, its mean value can be determined statistically by using a model. The model predicts an observed distribution of θ_B in which the ratio $B_{\perp,0}/\sigma_{B_{\perp}}$ is the only free parameter. In most cases, the expected distribution is well-represented by a Gaussian, and σ_{θ_B}, the 1-σ width of the distribution, is a function only of the ratio $B_{\perp,0}/\sigma_{B_{\perp}}$ and N, the number of independent samples of B_r along the line of sight. The model also relates $\sigma_{B_{\|}}$ to $\sigma_{B_{\perp}}$. The resulting $B_{\perp,0}$, together with $B_{\|,0}$ and $\langle \theta_B \rangle$, provide $\vec{B}_0$ within a two-fold directional ambiguity. $\sigma_{B_{\|}}$ and $\sigma_{B_{\perp}}$ provide B_r, and combining B_0 and B_r in quadrature yields B.

Models include one for which $\vec{B_r}$ is essentially two-dimensional, resulting from Alfvén waves (Zweibel 1990; MG90), and a 'turbulent' field model, in which $\vec{B_r}$ is distributed isotropically in three dimensions (MG90). The isotropic model may be unrealistic because transverse fluctuations undergo weaker nonlinear damping than do aligned or longitudinal fluctuations. An ideal model would incorporate both density and magnetic field fluctuations using the concepts first enunciated by Chandrasekhar and Fermi (1953). In any model there is the practical problem of separating fluctuations from systematic gradients.

We mention two important caveats. First, the Zeeman and polarization observations highlight different portions of the cloud unless care is exercised in selecting the statistical samples. Zeeman observations emphasize the regions of high $B_{\parallel}$, which tend to be those of high n; polarization observations are most easily obtained in regions of low extinction, which tend to have low n. Second, the result depends on the model used to describe the fluctuations in $\vec{B}$. Magnetic fluctuations are not inherently isotropic, so the relation between $\sigma_{B_{\parallel}}$ and $\sigma_{B_{\perp}}$ depends on $B_{\parallel,0}/B_0$, which is itself derived from the model. Finally, in the best of worlds the model is consistent both with basic physical principles and with their application to the specific physical conditions in the region.

In the dark cloud Lynds 204 (L204), Heiles (1988) has made H I Zeeman measurements on a grid of 27 positions covering approximately a 6 by 15 pc area. McCutcheon *et al.* (1986) have made well-sampled optical polarization measurements. These data are sufficient to apply the technique. We quote results for the $N = 1$ case and the self-absorption component only, which should be representative of the denser parts of the cloud.

MG90 find $B_{\parallel,0} = 7.6$ μG and $\sigma_{B_{\parallel}} = 4.7$ μG. They use the isotropic model and find $B_{\perp,0} = 16.9$ μG, $B_0 = 18.5$ μG, $B_r = 8.1$ μG, and $B = 20.2$ μG. L204 is in approximate virial equilibrium, and this field value is in reasonable agreement with the expectations from Paper II (Heiles 1988). Heiles finds two correlations, one between the cloud shape and line-of-sight velocity and one between $B_{\parallel}$ and line-of-sight velocity, which allow a resolution of the two-fold ambiguity in $\widehat{B}_{0,\perp}$ and $\widehat{B}_0$. The correlations also imply the presence of large-amplitude Alfvén waves, which may make the isotropic model inapplicable.

4. OBSERVATIONAL EXAMPLES: THE VIRIAL THEOREM AND CLOUDS

4.1. The virial theorem: observational considerations

Many observational treatments of molecular clouds rely on the virial theorem, written for a spherical cloud of uniform density and mass M:

$$|W| + 3P_0 V = 2T + \mathcal{M} \ , \tag{4.1}$$

where W is the gravitational potential energy $-3GM^2/5R$, $3P_0V$ is the external pressure term including turbulent pressure, T the total kinetic energy $0.27M\,\Delta v^2$ ($2T$ is $3\bar{P}V_{cl}$), and $\mathcal{M}$ the total *net* magnetic energy (including surface terms) $0.1B^2R^3$, where the factor 0.1 comes from Tomisaka, Ikeuchi, and Nakamura (1988). Here we use the observationally-oriented Δv, which is the line width at half peak intensity; it is related to the theoretically-oriented one-dimensional velocity dispersion, σ, by $\sigma^2 = 0.18\,\Delta v^2$.

This allows the overall equilibrium to be described in terms of global cloud parameters, which can be derived directly from observational data. This is a 'minimalist' approach, because it avoids dealing with the internal structure of clouds. The problem is the uncertainties in the virial terms. This problem is well exemplified by our discussion of the B1 cloud below in section 4.3 and, apart from matters such as surface terms and nonuniform cloud structure, involves three specific observational quantities: the assumed distance, the conversion of $B_\parallel$ to B, and the determination of the true H_2 content.

The distance uncertainty can be very serious and should never be ignored or minimized.

The second problem is a matter of projection angle. We usually measure $B_\parallel$ instead of B. For a large sample $\langle B \rangle = 2B_\parallel$. However, in any individual case we can only be sure that $B \geq B_\parallel$. The matter is important, because the magnetic term in the virial theorem $\mathcal{M} \propto B^2$, and the arbitrary use of the relation $B = 2B_\parallel$ increases $\mathcal{M}$ by a factor of 4 over its minimum value.

The third problem involves determination of the volume density n_{H_2} or, equivalently, the column density N_{H_2} or cloud mass M. Derived values of neither volume nor column density are generally accurate within a factor of two or three. Turner and Ziurys (1988) provide a contemporary summary of this problem.

Factors of two are good from the standpoint of astronomical accuracy but inadequate for a definitive discussion involving the virial theorem, as we shall see in our discussion of B1 (section 4.2) below. If $\mathcal{M}$ is important, then all terms in the virial theorem are likely to be comparable, and the uncertainties allow virial equilibrium whether $\mathcal{M}$ is included or not.

4.2. B1: Observations.

B1 is, to our knowledge, the dark cloud about which most is known from the combined standpoints of structure and magnetic field, and represents the best example for application of the virial theorem. B1 is a well-defined cloud in the Perseus region which has been mapped in several molecular transitions, including ^{13}CO with 4.4 arcmin resolution by Bachiller and Cernicharo (1984); OH with 3 arcmin resolution by Goodman *et al.* (1989; hereafter GCHMT); and NH_3 with 40 arcsec angular resolution by Bachiller, Menten, and del Río-Alverez (1990, hereafter BMdR). For these molecules—^{13}CO, OH, and NH_3—the sizes decrease and the line widths decrease monotonically. This is a reasonable result, because the three molecules trace increasingly dense regimes of H_2 volume density, and according to 'Larson's laws' (Larson 1981; Solomon *et al.* 1987), size $\propto n_{H_2}^{-1}$ and line width $\propto n_{H_2}^{-1/2}$.

The appearance of B1 in these three molecules is depicted in Figure 2. As the angular scale gets smaller B1 gets more complicated. We define the cloud *core* as the $\sim 3'$-diameter portion of the cloud highlighted by the NH_3 lines. The core contains an IRAS source that exhibits an apparent outflow, and thus is a protostar or newly-formed star. The cloud core consists of two or three condensations, which

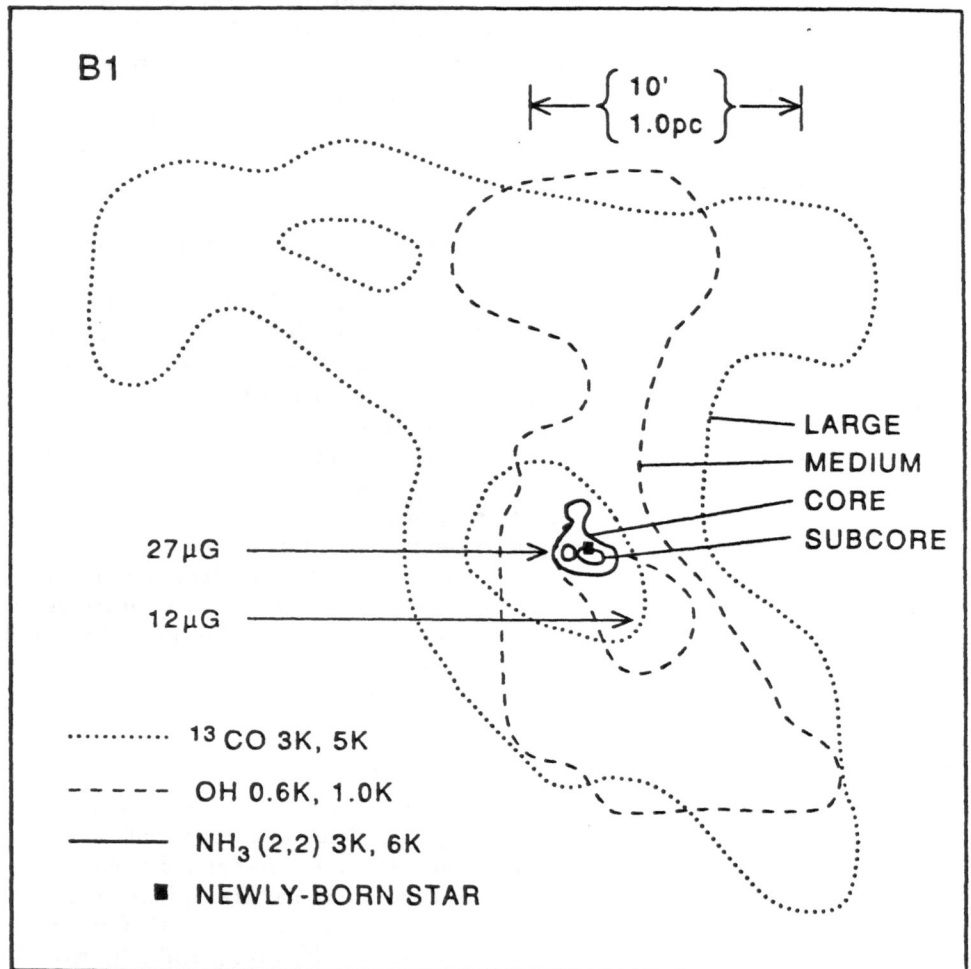

Figure 2. B1 as observed in ^{13}CO, OH, and NH$_3$. The three molecules highlight increasingly smaller size scales (see text). Only two contours are shown for each molecule, one near the peak and one about half the peak intensity. The four size ranges, ranging from 'large' to 'subcore', are indicated, as are the two $B_\parallel$'s derived from OH Zeeman splitting. There are several young stars in the region; we show only the one located in the core.

we define as *subcores*, separated by about 1.2 arcminutes, in which $n_{H_2} \sim 4 \times 10^4$ and 8×10^4 cm^{-3}. If these subcores themselves have smaller substructure, it would not have been resolved by existing observations.

Zeeman splitting of the OH 1665 and 1667 MHz lines was measured by GCHMT and Goodman (1989). On the core, $B_\parallel \approx -27$ μG, while 4' to the southwest $B_\parallel \approx -12$ μG (Figure 2). Goodman's (1989) off-core result of only -12 μG is less than half the field strength measured on the core, which implies that the field strength increases in the core. Further, if we were able to sample exclusively the dense core gas, then the measured field strength might be even greater than the -27 μG obtained from the OH, because as discussed in section 1.1 the OH may not sample the densest portions of the core.

The question of just how well OH traces high-density regions is a major one and deserves intense observational and theoretical attention.

4.3. The virial theorem applied to B1.

Figure 2 depicts four size ranges of B1: 'large', 'medium', 'core', and 'subcore'. We have reasonably complete information on physical conditions for each size range, which allows us to evaluate the virial terms in equation (4.1) and other quantities. Unfortunately, we do not have a complete sample of $B_\parallel$ in the four size ranges, and there is always the annoying matter of converting $B_\parallel$ to B. We assume $B = 2B_\parallel$. For the medium size and core the observations provide $B_\parallel = -12$ and -27 μG, respectively. For the subcore we assume that $B_\parallel$ is the same -27 μG that it is for the core. This may well be incorrect, because $B_\parallel$ increases as we go from the medium size to the core, but it serves as a reasonably interesting example.

The external pressure term is not usually included in observational discussions, but it is important for the three smallest size ranges. Let P_0 be the external pressure at the boundary of each size range. At the boundary of the largest size range, P_0 should equal the ambient interstellar pressure, for which we adopt $P_0 = 1.6 \times 10^4 k_B$ (Paper II). For the smaller size ranges, we set P_0 equal to the total pressure of the gas in the next larger size, calculated from the line width and density. That is, we take $P_0 = \rho\sigma^2 = 0.18 \times 10^{10} \rho \Delta v_s^2$, which amounts to assuming that the velocity field is 'microturbulent'. This assumption is theoretically justified, because in the subcore the damping length of MHD turbulence is about 0.005 pc (equation II 4.2), which is 20 times smaller than the size of the subcore.

Table 1 gives numerical values for the virial terms. We emphasize that the accuracies are low. Virial equilibrium is easily achieved for any size by changing the mass by a factor of less than two, which is within the observational uncertainties. For all size ranges, the virial theorem is satisfied as well by excluding $\mathcal{M}$ as by including it, which is the inevitable consequence of the uncertainties and having all terms in the virial theorem comparable.

We now we forge ahead and discuss the stability of the virial equilibrium, temporarily assuming absolute accuracy. We do this with ratios of the mass to three different critical masses. Table 1 lists the cloud mass M in terms of the magnetic critical mass M_Φ (equation II 2.16), the Jeans (gravitational) critical mass M_J (equation II 2.12), and the combined magnetic/gravitational critical mass M_{cr} (equation II 2.19). All sizes have $M/M_\Phi > 1$ and are magnetically supercritical, which means that without gravity they would expand. Nevertheless, the magnetic field is crucial for the stability of the medium size and core. These have

TABLE 1

B1 AT VARIOUS SIZE SCALES

SIZE	$n(H_2)$ cm^{-3}	B μG	MASS M$_\odot$	$-W$ erg	$3P_0V$ erg	$2T$ erg	$\mathcal{M}$ erg	M/M_Φ	M/M_J	M/M_{cr}
Large[a]	900	-10^b	630	1.5(46)	.20(46)	3.0(46)	.07(46)	4.5	0.43	0.34
Medium[a]	2600	-24^c	120	1.5(45)	2.0(45)	2.3(45)	.30(45)	2.2	2.2	1.13
Core[a]	20000	-54^c	10	4.3(43)	2.4(43)	5.3(43)	1.6(43)	1.64	2.3	0.99
Subcore[d]	70000	-54^b	2.5	6.4(42)	3.8(42)	13.0(42)	1.1(42)	2.4	0.85	0.65

a. From GCHMT, taking R in their Table 2 as the cloud diameter.

b. Assumed. For the core, where B probably increases, this value is probably a lower limit.

c. Twice the measured value, to account for orientation to line of sight.

d. From Bachiller, Menten and del Rio-Alvarez (1990); these parameters apply to a single NH$_3$ subcore, and differ from the parameters given by those authors, which apply to the whole NH$_3$ core.

$M/M_J \approx 2.2$, which means that in the absence of the magnetic field they would be unstable to collapse. But M_{cr} includes the magnetic support, and these two sizes have $M/M_{cr} \approx 1$, which either removes the instability or makes it much less pronounced.

We emphasize again that the statements in the above paragraph are based on uncertain parameters. If we had used a more conventional definition of cloud radius, or different values for n_{H_2}, we would have reached different conclusions. Nevertheless, we are encouraged by the result $M/M_{cr} \approx 1$ for the core. The core has split into three subcores, and it also contains a protostar or new star. These are *observational* indications that the core of B1 is on the verge of instability. With $M/M_{cr} \approx 1$, we have a *theoretical* indication that it is close to instability. This argues that our numerical estimates in Table 1 are reasonably accurate.

ACKNOWLEDGMENTS

We thank Rafael Bachiller, Vladimir Burdyuzha, Dick Crutcher, Rolf Güsten, Roger Hildebrand, Terry Jones, Karl Menten, Mark Reid, and Tom Troland for providing unpublished material; and John Bally, Gary Fuller, Miller Goss, Roger

Hildebrand, Terry Jones, Nick Kylafis, Phil Myers, Barry Turner, and Paul van der Werf for instructive discussions. CH was supported in part by NSF grant 443836-21705. AG is supported in part by a President's Fellowship at the University of California, Berkeley. CFM's research is supported by NSF grant AST-8918573; his research on star formation is supported in part by a NASA grant to the Center for Star Formation Studies. EAZ was supported in part by NSF grant ATM85-60032 and the Institute for Theoretical Physics NSF grant PHY89-04035.

REFERENCES

Aitken, D.K., Bailey, J.A., Roche, P.F., and Hough, J.M. 1985, *M.N.R.A.S.*, **215**, 815.

Bachiller, R., Menten, K.M., and del Río-Alverez, S. 1990, *Astron. Ap.*, **000**, 000.

Bachiller, R. and Cernicharo, J. 1984, *Astron. Ap.*, **140**, 414.

Balick, B., Gammon, R.H., and Hjellming, R.M. 1974, *Pub. Astr. Soc. Pacific*, **86**, 616.

Bally, J., Langer, W., Stark, A.A., and Wilson, R.W. 1987, *Ap. J. (Letters)*, **312**, L45.

Barvainis R., Clemens, D.P., and Leach, R. 1988, *A.J.*, **95**, 510.

Bel, N. and Leroy, B. 1990, to appear in *Galactic and Extragalactic Magnetic Fields*, eds. R. Beck, P.P. Kronberg, and R. Wielebinski, (Dordrecht: Kluwer).

Burdyuzha, V.V. and Varshalovich. D.A. 1972, *Astron. Zh.*, **49**, 727.

Chandrasekhar, S. and Fermi, E. 1953, *Ap. J.*, **118**, 113.

Crutcher, R.M. 1979, *Ap. J.*, **234**, 881.

Crutcher, R.M. 1988, in *Molecular Clouds in the Milky Way and External Galaxies*, ed. Dickman, R.H., Snell, R., and Young, J., p. 105.

Deguchi, S. and Watson, W.D. 1984, *Ap. J.*, **285**, 126.

Deguchi, S. and Watson, W.D. 1990, *Ap. J.*, **354**, 649.

Draine, B.T. and Katz, N. 1986, *Ap. J.*, **310**, 392.

Elitzur, M., Hollenbach, D.J., and McKee, C. 1989, *Ap. .J.*, **346**, 983 (EHM).

Erickson, N.R., Goldsmith, P.J., Snell, R.L., Berson, R.L., Hugukenin, G.R., Ulich, B.L., and Lada, C. 1982, *Ap. J..*, **261**, L103.

Fiebig, D. and Güsten, R. 1989, *Astron. and Astrophys.*, **214**, 333.

Genzel, R., Reid, M.J., Moran, J.M., and Downes, D. 1981, *Ap. J.*, **244**, 884.

Goldreich, P., Keeley, D.A., Kwan, J.Y. 1973, *Ap. J.*, **179**, 111.

Gonatas, D.P., Hildebrand, R.H., Platt, S.R., Wu, X.D., Davidson, J.A., Novak, G., Aitken, D.K., and Smith, C. 1990, *Ap. J.*, submitted.

Goodman, A.A. 1989, Ph. D. Thesis, Harvard University.

Goodman, A.A., Crutcher, R.M., Heiles, C., Myers, P.C., and Troland, T.H. 1989, *Ap. J.*, **338**, L61.

Gredel, R., Lepp, S., Dalgarno, A., and Herbst, E. 1989, *Ap. J.*, **347**, 289.

Güsten, R. and Fiebig, D. 1990, to appear in *Galactic and Extragalactic Magnetic Fields*, eds. R. Beck, P.P. Kronberg, and R. Wielebinski, (Dordrecht: Kluwer).

Heiles, C. 1988, in *Galactic and Extragalactic Radio Astronomy*, eds. G.L. Verschuur and K.I. Kellerman, (New York: Springer-Verlag), p. 171.

Heiles, C. 1990, to appear in *Galactic and Extragalactic Magnetic Fields*, eds. R. Beck, P.P. Kronberg, and R. Wielebinski, (Dordrecht: Kluwer).

Heiles, C. 1988, *Ap. J.*, **324**, 321.

Heiles, C., Goodman, A.A., McKee, C.F., and Zweibel, E.G. 1990, in *Protostars and Planets III*, in press.

Heiles, C. and Stevens, M. 1986, *Ap. J.*, **301**, 331.

60

Heiles, C. and Troland, T.H. 1982, *Ap. J.*, **260**, L23.
Herbst, E. and Klemperer, W.B. 1973, *Ap. J.*, **185**, 505.
Hildebrand, R.H. 1988, *Q. Jl. R. Astr. Soc.*, **29**, 327.
Johnston, K.J., Migenes, V., and Norris, R.P. 1989, *Ap. J.*, **341**, 847 (JMN).
Knowles, S.H. and Batchelor, R.A. 1978, *Mon. Not. Royal Astr. Soc.*, **184**, 107.
Kylafis, N.D. 1983a, *Ap. J.*, **267**, 137.
Kylafis, N.D. 1983b, *Ap. J.*, **275**, 135.
Kylafis, N.D. and Shapiro, P.R. 1983, *Ap. J.*, **272**, L35.
Larson, R.B. 1981, *M.N.R.A.S.*, **194**, 809.
Lepp, S., Dalgarno, A., and Sternberg, A. 1987, *Ap. J.*, **321**, 383.
Leung, C.M., Herbst, E., and Huebner, W.F. 1984, *Ap. J. Suppl.*, **56**, 231.
Lis, D.C., Goldsmith, P.F., Dickman, R.L., Predmore, C.P., Omont, A., and Cernicharo, J. 1988, *Ap. J.*, **328**, 304.
McCutcheon, W.H., Vrba, F.J., Dickman, R.L., and Clemens, D.P. 1986, *Ap. J.*, **309**, 619.
McKee, C.F., Zweibel, E.G., Heiles, C., and Goodman, A.A. 1990, in *Protostars and Planets III*, in press.
Mitchell, G.F. and Watt, G.D. 1985, *Astron. Ap.*, **151**, 121.
Myers, P.C. and Goodman, A.A. 1990, in preparation (MG90).
Myers, P.C., Ho, P.T.P., Schneps, M.H., Chin, B., Pankonin, V., and Winnberg, A. 1978, *Ap. J.*, **220**, 864.
Nedoluha, G.E. and Watson, W.D. 1990, *Ap. J.*, **354**, 660.
Norris, R.P. 1984, *Mon. Not. Royal Astr. Soc.*, **207**, 127.
Novak, G., Gonatas, D.P., Hildebrand, R.H., Platt, S.R., and Dragovan, M. 1989, *Ap. J.*, **345**, 802.
Novak, G., Predmore, C.R., and Goldsmith, P.F. 1990, *Ap. J.*, **355**, 166.
Plambeck, R.L., Wright, M.C.H., Welch, W.J., Bieging, J.H., Baud, B., Ho, P.T.P., and Vogel, S.N. 1982, *Ap. J.*, **259**, 617.
Reid, M.J. and Moran, J.M. 1981, *Ann. Rev. Astron. Ap.*, **19**, 231.
Reid, M.J. and Silverstein, E.M. 1990, *Ap. J.*, submitted.
Solomon, P.M., Rivolo, A.D., Barrett, J., and Yahil, A. 1987, *Ap. J.*, **319**, 730.
Tielens, A.G.G.M. and Hollenbach, D., 1985, *Ap. J.*, **291**, 722.
Tomisaka, K., Ikeuchi, S., and Nakamura, E. 1988, *Ap. J.*, **335**, 239.
Troland, T.H. 1990. To appear in *Galactic and Extragalactic Magnetic Fields*, eds. R. Beck, P.P. Kronberg, and R. Wielebinski, (Dordrecht: Kluwer).
Troland, T.H., Crutcher, R.M., and Kazès 1986, *Ap. J.*, **304**, L57. Troland, T.H., Heiles, C., and Goss, W.M. 1989, *Ap. J.*, **337**, 342.
Turner, B.E. and Ziurys, L.M. 1988, in *Galactic and Extragalactic Radio Astronomy*, ed. G.L. Verschuur and K.I. Kellermann, 200.
van Dishoeck, E.F. and Black, J.H. 1986, *Ap. J. Suppl.*, **62**, 109.
Vogel, S.N., Wright, M.C.H., Plambeck, R.L., and Welch, W.J. 1984, *Ap. J.*, **283**, 655.
Wannier, P.G., Scoville, N.Z., and Barvainis, R. 1983, *Ap. J.*, **267**, 126.
Werner, M.W., Dinerstein, H.L., and Capps, R.W. 1983, *Ap. J.*, **265**, L13.
Wright, M.C.H., Plambeck, R.L. Vogel, S.N., Ho, P.T.P., and Welch, W.J. 1983, *Ap. J.*, **267**, L41.
Zweibel, E. 1990, *Ap. J.*, in press (20 Oct 1990).

STRUCTURE IN THE GAS AND MAGNETIC FIELD IN S106

RICHARD M. CRUTCHER
Astronomy Department
University of Illinois
Urbana, IL 61801
U.S.A.

ABSTRACT. BIMA molecular-line observations show evidence for an expanding molecular ring around IRS 4, a newly formed massive star at the center of the bipolar nebula S106. VLA observations of the Zeeman effect in the OH 1665 MHz line show that the magnetic field strength is about 1 mG and that it reverses direction from one lobe of the bipolar nebula to the other.

1. Introduction

S106 is a biconical H II region and a site of recent star formation; it consists of a dark lane which separates two lobes of H II flowing away from a newly formed massive star, IRS 4. The entire complex is embedded in the placental molecular cloud from which it formed. We have mapped the small scale structure of the gas and of the magnetic field in this region. The Berkeley-Illinois-Maryland millimeter-wave Array was used to map the 3-mm emission lines of HCO^+, $H^{13}CN$, and CS. The VLA was used to map the Stokes parameter I and V spectra of the OH 1665 MHz absorption line.

2. BIMA Observations

The BIMA observations of the $\lambda = 3$ mm lines of HCO^+, $H^{13}CN$, and CS produced maps with an angular resolution of about 8 arcseconds and a velocity resolution of 1 km s^{-1}. All maps were centered on IRS 4 [$\alpha(1950) = 20^h 25^m 33.8^s$, $\delta(1950) = 37° 12' 50"$] and cover the full-width-at-half-power diameter of the 6-m BIMA antennas, about 2 arcminutes. Figure 1 shows the HCO^+ data summed over the velocity extent of the line. The brightest peak in the HCO^+ emission lies at the northwest corner of the southern ionized lobe. Other peaks occur near each of the other three corners of the ionized lobes; on the eastern side of the H II, there is a north-south ridge of HCO^+ emission. IRS 4 lies in a relative minimum of emission, with emission extending east-west just to the north and to the south of IRS 4. To investigate the kinematics of the dark lane region between the ionized lobes, we made a position-velocity cut (figure 2) through IRS 4 parallel to the dark lane separating the two ionized lobes. The dark lane extends ±20" from IRS 4. Outside this region, there is little or no velocity gradient; the line width is about 2 km s^{-1}. Within the dark lane region, there is an incomplete velocity ellipsoid, with the positive velocity gas to the west

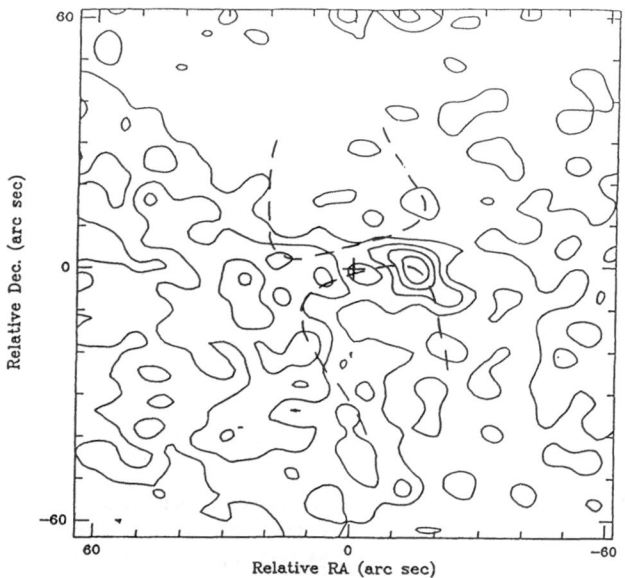

Figure 1. Contours show HCO$^+$ emission summed over velocity. Coordinates are with respect to IRS 4, which is shown as a cross at the center. Dashed contours mark the outer boundary of the H II region as delineated by VLA continuum observations.

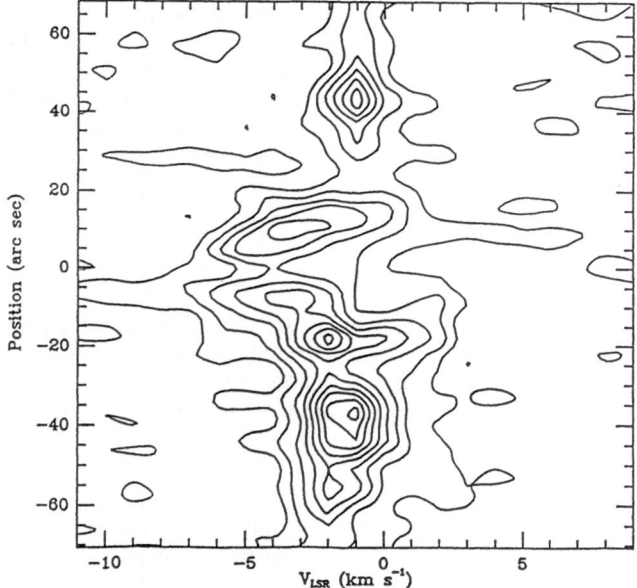

Figure 2. Position-velocity cut through IRS 4 from east to west at position angle 115°.

being missing. Also, there are higher velocity clumps of gas, seen at $(+10", >3$ km s^{-1}) and at $(-5", <-6$ km s^{-1}) in figure 2; additional such clumps are seen at other positions north and south of the dark lane.

Our interpretation of these data is that there is a clumpy molecular ring around IRS 4. The radius of the ring is about 2×10^{17} cm and its mass is ≤ 10 $M_\odot$. The ring is not rotating, but it is expanding with a velocity of about 2 km s^{-1}. This ring may be the residual of a disk of cold, dense material which developed during the formation of IRS 4 and which initially directed the ionized outflow from IRS 4 into a bipolar morphology. The star has begun to dissipate its surrounding placental material, so that the disk is now almost gone. The clumps of high velocity material which are observed may have been ablated from the ring and entrained in the ionized bipolar outflow.

The molecular cloud surrounding the S106 region is also very clumpy, although unlike the dark lane region there is very little velocity structure. The maps suggest that gas traced by CS and H^{13}CN emission is relatively quiescent and located in a cocoon-like structure surrounding the ionized lobes. On the other hand, HCO$^+$ emission peaks closer to the ionized regions than CS and HCN. The HCO$^+$ is enhanced just outside continuum peaks where the ionized outflow appears to encounter the surrounding molecular cloud and to be deflected northward and southward into the lobes. The HCO$^+$ enhancement may be a result of shock chemistry.

3. VLA Observations

The VLA was used to map Stokes parameter I and V spectra for the 1665 MHz line of OH in absorption toward the S106 continuum. A continuum map was produced from spectrometer channels with no line signal. Brief observations were made of the 1720 MHz maser. The angular resolution was about 10" and the velocity resolution 0.5 km s^{-1}. Maps of the optical depth of the OH line were made from the continuum and I data. The line-of-sight component of the magnetic field strength and its direction were inferred from the V spectra.

The OH absorption is strongest at about 0 km s^{-1}, with a gradually decreasing strength out to about -10 km s^{-1}. The ring is seen in absorption in OH with a velocity which shows that the ring is expanding. OH absorption against the ionized lobes is very clumpy; these clumps are responsible for the long negative velocity tail to the OH line profile. This OH appears to be confined to the interface between the ionized lobes and the surrounding molecular gas, and is interpreted as material being entrained in the bipolar outflow.

The magnetic field in S106 measured by single-antenna Zeeman experiments is about 0.1 milligauss. The VLA results are shown in figure 2; they show that the field reverses direction from one lobe to the other, with a typical magnitude of about 1 milligauss. This result is consistent with the magnetic field having an hourglass morphology, with the maser behind the H II of the northern lobe and the OH absorption tracing the field at the front interface between the H II and the surrounding molecular cloud. This morphology is that expected for a centrally condensed cloud which has contracted with flux freezing. The magnetic field strength is clearly large enough to be important in the dynamics of the region. Since the molecular ring is not observed to be rotating, magnetic fields could be playing a dominant role in its support.

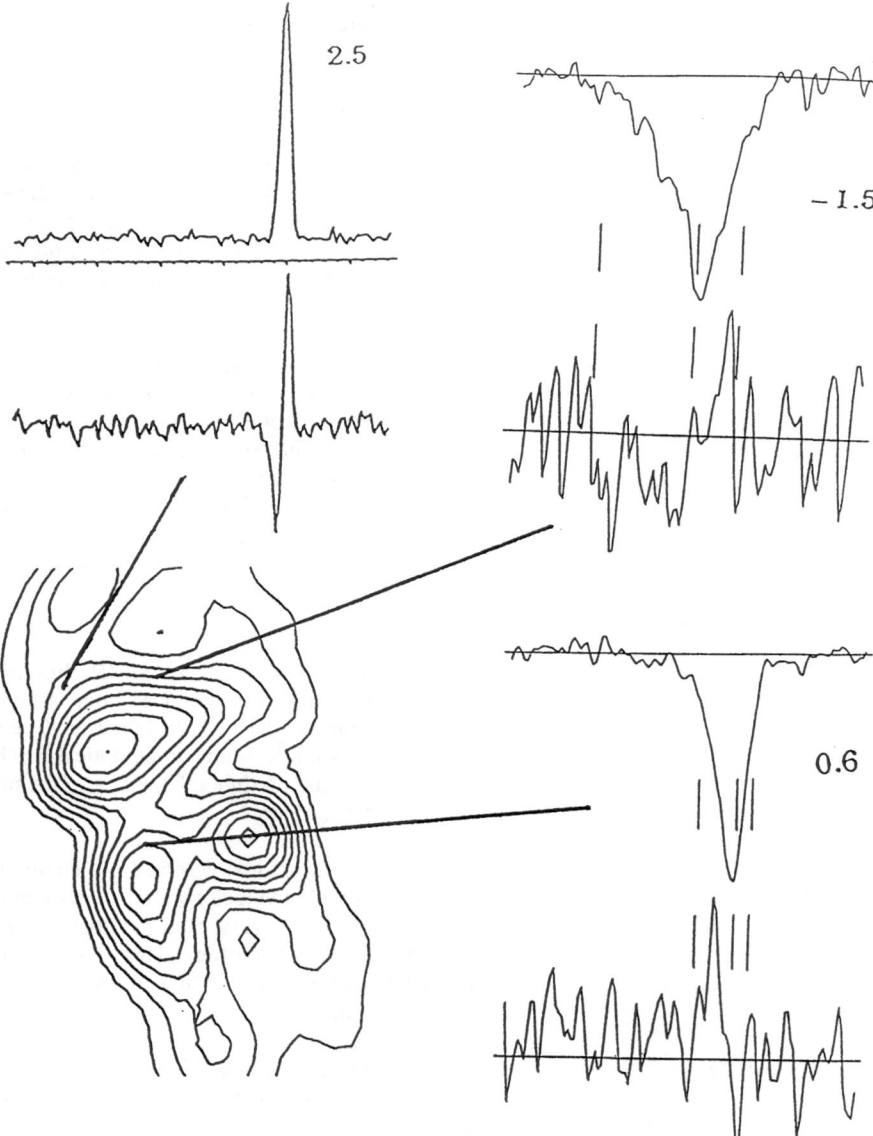

Figure 3. The contours at bottom left show our 18 cm VLA continuum map. Stokes parameter I (upper) and V (lower) profiles are shown for the 1720 MHz maser (upper left) and for two positions for the 1665 MHz absorption (right). Lines show the positions on the continuum where each of the profiles was obtained. Magnetic fields are +2.5 mG for the maser, -1.5 mG for the northern lobe in OH absorption, and +0.6 mG for the southern lobe in absorption.

THE INVESTIGATION OF THE HIERARCHICAL STRUCTURE
OF THE INTERSTELLAR MAGNETIC FIELDS

Lazarian A.L.

P.N.Lebedev Physical Institute, U S S R

Abstract. A new method of investigation of the hierarchical structure of the interstellar magnetic fields is discussed. Synchrotron intensity fluctuations are used. Statistical characteristics of the irregular magnetic field are available. The method was successfully tested.

1. Introduction

It is well known, that magnetic field is very important for the processes of fragmentation of molecular clouds and processes of star formation. But it is rather difficult to investigate the magnetic field itself. The problem is that the field has a very complicated spartial structure. The interstellar magnetic field can be approximated by a superposition of a regular "H" and a random "h" magnetic fields. Correlation properties of "h" determine the hierarchical structure of the interstellar magnetic fields on the scales much smaller than the size of the Galaxy. Using traditional methods it is impossible to determine correlation functions of the interstellar magnetic field. That is why a new method of investigation was developed [1].

The intensity of synchrotron radio emission depends on the number density of relativistic electrons and on the strength of the magnetic field component normal to the line of sight. In light of observational evidence for cosmic–ray isotropy one may start by assuming that relativistic electrons are distributed uniformly in the Galaxy. But the magnetic field is not homogeneous. Hence the synchrotron radiation observed in different directions also ought to be nonuniform. The problem of the determination of correlation properties of magnetic field is caused by the fact that the radiation is accumulated along the lines of sight. Different hierarchical structures of the interstellar magnetic field contribute to the resulting intensity. Under these conditions the information about statistical properties of a random magnetic field seemed to escape. Working in collaboration with Chibisov G.V. we have shown that it was not the case and it was possible to determine the correlation functions of the chaotic magnetic field on the basis of sycnhrotron fluctuations.

2. Materials

As it was shown back in 1980 [2] a correlation function of intensity of Galaxy's synchrotron radiation $K(\theta,\varphi)$ consists of an isotropic and anisotropic parts:

$$K(\theta, \varphi) = K_0(\theta) + K_2(\theta)\cos^2\varphi \qquad (1)$$

where θ is the angle between the lines of sight and φ is the angle setting the direction of the regular field component normal to the line of sight H. The appearance of the anisotropic part is caused by the regular magnetic field and it is possible to find the direction of H by an abalysis of correlation function of synchrotron intensity. Our research [3] has shown that the expression for $K_2(\theta)$ can be transformed into Abel's equation and thus the reversed task can be solved for $K_2(\theta)$. Correlation functions of the random magnetic field as well spectrum $E(k)$ are available:

$$E(k) = -\frac{\chi}{4\pi} kR \int_0^{\theta_{max}} d\theta \frac{d^2}{d\theta^2} \left[K_2(\theta) \cdot \theta^2 \right] \cdot J_1(kR\theta) \qquad (2)$$

where $J_1(x)$ is Bessel function, χ is a coefficient, R is the size of emmiting region. This size can be found using HII regions as distance indicators in the framework of the cross–correlation method [3].

3. Results

According to observational data ([4], [5]) the hierarchical structure of the Galaxy's interstellar magnetic fields can be described as follows: regular field with the intensity comparable with square root of the dispersion of the random field exists. The random magnetic field is formed by the hierarchy of turbulent magnetic vertices. The characteristic angular scale of the turbulence discussed is

about 3°. In order to estimate the corresponding linear scale L we have to know the effective depth of the emission layer. The preparations for the measurements of this value using cross–correlation method are under way at the decameter radiotelescope in vicinity of Kharkov. For the rough estimate we can take 1 kpc as synchrotron–emissivity depth. In this case we would then obtain L~50 pc as the size of typical magnetic inhomogeneities.

4. Discussion

The initial results from the study of the intensity variations of the diffuse radiation by means of correlation functions show this approach to be promising one for investigating the Galaxy's magnetic field.

The method is progressing. Now it is clear that not only synchrotron intensity fluctuations, but also fluctuations of different parameters (polarization, distribution of molecular line's profiles etc.) can be used to find statistical characteristics of the interstellar medium [6]. This information is essential for better understanding of processes taking place in the interstellar medium, including the processes of fragmentation of molecular clouds and star formation.

5. References

1. Chibisov G.V., Lazarian A.L. Preprint 283, Moscow, FIAN, 1987.
2. Chibisov G.V., Ptuskin V.S. Proc. 17–th Int. Cosmic Ray Conf. (Paris) 2, 233 (1981).
3. Chibisov G.V., Lazarian A.L. Preprint 191, Moscow, FIAN, 1989.
4. Oagkesamanskii R.D., Shutenkov V.R. (1987) "Radio background fluctuations and the structure of the galactic magnetic field". Sov. Astron. Lett. 13 (2), 73–76.
5. Lazarian A.L., Shutenkov V.R., Preprint 180, Moscow, FIAN, 1989.
6. Lazarian A.L., Preprint 190, Moscow, FIAN, 1989.

LOSS OF MAGNETIC FLUX AND ANGULAR MOMENTUM FROM MOLECULAR CLOUDS

Takenori Nakano
Nobeyama Radio Observatory, National Astronomical Observatory
Nobeyama, Minamisaku, Nagano 384-13, Japan

1. Introduction

The magnetic field and the angular momentum are major obstacles against cloud contraction. I will review recent results on magnetic flux loss rate and angular momentum loss rate and will investigate a gross feature of cloud contraction.

2. Magnetic Flux Loss

2.1. MAGNETIC FLUX PROBLEMS

In an oblate cloud penetrated by magnetic fields parallel to its minor axis , the mean magnetic force is weaker than the self-gravitational force only when its magnetic flux Φ is less than a critical flux

$$\Phi_{cr}^{(f)} \approx 2\pi G^{1/2} M, \qquad (1)$$

where G is the gravitational constant and M is the cloud mass (*e.g.*, Strittmatter 1966). Thus, a cloud with $\Phi > \Phi_{cr}^{(f)}$ must lose some of its initial flux before it can begin dynamical contraction (for more accurate criterion, see §4). Moreover, the magnetic flux to mass ratio for an interstellar cloud is estimated to be several 10^2 to 10^5 times greater than the ratio for a magnetic star (Nakano 1983, 1984). Therefore, clouds, or more exactly, cloud cores, must lose most of their initial flux at some stage of star formation.

2.2. PROCESSES OF MAGNETIC FLUX LOSS

As is well known, ohmic dissipation is quite inefficient in ordinary clouds. Mestel and Spitzer (1956) pointed out that in a slightly ionized cloud the magnetic field and ions, which are frozen to the magnetic field, drift in a sea of neutral gas, and magnetic flux decreases gradually. Elmegreen (1979) investigated the effect of charged grains, which may be only weakly coupled to the magnetic field, on magnetic flux loss. At very high densities even ions are not frozen to field lines. The transition from strong coupling to weak coupling of particles with magnetic fields occurs gradually, and the density at which this transition effectively occurs depends sensitively on the kind of particles, being especially influenced by the particle mass. Since various kinds of charged particles are contained in a cloud and their relative abundances change greatly with density (see §2.4 below), the study of magnetic field dissipation requires careful treatment. Nakano (1984) and Nakano and Umebayashi (1986a) introduced a formalism with which magnetic field dissipation can be treated quite generally.

2.3. CLOUDS WITH REALISTIC GRAIN SIZE DISTRIBUTION

With this formalism Nakano (1984), Nakano and Umebayashi (1986a, b), and Umebayashi and Nakano (1990) investigated magnetic field dissipation assuming that all grains have the same radius, typically $a = 0.1$ μm. In reality, however, interstellar grains have a wide size distribution. For example, Mathis, Rumpl, and Nordsieck (1977, referred to as MRN hereafter) deduced a power law size distribution for graphite and silicate grains,

$$\frac{dn_{gr}}{da} = An_H a^{-3.5}, \quad a_{min} < a < a_{max}, \tag{2}$$

where n_H is the gas density by hydrogen number, $a_{min} \lesssim 100$ Å, $a_{max} \approx 2500$ Å, and $A \approx 1.5 \times 10^{-25} cm^{2.5}$ (Draine and Lee 1984; Mathis 1986). In addition, recent observations suggest the existence of very small "grains" with a as small as 3Å (Sellgren 1984; Leger and Puget 1984; Draine and Anderson 1985).

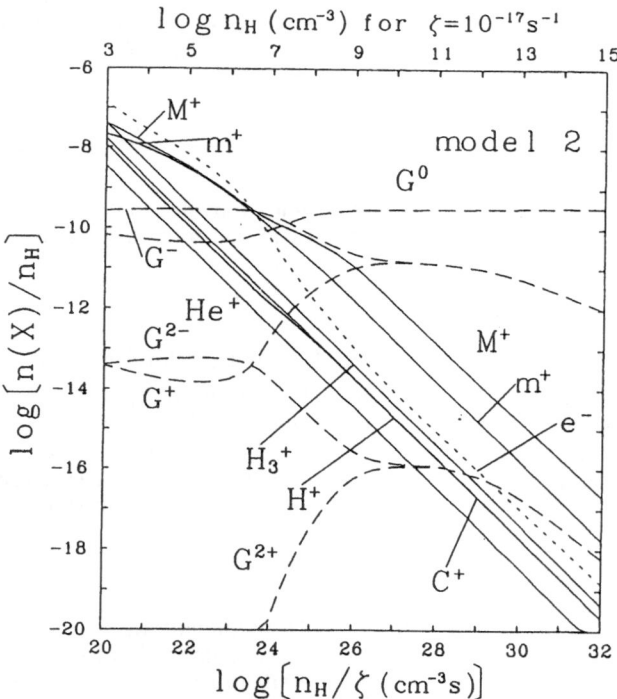

Figure 1. Abundances of various charged particles in clouds with ice-mantled grains (model 2) as functions of the density, n_H. Grains are denoted as G with a superscript representing their electric charge.

Tiny grains have two effects on magnetic field dissipation. First, the large total surface area of tiny grains allows efficient recombination of ions and electrons on grain surfaces. This suppresses the ion density in clouds and has an effect of enhancing field dissipation. Second, most grains are electrically charged, and smaller grains couple more strongly with magnetic fields. This has the effect of suppressing field dissipation. Since both effects are large but operate in opposing directions, a detailed investigation is needed in order to determine which effect is dominant and how efficient field dissipation can be. The following is a summary of the recent work by Nishi, Nakano, and Umebayashi (1990).

We adopt the following four grain models: (1) the standard MRN distribution ($50\text{Å} < a < 2500\text{Å}$); (2) the MRN distribution with ice-mantles ($90\text{Å} < a < 4500\text{Å}$); (3) the extended MRN distribution without ice-mantles ($3\text{Å} < a < 2500\text{Å}$); (4) the standard MRN distribution plus an additional population of grains of size $a = 4\text{Å}$ with number density $2 \times 10^{-7} n_H$. Although a particle as small as 3–4Å may have properties quite different from ordinary grains, we neglect the effects due to such properties. Models 3 and 4 should be regarded as extreme cases for the effects of tiny grains mentioned above.

2.4. DENSITIES OF CHARGED PARTICLES

In clouds shielded from ultraviolet radiation, ions are first formed by ionization of hydrogen molecules and helium atoms by cosmic rays. Many of molecular ions, denoted as m^+, recombine with electrons dissociatively. But some of them undergo charge-transfer reactions with metal atoms, forming metal ions, denoted as M^+. Metal ions recombine radiatively with electrons. All ions can recombine with electrons, which have been adsorbed on grain surfaces. This reaction scheme is nearly in steady state. Figure 1 shows steady state abundances of various charged particles for model 2 (ice-mantled grains). In this model, molecular ions and metal ions are dominant charged particles at densities $n_H \lesssim 10^7 \text{cm}^{-3}$. At higher densities, however, charged grains are more abundant than ions.

2.5. MAGNETIC FLUX LOSS FROM CLOUD CORES

Quasistatic contraction of a cloud due to magnetic flux loss is highly nonhomologous. Nakano (1979) showed that a high-density core appears, which subsequently contracts much faster than the other parts of the cloud. The timescale of such nonhomologous contraction is nearly equal to the timescale of magnetic flux loss from the core whose size across field lines is nearly equal to the size along field lines. Figure 2 shows the timescale, t_B, of magnetic flux loss from cloud cores as a function of the core density, n_H, for each of the four grain models. We have assumed that the magnetic flux, Φ, of the core is nearly equal to the critical flux $\Phi_{cr}^{(f)}$ given by equation (1). The flux loss time, t_B, is 10 to 10^2 times the free-fall time, t_f, at densities lower than a critical density $n_{cr}^{(1)}$. For model 2, $n_{cr}^{(1)} \approx 10^9 \text{cm}^{-3}$, for example. The flux loss time is smaller than t_f only at densities higher than another critical density $n_{cr}^{(2)}$. For model 2, $n_{cr}^{(2)} \approx 10^{10} \text{cm}^{-3}$. Since a cloud cannot contract faster than free fall, the magnetic field is decoupled from the gas at $n_H \gtrsim n_{cr}^{(2)}$. We find that charged grains are more efficient than ions in preventing the drift of magnetic fields at densities $n_H \gtrsim 2 \times 10^4 \text{cm}^{-3}$ for model 2. Thus, even at ordinary densities of molecular clouds, grains are important in preventing magnetic flux loss.

When $\Phi < \Phi_{cr}^{(f)}$, t_B is proportional to Φ^{-2} at $n_H \ll n_{cr}^{(1)}$, and is almost independent of Φ at $n_{\rm H} \gtrsim n_{\rm cr}^{(2)}$. In the latter density range, ohmic dissipation is dominant.

Since a cloud with $\Phi < \Phi_{\rm cr}^{(f)}$ contracts dynamically, or nearly freely, assuming that the cloud mass is much greater than the Jeans critical mass, and also $t_B \gg t_f$ at $n_{\rm H} \lesssim n_{\rm cr}^{(1)}$, flux loss down to far below $\Phi_{\rm cr}^{(f)}$ does not occur at densities $n_{\rm H} \lesssim 10^9 {\rm cm}^{-3}$ for any of the four grain models.

2.6. FRAGMENTATION OF CLOUDS DUE TO MAGNETIC FLUX REDISTRIBUTION

As shown in §2.5, the flux loss time, t_B, is a decreasing function of the density, $n_{\rm H}$. This means that a part of a cloud with higher density loses magnetic flux faster, and then contracts faster, than a part with lower density, assuming that the high-density part has a mass exceeding the Jeans critical mass. Therefore, a density fluctuation in a cloud is amplified by magnetic flux redistribution, and the cloud breaks into fragments. This process was first pointed out by Nakano (1976) using his old results, and also works with our new results.

3. Angular Momentum Loss

The most efficient process of removing cloud's angular momentum is transport of it to an ambient medium via magnetic torque. Magnetic braking of cloud's rotation has been investigated for some cases of magnetic configuration around a cloud.

3.1. CLOUDS WITH UNIFORM MAGNETIC FIELD

The first investigation of magnetic braking was for a cloud penetrated by a uniform

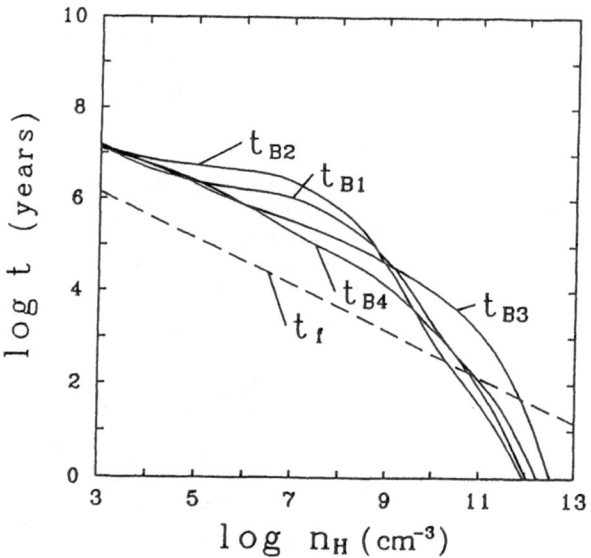

Figure 2. The magnetic flux loss time for cloud cores in which magnetic force nearly balances the self-gravitational force as a function of the core density, $n_{\rm H}$, for each of four grain models.

magnetic field with a rotation axis parallel to field lines. The angular momentum of such a cloud is lost in a time taken by a torsional Alfvén wave to cross a domain in the ambient medium whose column density is equal to the column density of the cloud (Ebert *et al.* 1960; Mouschovias and Paleologou 1980). However, such a magnetic configuration is realized only around a cloud with strong magnetic field ($\Phi \gg \Phi_{cr}^{(f)}$), or with weak gravity, where field configuration is hardly affected by cloud's gravity. Stars hardly form in such a cloud.

3.2. CONTRACTED CLOUDS

The magnetic field around a cloud with strong gravity is distorted considerably. Gillis, Mestel, and Paris (1974, 1979) investigated angular momentum loss from such a cloud assuming that the poloidal component of magnetic field is fixed. However, the assumed poloidal field is not force-free, and magnetic configuration changes in an Alfvén-crossing time, which is usually much shorter than the spin-down time of the cloud. These pioneering works are insufficient in this sense.

3.3. CLOUD CORES

As mentioned above, contraction of a magnetized cloud is highly nonhomologous and only a core finally contracts to form a star. Therefore, angular momentum loss from a cloud core is essential in star formation. Let us consider an axisymmetric cloud with a core at the center. Although the magnetic field may be nearly uniform very far from the cloud, there is an intermediate zone where field lines anchored to the core are nearly radial and decline only slightly from the symmetry axis. We can take the field in this zone nearly force-free. Assuming that the intermediate zone is sufficiently large, Nakano (1989) investigated the transport of core's angular momentum to this zone. The density in the intermediate zone is assumed to be proportional to r^{-n}, where r is the distance from the cloud center. It is convenient to introduce $\nu \equiv 1/(6 - n)$. The results are as follows.

For $\nu = 1/2$ ($n = 4$), propagation of torsional Alfvén waves can be solved analytically, and the rotation velocity of the core, Ω_{cc}, is given by

$$\Omega_{cc}(\tau) = \Omega_i \exp(-s_{1/2}\tau), \tag{3}$$

where τ is a dimensionless time, Ω_i is the initial rotation velocity of the core, and s_ν is a quantity determined by the ratio of the density just outside the core and the mean core density.

For ν between 0 and $1/2$ ($n < 4$), the solution is obtained semianalytically as

$$\Omega_{cc}(\tau) \approx \frac{\Omega_i}{1-\nu} \exp\left(s_\nu\tau \cos\frac{\pi}{2-2\nu}\right) \cos\left(s_\nu\tau \sin\frac{\pi}{2-2\nu}\right) - \frac{\Omega_i\nu/(1-\nu)}{(1+a_\nu s_\nu\tau)^{2-2\nu}}, \tag{4}$$

where a_ν is a constant of order unity. The first term oscillates with time with the amplitude decreasing exponentially. The second term decreases with time only by a power law and finally prevails over the first term.

In all cases most of the initial angular momentum is lost in a timescale equal to s_ν^{-1}. This is the time taken by a torsional Alfvén wave to cross a domain whose moment of inertia is equal to the moment of inertia of the cloud core.

3.4. DISTRIBUTION OF ROTATION VELOCITY IN CLOUDS

In cases other than $n = 4$, spin-down of a core is rather complicated. The $n = 4$ case is special; the Alfvén velocity is constant along field lines, and the moment of inertia of the gas in a magnetic tube with unit length is also constant along field lines. This is why waves propagate smoothly and the core loses angular momentum smoothly. In other cases wave propagation is rather complicated. Reflection of waves occurs everywhere. This is why the core behaves in a complicated way.

The second term of equation (4) represents a rotation reverse to the initial rotation. Because a core has a much smaller moment of inertia than the outer part of the cloud, the core loses angular momentum faster than the outer part. Therefore, after a few spin-down timescales, the core rotates reversely to the outer part, though this retrograde rotation gradually decays afterwards. Thus, there must be a stage in which a core rotates reversely to the outer part. With minute observations we may be able to find clouds with such rotational structure.

4. Contraction of Rotating Magnetized Clouds

A gross feature of contraction of rotating magnetized clouds can be obtained by comparing the contraction time with the magnetic flux loss time and with the angular momentum loss time, as done by Nakano (1990). The following is a summary of the results.

First, let us consider a cloud core whose initial magnetic flux, Φ_i, is greater than a critical flux, $\Phi_{cr}^{(d)}$, defined by equation (6) just below. The cloud is embedded in a medium which has uniform magnetic field and uniform density far from the cloud. The timescale of quasistatic contraction of the core with flux Φ is given by

$$t_{cont} \equiv \left(\frac{d \ln R_{cc}}{dt}\right)^{-1} \approx \frac{1}{2}\left[1 - \left(\frac{\Phi_{cr}^{(f)}}{\Phi}\right)^2\right] t_B, \tag{5}$$

where R_{cc} is an equilibrium radius of the cloud core, which we determine from the virial theorem, and t_B is the magnetic flux loss time investigated in §2. Contraction can be regarded quasistatic only when $t_{cont} > t_f$, or Φ is greater than

$$\Phi_{cr}^{(d)} \approx \Phi_{cr}^{(f)}\left(1 + \frac{t_f}{t_B}\right). \tag{6}$$

When Φ decreases to $\Phi_{cr}^{(d)}$, dynamical contraction sets in as long as the core mass, M_{cc}, is considerably greater than the Jeans critical mass, M_J.

An outline of contraction is shown in Figure 3. In an early stage of a core with $\Phi_i > \Phi_{cr}^{(d)}$, the angular momentum loss time, t_L, is much smaller than t_{cont}, and so the core loses angular momentum, L, rapidly, and soon settles down in corotation with the ambient medium. After this, contraction is induced by magnetic flux loss, and L decreases with contraction keeping corotation with the ambient medium because $t_L < t_{cont}$. As contraction proceeds, the ratio t_L/t_{cont} increases. When the Alfvén velocity, $V_A^{(0)}$, in the ambient medium is greater than the sound velocity, V_s, in the core, a condition $t_L > t_{cont}$ is realized and angular-momentum-conserving contraction sets in while the core is still contracting quasistatically. Afterwards, Φ decreases to $\Phi_{cr}^{(d)}$, and dynamical contraction sets in. When $V_A^{(0)} \lesssim V_s$, dynamical contraction begins while the core is still in corotation with the ambient medium. Afterwards, the core shifts to angular-momentum-conserving contraction.

A cloud with Φ_i slightly smaller than $\Phi_{cr}^{(d)}$ contracts dynamically from the beginning losing angular momentum because t_L is nearly equal to the magnetically diluted free-fall time, t_f', assuming the cloud mass $M \gg M_J$. However, it is unclear whether corotation with an ambient medium is realized or not.

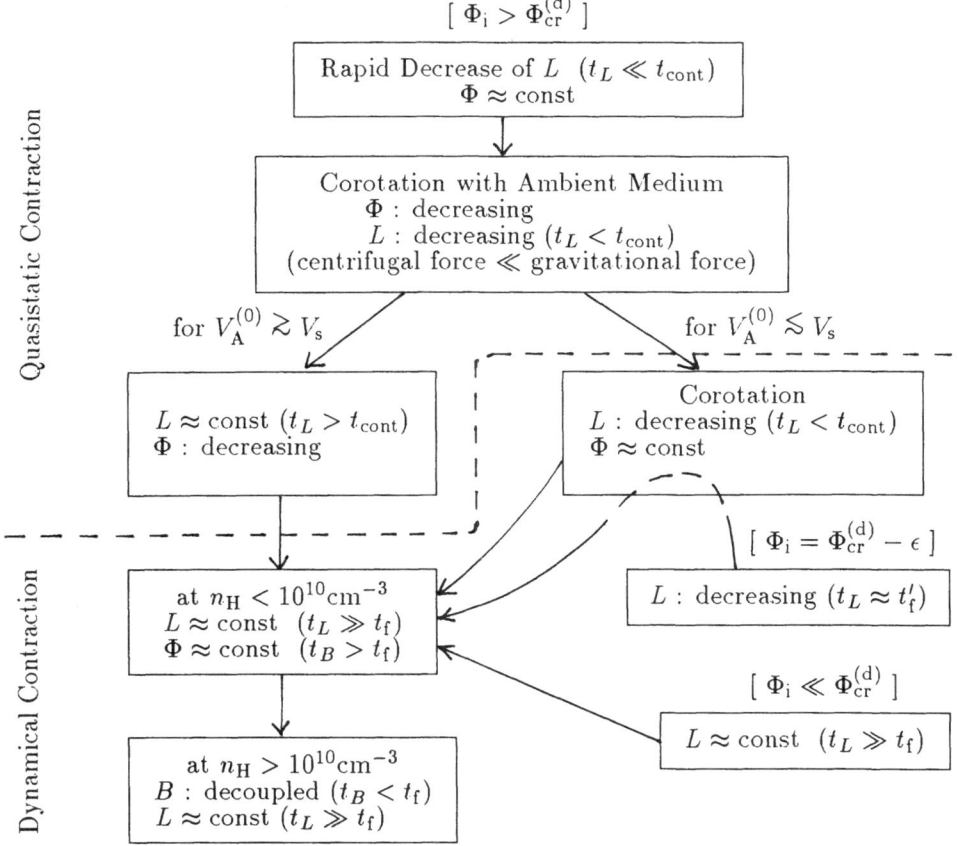

Figure 3. Gross feature of contraction of rotating magnetized clouds. The dashed line corresponds to the stage of magnetic flux $\Phi = \Phi_{\mathrm{cr}}^{(d)} \approx \Phi_{\mathrm{cr}}^{(f)}(1 + t_{\mathrm{f}}/t_B)$. In stages above this line the clouds contract quasistatically, and below this line they contract dynamically.

A cloud with $\Phi_{\mathrm{i}} \ll \Phi_{\mathrm{cr}}^{(d)}$ contracts dynamically from the beginning, as long as $M > M_{\mathrm{J}}$, nearly conserving L.

In dynamically contracting clouds, magnetic flux Φ is nearly conserved at n_{H} less than the critical density $n_{\mathrm{cr}}^{(2)}$, which was obtained in §2 and is $10^{10}\mathrm{cm}^{-3}$ for model 2 (ice-mantled grains). At $n_{\mathrm{H}} \gtrsim n_{\mathrm{cr}}^{(2)}$ the magnetic field is decoupled from the neutral gas and Φ decreases rapidly. Angular momentum is conserved in this stage.

Let us go back to the case of $\Phi_{\mathrm{i}} > \Phi_{\mathrm{cr}}^{(d)}$ with $V_{\mathrm{A}}^{(0)} \gtrsim V_{\mathrm{s}}$. The angular-momentum-conserving contraction effectively sets in at a stage $t_L \approx t_{\mathrm{cont}}$, and angular momentum of a core is fixed at this stage. If binary stars form from this core, we can predict the orbital period of the binary stars. Assuming that stars of equal mass are formed, Nakano (1990) found that the orbital period is

$$P_b \approx 3\text{yr}\left(\frac{\Omega_0}{2 \cdot 10^{-15}\text{rad s}^{-1}}\right)^3 \left(\frac{10^5\text{cm}^{-3}}{n_L}\right)^2 \approx 3\text{yr}\left(\frac{\Omega_0}{2 \cdot 10^{-15}\text{s}^{-1}}\right)^3 \left(\frac{10^2\text{cm}^{-3}}{n_0}\right)^2, \quad (7)$$

where n_L is the core density at the stage $t_L \approx t_{\text{cont}}$ and is about 10^3 times the density, n_0, of the ambient medium, and Ω_0 is the rotation velocity of the ambient medium. If we take n_0 as a mean density of a molecular cloud complex whose rotation velocity Ω_0 is nearly equal to the shear of galactic rotation, we obtain observed orbital periods of binary stars. Thus, the magnetic flux loss rate and the angular momentum loss rate shown above are consistent with orbital periods of binary stars.

References

Draine, B. T., and Anderson, N., 1985. *Astrophys. J.*, **292**, 494.
Draine, B. T., and Lee, H. M., 1984. *Astrophys. J.*, **285**, 89.
Ebert, R., Hoerner, S. von & Temesvary, S., 1960. *Die Entstehung von Sternen durch Kondensation diffuser Materie*, p. 315, Springer-Verlag, Berlin.
Elmegreen, B. G., 1979. *Astrophys. J.*, **232**, 729.
Gillis, J., Mestel, L. & Paris, R. B., 1974. *Astrophys. Space Sci.*, **27**, 167.
Gillis, J., Mestel, L. & Paris, R. B., 1979. *Mon. Not. Roy. Astr. Soc.*, **187**, 311.
Leger, A., and Puget, J. L., 1984. *Astr. Astrophys.*, **137**, L5.
Mathis, J. S., 1986. *Astrophys. J.*, **308**, 281.
Mathis, J. S., Rumpl, W., and Nordsieck, K. H., 1977. *Astrophys. J.*, **217**, 425.
Mestel, L. & Spitzer, L., Jr., 1956. *Mon. Not. Roy. Astr. Soc.*, **116**, 503.
Mouschovias, T. Ch. & Paleologou, E. V., 1980. *Astrophys. J.*, **237**, 877.
Nakano, T., 1976. *Publ. Astr. Soc. Japan*, **28**, 355.
Nakano, T., 1979. *Publ. Astr. Soc. Japan*, **31**, 697.
Nakano, T., 1983. *Publ. Astr. Soc. Japan*, **35**, 209.
Nakano, T., 1984. *Fund. Cosmic Phys.*, **9**, 139.
Nakano, T., 1989. *Mon. Not. Roy. Astr. Soc.*, **241**, 495.
Nakano, T., 1990. *Mon. Not. Roy. Astr. Soc.*, **242**, 535.
Nakano, T. & Umebayashi, T., 1986a. *Mon. Not. Roy. Astr. Soc.*, **218**, 663.
Nakano, T. & Umebayashi, T., 1986b. *Mon. Not. Roy. Astr. Soc.*, **221**, 319.
Nishi, R., Nakano, T., and Umebayashi, T., 1990. submitted to *Astrophys. J.*
Sellgren, K., 1984. *Astrophys. J.*, **277**, 623.
Strittmatter, P. A., 1966. *Mon. Not. Roy. Astr. Soc.*, **132**, 359.
Umebayashi, T., and Nakano, T., 1990. *Mon. Not. Roy. Astr. Soc.*, **243**, 103.

MAGNETIC FIELDS AND THE DYNAMICS OF MOLECULAR CLOUDS

J.L. PUGET
Institut d'Astrophysique Spatiale,
Bat. 120, Université Paris Sud
F-91405,Orsay, France

ABSTRACT: Magnetic fields are believed to play an important role in the star formation process. Correlations in the velocity field in molecular filaments are indicative of dynamical interactions between clouds and parts within a cloud. The magnetic field is a likely candidate as the vector of such interactions. Perturbations of the field at large scales can feed the velocity dispersion within condensations at small scale. This mechanism is discussed in the framework of two simple analytical approximations describing transverse waves fed into plane parallel slabs.
KEYWORDS: Molecular clouds evolution; Alfvèn Waves; Magnetic field; Filamentary structure; Turbulence .

1. INTRODUCTION

Interstellar gas is cycled through a system which can be very schematically described by a three steps cycle (molecular clouds are observed in these different states, the evolution from the one to the next is only a reasonable speculation at this stage).

1- a large system of low density gas and HI clouds in quasi pressure equilibrium undergoes a gravitational and/or thermal instability (Elmegreen 1985, Lioure and Chièze 1990) and gathers into a marginally bound complex.

2- the large scale kinetic energy in this complex is dissipated and the complex slowly contracts. This stage is best illustrated by large quiescent molecular complexes observed in the solar vicinity like the one observed by Maddalena and Thaddeus (1985) or Pérault et al. (1985).
Low mass stars are forming within the densest molecular clouds in such a complex. The Taurus cloud belongs to this phase.

3- as more kinetic energy is dissipated, the central part of the complex becomes more condensed and more and more massive stars are formed in this core. The ρ Ophiuchus and the Orion molecular clouds belong to this third phase.

In this last phase, massive stars produce strong winds, HII regions and supernovae.Part of the central core is photodissociated and the complex is dispersed. In this phase the potential energy of the whole cloud system is restored to be dissipated in the next contraction phase.
Low mass stars also have strong winds and outflows during the T Tauri phase and also inject in the complex an amount of kinetic energy which could be non negligible in some phases but smaller than that coming from massive stars (Puget 1985).

The large scale kinetic energy plays an essential role in the equilibrium of the complex at least on large scales. It probably also contributes to create density structures as the flows are highly supersonic and thus compressible. The star formation process cannot be described by a simple Jeans analysis of gravitational instability in a uniform medium where thermal pressure is the main force acting against gravity. Bonazzola et al. (1987, 1990) have discussed some of the complicated processes which control star formation when the supersonic turbulence might stabilize the large scales more than the small scales and at the same time creates density fluctuations. In such a system the star formation process is not any more controlled by local parameters (like temperature, gas density, magnetic field strength, and ionization state), but also, and possibly mainly, by the dynamical state of the gas which depends, through turbulent cascades, on the large scale environment of the cloud considered
It is thus essential to understand the physical mechanisms controlling this turbulent cascade in the interstellar medium to learn what controls the star formation rate and the initial mass function.

It has been stressed for several years that the magnetic field has an intensity such that its dynamical role is probably important. Recent observations have confirmed this point of view (see for example the review by Heiles at this conference).
The role of the static magnetic field on the gravitational stability of cold interstellar gas has been studied in detail. Secular evolution through ambipolar diffusion has also been studied and dynamical effects during the collapse phase to transfer out the angular momentum have also been addressed (see Mc Kee et al. 1991 for a review).

This paper concentrates on the less studied role of the magnetic field on the dynamics of the interstellar gas before any protostellar collapse takes place, and more specifically on the turbulent cascade in molecular complexes. We thus stress the role of the perturbations of the average field and the associated magneto hydrodynamical (MHD) waves on the stability of clouds more than the static role of the average field.

To separate factors as much as possible, we concentrate on complexes with low star formation rate or on part of complexes far away for massive stars.

2. OBSERVATIONAL EVIDENCE

The quasi virialized state of quiescent molecular complexes is well established. Furthermore the average density is low and thus the dense molecular material has a low volume filling factor in those complexes: 5 to 10 % (Pérault et al., 1985).
Most of the volume of the complex is filled with rather low density gas likely to be warm HI (Falgarone and Puget, 1986). Nevertheless the molecular clouds do not behave in the complex like stars in a bound cluster. The clouds are gathered into substructures of the complex in a hierarchical structure in which no preferred scale emerges (Pérault et al 1986, Scalo 1987, Falgarone and Phillips this conference). This is a strong indication that, despite the large density contrasts, the clouds in the complex interact strongly with each other. The large density contrast makes it very unlikely that they interact through pure hydrodynamical effects (in a supersonic turbulence large density contrasts can be generated but the dense phase contains only a small fraction of the total mass).

Correlations in both the spatial and the velocity distribution of molecular material in complexes is in fact directly observable in form of long filaments. Well known examples can be found in the Taurus-Auriga-Perseus complex (Ungerechts and

Thaddeus, 1987), the ρ Ophiuchus complex (Loren, 1989), the Northern Coalsack (Pérault et al., 1985), and L204 (Heiles 1988). These structures are very elongated. Along these filaments, the velocity is very well correlated, changing smoothly. The velocity gradient is much larger than the velocity dispersion of the individual clumps making the filament (see for example the case plotted in figure 1). This suggests that in these filaments a physical link maintains the spatial correlation to keep them from dispersing in a few million years. Pure hydrodynamical turbulence could create such correlations as long as the specific inertia in the surrounding medium is not small with respect to that of the filament itself, but this is probably not the case in the low density complexes.

The magnetic field is an obvious candidate to play that role. Nevertheless the direction of the field as revealed by star light polarization is not always that of the filament as for example in L204 (Heiles 1988).Neverless a wave pattern is present both in projection on the sky and velocity. We searched for similar patterns along other filaments (see also Fukui, this conference).

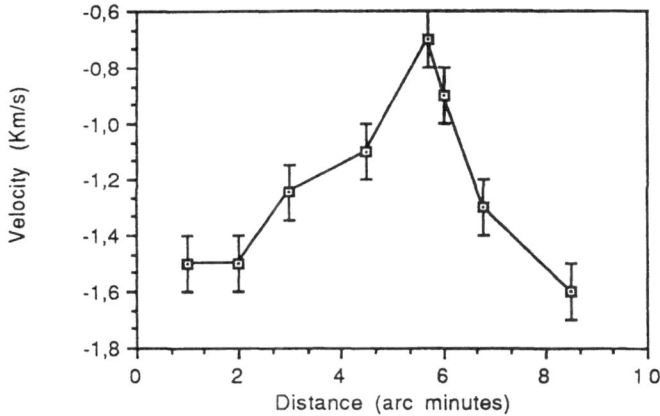

Figure 1: Radial velocity as a function of distance along the C18O ridge
At a distance d=200 pc, 1 arc minute is 0.05 pc

Assuming that the field is strong enough to dynamically link efficiently the clouds together along a filament, we can extract the basic characteristics of transverse MHD waves along these filaments which would account for the velocities observed. The values obtained that way and reported in Figure 2 are interesting indicators but it should be remembered that the assumption of pure tranverse waves is certainly too simple.

To test further this hypothesis, Falgarone et al. (1991) have investigated the substructure of some of these filaments. In Auriga a dense ridge 0.1 pc wide is observed in C18O which is aligned with the large filament (5 pc wide) in which it is embedded. Along this ridge several condensations (0.02 pc dense cores) can be identified. Figure 1 shows the radial velocity as function of distance along the ridge. This reinforces the idea that the magnetic field is needed even if the filaments are generated by a wind or a strong photon flux acting on the surface(Bally, this conference), because it shows that the inner small scale and the outer large scale are dynamically coupled.It is quite striking than the periods involved in these waves are not changing fast with scale within a given filament.

Because projection effects affect the observed amplitudes of transverse waves in opposite directions for velocity and displacement, one must rely on the upper envelopes in Figure 2. They can be approximated by

$$\Delta v = 1.22 \ \lambda_{pc}^{0.2} \ \text{km/s and } \Delta l = 0.26 \ \lambda_{pc}^{0.8} \ pc$$

leading to $\omega = 4.75 \ \lambda_{pc}^{-0.6} \ \text{Myr}^{-1}$ and $v_A = 0.75 \ \lambda_{pc}^{0.4} \ \text{km/s}$.

This behaviour of the Alfvèn velocity corresponds to a magnetic field of 7 µG on large scale (100 pc, $n_{H2} \sim 5 \text{cm}^{-3}$) and 20 µ G on the smallest scales (0.1 pc, $n_{H2} \sim 10^4 \text{cm}^{-3}$)

3. ROLE OF MAGNETIC FIELDS IN INTERSTELLAR TURBULENCE

Clifford and Elmegreen (1983) and Arons and Max (1975) have shown that molecular clouds can interact through the magnetic field threading them and that this limits their mean free path. Falgarone and Puget (1986) have used the same analytical approach to compute the kinetic energy injected in a cloud by the magnetic interactions with its neighbours. They show that this mechanism generates trapped waves in the clouds with most of the power at scales comparable to the cloud size and with a level such that the waves are becoming non linear (perturbations of the field not very small compared to the average field).This stabilisation mechanism has been further investigated by Pudritz and Carlsberg (see Pudritz, this conference).

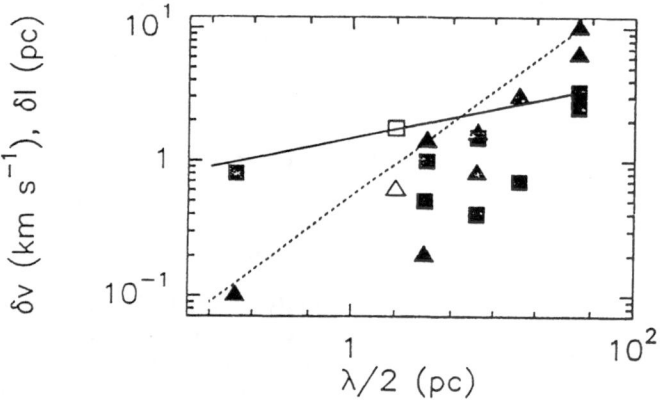

Figure 2: Amplitudes of waves along filaments as a function of half the wavelength. Triangles are velocities, squares are displacements. The full symbols refer to cases studied by Puget et al .(1991), the empty symbols are from Heiles (1988). The upper envelopes are given by the full line (displacement) and dashed line (velocity).

This has two important consequences : it could explain how the turbulence is fed in molecular clouds even when no detectable star is forming within these clouds. Futhermore it leads to a field configuration in which the support against gravity in the direction of the average field might be comparable to the support in the direction perpendicular to it. Finally it efficiently feeds energy at scales smaller but comparable to the cloud size: the spectrum of the injected energy in the cloud is steep ($P(k) = k^{-\alpha}$, $\alpha > 2$). This is what is required to explain why small scales seem to be more unstable than large ones contrary to what the classical Jeans criterion would indicate (Bonazzola et al. 1987,1990).

The kinetic energy injected by this mechanism would cascade to smaller scales through non linear interactions and dissipate through wave steepening (long wavelengths) and ohmic dissipation (short wavelengths) (Zweibel and Josafatsson , 1983).
The resulting MHD turbulence in the cloud cannot be evaluated simply and requires numerical simulations.
As a first approach to the question of stability, Puget and Falgarone (1991) develop a very simple model to investigate how the energy injected scales with density.
The cloud can be modeled as a dense slab which receives a spectrum of transverse Alfvèn waves. If the ratio of the Alfvèn velocity inside the slab to the Alfvèn velocity outside is noted $\varepsilon = (\rho_{in}/\rho_{out})^{-1/2} \ll 1$, the amplitude of the waves inside the slab (of thickness d) as a function of frequency shows resonances for :

$$\omega = n \frac{\pi}{d} v_A \text{ (n being an integer)}$$

Table 1 gives amplitude and energy density for resonant waves after integrating over frequency assuming a white incident spectrum. This shows that the energy density inside the slab is dominated by the resonant modes and that it **increases with density** like $1/\varepsilon$. The finite size of the clouds is thus essential to the feeding mechanism (the energy fed a semi infinite medium **decreases with density**).

This very simple linear analysis suggests that a two dimensional slab can be stabilized by the wave energy fed into it by this mechanism.
The problem is of course much more complicated (it is non stationary, non linear). The spectrum of waves incident on a cloud certainly does not have a flat spectrum but more likely a steep spectrum.

The longest wavelengths just generate motions of the cloud as a whole. Only can the short wavelengths perturbations with smaller amplitude feed the inside of the slab in the linear regime. Falgarone and Puget (1986) suggest that in the non linear regime low frequency but large amplitude perturbations of the field will excite all the resonant modes with a spectrum steeper than k^{-2}.

Of course a molecular cloud, even if it has sharp edges, does not have a simple geometry and thus no sharp resonant modes. Furthermore, density fluctuations will scatter and reflect the waves inside the slab (Li and Zweibel, 1987). The large injection of energy at the resonant frequencies is due to the reflected waves from the other interface coming back with the right phase. In a realistic cloud geometry , there will not be plane reflected waves but there will still be perturbations of the field inside the cloud boundary due to reflected waves on other cloud boundaries and thus one might expect that the resonant peaks are washed out but that the integral of the injected energy is conserved.

This can be checked to be true for multiple layers with different densities. The resonant pattern becomes much more complicated but in each slab the behaviour of the energy density of the waves with density remains those given in Table 1 (after averaging over frequency). A three dimensional MHD numerical simulation is needed to check this point under more general conditions.

In summary, the non linear waves generated in the cloud have large enough magnetic energy gradients to generate density fluctuations (see also B. Elmegreen at this conference). The question of the density and velocity dispersion distribution generated by non linear outside perturbations and the stability of the resulting configuration against gravitational collapse is an open question which requires three dimensional numerical simulations with a large dynamical range and non periodic boundary conditions in one dimension (Gispert, Puget, Bonazzola 1991).

Quantities measured in the outside medium		Scaling factor for same quantities inside the slade		
		non resonant waves	resonant waves	averaged over ω
Wavelength	$\lambda = \dfrac{2\pi}{\omega} v_{Aout}$	ε		
Amplitude $V_\perp$	ωL	$1.4\,\varepsilon$	1.4	$\varepsilon^{1/2}$
Amplitude $B_\perp$	$\dfrac{v_\perp B_0}{v_{Aout}}$	1.4	$\varepsilon^{1/2}$	$\varepsilon^{-1/2}$
Magnetic Energy density	$B_\perp^2 / 8\pi$	2	$0.5\,\varepsilon^{-2}$	ε^{-1}
Kinetic Energy density	$\rho\, v_\perp^2$	2	$0.5\,\varepsilon^{-2}$	ε^{-1}
Energy flux	$\dfrac{B_\perp^2}{8\pi} v_{Aout}$	$4\,\varepsilon^2$	1	$2\,\varepsilon$
Pressure gradient	$\dfrac{B_\perp^2}{8\pi}\dfrac{1}{\lambda}$	$1.4\,\varepsilon^{-1}$	ε^{-3}	ε^{-2}

Table 1

REFERENCES:

Arons J.and Max, C.E., 1975, Ap.J. Lett. **196**,L77.

Bonazzola S., Falgarone E., Heyvaerts J., Pérault M., and Puget J.L. 1987, Astron. Astrophys. **172**, 293

Bonazzola S., Pérault M., Puget J.L., Heyvaerts J., Falgarone E., 1990, submitted to Journal of Fluid Mechanics.

Clifford, P. and Elmegreen B.G. 1983, M. N. R. A. S. , **202**,629

Elmegreen B.G. 1985, in Birth and infancy of stars ed. R. Lucas, A. Omont,and R. Stora , North Holland Physics Pub.,p.215

Falgarone E., Puget J.L. 1986, Astron. Astrophys. **162**,235

Falgarone E., Puget J.L., Pérault M., 1991, in preparation.

Gispert R., Puget J.L., Bonazzola S., 1991, in preparation.

Heiles C , 1988, Ap.J. **324**,321

Li H.S. and Zweibel E.G., 1987, Ap.J. **322**,248

Lioure A. and Chièze 1990, Astron. Astrophys. **235**, 379.

Loren R.B. 1989, Ap.J. **338**, 925.

Maddalena R.J., and Thaddeus P.,1985,Ap.J.,**294**,231.

McKee C., Zweibel E.G., Goodman A.A. and Heiles C., 1991, to be published in Protostars and Planets III.

Pérault M., Falgarone E. and Puget J.L. 1985, Astron. Astrophys., **152**,371

Pérault M., Falgarone E. and Puget J.L. 1986, Astron. Astrophys.,**157**,139

Puget J.L., 1985, in Birth and infancy of stars, R.Lucas, A.Omont and R.Stora Eds.,North Holland Physics Pub.

Puget J.L., Falgarone E., 1991, in preparation.

Scalo J.,1987, Interstellar Processes, eds. D.J. Hollenbach and H.A. Thronson.

Ungerechts, H. and Thaddeus, P., 1987, Ap. J. Suppl. **63**,645.

Zweibel E.G., and Josafatsson K. 1983, Ap.J. **270**,511

TURBULENCE AND MAGNETIC FIELDS IN MOLECULAR CLOUDS

R.N. HENRIKSEN

service d'astrophysique, batiment 528
c.e.n. saclay, 91191 gif sur yvette
cedex, france
E–mail @32779 : : henriksen.

Keywords : Gravitation, Turbulence, Magnetic, Molecular, Clouds
Abstract. In this paper I first review some of the simple structural concepts associated with compressible turbulence. In particular the hierarchical or self-similar fractal structure to be expected is formulated in a manner readily compared to the observations, and to previous work. In the next section I present the first results of a wavelet analysis on molecular clouds, which seem to comfirm the hierarchical scaling. I conclude with an extention of the theory to include magnetic fields. This latter theory represents an alternative to the more conventional dynamo theory.

1. Compressible Turbulence and Fractal Hierarchies

All quantitative science depends on a quantitative description of the phenomenon in question. In the case of the interstellar molecular clouds various authors (Zinnecker,1984; Henriksen and Turner, 1984; Henriksen, 1986 ; Elmegreen, 1989) have suggested that the phenomenon is sufficiently complex and hierarchical to be termed 'turbulent'. In such a framework the concept of fractal geometry is known to permit a precise and quantitative description, although in general an infinite set of parameters is required. Fortunately these can be greatly restricted by selection on the basis of their physical significance. I shall begin by introducing the parameters that I have tried to use consistently in my own descriptions of the molecular cloud turbulence, taking care to emphasize their physical significance.

The fractal capacity or Hausdorf dimension D is defined, in a discrete model , in terms of the number of objects of scale ℓ in the structure say $n(\ell)$. Thus if two scales ℓ and L have the ratio $r = \ell/L$, then the number of fragments of size ℓ per fragment of size L is given by

$$n(\ell)/n(L) = r^{-D}, \tag{1}$$

where for a self-similar structure r is constant over all the various stages. Notice that the pieces may be irregular in detailed shape, but that in a statistical average they should be characterizable by one scale in this simple picture. Otherwise anisotropic fractal structure is required.

It is important to note that on passing to a continuum model of the structure wherein $N(L)$ gives the total number of fragments having $\ell \leq L$, the correspondence with the discrete $n(\ell)$ is the logarithmic derivative of N with respect to ℓ. For we clearly must associate equal ratios with equal differences in the logarithmn. Thus

$$\frac{dN}{d\ln(\ell)} \propto \ell^{-D}, \tag{2}$$

for the continuum model.

Now let us in addition introduce a dimension (Henriksen, 1986) that gives the variation of the mean density on a scale ℓ with ℓ. By mean density we shall mean here $\overline{\rho}(\ell)$ such that the mass of an element of scale ℓ is

$$M(\ell) = \overline{\rho}(\ell)\ell^3. \tag{3}$$

Previously, I have used the density of a subscale fragment (Henriksen 1986) so that 3 in equation (3) is replaced by a fractal dimension. I hereby renounce that somewhat confusing procedure! Then I introduce D_ρ so that

$$\overline{\rho}(\ell) \propto \ell^{-D_\rho}. \tag{4}$$

These two dimensional indices allow us to construct the dependence of various significant physical quantities on scale. Thus the total mass on a scale ℓ, $n(\ell)M(\ell)$ is related to that on a superscale L as

$$n(\ell)M(\ell) = n(L)M(L)\, r^{(3-(D+D_\rho))}. \tag{5}$$

It is convenient to introduce an 'aggregation index'

$$i_1 \equiv 3 - (D + D_\rho),$$

such that $i_1 = 0$ in a pure aggregation or fragmentation hierarchy where each superscale is composed solely of fragments of subscale ℓ. In this case D and D_ρ are not independent. However such a case is rather idealised for the molecular clouds where one expects there to be lower density material filling the voids between the subscale density peaks, that is 'fragments'. Such a 'smoothed' aggregation hierarchy requires the two independent dimensions, and the difference between i_1 and 0 is a measure of the deviation from a pure geometric hierarchy, with $i_1 > 0$ implying that there is mass between the subscale fragments (in the uniform limit $i_1 = 3$).

Moreover, following Henriksen and Turner (1984), the collision time between two fragments of scale ℓ moving in a volume L^3 is a measure of the dynamical coupling between the scales when compared to the dynamical time on the scale L. An interesting example is afforded by a simple binary hierarchy, wherein a resonance between the period of one binary and the time scale of the encounter between two such binaries on the next superscale can be expected to produce the most pronounced exchange of angular momentum and energy between the scales (see e.g. Lattanzio and Henriksen,1988). We calculate this quantity in a straightforward way as (v_L is the velocity dispersion on the scale L)

$$t_c(\ell) \equiv \frac{1}{\left(\frac{n(\ell)}{n(L)L^3} \times \ell^2 v_L\right)},$$

or on using equation (5) and the definitions above and after multiplying this expression top and bottom by $M(\ell)$, one obtains

$$t_c(\ell) = \frac{L}{v_L} \times r^{(D-2)}. \tag{6}$$

In this way one defines a 'dynamical coupling index' i_2 such that

$$i_2 \equiv D - 2. \tag{7}$$

When $i_2 = 0$, we see that there is strong dynamical coupling wherein $t_c = L/v_L$. When $i_2 < 0$, the subscale collision time is longer than the superscale dynamical time, which reduces the resonance between the scales, while when $i_2 > 0$ the collision time is relatively short and one expects collisional damping to terminate the dynamical cascade ultimately. In Henriksen and Turner (1984) $i_1 = 0$ so that a pure aggregation hierarchy was assumed implicitly, and thus $i_2 = 1 - D_\rho$. Subsequently the imposition of the dynamical condition $t_c = L/v_L$, so that $i_2 = 0$ allowed HT to deduce that $D_\rho = 1$ and that the velocity scaling was $\propto \ell^{1/2}$. They did not use the index D, but in Henriksen (1986) we now see that D is constrained to be 2 *if* the pure aggregation hierarchy is maintained. For general D and D_ρ, neither one of the indices i_1 and i_2 are 0, and the scaling in the turbulence depends only on the constants $\lambda \equiv \bar{\rho}\ell^{D_\rho}$ and Newton's constant G. This general predicted scaling (Henriksen, 1990) is

$$
\begin{aligned}
v(\ell) &= \sqrt{G\lambda}\ell^{(1-D_\rho/2)}u(\xi), \\
\rho(\ell) &= \lambda\ell^{-D_\rho}\mu(\ell), \\
B &= \sqrt{G\lambda^2}\ell^{(1-D_\rho)}b(\xi), \\
p &= G\lambda^2\ell^{2(1-D_\rho)}P(\xi), \\
\xi &= \frac{G\lambda t^2}{\ell^{D_\rho}}.
\end{aligned}
\tag{8}
$$

The indices i_1 and i_2 remain useful as indicators of the respective physical hierarchies discussed above.

One of the interesting applications of these ideas is the predicted relation (Henriksen, 1986) between these indices and the index of the scaling portion of the IMF defined as

$$\alpha_* \equiv -\log(d\,N/d\log M_*).$$

As in Henriksen (1986) but using now $M(\ell) \propto \ell^{3-D_\rho}$, one obtains directly from equation (2) that for the clouds

$$\alpha = D/(3 - D_\rho). \tag{9}$$

Subsequently Henriksen(1986) assumed that the mass of a star formed in a cloud would be proportional to the mass of the cloud. However at this meeting Larson has argued that the relation may be closer to

$$M_* \propto M^{0.43}.$$

In general we may want to introduce a degree of freedom here (see e.g. Zinnecker 1989) and call this power f. Then we have simply that

$$\alpha_* = \alpha/f = D/(f(3 - D_\rho)).$$

If $\alpha = 0.6$ as suggested at this conference by Elisabeth Lada, then $\alpha_* = 1.4$ with $f = 0.43$. In such a case $D = 1.2$ from (9) if $D_\rho = 1$. This does not however agree with more direct measures of the dimension D (see the next section). At the other extreme, $f = 1$, $D_\rho = 1$, and $D = 2.7$ give $\alpha = 1.35$, the Salpetre value, which dimension was the suggestion of Henriksen(1986). This value however appears to overestimate the importance of the clouds of small mass relative to the value found by Lada and others.

2. Wavelet Analysis

In this section I wish to present what I believe is the first application of wavelet transform theory to a description of molecular clouds. The full description of this work is to be found in a paper by Gill and Henriksen (1990). Our intention is to use the velocities characterising emissivity peaks as the third coordinate for these peaks orthogonal to the plane of the sky. In this $\Delta V - \ell - b$ 'phase' space the wavelet analysis allows us to measure a dimension directly. Even without further interpretation of the velocity-scale relation, a fractional but well defined value provides evidence for a scaling fractal hierarchy in phase space indicative of turbulent processes.

I proceed by presenting briefly the wavelet analysis in its simplest form. A more general discussion can be found for example in Argoul et al. (1989). One works either in the space of 'scales' a ($1/a$ is the magnification) or in position space $\vec{b}$ corresponding to wave vector space and position space of ordinary Fourier analysis. Here we have supposed that the scaling is isotropic so that it is characterised by a single parameter, and moreover we do not rotate our spatial axes from point to point, consistent with this presumed isotropy. One would not expect to obtain a convergent result if either of these constraints are badly broken. We have in mind the representation of real functions over the plane of the sky in terms of carefully chosen base functions (the analysing wavelets) $g(\vec{y})$. The wavelet that we consider best for our purposes to date is the so-called Mexican Hat wavelet which has the form

$$g_M(\vec{y}) = (2 - \frac{\vec{y}^2}{a^2})e^{-\left(\frac{\vec{y}^2}{2a^2}\right)}. \tag{10}$$

In order that the transform theorem hold, the wavelet must satisfy in practice

$$\int g(\vec{y}) \, d\vec{y} = 0,$$

which is the case for the Mexican Hat. In addition it is worth noting that $g_M = 0$ where $\vec{y} = \sqrt{2}a$, and that it is 2 at the origin but $-2/e^2$ at the position of the negative maximum $2a$. The simplest statement of the wavelet transform theorem now reads for a properly behaved function $\rho(\vec{x})$;

$$T_g(a, \vec{b}) = \int g\left(\frac{\vec{x} - \vec{b}}{a}\right) \rho(\vec{x}) \, d\vec{x}, \tag{11}$$

$$\rho(\vec{x}) = K \int T_g(a, \vec{x}) \, a^{-3/2} \, d a. \tag{12}$$

Here K is a numerical constant which need only be calculated once.

For our purposes we work with the wavelet transform $T_g(a, \vec{b})$ itself since it possesses the important property of scaling with the analysed function. That is if $\rho(\lambda \vec{x}) \propto \lambda^\alpha \rho(\vec{x})$ then

$$T_g(\lambda a, \vec{b}) \propto \lambda^\alpha T_g(a, \vec{b}). \tag{13}$$

Thus a plot of $\log T_g$ versus $\log a$ will yield a straight line of slope α if indeed the function being analysed scales in the assumed manner. Such a demonstration of scaling is important in itself, although the physical interpretation is more problematical and clearly dependent on the function being analysed.

However, a second remarkable property of the wavelet transform is that some positional information is retained *even at a single value of the scale a*. In fact in Gill and Henriksen (1990) it is shown that if the maximum scale a_d to which the well defined straight line in the $\log T_g - \log a$ extends is used, then $T_g(a_d, \vec{b})$ maps the dominant structure of the analysed function over the plane of the sky reasonably well.

We have applied this technique to ^{13}CO spectra of the $L1551$ star forming region kindly loaned to us for this purpose by Moriarty-Schieven and Snell (1988). Approximately one half of the some 1500 spectra had sufficiently good S/N ratios that the velocity at the emissivity peak could be identified both by eye and by fitting a Gaussian. There were essentially no incidences of well defined multiple peaks in the data, so that on average one emissivity peak was observed per line of sight.

The spectra were distributed over a 42×44 step grid which included the region of ^{12}CO molecular outflow and the source IRS 5 itself. The high velocity cloud that overlaps the region was removed from the velocity data as in Moriarty-Schieven and Snell (1988), and the mean velocity of the remaining $L1551$ cloud was also subtracted from the peak velocity data. The resulting set of corrected peak velocities constituted the function $\rho(\vec{x})$ that was analysed using equations (10) and (11). We used an average over all points $\vec{b}$ on the grid for each value of a (see figure caption) so that we postulate a homogeneous scaling on the grid (or subgrid). The calculation was actually done using a Fast Fourier Transform over $\vec{b}$ for each scale a, making use of the fact that equation (11) is recognisably a convolution integral. Finally the transform was inverted and the result is shown in figure 1.

One sees that these emissivity peak velocities do show a scaling law indicative of hierarchical structure. If moreover we adopt the view that the peak velocity is a measure of the position of the region of dominant emission orthogonal to the plane of the sky (see Gill and Henriksen, 1990 for a justification), then the transform makes a statistical estimate of the rate at which the number of such emissivity peaks in a volume increases with the characteristic scale a. That is, we obtain a direct estimate of the fractal capacity or Hausdorf dimension. This gives $D = 2.42 \pm 0.01$ in the notation used above in section 1 if we adopt the value found for the region outside the molecular outflow only, and $D = 2.35 \pm 0.01$ for the average over the entire cloud.

On referring back to our discussion of section 1, we see that this value of D gives (equation 9) an α of either 1.21 or 1.17 respectively, if $D_\rho = 1$, which is roughly twice the value reported at this conference and previously from direct cloud counts. Moreover if indeed f is significantly different from 1, then α_* becomes ultimately too large (a value as

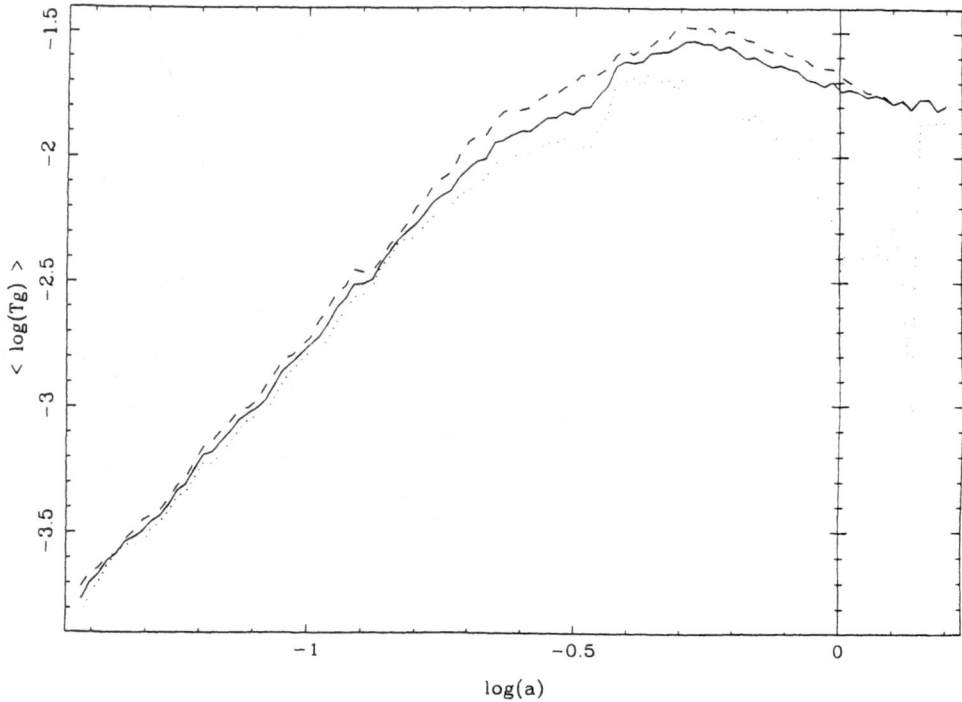

Fig. 1. The Mexican Hat wavelet transform of the peak velocity fluctuations of the ^{13}CO spectra in the $L1551$ region. Positive values of T_g are logarithmically averaged with a S/N weighting scheme. The solid line is averaged over the entire cloud, the dashed line is averaged over the region outside the ^{12}CO molecular outflow while the dotted line is averaged over the outflow region alone. The scale parameter a is defined such that $\log(a) = -0.6 \leftrightarrow 0.5pc$ in $L1551$. The slopes for $\log(a) < -0.6 = \log(a_d)$ are respectively, $2.35 \pm 0.01, 2.42 \pm 0.01$, and 2.29 ± 0.01.

small as .7 is however acceptable). Thus this method of determining the index D would be highly suspect were it not for the measurement of $D_p = 1.36 \pm 0.02$ reported here by Falgarone as the fractal dimension in the plane of the sky. If the ^{12}CO measurements are optically thick, so that the dimension obtained is essentially that of a planar intersection with the true spatial distribution, then indeed $D_p = D - 1$ and the agreement between these two highly independent measurements is essentially perfect. But to obtain this agreement, one must conclude that indeed $\Delta V \propto a$ in the region. This latter 'Hubble' law is an indication of turbulence, but is not the form usually quoted for molecular clouds $(\Delta V \propto a^{1/2})$. Perhaps the velocity field throughout the region has been disturbed by the molecular outflow. In this connection I note that Prasad and Sreenivasan (1990) give $D = 2.36 \pm 0.05$ for the Kolmogorov range of the turbulence associated with a water jet in quiet surroundings.

3. Magnetic Turbulence

If the indications of dynamical turbulence discussed in the preceeding sections are taken seriously, then a corresponding treatment of the magnetic field is necessary. In this section we wish to generalize the equations of Chaboyer and Henriksen (1990) for stochastic magnetic fields to include a Hubble type cloud divergence or contraction, as well as a feedback loop to the dynamical equations. This latter modification opens the question of true 'stochastic dynamos', although I do not pretend to close it here! The fundamental physical picture is that although the magnetic field may be ordered on the large scales, and again on the small protostellar scales, it is likely to be turbulent due to many competing dynamical processes on intermediate scales. On these scales then, the best global predictions for the magnetic field are probably in terms on the ensemble averaged two-point correlation functions in the spirit of Hoyng (1987a,b; 1988) and of Chandrasekhar (1955a,b; 1957). Observational evidence bearing on such a stochastic hierarchical image of magnetic stuctures is discussed in Chaboyer and Henriksen (1990), and recent results were presented by Heiles at this conference.

The fundamental approach to the description of the turbulence follows the self-similar model introduced by Henriksen and Turner (1984) following earlier work by Sedov (1982) and developed further in Henriksen (1986) and Henriksen (1988). The basic equations are the MHD equation with a non-constant turbulent resistivity η as;

$$\partial_t B^i = B^k \nabla_k v^i - \nabla_k (v^k B^i) - \nabla_k \left(\frac{\eta}{4\pi} \nabla^i B^k \right) + \nabla_k \left(\frac{\eta}{4\pi} \nabla^k B^i \right), \tag{14}$$

where the velocity should be taken as

$$v^i = u^i + H x^i,$$

and $H \equiv \dot{a}/a$ in terms of the overall scale factor $a(t)$. The coordinates x^i are Eulerian. In the same notation the fundamental dynamical equation is

$$\partial_t u^i = -\frac{1}{\rho} \nabla^i P - u^k \nabla_k u^i + F^i - H u^i - H(x^k \partial_k) u^i$$
$$+ \nu \nabla_k \sigma^{ik} + \zeta \nabla_k (g^{ik} \nabla_a u^a) + \sigma^{ik} \frac{\nabla_k (\rho \nu)}{\rho} + g^{ik} (\nabla_a u^a) \frac{\nabla_k (\rho \zeta)}{\rho}, \tag{15}$$

where

$$\sigma^{ik} \equiv \nabla^i u^k + \nabla^k u^i - (2/3)(\nabla_a u^a) g^{ik},$$
$$P \equiv p + \frac{B^2}{8\pi},$$
$$F^i \equiv f^i + \frac{B^k \nabla_k B^i}{4\pi \rho},$$
$$f^i \equiv -\nabla^i \Phi,$$
$$\Phi \equiv \phi - \frac{\ddot{a}}{2a} (\vec{R})^2.$$

Finally the continuity equation takes the form

$$\partial_t \rho + \vec{\nabla}.(\rho \vec{u}) = -\vec{\nabla}.(H \rho \vec{R}).$$

The analysis proceeds in terms of ensemble averaged second and third moments of these equations (see e.g. Henriksen 1988) expressed at two different points, O and O'. The results are written in terms of the two-point tensors

$$U^{i,j} \equiv \langle B^i B'^j \rangle,$$

$$F^{ik,j} \equiv \left\langle \frac{B^i B^k \rho' u'^j}{4\pi \rho \rho'} \right\rangle,$$

$$S^{ik,j} \equiv \langle u^i B^k B'^j \rangle,$$

$$R^{i,j} \equiv \langle u^i u'^j \rangle,$$

$$R^{ij,k} \equiv \langle u^i u^j u'^k \rangle,$$

together with their various contractions. One has also to use the properties of homogeneous and isotropic tensors (Henriksen,1988; Chaboyer and Henriksen,1990) which we shall not repeat here in the interests of brevity. When specific components of a tensor are given they refer to a right-handed reference system with the 1 axis along the line OO', and the $2-3$ axes in the orthogonal plane. An asterisk subscript emphasizes this frame specificity. One has no need to specify the absolute orientation of this system because of the assumed isotropy. The fundamental equations that we have derived for the stochastic dynamo are now;

$$\partial_t \tilde{U} = \frac{\eta}{2\pi} \nabla_k \nabla^k \tilde{U} + \frac{4}{L^2} \partial_L (L^2 \tilde{S}) + H L \partial_L \tilde{U}, \tag{16}$$

$$\partial_t \tilde{R} = 2\nabla_k \left(\delta_{i,j} \tilde{F}^{ik,j} \right) - 2\nabla_k \left(\delta_{ij} \tilde{R}^{ik,j} \right) + 2\nu \nabla_k \nabla^k \tilde{R} + 2\nu \nabla_k \langle \ln \rho \rho' \rangle \nabla^k \tilde{R}$$

$$+ 2\tilde{F}^{ik,}_{i} \nabla_k \langle \ln \rho \rho' \rangle + H L \partial_L \tilde{R}, \tag{17}$$

$$\langle \ln \rho \rho' \rangle = -6a - 2D_\rho \ln L. \tag{18}$$

In these equations L stands for the distance between the pairs of points considered (i.e. the spatial lag), the notation $\tilde{()}$ indicates $a^2 \times ()$, the index D_ρ is defined in a manner closely analagous to that of the preceeding sections so that $\rho(L) \equiv \exp \langle \ln \rho \rho' \rangle / 2 \propto L^{-D_\rho}$, and S is a source term depending on the lack of mirror symmetry as

$$S \equiv S_*^{12,2} - S_*^{21,2}. \tag{19}$$

Now we recall that the assumption of isotropy implies that the traces of the tensors $R^{i,j}$, and $U^{i,j}$ suffice to define these tensors entirely (Henriksen,1988). Thus our equations would be complete were it not for the presence of the third order tensors, which is the familiar closure problem common to all such moment equations. The tensor $R^{ij,k}$ introduces one additional unknown $R_*^{11,1}$ say, as $F^{ij,k}$ introduces $F_*^{11,1}$. In addition one has both S and $S_*^{1,11}$, although the latter does not yet appear explicitly. This means only that we have not yet specified the 'source' tensor completely.

The process of self-similar closure advocated in Henriksen(1988) is capable of relating S and U as well as R, $R_*^{11,1}$ and $F_*^{11,1}$. However one additional hypothesis is required in general. As an example, one posible choice would be to write the symmetric and anti-symmetric parts of $S^{ik,j}$ on ik as

$$S^{ik,j} = constant \times \rho(L)F^{ik,j} + S\left(\frac{\delta^{kj}L^i}{L} - \frac{\delta^{ij}L^k}{L}\right),$$ (20)

although the physical implications must be explored in order to justify such a hypothesis. In any case additional use of the self-similar closure must be made to finally close the equations.

In Chaboyer and Henriksen(1990) the feed-back to the dynamical equation was ignored completely so that equation (17) was not used. The magnetic field behaviour is however assumed to be dictated by the velocity field in that the self-similarity symmetry of equation (16) is fixed by using the same global constants that appear in the dynamical treatment of non-magnetic turbulence (Henriksen,1988). This essentially supposes equipartition magnetic fields in the ensemble average, since otherwise there would be no reason for such self-consistency.

The details of the solution do not concern us here, but it is worth reporting that the solution for U (neglecting the divergence term) that is consistent with the scaling laws discussed in section 1 for $D_\rho = 1$ is

$$U = const + \frac{t^2}{L}.$$ (21)

This indicates that at large separations the 2-point magnetic correlation tends to a constant, and that it grows in time at a fixed separation presumably as a result of stochastic diffusion from the small or protostellar scale where the simple theory has it diverging with the density. In fact we deduce that

$$B_{rms} \propto \sqrt{\rho},$$ (22)

on the small scales. This behaviour indicates an outward diffusion of magnetic flux consistent with the outward diffusion of angular momentum, and of heat predicted dynamically in the turbulent theory, and necessary for the formation of stars. However the effect found here changes the magnetic Jean's mass only $\propto \rho^{-1/2}$ which is too slow if the magnetic field is in equipartition on the large scales. This contrasts with the rate of reduction of the Jean's mass and of the 'centrifugal mass' which both vary as ρ^{-2} in the theory. Thus we must conclude that if the field is indeed in equipartition on the large scales then rather dramatic processes of magnetic dissipation must occur on the protostellar scale. However the conclusion is very sensitive to the magnitude of the mean field in molecular clouds (the magnetic Jean's mass $\propto B^3$).

Acknowledgements

The author wishes to express his thanks to the Service d'Astrophysique du C.E.A. Saclay for their hospitality and support while this work was completed.

References

Argoul,F.,Arneodo, A.,Elezgaray,J.,Grasseau,G. and Murenzi,R. : 1989,"Wavelet Transform of Fractal Aggregates." *Physics Letters A*, **135**,327–336.

Chaboyer,B. and Henriksen,R.N. : 1990,"Turbulent Magnetic Fields I" *Astron. Astrophys.*,in press.

Chandrasekhar, S. : 1955a,"Homogeneous Turbulence I" *Proc. Roy. Soc. A*,**233**,322.

Chandrasekhar,S. : 1955b,"Homogeneous Turbulence II" *Proc. Roy. Soc. A*, **233**,330.

Chandrasekhar,S. : 1957,"Homogeneous Turbulence III" *Annals of Physics*,**2**, 615.

Elmegreen, B.G. : 1989,"A Pressure and Metallicity Dependence for Molecular Cloud Correlations and the Calibration of Mass." *Astrophys. Journal*,**338**,178.

Gill A. and Henriksen,R.N. : 1990,"A First Use of Wavelet Analysis for Molecular Clouds" *Astrophys. Journal*, submitted.

Henriksen, R.N. and Turner,B.E. : 1984,"Star Cloud Turbulence" *Astrophys. Journal*,**287**,200–207.

Henriksen,R.N. : 1986,"Star Formation in Giant Molecular Clouds." *Astrophys. Journal*,**310**.189.

Henriksen,R.N. : 1988,"Inhomogeneous Turbulence" *Astrophys. Journal*,**331**,359–369.

Henriksen,R.N. : 1990,"On Molecular Cloud Scaling Laws and Star Formation" *Astrophys. Journal*, submitted.

Hoyng, P. : 1987a,"On Magnetic Dynamos I" *Astron. Astrophys.* , **171**,348.

Hoyng, P. : 1987b,"On Magnetic Dynamos II" *Astron. Astrophys.* ,**171**,357.

Hoyng, P. : 1988, "On Magnetic Dynamos III" *Astrophys. Journal*, **332**,857.

Lattanzio, J.C. and Henriksen, R.N. : 1988, "Collisions Between Rotating Interstellar Clouds" *Mon. Not. R. astr. Soc.*, **232**, 565–614.

Moriarty-Schieven, G.H. and Snell, R. : 1988 "The Star Forming Region L1551" *Astrophys. Journal*,**332**, 364.

Zinnecker, H. : 1984 "Star Formation from Hierarchical Cloud Fragmentation : A Statistical Theory of the log-normal IMF." *Mon. Not. Roy. astr. Soc.*,**210**,43–56.

Zinnecker, H. : 1989 "Star Formation in Galaxies" in *Evolutionary Phenomena in Galaxies*, (J. Beckman,Ed.), 115, Cambridge University Press.

DYNAMICAL CONDITIONS OF
DENSE CLUMPS IN DARK CLOUDS :
A STRATEGY FOR ELUCIDATION

SHEO S. PRASAD

Lockheed Palo Alto Research Laboratory
3251 Hanover Street (MC : O/91-20; B255)
Palo Alto, CA 94304; USA
and
Departments of Physics and Astronomy
University of Southern California
Los Angeles, CA 90089-1341; USA

Abstract. Chemical considerations and simplified dynamical modeling suggest that dark cloud cores may be incessantly evolving such that the time spent at high core densities decreases as the density increases. After reaching a high density, gravitationally contracting dark cloud cores may either form stars or expand to states of lower densities. Cloud mass and initial density are amongst the factors that may control the evolutionary fate of the core. This view is diametrically opposite of the common belief that dense cores may be in near mechanical equilibrium. Mutually consistent end-to-end modeling of the spectral line profiles and intensities is needed to discern the reality.

1. Introduction

The dynamical conditions of the dense clumps or cores in molecular clouds are discussed from the perspective of chemistry. This discussion suggests the possibility that dynamical evolution, rather than dynamical equilibrium, may be the norm for these clumps at least for dark clouds.

Reviews presented in this IAU Symposium leave little doubt that molecular clouds are very clumpy (e. g., Thaddeus 1990, Solomon 1990). This conclusion has been reached by panoramic surveys in CO lines (Thaddeus 1990, Solomon 1990), and has been substantiated by cloud specific surveys in the lines of other molecules (e. g., CS, NH_3) capable of acting as high density tracers (e. g., Swade 1990). A precise definition of clumps is, however, not available at the present time (Myers 1990). Indeed, the words clumps and cores have been used interchangeably. Nevertheless, in general terms, clumps or core are regions of enhanced density surrounded by regions where the density drops significantly by two to three orders of magnitudes. The number and sizes of the clumps in a given cloud may vary significantly from cloud-to-cloud. According to a model by Tauber and Goldsmith (1990), the OMC may contain thousands of small clumps of at best a few solar mass. In contrast, L134N in the Taurus region may have only a few massive clumps of about 25 solar mass. The cause of this vast diversity is not known. However, it is noteworthy that clouds modeled by Tauber and Goldsmith (1990) are associated with active star formation while L134N appears to be quiescent and perhaps only on the threshold of star formation.

While panoramic surveys indicate the presence of clumps, these maps of line intensities provide no information about the dynamical conditions in the clumps or cores. The dynamical conditions have, therefore, been inferred from the width and shapes of the spectral lines. This is possible because $\Delta\nu$, the line width in the frequency domain, is directly related to σ, the velocity dispersion. Indeed, the line widths are usually expressed as Δv where v is the velocity.

Molecules can also be used to infer the dynamical conditions in the dense clumps or cores because the abundances and the line intensities of at least some of them are sensitive to the dynamical states. For example, the abundances of "complex molecules" (e. g., C_2H, C_3H, C_3H_2) are much smaller in the equilibrium models, compared to the models in which the chemistry is prevented from equilibrating by the dynamics. By comparing the observed abundances (preferably line intensities) of the sensitive molecules with the theoretical predictions for the equilibrium and evolutionary models, it should be possible to form a consensus about the dynamical states of the targeted cloud cores.

2. Contrasting Views About the Dynamical States of the Clumps :

According to one school, the dense clumps or cores are thought to be in near mechanical equilibrium and probably magnetically supported (Myers 1983), because they approximately obey the relations of virial balance ($\sigma^2 \simeq GM/R$), and (Larson's) power law relation ($\sigma \simeq R^{0.5}$) and ($n \simeq R^{-1}$). In these relations, n, M and R are the cloud density, mass and size, and σ is the velocity dispersion which is related to the spectral line widths. If these objects were collapsing on a free fall time scale, it is argued, then the star formation rate in the galaxy would be too high (Zuckerman and Palmer 1974).

A diametrically opposite state of dark cloud cores is, however, suggested by their observed molecular abundances.

If these cores are in near equilibrium, then their molecular abundances should reflect chemical equilibrium at fixed conditions of density, temperature and visual extinction. The observed abundances show serious disagreement with equilibrium chemistry abundances (Herbst and Leung 1986a,b, 1989). The equilibrium chemistry tends to under-estimate the abundances of long carbon chain molecules (complex molecules) and over-estimates the abundances of H_2O and O_2 (Chièze and Pineau des Forêts 1989).

One possible solution of this chemical dilemma is to assume the existence of mixing currents which bring core material to the envelop at short time intervals and limit the time available to chemistry to equilibrate at any fixed conditions of density, temperature and visual extinction (Boland and de Jong 1984, Chièze and Pineau des Forêts 1989). Williams and Hartquist (1984) have, however, pointed out significant difficulties with this solution. These difficulties are all the more accentuated in dark cloud cores, where turbulent line widths are quite small (Swade 1990). Penetration of uv due to extreme clumping (Boissé 1990) may work only for cores near star formation in giant molecular clouds. The other possible solution is that dark clouds may be gravitationally evolving, so that the cloud cores spend only a limited time at any given core density, and this time decreases as the core density increases (Tarafdar et al 1985, Prasad 1987, Prasad et al 1987). In these models the time available for dark cloud core chemistry to equilibrate is severely restricted, and the modeled chemical composition of these cores are in better agreement with observations (Tarafdar et al 1985). These early evolutionary models were, however, susceptible to criticism that they imply excessive star formation rate.

3. Improved Evolutionary Models

Recent improved evolutionary models have the potential to avoid the above-mentioned conflicts with the observed star formation rates. This follows from the theoretical result that all gravitationally contracting clouds may not form star; some may expand after attaining high core densities and revert to a diffuse state.

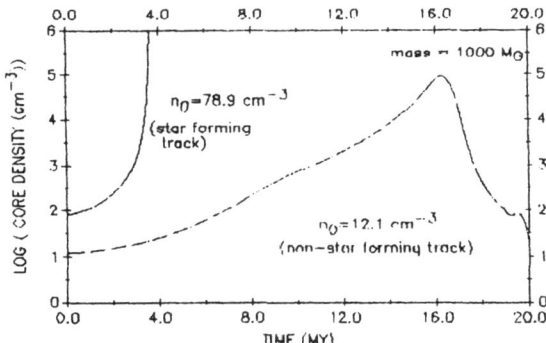

Fig. 1.   Time evolution of the core density in a gravitationally contracting cloud of $1000 M_\odot$.

Improved evolutionary models span lower initial densities and also include non-gravitational forces that may oppose gravity. Modeling details have been presented by Prasad et al (1990). Here we discuss the results only. Figure 1 shows the time evolution of the core density in a model cloud of $1000 M_\odot$ gravitationally contracting from initial densities of 12 and 79 atoms cm^{-3}. When the initial density was high (e. g., $n_0 \geq 78$ cm^{-3}) the cloud collapsed monotonically on a star-forming track. It attained a core density of $10^3 cm^{-3}$ in about 3 MY. After that the core density increased rather rapidly and very soon the cloud was on the threshold of star formation. In sharp contrast, the cloud evolved on a non-star-forming track when the initial density was low, say, only 12 cm^{-3}. In this case, the initial evolution

96

towards the higher core densities was very slow. It took 12 MY to reach the core density of $10^3 \mathrm{cm}^{-3}$. Thereafter the core density increased relatively more rapidly, so that a 100 fold increase in the core density took place in only 4 MY. At this stage, however, an interesting phenomenon occurred. The cloud began to expand and its core density started to decrease. In the course of the initial contraction, the pressure gradient forces increased more rapidly than the gravity. This should explain the subsequent expansion by which the cloud reverts to its initial diffuse state.

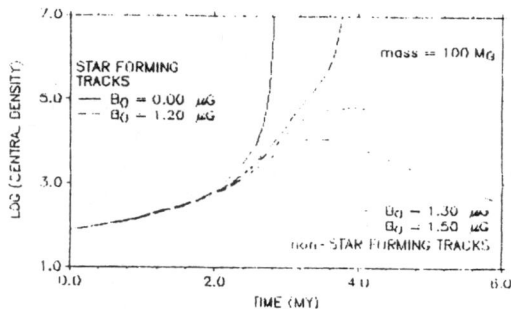

Fig. 2.   Time evolution of the core density in model clouds of $100 M_\odot$. The clouds differed in the values of the initial strength of the tangled frozen-in magnetic field.

Similar behaviors were exhibited by model clouds with non-gravitational forces capable of opposing gravity. In the present study, those forces were mimicked by randomly oriented frozen-in magnetic fields which increased with density according to a power law $B = B_0(\rho/\rho_0)^k$ where $k = 2/3$ was assumed. It is recognized that the existence of tangled frozen-in magnetic field is by no means certain. Even so, it has been used here as a simplified surrogate for a number of physical processes that may oppose gravitational contraction in such a manner that the opposition increases as the core density increases. One example is the progressive removal of coolants (e. g., CO) from the gas phase by condensation onto the grains which might produce a warmer core and thereby a stronger thermal pressure gradient force opposed to gravity. Figure 2 shows the results for a cloud of $100 M_\odot$ with an uniform initial density of $79 \mathrm{cm}^{-3}$. The higher initial density was chosen because, with reference to the data presented in Fig. 1, clouds with this high initial density have a greater tendency to follow star-forming evolutionary track. Thus, in the non-magnetic case with $B_0 = 0$ μG the cloud followed the typical star-forming evolutionary track. The same was the case, albeit on a slower time scale, for the cloud with $B_0 = 1.2$ μG. For larger B_0 ($= 1.3$ μG), the cloud followed the non-star-forming track. The reversal of the gravitational collapse in this case was the result of the increase in the magnetic pressure gradient force acting outward. The magnetic pressure gradient force increased because the magnetic field in the core increased, in response to the increase in the core density, according to the assumed scaling law for the field strength.

4. Important Implications

The possibility of the reversal of the gravitational collapse in the model clouds has important ramifications. Generally, gravitationally contracting cloud cores are thought to be a rarity, because their common occurrence are interpreted as leading to star formation rate in excess of the observed. Cloud cores were, therefore, thought to be in near equilibrium. These general believes could now reverse, because gravitationally evolving clouds may not always form stars. Most of the time they may be on non-star-forming evolutionary tracks. As a corollary, it appears possible that cloud cores in dynamical equilibrium may be exception rather than the norm.

The reversal of gravitational collapse after the formation of a dense core is equivalent to a mechanism for dissipating dense cores and cycling of interstellar gas between dense and diffuse phases. Shocks and winds from nascent stars may also disperse dense cores (Williams 1986, 1987). These mechanisms are, however, clearly limited to cores in the vicinity of active star formation. In contrast, the dissipation of dense cores through the reversal of gravitational contraction has the potential to be effective in both star forming and quiescent regions.

Possible conflict with the observed star formation rate having been eliminated, dynamically evolving clouds now seem to be a viable (or, perhaps the preferred) explanation for why the "early-time" molecular abundances from classical (pseudo time-dependent) models appear to agree with the observations better than the abundances predicted for the "late-time" near equilibrium conditions. The dynamically evolving models have the property that the time spent at any given core density decreases as the density increases. This property is clearly seen in the Figures 1 and 2. It introduces characteristic times that limit the time available to chemistry for equilibrating at any given high density. The explanation in terms of turbulent circulation currents is at best empirical at the present, and wind driven processes may operate in star forming region only.

By virtue of these positive attributes, dynamically evolving clouds become serious alternatives to cores in dynamical equilibrium. It is, therefore, imperative to examine the very basics of the two models in order to determine the reality.

5. The Search for Reality

At present the foundations for believing in either the evolutionary or the equilibrium models of dense cloud cores are not firm. We will now outline the studies needed to elucidate the reality.

The equilibrium models have put great reliance on the approximate (noisy) obedience, by the cores, of the virial balance equations ($\sigma^2 \simeq GM/R$) and Larson's relations ($\sigma \simeq R^{0.5}$ and $n \simeq R^{-1}$). While these relations have been used extensively, their utility as diagnostics of equilibrium has not been evaluated quantitatively. For example, a modeling study by Villere and Black (1982) suggests that gravitationally contracting clouds may also exhibit line shapes consistent with the observations. This study should be revived and extended to include clouds in dynamical equilibrium and the expanding phase of the dynamically evolving clouds. Villere and Black's (1982) studies were limited to CO lines. New studies should include lines of high density tracer molecules such as CS, and NH_3. Relations between σ, M/R, n and R exhibited by dynamically evolving model clouds can then be compared with those shown by model clouds in dynamical equilibrium for a proper diagnosis of the dynamical conditions in cloud cores.

Chemical diagnostic of the dynamical conditions in dense cores utilize the observed molecular abundances. Unfortunately, the abundances are not the primary observed quantities. They are the derived quantities. Astronomers observe only the line intensities and line profiles. Observed line intensities are then converted into abundances using simplifying assumptions about the excitations (e. g., LTE), radiative transfer (e. g., slab geometry), and physical-chemical conditions (e. g., uniformity in the emitting or absorbing region). This conventional practice suffers from serious drawbacks. The conditions of ambient density, temperatures and velocity field assumed by the observers in reducing their data may significantly differ from those in the theoretical models. Prasad et al (1990) have presented several examples of these differences which are dramatic for C_2 and C_2H. We need complete, end-to-end, mutually consistent, models to predict line intensities for comparison with the observational data. These models would start with the results of ab-initio chemical-dynamical or chemical equilibrium calculations. They should then proceed to statistical equilibrium calculations and end with radiative transfer calculations for the emergent line intensity and profiles for direct comparison with the primary observational data. Only then the molecular chemistry diagnostics of the dynamical conditions in the dense cores will be on a firm foundation.

6. Summary

Observed molecular abundances and chemical dynamical modeling results suggest that dense cloud cores may not be in equilibrium. Instead, they may be dynamically evolving in such a manner that the time spent at high densities decreases as the density increases. After attaining a high density, a gravitationally contracting core may either expand to states of lower core densities or may proceed to star formation. Cloud mass and initial density constitute some of the factors that affect the evolutionary track. Mutually consistent end-to-end modeling of the spectral line intensities and profiles may have the potential to provide the data needed to discern the reality.

Acknowledgements

This research was supported by the National Aeronautics and Space Administration (NASA), through a research grant to the University of Southern California. Grateful thanks are also due to the organizers of the Symposium, particularly Dr. Edith Falgarone, and to the Lockheed Palo Alto Research Laboratory whose supports made it possible for me to participate in the Symposium.

References

Boissé, P. 1990, *Astron. Astrophys.*, **228**, 483.

Boland, W., and de Jong, T. 1982, *Ap. J.*, **261**, 110.

Chièze, J. P., and Pineau des Forêts, G. 1989, *Astron. Astrophys.*, **221**, 89.

Herbst, E., and Leung, C. M. 1986, *M. N. R. A. S.*, **222**, 689.

Herbst, E., and Leung, C. M. 1986b, *Ap. J.*, **310**, 378.

Myers, P. C. 1983, *Ap. J.*, **270**, 105.

Prasad, S. S. 1987, in *Astrochemistry, Proceedings of IAU Symposium 120* (M. S. Vardya and S. P. Tarafdar, Eds.), p. 259. D. Reidel Publishing Co.

Prasad, S. S., Tarafdar, S. P., Villere, K. R., and Huntress, W. T. Jr., 1987, *Interstellar Processes* (D. J. Hollenbach and H. A. Thronson, Eds.) p. 631, D. Reidel Publishing Co.

Prasad, S. S., Heere, K. R., and Tarafdar, S. P. 1990. "Dynamical Evolution and Molecular Abundances of Interstellar Clouds". To appear in *Ap. J.*

Solomon, P. 1990. "Fragmentation and clumpiness in inner-Galaxy giant molecular clouds". In these *Proceedings*.

Swade, D. A. 1989, *Ap. J.*, **345**, 828.

Tarafdar, S. P., Prasad, S. S., Huntress, W. T., Jr., Villere, K. R., and D. C. Black. 1985, *Ap. J.*, **289**, 220.

Tauber,J. A., and Goldsmith, P. F. 1990, *Ap. J.*, **356**, 163.

Thaddeus, P. 1990. "The Hierarchial structure of molecular clouds (gas and dust). Connection between observations in our galaxy and nearby galaxies." In these *Proceedings*.

Villere, K. R., and Black, D. C. 1982, *Bull. Am. Astron. Soc.*, **14**, 970.

Williams, D. A. 1986, *Quart. J. Roy. Astr. Soc.*, **27**, 64.

Williams, D. A. 1987 in *Rate Coefficients in Astrochemistry*, (T. J. Millars and D. A. Williams, Eds.), Kluwer Academic Publishers, p. 281.

Williams, D. A., and Hartquist, T. W. 1984, *M. N. R. A. S.*, **210**, 141.

Zuckerman, B., and Palmer, P. 1974, *Ann. Rev. Astron. Astrophys.*, **12**, 279.

NUMERICAL SIMULATIONS OF TURBULENT COMPRESSIBLE FLOWS

A. Pouquet[1], T. Passot[1,2] and J. Léorat[3]
[1] *Observatoire de la Côte d'Azur, BP 139, Nice cedex 06003, France*
[2] *Department of Mathematics, University of Arizona, Tucson AZ, 85721*
[3] *Observatoire de Meudon, 92 Meudon, France*

ABSTRACT. We give an overview of the use of numerical simulations in the modeling of turbulence in molecular clouds.

1. Introduction

Observations of spectral lines in molecular clouds reveal the existence of supersonic motions whose origin has not been clarified. Scaling laws relating velocity dispersion and cloud size can be attributed to turbulent motions. The physical properties of this turbulence and its feeding mechanisms are still unknown, but independent observations seem to confirm the existence of very irregular and hierarchical structures. Magnetic fields are probably dynamically important in most of these objects. They can help support the cloud against gravity and, through Alfvén wave turbulence, give an alternative explanation for the observed molecular linewidths.

Numerical simulations have become an important tool for studying nonlinear dynamics and are helpful in deciding between competing physical models of molecular clouds. It is generally admitted that nonlinearities are at play in these media and that they are in part the source of our lack of success in analytical modeling. In the absence of an adequate theory of turbulent flows, our understanding of nonlinear complexity necessitates high resolution numerical simulations.

In this paper, we concentrate on homogeneous compressible flows, leaving aside problems which lead to more physically elaborate modeling, including for example radiative transfer.

We begin with a brief account of several basic concepts in turbulence. The next Section of the review is devoted to the case of neutral two and three–dimensional flows without self–gravity, discussing the general properties of supersonic turbulence. In particular we will describe the structures observed in physical space, the temporal evolution of large scale variables, and the scaling laws for the velocity correlations. One of the most striking results that seems to persist in three dimensions is the distribution of density in patches, within which it is filamentary and with small fluctuations.

101

The fourth Section is devoted to a brief account of recent numerical calculations of compressible two-dimensional MHD flows, focusing on overall aspects for different magnetic over kinetic energy ratios. A description will be given of the structures that develop, namely current sheets and bubbles of density.

In the following Section we show, on the basis of phenomenological arguments supported by two–dimensional numerical simulations, that supersonic turbulence can slow down and even stop the gravitational collapse.

Before concluding, we discuss the limit of small Mach numbers, relevant for both the large–scale interstellar medium and the sub–regions that develop within a supersonic flow.

2. Basic Concepts and Tools in Turbulence

Turbulence, as a strongly nonlinear phenomenon, is still lacking a definitive theoretical description, and a resort to a combination of theory, phenomenology, modeling, experiments and observations is needed to progress. In that light, it may appear bold to extend the analysis to more complex problems, involving coupling to rotation, magnetic fields, compressibility, convection and self–gravity, to name a few. But observational facts for one thing compel us in that direction. Also, and somewhat paradoxically, the problems at hand may become simpler, in that small parameters are introduced and at least some subsets become amenable to analytical treatment, *e.g.* through multiple–scale analysis : low Mach number, fast rotation, or strong magnetic fields.

The robustness of such regimes for the general case remains to be shown, and numerical experimentation has certainly become a primary way of investigation. Indeed some flows, for example at high magnetic Reynolds number $R^M = u_0 L_0 / \eta$ where u_0 and L_0 are characteristic velocity and length and η the magnetic diffusivity, or at high *rms* Mach number $M_a = u_0 / c_s$ where c_s is the sound speed of the medium, may not be feasible in the laboratory. Numerical experiments, on the other hand, do not allow to reach very high Reynolds numbers because of limitations in both memory and CPU time. Indeed, for a flow to be well resolved down to the dissipation length $\ell_D = (\nu^3 / \epsilon)^{1/4}$ where ν is the kinematic viscosity, ϵ the rate of energy transfer and dissipation, and where a Kolmogorov energy spectrum has been assumed, *viz* $E(k) = \epsilon^{2/3} k^{-5/3}$, the dynamics of the numerical simulation *ie* the ratio of the large–scale L_0 to the smallest resolved scale Δx must be of the order of the Reynolds number itself ($R^{3/4}$ in the incompressible case). In three dimensions, 10^6 modes will thus be needed to experiment on a flow with a Reynolds number of ~ 100. Herein lies the fundamental limitation of numerical simulations. By–passes exist. One can reduce the space dimensionality to two (cylindrical) or one (spherical). Or one can decide that the precise way by which the flow dissipates the energy is not fundamental and thus resort to a model of dissipation. Among such methods, the more popular ones use the Euler equations (viscosity identically zero) and add some *ad–hoc* dissipation in steep gradients and shocks in the compressible regime (Woodward and Colella, 1984; Moretti, 1987). However, when dealing with small–scale phenomena, such as the reconnection processes in current sheets that may be at the origin of the heating of the solar corona, care must be taken in the precise treatment of the internal structure of dissipative layers. In that case the spectral methods retain all their advantages (Gottlieb and Orszag, 1977) because of their exponential precision for smooth flows.

Finally, one should expect only very slow progress in this experimental–numerical approach to turbulence : a factor two in resolution represents an eightfold increase in memory and twice that in CPU time, to follow explicitly all time scales involved. Some speed–up will come from hierarchical grids (Dorfi, 1982), from dynamical grid–tightening (Landman et al., 1990), and from heavy parallelisation of codes on computers such as the successor to the Connection Machine (Boghosian, 1990).

The concept of a cascade of energy from the large–scale containing eddies to the small–scale dissipative ones, in an energy–conserving way through the inertial range, is well–known (Rose and Sulem, 1978; Leslie, 1973; Monin and Yaglom, 1971). Modifications to the Kolmogorov spectral index of this range to take into account either magnetic fields (Iroshnikov, 1963; Kraichnan, 1965; Grappin et al., 1983; Matthaeus and Zhou, 1989), or compressible effects (Moiseev et al., 1983) as well as intermittency have been proposed.

Possibly less familiar is the concept of **inverse** cascade, from L_0 to scales larger than L_0. In incompressible MHD, an inverse cascade of magnetic helicity $H^M = \int \mathbf{a} \cdot \mathbf{b} \, d^3\mathbf{x}$ where $\mathbf{b} = \nabla \times \mathbf{a}$ with $\mathbf{a}$ the magnetic potential, leads to large-scale helical magnetic fields (Horiuchi and Sato, 1989; Pouquet, 1990). Such magnetic helical structures may have been observed in the Sun (Berger, 1988) and in molecular clouds (Heiles, 1987; see also the discussion in Scalo, 1990), but more data analysis is needed (Heiles, private communication). Whether such large scales will persist in a sustained supersonic flow is an open question. The precise mechanism by which these instabilities grow and saturate can be recast in the framework of multiple–scale analysis (Gilbert and Sulem, 1990).

Coherent structures in flows are pre–eminent, and their origin through an inverse cascade or more esoteric mechanisms (Nicolaenko and She, 1989) is unclear. The topological approach to turbulent flows (Moffatt, 1989) may be a helpful way to encompass the three–dimensionality of structures and also allow for a substantial reduction in data storage and analysis (Perry and Chong, 1987), such as for separated flows.

A consequence of the **direct** cascade of energy to the small scales is the added dissipation that takes place in a turbulent flow. These effects are modelized through turbulent transport coefficients, the precise computation of which still remains a problem of current research (Dubrulle and Frisch, 1990). However, it has been conjectured (Moffatt, 1985) that nonlinear interactions may be self–defeating, in the sense that they themselves produce a flow in which they become negligible : Beltrami flows in which the kinetic helicity $H^V = \int \mathbf{u} \cdot \omega \, d^3\mathbf{x}$ (with $\omega = \nabla \times \mathbf{u}$ the vorticity) is maximal. In MHD, the flow could become either force–free (vanishing Lorentz force), or fully correlated (normalized $\int \mathbf{u} \cdot \mathbf{b} \, d^3\mathbf{x}$ maximal), or both. Numerical evidence in two dimensions seems to corroborate these ideas in MHD, but the 3D problem remains open.

Finally, mention should be made of chaos and intermittency, and the ensuing spatial complexity of the flow with a possible fractal structure (see Scalo, 1990 for a review in the context of molecular clouds and also Falgarone and Phillips, 1990).

Low–dimensional dynamical systems exhibit complex behavior, with a transition that is now well mapped. However, when the number of relevant modes increases substantially, the concepts developed in the framework of chaos do not readily apply. The difficulty may lie in the way to couple many temporal scales as well as spatial scales. The wavelet technique (Combes et al., 1988) that combines local spatial information and Fourier mode analysis has been recently applied to the identification of structures in turbulent flows (Argoul et al., 1989; Farge et al., 1989; Everson et al., 1990) and in galaxy counts (Slezak et al., 1990). This technique may prove useful in the analysis of well–resolved maps of molecular clouds, the Taurus cloud (from IRAS data) being a possible candidate (Henriksen, 1990).

Chaos, however, may be relevant in fully developed turbulence as well, for example at stagnation points of the velocity and in the dynamo problem. In the latter case, it was shown (Galloway and Frisch, 1986) that the emerging magnetic structures are elongated. Numerical simulations (Meneguzzi et al., 1981) also point to an intermittency of the magnetic field, but it is not clear whether this is a dynamical effect or simply due to the proximity from the cross–over in the magnetic Reynolds number, separating the non–magnetic from the magnetic regime. The question of whether intermittency will steepen the Kolmogorov spectrum or not is yet another open problem (Kraichnan, 1990). Intermittency may have been observed in molecular clouds. Its origin may vary (fluid, MHD, gravitation, or a combination). Falgarone and Phillips (1990) have shown that there is a systematic departure from a Gaussian profile in the wings of molecular lines, that they relate to an intermittent behavior. Indeed, Anselmet et al. (1984) and Gagne and Castaing (1990) have shown that the probability distribution function of the velocity field from wind–tunnel data, as well as for the derivatives of the velocity and for a passively advected temperature, all have exponential wings with a Gaussian core. This is also the case in numerical simulations of MHD (Biskamp, 1990).

3. Compressible Turbulence

With increasing resolution in the observations it appears that molecular clouds are agitated by turbulent flows. Models assuming distinct clouds and quasi-equilibrium should thus be modified to account for a more dynamical vision (Scalo 1990) as suggested by observations exhibiting an enormous variety of structures with irregularities at all scales, filaments, bubbles etc...(Bally et al. 1987, Bally 1989; Falgarone 1989, Fukui, 1990). A first step in this direction can be attempted by studying compressible turbulence and the effect of the nonlinear advection term of the momentum equation on the shaping of the flow. Although interstellar cloud turbulence certainly includes magnetic fields, stellar energy sources, radiative cooling and gravitation, nonlinear advection is a major common feature to take into account.

Homogeneous compressible turbulence has not been extensively studied, partly due to the fact that the incompressible case remains unsolved. Analytical studies pertain mostly to the weakly compressible regime, either concerning the acoustic part of the flow (its generation (Lighthill, 1954), or statistical properties (Zakharov et al., 1970)) or the extension of the incompressible phenomenology for small Mach numbers (Moiseev et al., 1983). General arguments (Kraichnan, 1953) and closure schemes (Chandrasekhar 1951a, Weiss 1979, Hartke et al. 1988, Marion 1988) have been developed which are restricted to the small Mach number regime (see also Passot and Pouquet (1987) for a review).

Feireisen (1981) studied numerically the effect of a weak compressibility on the statistics of 3D turbulent shear flows. Computations on homogeneous and turbulent supersonic flows have been performed in both two dimensions and in three dimensions for decay flows (Erlebacher et al., 1990; Passot and Pouquet, 1990, and references therein) and forced flows (Kida and Orszag, 1990). Large–Eddy Simulations (Erlebacher et al., 1987; Porter et al., 1990a) have also been implemented.

In the supersonic regime, a dominant feature is the presence of shocks which are the major cause of dissipation. In the case of decaying turbulence, the flow remains globally supersonic for short times (a few turnover times of the large–scale vortices) whatever the initial value of the Mach number. However, the trace of an initially supersonic flow is still visible at late times on hot spots of temperature (assuming no radiative leaks) and entropy production. Dissipation is found to be similar for the two and three dimensional cases during this first period of supersonic evolution, but different in the subsequent part of the evolution. Whereas in 2D the Mach number almost stabilizes at $r.m.s.$ values of about .6, possibly due to a remnant of the incompressible property of global squared vorticity conservation (Kraichnan and Montgomery, 1979), in the 3D case it keeps decreasing due to the usual nonlinear transfer of energy towards small scales.

When a strong pressure imbalance is present in the initial conditions it is found that the compressive component of the kinetic energy and the internal energy both oscillate periodically (in opposite phase) even for long times, indicating that large scale, large amplitude sound waves are still free to propagate into the system, being minutely influenced by the vortices interacting mostly with themselves. For rms Mach numbers smaller than .3, the flow can be considered incompressible, the interaction between rotational and compressive modes being weak, mostly consisting of sound production by large–scale vortices. For larger Mach numbers the compressive modes are fed more efficiently and contribute in a dominant way to the small–scale kinetic energy because of the presence of strong shocks, the large scales being mostly solenoidal. The ratio χ of the compressive over total kinetic energy is typically .2. The same transitional Mach number has been found in MHD.

The opposite interaction consisting of the production of rotational modes by the compressive ones has been observed in 2D during collisions of shocks or behind curved shocks where entropy gradients are not colinear to temperature gradients (Passot and Pouquet, 1987). It is intermittent and concentrated near the small dissipative scales. This interaction takes the form of small vortices created just behind the shocks or during the process of a Kelvin-Helmotz instability developing in contact discontinuities. It has not yet been observed in 3D.

The interaction between rotational and compressive modes is thus mostly concentrated in the large and small scales, the latter being only efficient for large Mach numbers. Consequently the flow presents a dual nature, consisting of a weakly compressible turbulence, retaining most of its characteristics of the incompressible case, on which is superimposed sound and shock waves. This weak interaction can also be observed when measuring spectra in 2D. The velocity correlation spectrum for the rotational modes still presents an inertial range whose slope is close to –3, a value observed in the incompressible case. The compressive modes, being dominated by shock waves, present a k^{-2} spectrum (Passot et al. 1988). Inertial ranges cannot be observed in 3D due to a lack of resolution, although time averaging will help. When visualizing the density field it appears that there are large patches in which fluctuations are mild. The local *rms* Mach number is small but both mean velocity and density may vary greatly between patches. The existence of such patches has also been observed in the 3D case (Porter et al., 1990b) at late times.

A supersonic turbulent flow also presents striking filamentary structures both in 2D and in 3D, but for possibly different reasons. In 2D the filaments observed on the density field are strongly correlated to entropy fluctuations, created by heating due to dissipation in shocks or vortex sheets. Being passively advected by the flow (Bayly et al 1990) these fluctuations naturally form ribbons, and accumulate as time evolves. These filaments pierce from one patch to the other, revealing once more the dual nature of the flow (Passot et al., 1988). In 3D the filaments which are observed at earlier times (Porter et al. 1990b) are more likely to be associated to shock collisions and intersections, as well as over–compressions in shock bendings or vortex tubes (Vincent and Meneguzzi, 1990). These filaments are striking in the compression field, vorticity and density, and may be a locus of star formation.

The most striking difference between a 2D and a 3D compressible flow is in the density contrast defined as $\Delta\rho = \rho_{max}/\rho_{min}$. Whereas in 2D $\Delta\rho \sim 4$, at similar Reynolds and Mach numbers in 3D $\Delta\rho \sim 100$ (Passot and Pouquet, 1990).

When recasting all these results in the framework of molecular clouds dynamics, several problems emerge. The most important is linked to the rate of dissipation of supersonic turbulence, too high in comparison with estimated energy injection rates, although the gravitational potential well is omnipresent. This problem may possibly be alleviated by the presence of a magnetic field as discussed in the next Section. The structures that obtain in such neutral flows are however not in contradiction with observations and it will be interesting to see how they will be modified by gravitation and magnetic fields.

4. Supersonic MHD Turbulence

This Section briefly reports on some recent calculations of supersonic magneto-hydrodynamic flows. Homogeneous compressible MHD flows have attracted little attention until very recently, previous works being mostly devoted to the study of reconnection processes (see e.g. Ugai, 1988 and Sonnerup 1988). The growth of correlations $\int \mathbf{v} \cdot \mathbf{b} \, d^n\mathbf{x}$ between the velocity field $\mathbf{v}$ and the magnetic field $\mathbf{b}$ which occurs in the incompressible case (see *e.g.* Pouquet, 1990) has also been shown to occur in compressible flows by Dahlburg and Picone (1988), and a study of turbulent relaxation (dynamic alignment versus selective decay) has been undertaken by Ghosh et al. (1988). Shebalin and Montgomery (1988) studied the pseudo–sound generation in an isentropic flow and more recently, Dahlburg and Picone (1989) described the influence of compressibility on the evolution of the Orszag–Tang (1979) vortex. These works are dealing with 2D subsonic flows (thus with a high value of the plasma β value, ratio of kinetic to magnetic pressure).

Even when studying the simple case of a conducting perfect gas in a two-dimensional periodic box, we are faced with a large free–parameter space. It has been chosen here to concentrate on a comparative study of the overall aspects of a supersonic flow when permeated with a random magnetic field at the same scale as that of the velocity field but with differing magnitudes.

When shocks are present, kinetic energy dissipates into heat within a few non-linear times. One of the goals of this work is to identify the regimes where strong shocks are inhibited but where supersonic linewidths could nevertheless be observed. Does this regime exist and does it correspond necessarily to a state of Alfvénic wave turbulence as suggested by Falgarone and Puget (1986) and Lizano and Shu (1987) ? Does this require a smooth large–scale magnetic field ?

New characteristic times appear in the evolution of an MHD flow. Certainly one of the most important is the Alfvén time $T_a = L/V_a$ where L is a characteristic scale and V_a is the Alfvén speed $B_0/\sqrt{4\pi\rho}$ which governs the propagation of transverse waves in the direction of the magnetic field B_0 in an average density field ρ. There are also characteristic times based on the speed of the compressive waves which propagate into the system (i.e. the slow and fast magneto-acoustic waves). For reasons of simplicity we will classify the flows with respect to their character (sub or super–sonic and sub or super–Alfvénic), comparing the rms velocity with the sound speed and the Alfvén speed respectively. This defines in turn the Mach number M_a and the Alfvénic Mach number M_{alf}. The Alfvén speed is based on the rms magnetic field since no mean field is considered here.

Before describing the new numerical simulations on supersonic MHD flows, it will be useful for the discussion to recall some properties of discontinuities in a compressible magnetic fluid. Beside transverse waves which are non–compressive and remain smooth, there are also magneto-acoustic waves which can steepen into shocks. Transverse shocks and contact layers (examples of which are current sheets) are not formed by steepening but have to result from topology, breaking of equi-librium, etc.... Current sheets can be present regardless of the properties of the velocity field. They do not propagate and in contrast to pure hydrodynamical con-tact discontinuities, there cannot be a non–zero component of the magnetic field perpendicular to the plane of the discontinuity, in which case only jumps in den-sity and entropy are allowed. Thus in 2D the center of a current sheet is a neutral line. These discontinuities are known to dissipate through various mechanisms, in particular because they are subject to internal instabilities, such as the tearing mode.

In contrast, the dissipation in a shock is only due to molecular transport coef-ficients and depends on its strength as measured by the entropy or pressure jump. Weak shocks necessarily propagate at the velocity of the corresponding linear wave. Stationary shocks in a medium of zero mean velocity are then moderately strong and we want to discuss their existence in some particular cases. In absence of magnetic fields, their existence requires a supersonic flow. Numerical simulations in two di-mensions (Passot and Pouquet, 1987) indicate that a supersonic velocity fluctuation produces strong shocks that dissipate rapidly until the rms Mach number attains values close to .6. It thus seems that within a factor of two the above mentioned criterion gives good approximations for the upper bound Mach number compatible with small dissipation. We refer here only to dissipation due to shocks, leaving aside the dissipation due to the cascade of eddies in the three-dimensional case or due to current sheets. In three dimensions, indeed, the Mach number may drop to substantially lower values (Porter et al., 1990b) due to the usual turbulent eddy viscosity. In the presence of a magnetic field, shocks propagate anisotropically. It is particularly true for slow shocks which propagate efficiently only in a cone centered on the local direction of the magnetic field. It will be useful to consider two limit cases of shocks. In the first one corresponding to the propagation perpendicular to a constant magnetic field, slow shocks degenerate into contact discontinuities and the existence of fast shocks requires the velocity of the fluid V to be greater than $\sqrt{B_0^2/(4\pi\rho) + c^2}$. In the case of the propagation parallel to the magnetic field, the pure gas limit is recovered but the evolutionary conditions are different since in presence of small perturbations transverse waves can be induced by the longitudinal magnetic field. In particular for strong shocks, the flow has to be super–Alfvénic behind the shock. What is important to point out is the general trend that in the presence of a magnetic field the strength of the shocks is reduced. A shock propagating in a medium permeated with a magnetic field parallel to its plane can eventually degenerate into a sound wave if the strength of the magnetic field increases beyond a critical value (Ferraro and Plumpton, 1966, p. 105). From the preceding discussion, it seems also that a sub–Alfvénic (but possibly supersonic) flow will develop less strong shocks and it is also generally admitted that turbulence is inhibited by magnetic fields.

The following description of some of our numerical simulations reveals indeed this tendency, but points out to some other interesting features of a magnetic turbulent flow. The three runs presented now are for decaying two–dimensional flows, starting with a *rms* Mach number of unity, with excursions up to 2. The velocity field has most of its energy at a length scale $L = \pi$ corresponding to one–half the computational box. Both the usual and magnetic Prandtl numbers are unity, $\gamma = 5/3$ and the Reynolds number is 150. All primitive variables are initialized randomly. We define $\chi = E^c/E^v$, with $E^c + E^s = E^v$ the kinetic energy decomposed into its solenoidal E^s and compressive E^c components. Initially, $\chi = 15\%$, a deliberate choice since we want to start our computation in a fully compressible regime.

The case without magnetic field is analogous to the one described in the previous Section, developing numerous elongated shocks. When taking initially a random magnetic field at the same scale and such that the ratio $r_m = E^m/E^v$ of the magnetic over kinetic energy is equal to .9, the shocks are already completely absent although the flow is not yet sub–Alfvénic. It has to be mentioned however that the compressibility of the flow measured by the ratio χ is enhanced, attaining values of 30% or 1.5 times the peak value of the neutral case. Density fluctuations are also higher. This is due to the increase in the total pressure gradients induced by the presence of the magnetic field. We see at this point that the effect of the magnetic field on the velocity will depend on its scale and not only on its magnitude, the smoothing mechanism being valid only if the field is at a scale larger than or comparable to that of the velocity.

Increasing the magnitude of the ratio r_m to 3.45 and leaving all other parameters constant, the aspect of the flow changes drastically. The smoothing mechanism does not persist when increasing the magnetic field strength, leaving its scale constant; indeed the initial desequilibrium in pressure is large enough to produce shocks

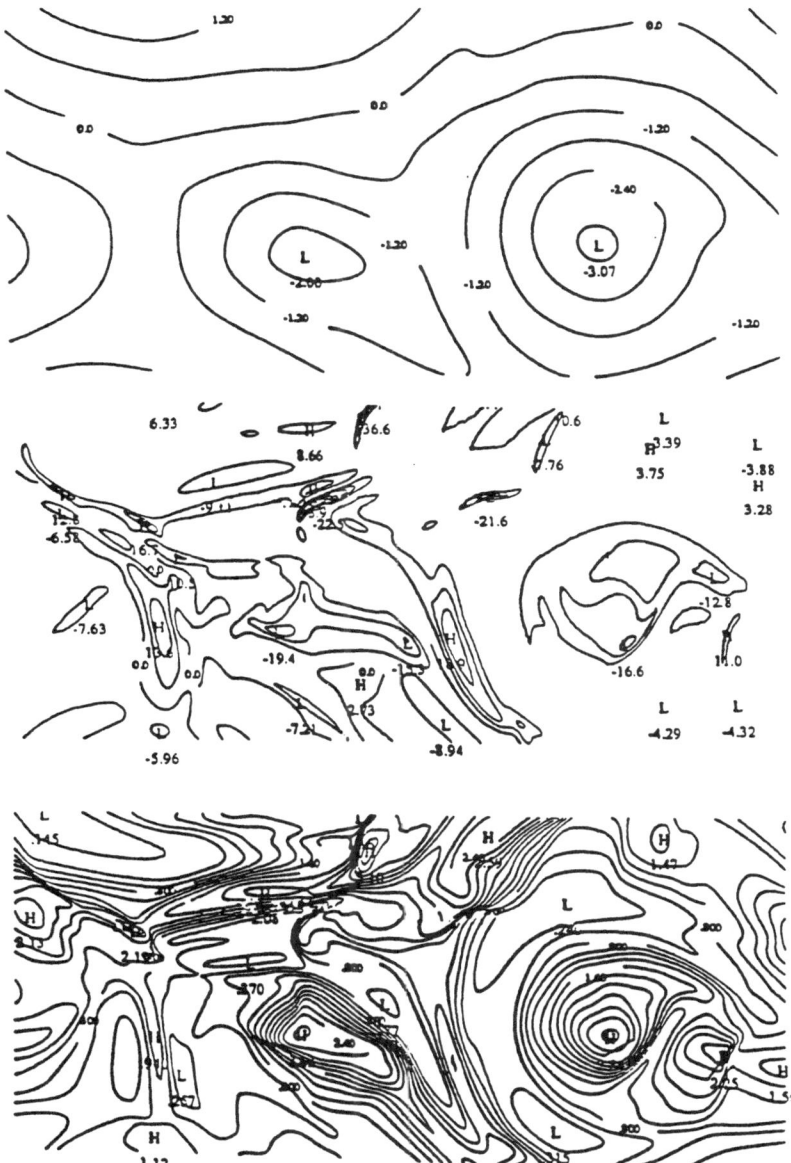

Figure 1: (top) magnetic potential, (middle) magnetic current, (bottom) density, at $t = 1.5$ for a flow with an initial Mach number of 1 and $r_m = 3.45$.

110

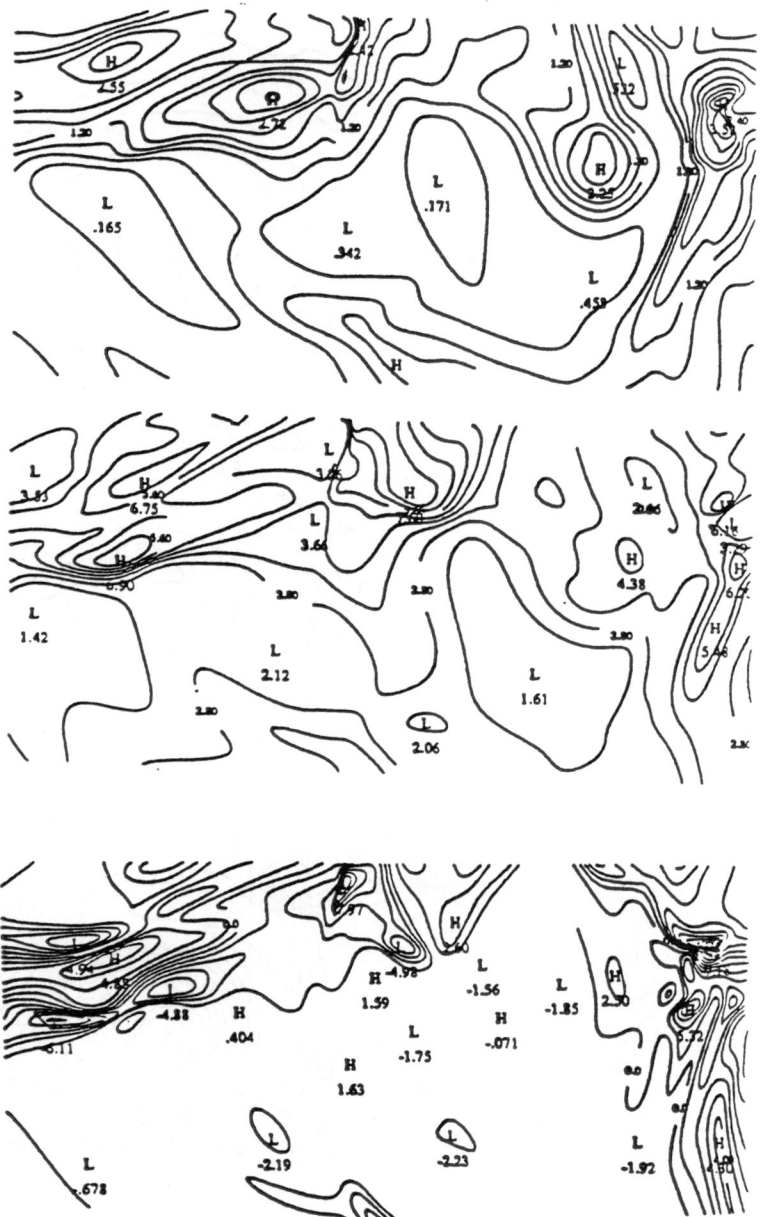

Figure 2: (top) density, (middle) temperature, (bottom) vorticity, at $t = 2.5$ for a flow with an initial Mach number of 2, and $r_m = 1.7$.

although the flow is now sub–Alfvénic. In a short time, violent processes occur whose dynamics is mostly governed by the magnetic field. There is initially a large transfer of energy from magnetic to kinetic. Shocks form and current sheets develop near X-type neutral points, releasing a large quantity of heat. This heat causes the fluid to expand, producing bubbles. When growing, these bubbles form circular shocks which smooth out rapidly while expanding. The discontinuities present in this flow have therefore a shape very different from the neutral case. Shocks are thin but much shorter, their life time being also much smaller as we could predict. Current sheets are more long-lived and thicker than shocks since they arise from a topological and not a dynamical constraint; they have a certain amount of magnetic flux to dissipate which can be quite large. We show in Figure 1 the vertical component of the magnetic potential (top), the current density (middle) and the density (bottom) for the lower half of the flow at $t = 1.5$. The current sheet, corresponding to the magnetic X–point (hyperbolic neutral point), is thicker than the filamentary structures observed both in the current and the density, associated with angular points of the potential. The bubbles in density correspond to previous current structures, and the thinner compressions to the magnetic shocks. The Joule heating mechanism, more important in MHD shocks than in current sheets, is now able to decorrelate efficiently the temperature field from the density thus increasing the strength of the baroclinic term $\frac{(\nabla p \times \nabla \rho)}{\rho^2}$ responsible for vorticity production; vorticity is also created by the purely magnetic term leading for example to quadrupolar structures centered on the current sheets, as in the incompressible case. Thus, a polytropic approximation becomes questionable when the Alfvén time is shorter than the characteristic radiative cooling time. Ambipolar diffusion will also provide an efficient dissipation mechanism.

Reconnection is seen to be more efficient than in the incompressible case. Current sheets are broken and dissipated rapidly by e.g. tearing modes, when high–speed flows converge onto it (fast reconnection). This is clearly observed in a simulation of the Orszag–Tang vortex performed at Mach one, with constant-initial pressure. In this case, the lateral (as opposed to central) current sheets are the strongest. This is due to the combined action of the centrifugal force (not compensated by the initial pressure) and the periodic boundary conditions which direct a flow onto the sheet and enforce reconnection in an intrinsic dynamical way.

Similar results obtain when other runs with initially smaller scales and/or higher Mach numbers are performed. In a run with an initial Mach number $M = 1.8$ and with $r_m = 1.7$, the Mach number is equal to .46 after ten turn–over times, as opposed to .4 when starting with $M = 1$. The ratio $\chi = 48\%$ is much larger in the high Mach number case, and yet the flow appears smooth at that Reynolds number (computation on a 256^2 grid). In Figure 2, we show the density (top), temperature (middle) and vorticity (bottom) for the higher half of the flow at $t = 2.5$ for the Mach 2 run. We see both the bubbles linked to current and thin structures in the density, we see the decorrelation between density and temperature, and we see the breaking of the vorticity in small often roundish quadrupoles.

Common to all these runs is the fact that the ratio r_m increases with time after an initial decaying transient and that the scale of the magnetic field seems to diminish. Here it is important to note that the initial conditions of these runs are violent. For later times, the flow is gentler and discontinuities almost inexistent. The dissipation is then much smaller. This is consistent with the observations on the evolution of the strength and scale of the magnetic field. Finally, we note that, in agreement with the results of Dahlburg and Picone (1988), the alignment factor $\frac{<V.B>}{<V^2>^{1/2}<B^2>^{1/2}}$ increases, starting from .33 and rising to .6.

The properties of the spectra are also noteworthy. There is quasi–equipartition between the solenoidal component of the kinetic energy and the magnetic energy in the small scales, with a slight excess for the latter as for incompressible flows. On the other hand, the correlation spectrum of the compressible part of the velocity itself dominates in the small scales both the E^s and E^m modes for sufficiently high Mach numbers.

Summarizing these results, we can say tentatively that in sub–Alfvénic but still supersonic flows, shocks are hindered. When the magnetic field is tangled, reconnection takes place, but this process dissipates less energy than would do shocks. If the flow is perpendicular to the magnetic field (possibly due to gravitation), we could imagine the flow to be less turbulent. Also, the formation of bubbles of density is striking.

It is planned to study these dissipation mechanisms with different topologies of the magnetic field and different initial conditions, trying to test the stability of regimes consisting of nonlinear Alfvén waves.

Magnetic fields may play an essential role in the collapse of molecular clouds (Mouschovias, 1987; Lada and Shu, 1990). The numerical treatment of a turbulent MHD compressible flow undergoing gravitational collapse, is scanty (Dorfi, 1982; Pouquet et al., 1990). In the next part we shall discuss the simpler case (although still not resolved) of the interplay of gravitation with a compressible turbulence in a barotropic flow. Works performed directly in the astrophysical context are reviewed for example in Scalo (1988).

5. Turbulence and Gravitation

In molecular clouds, large scales are found to be relatively stable over times long compared to the free-fall time of the cloud, whereas small scales are clumpy. The generalization of the Jeans' stability analysis to account for turbulent kinetic energy has been considered by Chandrasekhar (1951b), Sasao (1973), and by Bonazzola et al. (1987); magnetic fields (Lizano and Shu 1987; Pudritz, 1990) have also been taken into account. A phenomenological argument (Léorat et al. 1989,1990) which encompasses the interaction between turbulent eddies, the phase coherence of shocks, waves and gravitational collapse is briefly exposed below. It is confirmed by numerical simulations assuming an isothermal flow and bidimensionality.

Gravitation acts directly on the compressive modes and competes with pressure and turbulent dissipation. At a given scale, the comparative strength of these processes can be evaluated by estimating their characteristic times. For turbulence, the ratio $r_g = \tau_{tr}/\tau_{ff}$ of the transfer time of the compressible kinetic energy towards small scales τ_{tr} to the free-fall time τ_{ff} is the relevant parameter to consider. In the Jeans' case where only pressure is considered, the relevant parameter is obtained by replacing τ_{tr} by the acoustic time $\tau_{ac} = \ell/c_s$ where c_s is the sound speed. At small Mach number, the transfer time is obtained by considering sound wave coupling as in Zakharov et al. (1970). The parameter r_g increases with scale and eventually collapse will take place for a critical length, turbulence resulting in a shift of the Jeans' length towards larger scales (see also Chandrasekhar, 1951b).

However, r_g is found to be very sensitive to the value of the Mach number M. With increasing M, dissipation tends to occur mainly in shocks, with an equal strength for all scales. The concept of inertial range and energy cascading becomes meaningless and τ_{tr} must be modified accordingly. Taking the Burgers' equation limit, τ_{tr} is found independent of scale. Assuming a constant local mean density, the parameter r_g will thus also be scale–independent and reads in the simplest case :

$$r_g = \tau_{tr}/\tau_{ff} = (L/L_J)c_s/u_c$$

where L_J is the Jeans' length and u_c the compressive component of the velocity at scale L. There is a critical value $\tilde{r}_g$ of order unity below which the flow becomes stable. This leads plausibly to a global marginal equilibrium between turbulence and collapse at all scales. This equilibrium can be broken at scales smaller than the Jeans' length, when local peaks in density are formed by turbulence, e.g. in the vicinity of shocks. Large density enhancements can be expected in shocks when radiative transfer is included (Zeldovich and Raizer, 1966), which will favor this kind of local mechanism for collapse. Observations do support clumpiness of the medium on a wide range of scales. An unmagnetized medium will have planar shocks, and clumps might occur at their intersections, on filaments. Furthermore, in the MHD case, bubbles form in the density leading to arched–like shocks. Also, shocks may be unstable through a hernia–type effect.

Numerical simulations support the predictions based on the preceding phenomenology, with $\tilde{r}_g \sim 0.3$ in two dimensions. Although the Reynolds numbers of the computed flows are not large, and thus no complete inertial range is exhibited, it has nevertheless been observed that the large scales of the flow tend to tear the collapsing clumps into pieces and fragment them into stable entities. When the value of r_g is small enough, the gravitational collapse is stopped for at least ten free-fall times. For larger values of r_g, collapse occurs but strongly structured by turbulence. Clumps tend to form on filaments of denser matter (Passot, 1987). Their scale decreases with the Mach number and hierarchical structures are also found when several Jeans' masses are present in the cloud (Léorat et al., 1990). However, no successive fragmentation is observed, possibly due to a lack of numerical resolution. The hierarchical structures observed are formed one at a time through the influence of the turbulent velocity field; it happens also that small structures are formed first before collapsing together. In three dimensions, the situation is not clear. Chantry et al. (1990) find that density clumping tends to form initially at hyperbolic points of the velocity, where the vorticity is weak, through the centrifugal force. This effect is amplified when two vortices meet. However, as both the Reynolds number and the Mach number are increased and vortex interactions become embedded in a turbulent flow, the preceding coherent mechanism may be swamped by shock formation and ensuing strong density contrasts in their vicinity.

Finally, when coupling gravity and MHD, one may conjecture that pressure is modified by both the turbulent pressure, and the magnetic one (see also Pudritz, 1990). A somewhat *ad hoc* modification of the phenomenological argument presented above is given in Pouquet et al. (1990).

6. Conclusion

It has been mentioned in Section II, that in a supersonic flow, there are regions where density fluctuations as well as the local *rms* Mach number are small. It is thus of interest to consider the limit of quasi-incompressible flows. This has been studied rigorously for isentropic fluids (Klainerman and Majda 1981, 1982). Entropy fluctuations should nevertheless be considered as soon as forcing or dissipative processes are no longer negligible. We do not review this rather large subject but rather discuss the observations of Armstrong et al. (1981). They have measured the density fluctuations in the large scale interstellar medium, where the flow can be considered quasi–incompressible and found that they exhibit a power law close to $k^{-5/3}$. Assuming a barotropic fluid, density fluctuations are correlated with pressure fluctuations and thus should obey a $k^{-7/3}$ law if a Kolmogorov–like cascade is assumed. Recently Montgomery et al. (1987) and Matthaeus and Brown (1988) suggested that the presence of a magnetic field in equipartition with the velocity field could possibly explain a $k^{-5/3}$ law. By systematically deriving the incompressible limit for a gas whose equation of state depends on two thermodynamic variables, Bayly et al. (1990) showed both analytically and numerically that a $-5/3$ law can also be expected as soon as entropy fluctuations are non–negligible. The basis of the argument is that when taking the incompressible limit two parameters have to be taken to zero, namely the Mach number and the size of temperature fluctuations. Depending on the relative size of these two parameters, different physical limits can be identified. When the latter is larger than M^2, the density and temperature both obey at first order a passive scalar advection equation. This in turn can also explain the filamentary structures as discussed in Section II at long times, with relatively small density fluctuations within them.

Very few high resolution numerical simulations of compressible turbulence exist. Although turbulence *per se* is certainly still far from being solved, the results described here may lead to some insight in the dynamics of the interstellar medium. The overall aspect of the flow, when compared to observations, indicate that nonlinearities are certainly at play in the interstellar medium. Filaments are observed both in molecular clouds and numerical simulations, with density clumps along them. In the latter we saw three different occurrences for such structures, showing that they are general patterns produced by nonlinear advection terms. Sheets also occur in three dimensions, and bubbles obtain when strong magnetic fields reconnect. Finally, hierarchical structures appear with gravitational collapse. Virialisation of the medium may end–up being a consequence of its dynamical evolution when few density gradients occur and there is an equilibrium, albeit marginal, between turbulence and collapse. Furthermore, nonlinear build–up of intermittent structures and fractal behavior have been shown to take place in molecular clouds.

ACKNOWLEDGEMENTS

The three–dimensional computations with the PPM code reported here were done in collaboration with D. Porter and P. Woodward on the Minnesota Super–Computer

Institute (MSI) computers. The other numerical simulations were performed at the Centre de Calcul Vectoriel pour la Recherche (Palaiseau, France) at NCAR and at the Pittsburgh Supercomputing Center under N.S.F. grant D.M.S. 8703397. We received partial support from the DRET grant 89–236.

References

F. Anselmet, Y. Gagne and E.J. Hopfinger (1984) J. Fluid Mech. **140**, 63.

F. Argoul, A. Arneodo, G. Grasseau, Y. Gagne, E.J. Hopfinger and U. Frisch (1989) Nature **338**, 51.

J.W. Armstrong, J.M. Cordes and B.J. Rickett (1981) "Density Power Spectrum in the Local Interstellar Medium", Nature **291**, 561.

J. Bally, W.D. Langer, A.A. Stark and R.W. Wilson (1987) "Filamentary Structure in the Orion Molecular Cloud", Astrophys. J. **312**:L45.

J. Bally (1989) "The Structure of Molecular Clouds from Large Scale Surveys of CO and CS", in "Structure and Dynamics of the Interstellar Medium", IAU Symposium No 120, eds. G. Tenorio-Tagle, M. Moles and J. Melnick, Springer–Verlag.

B.J. Bayly, D.C. Levermore and T. Passot (1990) "Density Variations in Weakly Compressible Flows", submitted to Phys. Fluids. A.

M. A. Berger (1988) "Turbulent Relaxation in Compressible Two-dimensional Magnetohydrodynamics", in Proceeding of the workshop on Turbulence and Nonlinear Dynamics in MHD Flows, Cargèse, France, July 4-8 1988, p. 197, Eds. M. Meneguzzi, A. Pouquet, and P.L. Sulem.

D. Biskamp, H. Welter and M. Walter (1990) preprint.

B.M. Boghosian (1990) Computers in Physics Jan/Feb 1990, p. 14.

S. Bonazzola, E. Falgarone, J. Heyvaerts, M. Perault and J.L. Puget (1987) Astron. Astrophys. **172**, 293.

S. Chandrasekhar (1951a) "The Fluctuations of Density in Isotropic Turbulence", Proc. Roy. Soc. **A210**, 18.

S. Chandrasekhar (1951b) Proc. Roy. Soc. **A220**, 26.

G. Chantry, R. Grappin and J. Léorat (1990) this conference.

J.M. Combes, A. Grossmann and P. Tchamitchian (1988) "Wavelets", Springer Eds., Berlin.

R.B. Dahlburg and J.M. Picone (1988) "Growth of Correlation in Compressible Two–Dimensional Magnetofluid Turbulence", J. of Geophys. Res. **93**, 2527.

R.B. Dahlburg and J.M. Picone (1989) "Evolution of the Orszag-Tang Vortex System in a Compressible Medium. I. Initial Average Subsonic Flow", submitted to Physics of Fluids B.

E. Dorfi (1982) Astron. Astrophys. **114**, 151.

B. Dubrulle and U. Frisch (1990) Preprint Observatoire de Nice.

G. Erlebacher, M.Y. Hussaini, C.G. Speziale and T.A. Tang (1987) ICASE Report 87–20.

G. Erlebacher, M.Y. Hussaini, H.O. Kreiss and S. Sarkar (1990) ICASE Report 90–15.

R. Everson, L. Sirovich and K.R. Sreenivasan (1990) Physics Letters **145** , 314.

E. Falgarone (1990), "Turbulence in Interstellar Clouds", in IAU Symposium No 120 op. cit.

E. Falgarone and J.L. Puget (1986), Astron. Astrophys. **162**, 235.

E. Falgarone and T.G. Phillips (1990) "A Signature of the Intermittency of Interstellar Turbulence: The Wings of Molecular Line Profiles", Astrophys. J. **359**, 344.

M. Farge, M. Holschneider and J.F. Colonna (1989) "Wavelet Analysis of Coherent Structures in Two–Dimensional Turbulent Flows", Topological Fluid Mechanics, p. 777, Proceedings of the IUTAM Symposium, Cambridge, August 13-18, 1989, Ed. H.K. Moffatt and A. Tsinober.

W.J. Feireisen, W.C. Reynolds and J.B. Ferziger (1981) "Numerical Simulations of Compressible Homogeneous Turbulent Shear Flows", Report TF-13 (unpublished) Thesis, Stanford University.

V.C.A. Ferraro and C. Plumpton (1966) "An Introduction to Magneto-Fluid Mechanics" Oxford University Press.

Y. Fukui (1990) this conference.

Y. Gagne and M. Castaing (1990) Physica **D**, to appear.

B. Galanti, P.L. Sulem and A. Gilbert (1990) "Large–scale Instabilities and Inverse Cascades in MHD Flows", Preprint, Observatoire de Nice.

D. Galloway and U. Frisch (1986) Geophys. Astrophys. Fluid Dyn. **36**, 53.

S. Ghosh, W.H. Matthaeus and M.L. Goldstein (1988) "Turbulent Relaxation in Compressible Two-dimensional Magnetohydrodynamics", in the Proceedings of the workshop on Turbulence and Nonlinear Dynamics in MHD Flows, Cargèse, France, July 4-8 1988, p. 247, Eds. M. Meneguzzi, A. Pouquet, and P.L. Sulem.

A. Gilbert and P.L. Sulem (1990) Geophys. Astrophys. Fluid Dyn., to appear.

A. Gottlieb and S.A. Orszag (1977) "Numerical Analysis of Spectral Methods", SIAM Review, Philadelphia.

R. Grappin, A. Pouquet and J. Léorat (1983) Astron. Astrophys. **126**, 51.

G.J. Hartke, V.M. Canuto and C.T. Alonso (1988) "A Direct Interaction Approximation Treatment of Turbulence in a Compressible Fluid I. Formalism", Phys. Fluids **31**, 1034.

C. Heiles (1987) "Physical Processes in Interstellar Clouds", Eds. G.E. Morfill and M. Scholer, Dordrecht, Reidel, 429.

R.N. Henriksen (1990) this conference.

R. Horiuchi and T. Sato (1989) Phys. Fluids **31**, 1142.

P.S. Iroshnikov (1963) Astron. Zh. SSSR **40**, 742 and Soviet Astron. **7**, 566.

S. Kida and S.A. Orszag (1990) "Energy and Spectral Dynamics in Forced Compressible Turbulence" Preprint Kyoto University.

S. Klainerman and A. Majda (1981) Commun. Pure Appl. Math. **34**, 481.

S. Klainerman and A. Majda (1982) Commun. Pure Appl. Math. **35**, 629.

R.H. Kraichnan (1953) J. Acoust. Soc. of America, **25**, 1096.

R.H. Kraichnan (1965) Phys. Fluids, **8**, 1385.

R.H. Kraichnan (1990) "Models of Intermittency in Hydrodynamic Turbulence", Phys. Rev. Lett. **65**, 575.

R.H. Kraichnan and D. Montgomery (1979) Rep. Prog. Phys. **43**, 547.

C.J. Lada and F.H. Shu (1990) "The Formation of Sunlike Stars", Science **248**, 564.

M.J. Landman, G.C. Papanicolaou, C. Sulem, P.L. Sulem and X.P. Wang (1990) Physica D, to appear.

J. Léorat, T. Passot and A. Pouquet (1989) "Structure Formation in Self-Gravitating Flows", Topological Fluid Mechanics, p. 782, Proceedings of the IUTAM Symposium, Cambridge, August 13-18, 1989, Ed. H.K. Moffatt and A. Tsinober.

J. Léorat, T. Passot and A. Pouquet (1990) "Influence of Supersonic Turbulence on Self-Gravitating Flows", Mon. Not. R. Astr. Soc. **243**, 293.

D.C. Leslie 1973 "Development in the Theory of Turbulence", Oxford, Clarendon Press.

M.J. Lighthill (1954) Proc. Roy. Soc. **A 222**, 1.

S.L. Lizano and F.H. Shu (1987) "Formation and Heating of Molecular Cloud Cores" in Physical Processes in Interstellar Clouds, Eds. G.E. Morfill and M. Scholer, p. 173-193.

J.D. Marion (1988) Thèse, Ecole Centrale de Lyon.

W.H. Matthaeus and M.R. Brown (1988) "Nearly Incompressible Magnetohydrodynamics at Low Mach Number", Phys. Fluid **31**, 3634.

W.H. Matthaeus and Ye Zhou (1989) Phys. Fluids B **1**, 1929.

M. Meneguzzi, U. Frisch and A. Pouquet (1981) Phys. Rev. Lett. **47**, 1060.

H.K. Moffatt (1985) J. Fluid Mech. **159**, 359.

H.K. Moffatt (1989) "The Topological (as Opposed to the Analytical) Approach to Fluid and Plasmas Flow Problems", Topological Fluid Mechanics, p. 1, Proceedings of the IUTAM Symposium, Cambridge, August 13-18, 1989, Ed. H.K. Moffatt and A. Tsinober.

S.S. Moiseev, R.Z. Sagdeev, A.V. Tur, G.A. Khomenko, and V.V. Yanovskii (1983) Sov. Phys. JETP, **58**, 1149.

A. Monin and A.M. Yaglom (1971) "Statistical Fluid Mechanics", ed. by J.L. Lumley, MIT Press.

D. Montgomery, M.R. Brown and W.H. Matthaeus (1987) "Density Fluctuation Spectra in Magnetohydrodynamic Turbulence", J. Geophys. Res. **92**, 282.

G. Moretti (1987) Ann. Rev. Fluid Mech. **19**, 313.

T. Mouschovias (1987) "Physical Processes in Interstellar Clouds", Eds. G.E. Morfill and M. Scholer, Dordrecht, Reidel, 453.

B. Nicolaenko and Z–S She (1989) "Coherent Structures, Homoclinic Cycles and Vorticity Explosions in Navier–Stokes Flows", Topological Fluid Mechanics, p. 777, Proceedings of the IUTAM Symposium, Cambridge, August 13-18, 1989, Ed. H.K. Moffatt and A. Tsinober.

S.A. Orszag and C.M.. Tang (1979) J. Fluid Mech. **90**, 129.

T. Passot (1987) "Simulations numériques d'écoulements compressibles homogènes en régime turbulent : application aux nuages moléculaires", Thèse de Doctorat, Université de Paris VII.

T. Passot and A. Pouquet (1987) "Numerical Simulation of Compressible Homogeneous Flows in the Turbulent Regime", J. Fluid Mech. **181**, 441.

T. Passot and A. Pouquet (1990) "Numerical Simulations of Three-Dimensional Supersonic Flows", submitted to J. of Mechanics B/ Fluids.

T. Passot, A. Pouquet and P.R. Woodward (1988) "On the Plausibility of Kolmogorov-Type Spectra in Molecular Clouds", Astron. Astrophys. **197**, 228.

A.E. Perry and M.S. Chong (1987) "A Description of Eddying Motions and Flow Patterns Using Critical Points Concepts", Ann. Rev. Fluid Mech. **19**, 125.

D. Porter, A. Pouquet, and P.R. Woodward (1990a) "A Comparative Study of the PPM Method to a Navier–Stokes Flow", in preparation.

D. Porter, A. Pouquet, T. Passot and P.R. Woodward (1990b) "The Characteristic Structures of a Turbulent Three–Dimensional Compressible Flow", in preparation.

A. Pouquet (1990) in Les Houches Summer School on Astrophysical Fluid Mechanics, J.P. Zahn Ed., Gordon and Breach.

A. Pouquet and T. Passot (1988) "Numerical Simulations of Turbulent Supersonic Flows", Princeton Meeting on Compressible Flows, October 1988, T. Birmingham Ed.

A. Pouquet, T. Passot and J. Léorat (1990) "Turbulence and Gravitation, a Marginal Equilibrium", Third European Conference on Turbulence, Stockholm, July 1990.

R. Pudritz (1990) this conference.

H. Rose and P.L. Sulem (1978) J. de Physique **39**, 441.

T. Sasao (1973) Pub. Astr. Soc. Japan **25**, 1.

J. Scalo (1988) "Theories and Implications of Hierarchical Fragmentation" in Molecular Clouds in the Milky Way and External Galaxies, Eds. R.L. Dickman, R.L. Snell and J.S. Young (New-York: Springer Verlag), p. 201.

J. Scalo (1990) "Perception of Interstellar Structure: Facing Complexity", Preprint no. 95, Mc Donald Observatory, University of Texas.

J.V. Shebalin and D. Montgomery (1988) J. Plasma Phys. **39**, 339.

E. Slezak, A. Bijaoui and G. Mars (1990) Astron. Astrophys. **227**, 301.

B.U. Sonnerup (1988) "On the Theory of Steady State Reconnection", Computer Physics Communications **49**, 143.

M. Ugai (1988) "MHD Simulations of Fast Reconnection Spontaneously Developing in a Current Sheet", Computer Physics Communications **49**, 185.

A. Vincent and M. Meneguzzi (1990) "The Spatial Structures and Statistical Properties of Homogeneous Turbulence", CERFACS Preprint TR/IT/90/16.

J. Weiss (1979) PhD Thesis, University of New-York.

P. Woodward and P. J. Collela (1984) J. Comp. Phys. **54**, 115.

V.E. Zakharov and R.Z. Sagdeev (1970) Sov. Phys. Dokl. **15**, 439.

Ya. B. Zeldovich and Yu. P. Raizer (1966) "Physics of Shock Waves and High–Temperature Phenomena", Eds. W.D. Hayes and R.F. Probstein, Academic Press.

SIGNATURES OF TURBULENCE IN
THE DENSE INTERSTELLAR MEDIUM

E. FALGARONE[1], T.G. PHILLIPS[2]

[1] *Radioastronomie Millimétrique, Ecole Normale Supérieure,*
24 rue Lhomond, 75005 Paris, France

[2] *California Institute of Technology,*
320-47, Pasadena CA 91125, USA

Abstract. We present an ensemble of recent observational results on molecular clouds which, taken separately, could all be understood by invoking various unrelated physical processes, but taken all together form a coherent ensemble stressing the imprints of turbulence in the physics of the cold interstellar medium. These results are first, the existence of wings in the molecular line profiles, which can be interpreted on statistical grounds as the signature of the intermittency of the velocity field in turbulent flows, second the fractal geometry of the cloud edges, with properties reminiscent of those of various surfaces studied in turbulent laboratory flows, and third, the fact that the dense gas fills only a very small fraction of the space. The last points are supported by CO multitransition observations of a few fields in nearby molecular clouds. They show that the excitation conditions are the same for the gas emitting in the linewings and in the linecores and are also remarkably uniform over a large range (factor 10) of column densities. An attractive interpretation of the molecular line data is that most of the ^{12}CO(J=2–1) and (J=3–2) emissions arise in cold ($T_k \leq 10K$) and dense ($n_{H_2} \sim 10^4\, cm^{-3}$ or more) structures distributed on a fractal set with no characteristic scale size greater than about $1000\, AU$.

Keywords : Molecular clouds, Turbulence, Intermittency, Linewings, Fractal structure, Self-similarity

1. Introduction

There is considerable evidence from both theory and observations, that gas flows in molecular clouds are turbulent. By this, we mean that in the equations of motion, the non-linear advection term is orders of magnitude larger than the dissipation term. The two Reynolds numbers (hydrodynamic and magnetic, which measure the importance of the non-linear terms in the equations of motion and evolution of the magnetic field respectively) are so large in interstellar clouds that all the characteristics of magnetohydrodynamic turbulence can be expected to be found in their evolution. While the mathematical behavior of the solution of the equation of motion is not yet well understood, the specific manifestations of turbulence in atmospheric and laboratory flows are numerous and it is tempting to view them as diagnostics of turbulence within interstellar clouds (eventhough the effects of magnetic fields and compressibility may be important in the interstellar medium).

Some of these manifestations can be found in the statistics of the velocity field, some others are purely geometrical, all of them are related. In particular, it is now understood that turbulence is highly intermittent, ("the element of surprise in the time history of the velocity field", quoting Dutton and Deaven, 1969), especially on the smaller scales, and that many properties, including velocity increments, exhibit distributions that depart significantly from Gaussian in the wings (see references in Falgarone and Phillips 1990). Also, turbulence is not space filling. The subset of the Euclidian space on which the turbulent activity develops is a fractal (Mandelbrot, 1982). Recent high sensitivity measurements in atmospheric and laboratory flows (Anselmet, Hopfinger and Gagne, 1984; Méneveau 1989) allow a quantitative approach to the fractal structure of the subset of space on which the dissipation of turbulence is concentrated and connect it to the intermittency of the velocity field.

These observable signatures of turbulence are present in molecular clouds. In this paper, we discuss first the information relevant to the velocity field contained in the molecular lineshapes and second that contained in the spatial and density structure of molecular clouds.

2. Molecular lineshapes

2. 1. INTERSTELLAR LINEWINGS

It has been clear for a long time that the linewidths of molecular transitions, as seen in emission from regions of the dense interstellar medium, are too large to be simply due to the thermal motions of the molecules. Typical linewidths from molecular clouds are in the 0.5 to $2\,km\,s^{-1}$ range for the relatively quiescent regions and of course can be much greater for disturbed regions. However, the thermal width for the CO molecule is only about $0.05\,km\,s^{-1}$ for a typical kinetic temperature of $10-20\,K$. Zuckerman and Evans (1974) suggested that this might be due to the effect of turbulence, but the characteristic signature of turbulence was not fully understood at that time. A major step forward was made by Larson (1981) who showed that the linewidths scale with the size of the region under study, roughly as predicted by Kolmogorov for the case of incompressible turbulence. This is suggestive of a turbulent origin to the linewidths, but there could be many other effects which would give a comparable scaling of the linewidth with size, including statistical effects (see the review of Scalo, 1987). An additional characteristic of the kinematics of molecular clouds is the existence of broad wings in the molecular line profiles often observed in the direction of clouds not associated with molecular outflows. The wing emission has been shown to be optically thick (Magnani, Blitz and Wendel 1988) and has received several interpretations. Keto and Lattanzio (1989) proposed that it could be generated by cloud collisions, Magnani et al. (1990) claimed that it arises in gas less dense than $10^3\,cm^{-3}$, and Elmegreen (1991) invokes magnetosonic waves in molecular clouds with the broad wings originating in regions of low average density where the Alfvén velocity increases.

Our approach to this problem has been to study, in a statistical way, the full profiles of the molecular lines to extend the information beyond the behavior of just the linewidths and propose an interpretation which is scale-independent (Falgarone and Phillips, 1990).

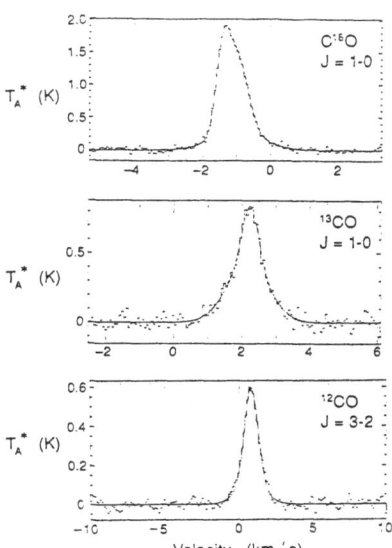

Fig. 1. Spectra of CO emission for various isotopes and ladder lines in non-starforming regions of *(a)* Auriga, (Falgarone, Puget and Pérault 1991), *(b)* Ursa Major (Falgarone and Pérault 1988), and *(c)* Taurus-Perseus (Falgarone, Phillips and Walker 1991). The lines have excess strength in the wings as compared to a Gaussian profile, but very high signal to noise ratios are needed to discriminate the wide wing structure in these relatively weak lines. The continuous lines are least square fits of a narrow Gaussian (two for the $C^{18}O$ spectrum) plus a weaker broad Gaussian.

The essential point is that turbulence studies show that the velocity distributions are non-Gaussian to some degree. There is a fundamental property of turbulence called intermittency which predicts that there is an excess of large velocity deviations in a given region as compared with a Gaussian distribution of velocities. For the interstellar medium this means that, if we consider the line profiles as the tracers of the probability distribution of the velocity differences, the linewings are enhanced. Some examples of non-Gaussian spectra in typical non-starforming regions are shown in Figure 1. In this conference paper and in Falgarone and Phillips (1990) we attempt to describe the nature of intermittency and to relate molecular lineshapes to the predictions of turbulence models.

2. 2. THE TURBULENT CASCADE

In 1941, Kolmogorov predicted the self-similar behavior of the velocity field in incompressible turbulent flows by postulating the existence of a local dissipationless cascade of kinetic energy from large scales to small scales in which large 'eddies' become unstable and split into smaller and smaller 'eddies'. L_0 is the large scale size at which the energy is injected into the cascade and λ_0, the small scale size at which the kinetic energy is dissipated and is transferred to another process such as heating of the fluid. He predicted that at scales $\lambda_0 << \ell << L_0$, the distribution of the velocity field is determined only by the mean dissipation rate of specific kinetic energy $\overline{\epsilon}_d$. Dimensional arguments give the scaling with ℓ of the nth-order moments of the increments of the velocity field

$$M_n = <| \mathbf{v}(\mathbf{r} + \ell) - \mathbf{v}(\mathbf{r}) |^n > \propto (\overline{\epsilon}_d \ell)^{n/3} \qquad (1)$$

This relation is statistical, not deterministic, and the brackets hold for an ensemble average.

However, any model for turbulence, or whatever the semantic description of the highly non-linear process may be, must be subject to experimental test. Kolmogorov's prediction was later modified (Landau and Lifchitz 1959; Kolmogorov 1962; Oboukhov, 1962) when laboratory experiments showed that the dissipation rate of specific kinetic energy, ϵ_d, was not uniform, but concentrated in an intermittent fashion into the limited regions of space and time where velocity gradients reach large values or diverge.

Much work has been done in this area in laboratory duct flows and in the Earth's atmosphere. In particular Dutton and Deaven (1969) carried out well defined measurements of the velocities in turbulent flows where they could monitor the velocity differences, as a function of time, between two points separated by a variable distance r along the flow. They discovered an excess of large velocity differences as compared to a Gaussian distribution, which they ascribe to the intermittency of turbulence. Figure 2 shows more modern measurements by Van Atta and Park (1971) of the probability density of the velocity differences in the atmosphere above the Ocean, clearly indicating the excess of large velocity dispersion events.

2. 3. INTERMITTENCY

In order to take into account the clearly modified scaling due to the intermittency of ϵ_d, Kolmogorov introduced a modified scaling law for the moments

$$M_n \sim \bar{\epsilon}_d{}^{n/3} \ell^{\zeta_n},$$

where ζ_n was permitted to deviate from $n/3$. Several theoretical approaches predict simple analytical forms of ζ_n. For a log-normal distribution of the energy dissipation rate averaged over a volume of size ℓ (Kolmogorov, 1962), $\zeta_n = \frac{n}{3} - \frac{\mu}{18} n(n-3)$. Alternatively, for the β-model of Frisch et al. (1978) in which the eddies are less and less space filling as their size decreases, $\zeta_n = \frac{n}{3} - \frac{1}{3}\mu (n - 3)$.

The constant μ is called the intermittency parameter. Neither model is correct although they are quite a good approximation for most of the low order moments ($n \leq 6$). More recent models of greater complexity (e.g. the multifractal formalism, Méneveau 1989) differ again in the algebraic form, but the concept of a small n-dependent modification to the scaling, whose magnitude is a measure of the intermittency, is still retained.

It is particularly important to note that the high order moments contain the information on the intermittency and that intermittency exists at all scales, being relatively more pronounced, as the sizes decrease.

2. 4. RELATION OF INTERSTELLAR SPECTRA TO THE MOMENT ANALYSIS AND ATMOSPHERIC FLOWS

The observation of an interstellar line involves the integration of emission from a cylindrical volume of diameter d and length D, representing the telescope beam

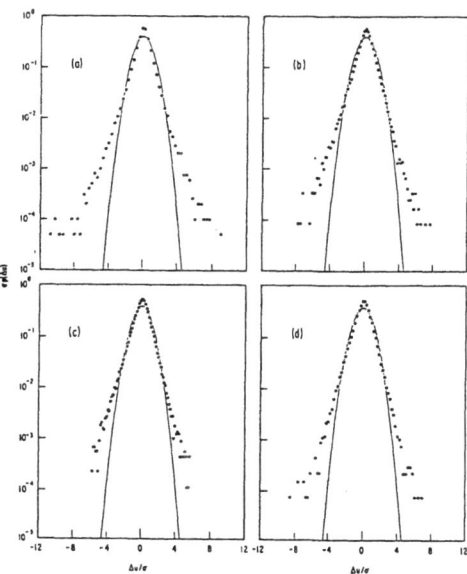

Fig. 2. Probability densities of the velocity difference for various values of the separation r : (a) r=1.38cm, (b) r=4.14cm, (c) r=9.67cm, (d) r=20.7cm. The distribution is clearly in excess of Gaussian (full curves) for large values of the relative velocity difference. The logarithmic scale for the probability distribution emphasizes the behavior at large velocity separations (from van Atta and Park 1971).

through the molecular cloud. We expect to see through the cloud, because the velocity gradient is high in regions of intermittency. The average separation of points within the tube is :

$$\bar{\ell} = \frac{D}{2\sqrt{3}} \left(1 + \frac{2d^2}{D^2} \right)^{1/2}$$

Returning to the moment description we can write :

$$M_n = \int p(\delta v) \, \delta v^n \, d(\delta v),$$

where the probability density is given in terms of the emission strength, T_A^*, by

$$p(\delta v) = T_A^* (\delta v) / \int_{\text{line}} T_A^* \, dv.$$

In principle this allows the computation of the moments and a test of the scaling laws. We make the simplifying assumption that any line profile is associated with the separation $\bar{\ell}$. However, the data set must fulfill a number of conditions, such as *(i)* a very large range in the probability density, or extremely high signal to noise ratios, particularly in the far line wings, *(ii)* a very large range in spatial separations $(\bar{\ell})$, i.e. in beam and cloud sizes, *(iii)* if possible (and at least for nearby clouds) it must avoid star-forming regions, to eliminate extra zones of injection of energy into the cascade.

An ideal data set would be a fully sampled, very large map of a nearby non-starforming cloud, with very high signal to noise in each pixel. Variation in $\bar{\ell}$ would be obtained by averaging areas of the cloud ranging from one pixel to the whole cloud. Since this is beyond the capabilities of current instrumentation, we have used a set of 20 molecular line spectra, built on data from the literature, ongoing work (Falgarone, Puget and Pérault 1991), unpublished spectra from M. Pérault and H. Ungerechts, and our own measurements. It involves an ensemble of clouds of various sizes at various distances, which gives a range for d from less than $0.1\,pc$ to more than $100\,pc$, or more than 3 orders of magnitude. The data is summarized in Table 1 of Falgarone and Phillips (1990).

2. 5. ANALYSIS OF THE DATA

In an ideal data set it would be possible to analyze each spectrum for the various moments. In fact, for reasons of finite signal to noise, we have to characterize the profiles with an analytic form in order to make progress. We have chosen to use a two Gaussian fit (core and wing Gaussians of velocity dispersions σ_c and σ_w respectively) as shown in Figure 3. This is physically reasonable and also has the necessary property that high order moments converge. A Gaussian core and exponential wing is another possibility. It is a slightly worse fit, but if adopted, does not change the conclusions. The result of the fitting procedure is that we observe a constant scaling with $\bar{\ell}$ such that

$$\frac{\sigma_w}{\sigma_c} = 3.3\,(\pm 0.2)$$

over the full range of $\bar{\ell}$. In other words, the existence of wings is an intrinsic property of the clouds over scale sizes less than $0.1\,pc$ to more than 100 pc and the wings and cores scale similarly with the size. The entire line profile is therefore self-similar.

The moments can be obtained from the two Gaussian fits. The first moment, M_1, is related to the linewidth and we find that the slope, ζ_1, is close to the value 0.33 required by the Kolmogorov scaling. Care is needed in deducing the slope of the data shown in Figure 4 because of the need to know the errors on both axes. For our best estimate of the shapes of the error boxes we find $\zeta_1 = 0.33$, but with a possible range extending from .28 to 0.37.

Information on the intermittency parameter is contained in the higher order moments. If we can rely on the assumption that the Gaussian (or exponential) fits to the wings are valid to $\Delta v = \infty$ then the higher moments are given by :

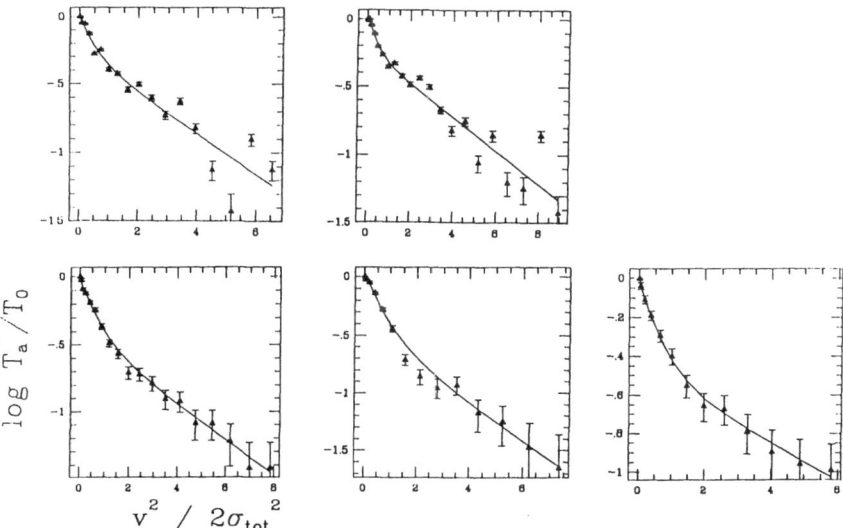

Fig. 3.   Linear-log display of the five (out of six) half-profiles of Figure 1 which present linewing excess emission. They have been normalized to the peak temperature and $\pm 1\sigma$ error bars are shown. The choice of coordinates is such that Gaussians appear as straight lines.

$$M_n \propto \frac{a_w \sigma_w^{n+1}}{\int_{\text{line}} T_A^*(v)\,d\,v}$$

where a_w is the amplitude of the wing Gaussian, and

$$\zeta_n = \frac{d\,\ln(M_n)}{d\,\ln(\bar{\ell})}.$$

We can plot the $\bar{\ell}$ dependence of a_w, σ_w and the line integral to deduce μ. If we compare with the model of Frisch et al. (1978) we get $\mu \sim 0.1$. The values found for atmospheric measurements are ~ 0.2 (Anselmet, Gagne and Hopfinger, 1984). However, the accuracy in the interstellar measurements is currently very poor and much higher signal to noise spectra are needed.

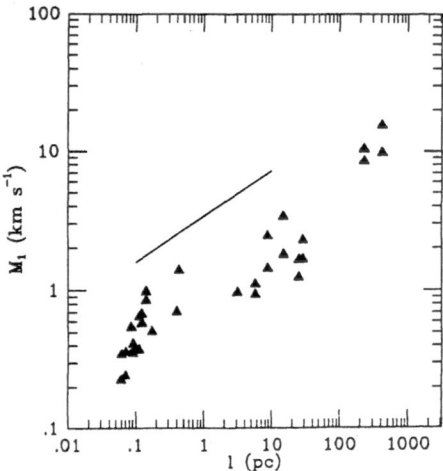

Fig. 4. Dependence of the first order moment of the line profiles on the characteristic dimension $\bar{\ell}$ adopted for each emitting region. The slope 1/3, predicted by Kolmogorov for incompressible turbulence, is shown for comparison.

3. The spatial structure of molecular clouds

3. 1. THE SELECTION OF THE OBSERVED FIELDS

The second part of the argument relies on a different data set consisting of an ensemble of maps of inactive parts of molecular complexes, taken at high angular resolution with the 10m antenna of the Caltech Submillimeter Observatory (CSO) in the ^{12}CO and ^{13}CO (J=2–1) and (J=3–2) transitions (Falgarone, Phillips and Walker 1991). The fields selected for observations are almost transparent when viewed at low angular resolution. *At the parsec scale*, they have a low average H$_2$ column density $N_{H_2} \leq 10^{21}$ cm^{-2}. The low *average* column density of matter ensures little crowding of emitting elements along the line of sight. This spatial transparency in principle allows a better view of the structure, its degree of convolutedness and its shape, because details do not blend with each other. In addition, the fields have been selected to include velocity components which belong both to the center and to the extreme wings of the velocity distribution of the entire complexes. It is this specific point which allows an elucidation of the connection between the velocity field at all scales within a complex and the density structure at very small scale, in the broad context of turbulence. The overall velocity distribution in a complex can be obtained with very good signal to noise ratio by summing all the individual spectra of a well-sampled map of this complex. The grand average spectra of the Taurus-Perseus-Auriga and Cygnus OB7 complexes, which would be those obtained with large beamsizes of $\sim 100\,pc$ and integration times of several hundreds hours, are shown in Figure 5. They were provided by H. Ungerechts and M. Pérault respectively, who integrated all the spectra of the large scale maps (Ungerechts and Thaddeus, 1987; Falgarone and Pérault, 1987). The velocity offsets of the gas components studied individually at high angular resolution are shown on these grand average spectra. Our targets therefore include gas representative of material emitting in the low brightness wings of complexes as well as in the line cores.

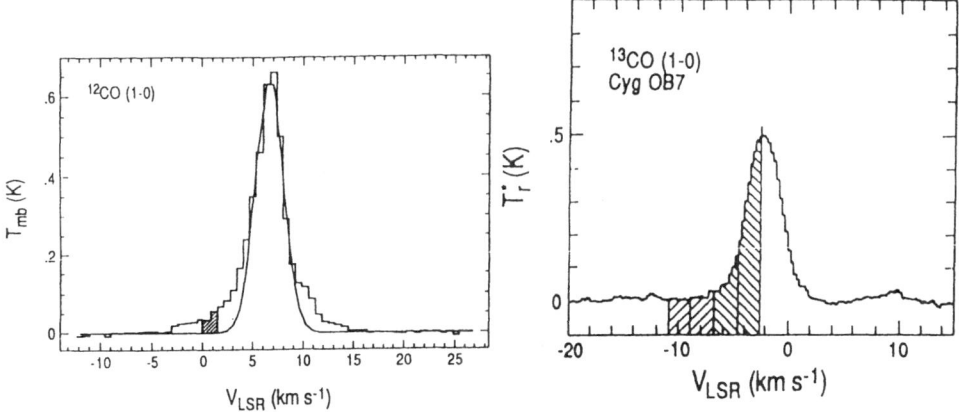

Fig. 5.   The grand average spectra. *(a)* ^{12}CO(J=1–0) integrated profile of the Taurus complex. The 3817 individual spectra of the $\sim$ 750 square degrees map have been added (H. Ungerechts, private communication). *(b)* ^{13}CO(J=1–0) integrated profile of the CygOB7 complex. About 3000 individual spectra of the $\sim$ 20 square degree map have been added (M. Pérault, private communication). The velocity coverage of the gas components studied individually at high angular resolution are hatched intervals.

It is nevertheless possible that the structure we see across the edges of molecular clouds is not representative of that of the bulk of the molecular material. In regions where UV photons are pervasive, the chemistry could be different and the CO abundances may not have reached the value they have in more shielded regions. There could be large fluctuations of molecular abundances from one position to another which would mimic variations of the total gas column density. Probably, this is not the case. In recent studies of molecular clouds, starting with the CI studies (Phillips and Huggins,1981; Keene et al., 1985; Genzel et al. 1988; Zmuidzinas et al., 1988) and reinforced by CII work (Stutzki et al. 1988; Boreiko, Betz and Zmuidzinas 1990), it has become clear that the interior regions of molecular clouds are more similar to the edge regions than might be expected. In particular, the similarity of the entire line profiles in the ^{13}CO and CI lines observed by Frerking et al.(1989) in ρ Ophiuchus strongly suggests that shielded and unshielded material in molecular clouds are intimately mixed with each other not only in space, but also in velocity space. This is also shown by the tendency for molecular species to be observed in interior regions where it might have been expected that they should be condensed on dust grains. In other words, from a chemical point of view, there is much in common between the interiors and the edges of clouds. A probable explanation is that the UV penetrates much deeper into the clouds than anticipated, due to what is usually described as 'clumpy structure' (Boissé, 1990). Thus, we feel confident in using the edges of the clouds in velocity and space as useful regions to probe the cloud material because of their transparency. The target selected in the Taurus complex is shown in Figure 6.

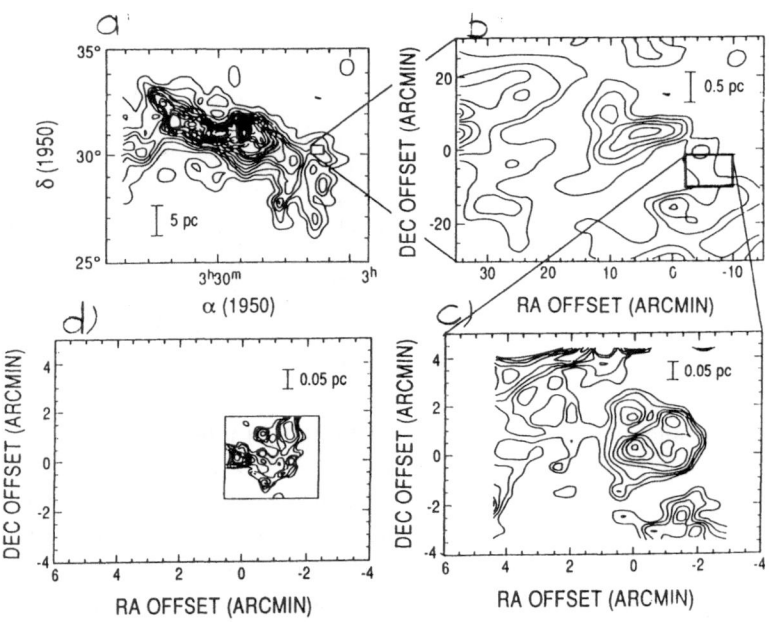

Fig. 6. (a) The field studied at high angular resolution, located on the appropriate fragment of the Ungerechts and Thaddeus velocity-integrated intensity map. First contour $0.5\,K\,km\,s^{-1}$, step $1.5\,K\,km\,s^{-1}$. (b) Velocity-integrated intensity of the ^{12}CO (J=2–1) emission (CSO, HPBW=30", sampling 5'). First contour $2\,K\,km\,s^{-1}$, step $2\,K\,km\,s^{-1}$. Reference position $\alpha(1950) = 3h\,07m\,27.4s$, $\delta(1950) = 30°\,05'$. (c) Velocity-integrated intensity of the ^{12}CO (J=2–1) emission (CSO, HPBW=30", sampling 30") within the area delineated in (b). First contour $1.8\,K\,km\,s^{-1}$, step $0.3\,K\,km\,s^{-1}$. Reference position $\alpha(1950) = 3h\,06m\,54.9s$, $\delta(1950) = 29°\,58'$. (d) Velocity-integrated intensity of the ^{12}CO (J=3–2) emission (CSO, HPBW=21", sampling 18"). First contour $0.85\,K\,km\,s^{-1}$, step $0.2\,K\,km\,s^{-1}$. Same reference position as in the previous map. The linear scales are computed for a distance to the cloud $d = 200\,pc$.

3. 2. RESULTS OF THE CO MULTITRANSITION OBSERVATIONS

All the data, in particular those from the remote fields at 750pc, are presented in Falgarone, Phillips and Walker (1991). The raw observational results are the following. First, for all the fields, and both transitions, spatial structure exists at all size scales down to the best angular resolution, which, in the case of the Taurus field and the high frequency beam, is $0.02\,pc$ or 4000AU (see Figure 6d).

Second, for all the fields, the maximum peak line intensities in the maps are comparable and low, $T_A^* \sim 3$ to $5K$ in both CO rotational lines, in spite of an anticipated variation of the beam dilution between the remote and nearby fields of $\left(\frac{750}{200}\right)^2 = 14$. In the simple picture of a clumpy medium, this result combined with the first one would imply that the emission arises in fragments much smaller than the smallest beam.

Third, the integrated intensity maps are remarkably self-similar. The maximum brightness contrast observed in a map scales with the size over which it is observed, and Taurus and Cygnus, although at very different distances from the Sun, exhibit the same brightness contrast over comparable distances.

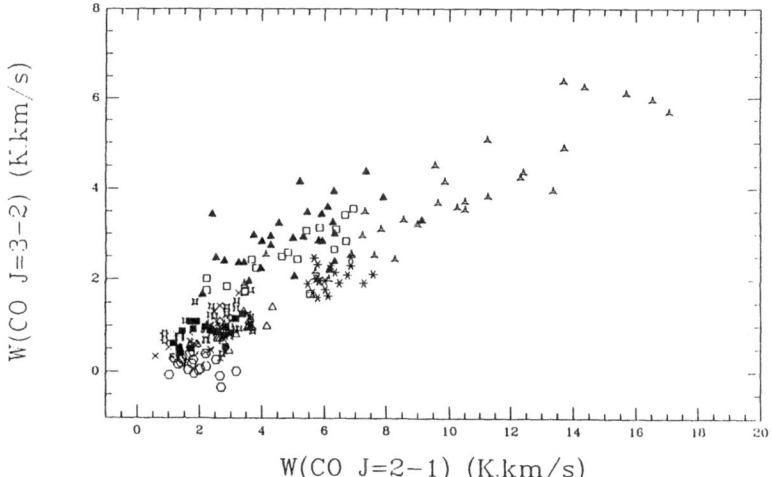

Fig. 7. ^{12}CO (J=3–2) versus ^{12}CO (J=2–1) integrated intensities (uncorrected for beam efficiency) for all the fields and velocity components. Solid symbols are for two core velocity components in CygOB7. Empty squares and triangles (starred or not) are for the four wing velocity components of these two fields. The data from the Taurus fields are the crosses and the asterisks. Empty hexagons indicate upper limits for the ^{12}CO (J=3–2) line. Note the absence of any trend for the solid symbols (linecore velocities) to be distinguished from the empty ones (linewings velocities). There is also no distinction between the remote (Cygnus OB7) and nearby (Taurus) fields.

Fourth, the (J=2–1) and (J=3–2) integrated intensities stay approximately proportional to each other over the entire range of values accessible to our observations (Figure 7). In particular there is no tendency for their ratio $R = \frac{W(3-2)}{W(2-1)}$ to increase as the ^{12}CO(J=2–1) integrated intensity, $W(2-1)$, increases by a factor as large as ~ 10, nor to vary with either the velocity range or the size of the emitting region. There is some significant scatter of the values of R from field to field, and component to component and even within components, but there is no indication of systematic variation with either the column density or the velocity. The ensemble of data is consistent with $R = 0.55 \pm 0.2$. Uncertainties affect the above average value R = 0.55 and are extensively discussed in Falgarone, Phillips and Walker (1991).

A fifth result is that the gas is optically thick in ^{12}CO (J=2–1). Our estimate, $2.5 < \tau_{CO(2-1)} < 6.5$, relies on the assumption of a similar excitation temperature for both isotopes, and an isotopic abundance ratio $[^{12}CO]/[^{13}CO] = 57$, (Langer and Penzias, 1990). Even for the lowest possible abundance ratio, the ^{12}CO (J=2–1) optical depth would remain larger than one.

Some immediate conclusions can be drawn from the raw observational results :
(i) the self-similarity of the maps, with structure at all scales down to the resolution size, suggests a fractal geometry for the CO emitting gas.
(ii) the uniformity of the ratio R (Figure 7) means that the observed gas has comparable excitation properties in all the fields, independent of their size and their velocity with respect to the centroid velocity of the complex.

3. 3. THE SPATIAL SELF-SIMILARITY

The self-similarity of the CO maps in non-starforming regions can be illustrated by the fact that it is impossible in general by simple inspection of a map of antenna temperature for which the contour levels are known, to guess its linear extent. It is possible to quantify this self-similarity by estimating the fractal dimension of the CO contours by the method of the area-perimeter relation which uses the property that for a curve of fractal dimension D_2 in a plane, the perimeter and area are related by $P \propto A^{D_2/2}$. Lovejoy (1982) applies this method to atmospheric clouds and rain areas of sizes ranging between 1km and 10^4km and measures a fractal dimension $D_2 = 1.35\pm0.05$. In the Taurus complex, Falgarone, Phillips and Walker (1991) have found the same dimension, $D_2 = 1.36$, for the large scale ($\sim 25°$) ^{12}CO(J=1–0) map of Ungerechts and Thaddeus (1987), for the $\sim 1°$ undersampled map in the ^{12}CO(J=2–1) line and for the $\sim 8'$ maps in the ^{12}CO(J=2–1) and (J=3–2) lines (Figure 8). It is remarkable that this dimension is the same for maps in three different rotational transitions of CO which in principle are sensitive to gas of different densities. It is also comparable to that found for other column density tracers of the cold interstellar medium, the IRAS $100\mu m$ emission in the Taurus complex (Scalo, 1990), in a high latitude cloud (Bazell and Désert, 1988) and for the HI emission in a high velocity cloud (Wakker 1990). The fact that whichever tracer is used, the fractal dimension is roughly the same suggests that lower density molecular gas and even atomic gas are distributed on a set which has the same fractal topology as denser gas, for instance that seen in ^{12}CO(J=3–2). This value is indicative of a possible link between the topology of the cold interstellar medium and the role of turbulence in structuring it.

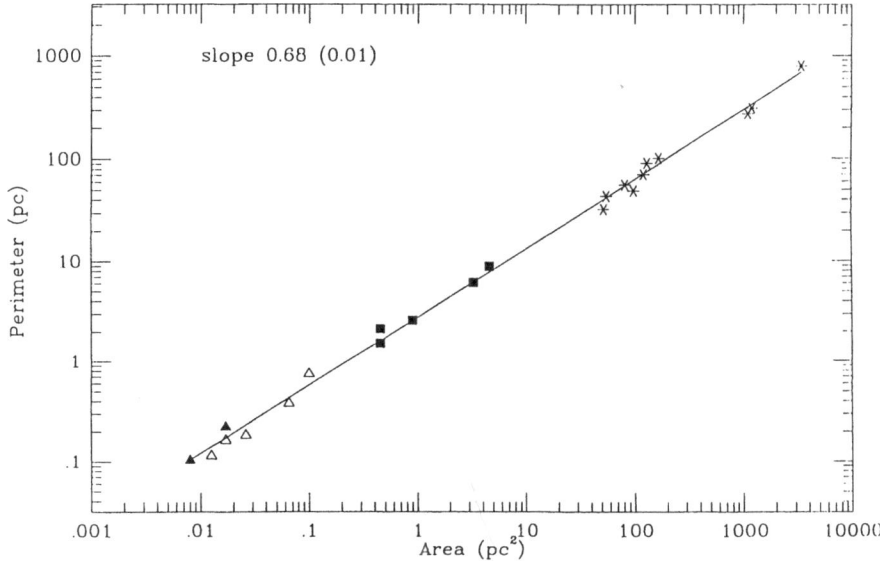

Fig. 8. Perimeter versus area of contours of integrated emission in the large scale ^{12}CO (J=1–0) map of Ungerechts and Thaddeus (1987) (asterisks), the undersampled ^{12}CO (J=2–1) map (empty squares) and the well sampled ^{12}CO (J=2–1) (empty triangles) and (J=3–2) (solid triangles) maps. The perimeters have all been scaled to the resolution $\epsilon = 30'$ of the (J=1–0) map, according to $P(\epsilon) \propto \epsilon^{1-D_2}$. The slope of the perimeter-area relation is $D_2/2 = 0.68$.

Sreenivasan and Méneveau (1986) measure the fractal dimension (D_2') of the curves obtained by 2-dimensional slicing of a variety of interfaces in turbulent flows. Their remarkable result is that, for several interfaces and surfaces (of isoconcentration or isodissipation of the kinetic energy), they find the same dimension, $D_2' = 1.36$. D_2' may differ from the dimension D_2 of the projection onto a plane of a fractal surface of dimension D_3, but limited experiments suggest that $D_2' \leq D_2$ (see Méneveau 1989). If isotropy is assumed, one can write $D_3 = D_2 + 1$. The same authors show that a dimension $D_3 = 7/3$ (in Euclidian space) can be predicted for any kind of interface in a turbulent flow in which the velocity scaling law can be written $\delta v \propto \delta \ell^{1/3}$.

Although it is not clear yet what conclusions could be drawn from the similarity of the dimensions among these various systems or what is due to mere coincidence, a remarkable consequence of the fractal geometry is that the quantities which describe the spatial distribution of matter are determined by expressions which differ notably from those of an ensemble of randomly distributed clumps. The number $\mathcal{N}(l)$ of structures of linear dimension l in a fractal of dimension D_3 scales as :

$$\mathcal{N}(l) = \left(\frac{L_0}{l}\right)^{D_3}$$

where L_0 is a normalization parameter determined by the observed number of structures of a given size in a map. From the maps in the Taurus field (Figure 6), we find $L_0 \sim 50\,pc$ by estimating that ~ 10 structures of $\sim 10'$ are present in the map. The volume filling factor of the ensemble of structures of size l is therefore

$$f_v(l) = \left(\frac{L_0}{l}\right)^{D_3-3} \tag{2}$$

and their surface filling factor

$$\eta_s(l) = \frac{L_0^{D_3-3}}{l^{D_3-2}} \times s \tag{3}$$

where s is the length along the line of sight of the volume occupied by the entire structure.

The paradoxical consequence of the fact that $2 < D_3 < 3$ is that the area filling factor increases as the structure size decreases while the volume filling factor keeps decreasing with l. For $D_3 = 2.36$, $\eta_s(l) \propto l^{-0.36}$ and $f_v(l) \propto l^{0.64}$. This implies that any beam is predominantly filled by the smallest structures. Another characteristic of such a structure is that the mass within any volume of projected size L scales as L^2, which seems to be the scaling law found for molecular clouds over a large range of sizes.

3. 4. THE PHYSICAL PARAMETERS OF THE GAS

Since the ratio R of the J=3-2 and J=2-1 transitions is a very sensitive measure of the relative excitation temperatures of the two transitions, the lack of systematic variation of this ratio with the column density and the relatively small amount of scatter (Figure 7), might indicate that a universal density, velocity and temperature structure, on size scales much smaller than the beam, conspires to give a uniform ratio. At present, we are not able to evaluate this situation in a computational way because it would require an analysis which includes the density-velocity correlation as developed by turbulence.

Fortunately, in trying to deduce the physical parameters of the gas distribution, it is possible to use the general properties of self-similarity of the molecular cloud structure as exemplified by the fractal nature of the cloud edges and the similarity of the maps in the Taurus and Cygnus fields. We do not attempt to derive a density structure from the observations but favor solutions in which all the $^{12}CO(J=3\text{--}2)$ and (most of) $(J=2\text{--}1)$ emission arises in cold and dense structures distributed on a fractal set. This solution, in which the excitation conditions are uniformly close to thermalization is preferred for the following reasons :

(i) quasi-thermalized cold gas naturally explains the low peak antenna temperatures without invoking beam dilution, or excitation temperatures, which would be systematically lower in nearer sources in order to reproduce the similarity of the observed peak values in fields at different distances. Low values of R imply $T_k \leq 10K$, since in this case R only depends on the kinetic temperature. Furthermore, it allows for values of the beam filling factor $\eta_s > 1$.

(ii) dense gas is needed to build up the CO column density required to reproduce the observed line integrated intensities in the smallest structures (Figure 6), without invoking unrealistically large CO abundances.

In this picture, small scale brightness fluctuations in the maps are due to variations of the crowding of structures much smaller than the beam size from one line of sight to the next. Variations of the line ratio R most likely trace those of the kinetic temperature. Also, we can use the dependence of η_s with l given in (3) to derive the maximum size l_{max} for which $\eta_s = 1$. It is :

$$l_{max} = 1000 AU \left(\frac{L_0}{50\,pc} \right)^{-1.8} \left(\frac{s_{max}}{2\,pc} \right)^{2.78} .$$

It is interesting to note that the volume filling factor of the ensemble of these structures in a molecular complex of size $L_0 = 50\,pc$ is $f_v = 2\ 10^{-3}$, which is remarkably small.

Our results on the uniformity of excitation conditions of dense gas in non-starforming regions are in harmony with recent observations which also suggest a remarkable homogeneity of the excitation conditions in quiescent molecular clouds and may support the idea that the cold interstellar medium is made of tiny dense structures. Values of the $CS(J=2\text{--}1)/CO(J=1\text{--}0)$ ratio observed in the Galaxy far from star forming regions and in external galaxies are shown in Figure 2 of Falgarone (1991). It is a compilation of data from Drdla et al. (1989), Mooney and Solomon (1990), Sage et al. (1989) and Mauesberger et al. (1989). There is no trend for this ratio, usually considered as a tracer of dense gas, to increase toward small scales. There is an important scatter in the ratios, but no trend of systematic variation with for example the CO integrated brightness Casoli et al. (1991) show that the ratio of the $CO(J=2\text{--}1)$ to $(J=1\text{--}0)$ emissions in the disks of a large variety of galaxies is low (between 0.5 and 0.8) and remarkably uniform and does not depend on the size of the emitting region between 50 pc and 25 kpc. These results are unlikely to be biased by optical depth effects since Falgarone, Puget and Pérault (1991) find also uniform and low values 4of the same ratio for the ^{13}CO lines in the direction of quiescent clouds.

The idea that the bulk of the CO emission of molecular clouds arises in cold thermalized gas is not new. In 1972, Penzias et al. derived from their CO observations of dark clouds that the J=1-0 emission is thermalized and that most clouds have kinetic temperature near 6K. More recently, Clemens and Barvainis (1988) find a sharp maximum at $T_R \sim 4K$ in the histogram of the $^{12}CO(J=2-1)$ peak temperatures of 248 small molecular clouds and conclude that most of the cataloged clouds are cool with kinetic temperatures $\sim 8K$. What is new is that the same results are now found over a very large range of sizes, with a spectacular self-similarity, a fact which cannot be easily dealt with in a simple clumpy structure. The fractal geometry, in turn, may allow us to understand why the previously believed isolated clouds are in fact connected by widespread emission at a very low level, as revealed by recent CO observations of large areas of the sky, down to unprecedented sensitivity limits (Heithausen and Thaddeus, 1990; Lee, Snell and Dickman, 1990).

4. Conclusions

We have interpreted several characteristics of the molecular line emission of the cold interstellar medium as a manifestation of the importance of turbulence in structuring the velocity field and the spatial distribution of dense gas, from the smallest to the largest sizescales.

Our analysis of the shapes of the molecular line profiles relies on an ensemble of 20 spectra of high signal to noise ratio, from emitting regions whose sizes range from less than 0.1pc to more than 100pc. We reach the following conclusions :

1. For all accessible scale sizes, a significant fraction of the mass of molecular clouds has unexpectedly high velocities relative to the cloud centroid velocity and provides the origin of the broad spectral wings. Considering the line profiles as tracers of the velocity probability distribution of the gas along the line of sight, we propose that the wings represent a direct observational signature of the intermittency of the velocity field, an intrinsic property of turbulence observed in atmospheric and duct flows.

2. Most of the line profiles can be described with a core and a wing Gaussian whose widths are both found to scale self-similarly with size. Kolmogorov scaling apparently applies to the entire profiles, with an exponent near 1/3, which is that predicted for incompressible turbulence. Our interpretation of the linewings implies the self-similarity of the entire lineshape with size.

3. It is hard to measure the value of the intermittency parameter and we find a smaller value ($\mu \sim 0.1$) than in atmospheric studies ($\mu \sim 0.2$). Line profiles extending to 20 times the half width are ideally required to accurately determine μ.

We then show that the spatial structure of clouds is amenable to observations if regions are chosen near the edges of both the spatial and velocity distribution of complexes, so that the information is not obscured by crowding of structures along the line of sight. Fields in Taurus and Cygnus OB7 were chosen with this criterion in mind. Our conclusions regarding the spatial distribution of gas are :

4. CO maps at various scale sizes and resolutions exhibit self-similarity. The fractal dimension D_2 is measured by means of the area-perimeter relation and is

about 1.36 in the Taurus complex, in large and small scale maps and whichever rotational transition of CO is used. This fractal dimension is comparable with that found earlier in HI maps of high velocity clouds, and $100\mu m$ maps of the dust emission of nearby clouds.

5. The ratio R of CO(3-2) to CO(2-1) intensities is relatively constant in all the fields, $R \sim 0.5$. It somewhat fluctuates from field to field or component to component but does not show any systematic variation with either the total column density of gas (over a factor of 10), or the size of the region, or its velocity relative to the bulk of the gas. The excitation conditions of the gas in non-starforming regions are therefore quite uniform over a large range of conditions.

6. The concept of fractal structure helps us to understand the density structure of molecular clouds. Together with a measure of the optical depth estimated from ^{13}CO measurements, we use it to show that the CO(J=2–1) and (J=3–2) emissions are close to thermalization and that the material dominating the emission is dense $(n_{H_2} \sim 10^4 \, cm^{-3})$, cold $(T_k \leq 10K)$ and distributed in structures much smaller than our best resolution $(\sim 1000AU)$.

The link between both sets of conclusions might be that surfaces of isodissipation of kinetic energy in laboratory flows are fractal and have area-perimeter dimensions comparable to that measured for molecular clouds. It is remarkable in particular that there is no difference between the excitation conditions of the gas emitting in the line wings and in the line cores. The important issue of the formation, lifetime and resilience of such a structure is presently an open question.

References

Anselmet, F., Gagne, Y., and Hopfinger, E. J. 1984, *J. Fluid Mech.*, **140**, 63.

Bazell, D. and Désert, F. X. 1988, *Ap. J.*, **333**, 353.

Boissé P. 1990, *Astron. Astrophys.*, **228**, 483.

Boreiko, R.T., Betz, A.L., and Zmuidzinas, J. 1990, *Ap. J.*, **353**, 181.

Casoli, F., Dupraz, C., Combes, F., and Kazes, I. 1991 *Astr. Ap.* submitted.

Clemens, D.P., and Barvainis, R. 1988, *Ap. J. Supp.*, **68**, 257.

Drdla K., Knapp G.R. and van Dishoeck E.F. 1989, *Ap. J.*, **345**, 815.

Elmegreen, B.G. : 1991, these proceedings.

Falgarone, E., and Pérault, M. : 1987, *Physical Processes in Interstellar Clouds*, eds. G.E. Morfill and M. Scholer, Kluwer Acad. Publ..

Falgarone, E., and Pérault, M. : 1988, *Astr. Ap.*, **205**, L1.

Falgarone, E., and Phillips, T. G. 1990, *Ap. J.*, **359**, 344.

Falgarone E. 1991, *From Ground-based to Space-borne Submillimeter Astronomy*, eds. N. Longdon and B. Kaldeich, ESA Publ..

Falgarone, E., Phillips, T. G., and Walker C. 1991, submitted to *Ap. J.*.

Falgarone, E., Puget, J.-L., and Pérault, M. 1991, in preparation.

Frerking M.A., Keene J.B., Blake G.A., and Phillips T.G. 1989, *Ap. J.*, **344**, 311.

Frisch, U., Sulem, P. L., and Nelkin, M. 1978, *J. Fluid Mech.*, **87**, 719.

136

Genzel, R., Harris, A.I., Jaffe, D.T. and Stutzki J. 1988, *Ap. J.*, **332**, 1049.

Heithausen, A., and Thaddeus, P. 1990, *Ap. J. Letters*, **353**, L49.

Jaffe D.T., Genzel R., Harris A.I., Howe J.E., Stacey G.J., Stutzki J. 1990, *Ap. J.*, **353**, 193.

Keene, J., Blake, G.A., Phillips, T.G., Huggins, P.J., and Beichman, C.A. 1985, *Ap. J.*, **299**, 967.

Keto, E.R., and Lattanzio, J.C. 1989, *Ap. J.*, **346**, 184.

Kolmogorov A.N. 1941 : *Dokl. Akad. Nauk.* **26**, 115.

Kolmogorov, A. N. 1962, *J. Fluid Mech.*, **13**, 82.

Landau, L. D. and Lifchitz, E. M. 1959, *Fluid Mechanics*, Addison-Wesley.

Langer, W.D., and Penzias, A.A. 1990, *Ap. J.* in press.

Larson, R.B. : 1981, *Monthly Notices Roy. Astron. Soc.*, **194**, 809.

Lee, Y., Snell, R.L. and Dickman, R.L. 1990, *Ap. J.*, **355**, 536.

Lovejoy, S. 1982, *Science*, **216**, 185.

Magnani L., Blitz L., and Wendel A. 1988, *Ap. J.(Letters)*, **331**, L127.

Magnani L., Carpenter, J.M., Blitz L., Kassim, N.E. and Nath B.B. 1990, *Ap. J. Suppl.*, **73**, 747.

Mandelbrot, B. B. 1982, *The fractal geometry of nature*, Freeman.

Mauersberger R., Henkel C., Wilson T.L. and Harju J. 1990, *Astron. Astrophys.*, **223**, 79.

Méneveau 1989, PhD. dissertation, Yale University.

Mooney, T.J., and Solomon, P.M. 1990 in preparation.

Oboukhov, A.M. 1962, *J. Fluid Mech.*, **13**, 77.

Penzias, A.A., Solomon, P.M., Jefferts, K.B., and Wilson, R.W. 1972, *Ap. J. Letters*, **174**, L43.

Pérault M., Falgarone E., and Puget J.L. 1985, *Astr. Ap.*, **152**, 371.

Phillips, T.G., and Huggins, P.J. 1981, *Ap. J.*, **251**, 533.

Sage L.J., Shore S.N. and Solomon P.M. 1990, *Ap. J.*, **351**, 422.

Scalo, J. : 1987, *Interstellar Processes*, eds. D.J. Hollenbach and H.A. Thronson.

Scalo, J.M. 1990 *Physical Processes in Fragmentation and Star Formation*, eds R. Capuzzo-Dolcetta et al., Kluwer Academic Publ. : Dordrecht.

Sreenivasan, K. R. and Méneveau, C. 1986, *J. Fluid Mech.*, **173**, 357.

Stutzki J., Stacey, G.J., Genzel, R., Harris, A.I., Jaffe, D.T., and Lugten, J.B. 1988, *Ap. J.*, **332**, 379.

Ungerechts, H. and Thaddeus, P. 1987, *Ap. J. Suppl.*, **63**, 645.

van Atta, C. W. and Park, J. 1971, *Statistical Models and Turbulence*, eds. M. Rosenblatt and C. van Atta : Springer.

Wakker, B.P., 1990, Ph.D. Dissertation, University of Leiden.

Zmuidzinas, J., Betz, A.L., Boreiko, R.T., Goldhaber, D.M. 1988, *Ap. J.*, **335**, 774.

Zuckerman, B. and Evans, N.J. 1974, *Ap. J.*, **192**, L149.

III - CHEMISTRY

CHEMISTRY AND SMALL–SCALE STRUCTURE OF DIFFUSE AND TRANSLUCENT CLOUDS

JOHN H. BLACK

Steward Observatory, University of Arizona,
Tucson, AZ 85721 USA[1]
and
Onsala Space Observatory, Chalmers University of Technology,
S-439 00 Onsala, Sweden

EWINE F. VAN DISHOECK

Sterrewacht Leiden, Postbus 9513,
NL-2300 RA Leiden, The Netherlands[1]
and
Division of Geological and Planetary Sciences,
California Institute of Technology, Pasadena, CA 91125 USA

Abstract. The small, thin diffuse and translucent molecular clouds are excellent laboratories for studying the ways in which small–scale structure and interstellar chemistry affect each other. Variations of density or column density and chemical stratification can be found on scales as small as 0.01 pc. The origin of such structures and the evolutionary states of small clouds remain elusive.

Keywords : interstellar chemistry; molecular clouds

1. Introduction

Diffuse and translucent molecular clouds present special advantages for the investigation of the coupling between interstellar chemistry and small–scale structure in interstellar clouds. They also represent important test cases for developing theoretical tools needed to characterize the chemistry in larger, denser, star–forming regions. By diffuse clouds, we mean regions with line–of–sight visual extinction of $A_V \approx 1$ mag or less. The translucent clouds have $A_V \approx 1 - 5$ mag and often possess substantial abundances of various molecules. Both types of clouds are thin enough to be observed by means of optical absorption lines superimposed on the spectra of background stars, but thick enough in many cases to show detectable emission lines of molecules like CO at mm and sub–mm wavelengths. Typically the diffuse and translucent clouds are not directly involved in star formation. Many of these clouds are quite nearby, with fairly well determined distances $D \approx 50 - 200$ pc, so that rather small linear scales are resolvable in emission lines. For example, a $30''$ antenna beam corresponds to 6000 AU (0.03 pc) at $D = 200$ pc. Absorption line measurements sample an even smaller scale dictated by the angular size of the background star : e.g., a star of radius 10 R$_\odot$ at 500 pc isolates a cylinder through a

---

[1] Permanent Address

cloud that is 0.04 AU across for a cloud distance of $D = 200$ pc. While the velocity resolution of radio measurements may routinely be 0.1 km s^{-1}, that of the optical spectra is rarely better than 3 km s^{-1}. This also raises basic questions about what is meant by an interstellar "cloud" : usually, a cloud is defined as a distinct velocity component. Counting the number of "clouds" along the line of sight may thus be seriously limited by instrumental resolution at optical wavelengths. In principle, the complementary information provided by radio and optical measurements may permit theories of molecule formation to be tested rigorously at the level of a factor of 2 or better in abundances. In practice, the interpretation can be complicated by subtle effects of small–scale structure. However, given enough observational information, we may be able to use the chemical information for indirect analyses of the internal structure.

The empirical definitions of diffuse and translucent clouds in terms of thickness (column density or extinction) allow quite different phenomena to be included in the same categories : small, isolated clouds; surface layers of giant molecular clouds; and high–latitude molecular clouds, for example. Translucent clouds are characterized by total hydrogen column densities $N_H \geq 10^{21}$ cm $^{-2}$. There may be a variety of origins and evolutionary states represented, which range from dissipating fragments of previous large clouds, to condensing pieces that are forming new larger structures. One unifying property is that ultraviolet starlight is likely to play an important role in the chemistry and thermal balance of all these clouds. Thus, translucent clouds embody most of the features that one expects in photodissociation regions (PDRs) in star–forming complexes where a molecular cloud surface is scorched by intense ultraviolet light from hot, young stars nearby (cf. Tielens and Hollenbach 1985), except that the ultraviolet fluxes and gas densities are typically lower. It is important to recognize that chemical stratification is a form of internal structure that can affect the thermal properties of the matter and that can sometimes mask physical stratification (variations in density and temperature).

Owing to the deluge of new observational results, only recent results will be reviewed here. Langer (1990) has discussed velocity and density structure in diffuse clouds. Reviews of the chemistry of diffuse and translucent clouds include Dalgarno (1988), van Dishoeck and Black (1988a), and van Dishoeck (1990a,b).

2. Chemical Structure

The basic *chemical* structure of a small molecular cloud is thought to be governed by the processes that form and destroy the hydrogen molecule. Until a large fraction of the hydrogen is in the form of H$_2$, very little molecular formation involving other elements can occur. H$_2$ molecules exposed to the general Galactic background of ultraviolet starlight are readily photodissociated on a time scale of the order of 10^3 years. The most efficient formation mechanism of H$_2$, association of H atoms on grain surfaces, is still slow in absolute terms, and the molecular fraction, $f = 2n(H_2)/n_H$, in unshielded regions will usually be very low. Because the photodissociation of H_2 is initiated by line absorptions (P. M. Solomon, as reported by Field *et al.* 1966), which begin to saturate at column densities as low as $N(H_2) \approx 10^{14}$ cm^{-2}, the molecules deeper into a cloud "feel" fewer line–core photons than do those near the boundary where the highest rates of absorption and dissociation occur. This effect is called *self–shielding* and has been described by Stecher and Williams (1967) and Federman *et al.* (1979). The line absorptions in H_2 are even more likely to lead to fluorescent excitation than dissociation (by a ratio of $\approx 6/1$), so that the destruction rate and abundance of H_2 are intimately linked to its excitation. A result of the depth–dependent balance between formation and destruction of H_2 is that the H_2/H abundance ratio varies over orders of magnitude through a diffuse or translucent cloud and the mean molecular weight, by approximately a factor of two. Detailed depth–dependent models of the excitation and abundance of H_2 in such clouds have been constructed by van Dishoeck and Black (1986) and Viala (1986). The transition from a fully atomic composition ($H/H_2 >> 1$) to a largely molecular gas ($H/H_2 \leq 1$) is predicted to occur at an extinction $\Delta A_V \approx 0.1$ into a diffuse cloud (mean density $\approx$ a few hundred cm^{-3}). This is substantiated by ultraviolet absorption observations of H_2 in diffuse clouds (Spitzer and Jenkins 1975, especially their Figure 4).

Carbon monoxide is one of the few other molecules whose abundance is limited by self–shielding ultraviolet line absorptions. Thus its abundance tends to be significant only when its column density exceeds the threshold for effective shielding. This threshold is only somewhat lower than the minimum column density for detection of its mm–wavelength emission lines. Because the abundances of carbon and oxygen in the interstellar gas are almost 10^4 times smaller than that of hydrogen, extinction by dust and absorption by overlapping lines of H_2 play a more important role in shielding the CO than does the dust in shielding the H_2. The dissociating lines of CO are intrinsically broader than those of H_2 so that their natural widths may dominate their line broadening in the interstellar medium and thus lead to a less sharp transition layer than for H_2. According to gas–phase theories of interstellar chemistry, CO arises from the basic oxygen cycle that generates OH, H_2O, and related ions, which react with carbon atoms or ions to produce CO$^+$, HCO$^+$, and CO; and from reactions that form simple hydrocarbons and hydrocarbon ions, CH, CH$_2$, CH$_3$, CH$_2^+$, CH$_3^+$, etc., which then react with oxygen to form HCO$^+$ and CO. The various isotopic varieties of CO can form similarly, but there is an additional, temperature–sensitive exchange reaction,

$$^{13}\text{C}^+ + {}^{12}\text{CO} \rightleftharpoons {}^{13}\text{CO} + {}^{12}\text{C}^+ + \Delta E \tag{1}$$

which enriches ^{13}C in ^{13}CO at low temperatures, because $\Delta E/k \approx 36$ K. The photodissociation of CO is also partly isotope–selective (Bally and Langer 1982, Glassgold *et al.* 1985, van Dishoeck and Black 1988b). Thus we expect the ^{12}CO/^{13}CO abundance ratio to vary with depth through a cloud, following the growth of attenuation of ultraviolet starlight and responding to any gradients in temperature.

Figures 11 and 12 of van Dishoeck and Black (1988) illustrate the depth–dependent variations of abundances of the isotopic varieties of CO in model clouds : the fractionation effects described above are significant at extinctions $\Delta A_V = 0.5 - 1$ mag, but depend sensitively on such parameters as the total density, the overall gas–phase carbon abundance, and the intensity of ultraviolet starlight. The growth of the total CO abundance is also sensitive to these parameters and is a steep function of depth. These effects emphasize the apparent structure in CO emission line maps : e.g., a 20% increase in H_2 column density can produce a factor of 2–3 increase in the column density of CO. If, in addition, there is a gradient in the density $n(H_2)$, the observable CO line intensity can be enhanced far out of proportion to the variations in density alone. Because the CO in translucent clouds is often just at the thresholds both of self–shielding and of detectable rotational line emission, it is very difficult to infer variations in physical parameters from apparent fluctuations in measurable intensities.

The reason for reviewing briefly this gas–phase chemistry is to reinforce the expectation that especially the small diffuse and translucent clouds are chemically stratified. Moreover, this chemical stratification is a form of small–scale internal structure, particularly as regards the abundances and excitation of molecules like ^{12}CO and ^{13}CO that are commonly used to trace the extent and internal structure of molecular clouds. Note also that this chemistry governs the abundances of the principal forms of carbon : C^+, C, and CO, which are major coolants of the gas and which include the major electron donor and a dominant ion, C^+. A thorough understanding of the chemistry and its variation with position inside a cloud is thus necessary for a complete description of the thermal properties of the gas (major coolants) and the coupling of the gas to magnetic fields (dominant ions and sources of electrons).

The identity of the principal ion and the coupling of gas, dust, and magnetic fields may be complicated by the chemical effects of large molecules such as polycyclic aromatic hydrocarbons (PAHs). These large molecules are probably very effective at capturing electrons and at neutralizing a variety of positive ions (Omont 1986, Lepp and Dalgarno 1988). In dense, dark clouds, the negative ion, e.g. PAH$^-$, may be the most abundant charged species. Even in diffuse clouds, it will affect the atomic ionization balance measurably (Lepp *et al.* 1988), if the interstellar PAH abundance is 1–10% of the total carbon. In the absence of PAHs, the principal ions would tend to be C^+, H^+, H_3^+, HCO^+, H_3O^+, etc., which all have charge/mass ratios at least 10 times larger than that of a 50-atom PAH$^-$ ion. The consequences of a high abundance of charged, large molecules like PAH$^-$ for coupling of gas to magnetic fields merit further consideration. The effects of PAHs on the chemistry and propagation of MHD shock waves in molecular clouds have been considered recently by Pineau des Forêts, Flower, and Dalgarno (1988) and Flower, Heck, and Pineau des Forêts (1989).

There are some fundamental, unresolved questions already lurking behind the facile discussion above. It is conventionally assumed that chemical abundances in diffuse and translucent clouds are in steady state. Is this a good assumption? It was pointed out above that an important limiting time–scale is the lifetime of H_2, which is only $\approx 10^3$ y at a typical cloud boundary; however, the effect of self–shielding can lengthen this to 10^7 y in the center of a diffuse cloud. The lifetime of a hydrogen atom against catalysis on a grain surface to form H_2 can also be long : $10^7(30/T)^{1/2}(200/n_H)$ y, where T and n_H are the kinetic temperature and total density of hydrogen nuclei, respectively. This latter chemical time–scale is comparable to the sound–crossing time for a cloud of total thickness $A_V \approx$ 1 mag : $5 \times 10^6 A_V (30/T)^{1/2}(200/n_H)$ y. These estimates suggest that chemical time–scales involving the dominant species can be long compared with relevant dynamical, evolutionary scales. However, once the H_2 abundance even approaches its steady–state value, most other molecular abundances rapidly achieve steady state on time scales of $10^3 - 10^4$ years. Wagenblast and Hartquist (1988, 1989) have considered time–dependent effects on the abundance and excitation of H_2 in diffuse clouds. Tarafdar et al. (1985) have performed elaborate computations of the combined chemical and dynamical evolution of diffuse clouds, but have started from the presumption that diffuse structures collapse to form denser clouds. In all cases, time–dependent effects on structure and chemistry and their consequences for the evolution of small clouds are just beginning to be explored in detail. It seems obvious that initial conditions must affect the future development of these structures, but it is not so clear that observable properties can provide clues to what those initial conditions were.

Some observations of translucent (cf. Jannuzi et al. 1988), high–latitude (Magnani, Blitz, and Mundy 1985; Keto and Myers 1986), and tiny (Knapp and Bowers 1988) molecular clouds indicate that the masses are too small for these clouds to be in virial equilibrium given their apparent sizes and velocity dispersions. The implications are that gravity does not completely control their structure and that they may be subject to rapid dissipation or other structural changes. Some observed features of the high–latitude clouds may have a natural explanation in terms of collisions between cloud condensations (Keto and Lattanzio 1989).

Even complicated theoretical models often make embarrassingly simple assumptions about the velocity field and about variations of density and temperature. Plane–parallel slabs or spheres with uniform (or uniformly varying) physical parameters are most common. How much difference does it make to our theoretical description of a cloud if, for example, it is filamentary with a turbulent velocity field and a non–Gaussian, high–velocity kinematic component? The chemical effects of "hot" atoms and ions (i.e., those with excess kinetic energy compared with mean thermal values) have been investigated in a few cases (Adams et al. 1984; Yee et al. 1987; Brown et al. 1990), but only in the microscopic sense in which the excess translational energy results from specific chemical reactions. The high abundance of widespread CH^+ in diffuse clouds remains a puzzle for interstellar chemistry, in part because C^+ ions do not react directly with H_2 unless they have excess translational energy of 0.4 eV or more. It should be noted that if a fraction 10^{-8} of all

C^+ in a diffuse cloud were translationally "hot", then the CH^+ abundance could be explained. It seems not completely unreasonable that a macroscopic component of hot atoms and molecules might exist in molecular clouds, especially in view of the observational evidence of intermittent turbulence on all scales (Falgarone and Phillips 1990). Whether or not such a hot component could affect observable abundances has not been explored yet. One way in which the turbulence could be chemically important is through mixing of dense and dilute gas (Boland and de Jong 1982, Chièze and Pineau des Forêts 1989).

This consideration raises another fundamental question : at what scale do microscopic and macroscopic phenomena overlap? This may be of practical importance if the characteristic scale of overlap is large enough to have observable effects. Almost all descriptions of chemical reaction rates in the interstellar medium assume Maxwellian velocity distributions and corresponding thermal reaction rate coefficients : for this assumption to be valid, particle motions must be able to be thermalized on a small, microscopic length scale. Consider, for example, the thermalization of photoelectrons in a diffuse cloud. The mean excess kinetic energy of the electron produced by

$$C + h\nu \rightarrow C^+ + \epsilon \qquad (2)$$

is $< E >= 0.87$ eV $>> kT$ in typical Galactic starlight. These electrons lose energy and become thermalized in collisions with thermal electrons and with abundant neutrals like H and H_2. At the densities $n(e) \approx 0.1$ and $n(H_2) \approx 300\text{--}1000$ cm^{-3} characteristic of translucent clouds (Black and van Dishoeck 1990), the thermalization time and corresponding mean–free path are $\approx 10^5$ s and $10^{12} - 10^{13}$ cm, respectively. This suggests that one cannot assume an ensemble of interstellar electrons to be fully thermalized on scales much smaller than 1 AU. If intermittent turbulent motions persist to such scales, they may add to the complexity of non–Maxwellian tails in the "microscopic" speed distributions. Similarly, highly reactive positive ions may always be more likely to undergo reactions before suffering enough elastic encounters to become thermalized.

The standard chemical models have been tested against detailed observations of specific lines of sight, such as the classical diffuse cloud toward ζ Oph. In general, it is found that the abundances of molecules like CH, CN, C_2 and CO are consistent with the quiescent gas–phase chemistry, but that CH^+ is most likely formed as a result of a MHD shock propagating into the cloud. An important recent advance has been the achievement of 0.5 km s^{-1} resolution in observations of the molecular absorption lines of CH, CH^+, and CN toward ζ Oph (Lambert, Sheffer, and Crane 1990), which permit *direct* tests of the relative importance of MHD shocks and quiescent chemistry : surprisingly, not only the CH^+, but also part of the CH may be produced in a (shock–)heated layer. There are many ways in which small–scale structures and chemistry can affect each other through the propagation of MHD shocks in predominantly neutral clouds. There is by now a vast literature on this subject, a review of which is beyond the scope of this paper (see reviews of

Draine 1987, and Hartquist *et al.* 1990). Such shocks have been suggested to play a major role in the chemistry of high–latitude clouds (Magnani, Blitz and Wouterloot 1988), but little observational evidence has been found to support this picture (de Vries and van Dishoeck 1988; Magnani and Siskind 1990, Penprase *et al.* 1990). In general, the chemistry in high–latitude clouds resembles very much that found in translucent clouds.

3. Observations and Interpretations

The small–scale structure of diffuse and translucent clouds can be revealed by high–resolution spatial and/or spectral observations. Recent observations of molecular absorption lines at optical wavelengths with velocity resolutions of 0.5–3.0 km s^{-1} have revealed much more complicated distributions of cloud components than was previously recognized (Jenkins *et al.* 1989; Lambert *et al.* 1990) : several components spaced less than 1 km s^{-1} are found in some cases, whereas weak components with nearly thermal line widths have been revealed in others. Such studies have been complemented by improved measurements of atomic constituents in many of the same clouds (e.g., Joseph and Jenkins 1990; Hobbs and Welty 1990). Emission line studies of diffuse and translucent clouds at mm/submm wavelengths have exposed complicated structures both on the plane of the sky and in radial velocity (e.g., Langer *et al.* 1987; Crutcher and Federman 1987; LeBourlot *et al.* 1989; Jannuzi *et al.* 1988; Magnani *et al.* 1985, 1990; Keto and Myers 1986; Knapp and Bowers 1988; van Dishoeck *et al.* 1991; Gredel *et al.* 1991). These measurements have spawned efforts to improve the interpretation and theoretical modeling (e.g., Jannuzi *et al.* 1988; van Dishoeck *et al.* 1991) and to refine the treatment of spectroscopic data in cases where intrinsic blending occurs at the same scale as astronomical velocity structure (Black and van Dishoeck 1988). It is important to note that optical absorption line measurements are much more sensitive to *very small* column densities ($N(H_2) \approx 10^{14}$ cm^{-2}) and thus can reveal a more diffuse, atomic component along the line of sight, whereas mm observations are limited to clouds with $N(H_2) \gtrsim 5 \times 10^{20}$ cm^{-2}.

Similar structure has been seen at the level of 0.02 pc in a molecular cloud of relatively low average column density by Falgarone and Pérault (1988). They interpret small–scale fluctuations in intensity of CO line emission as fluctuations of density. These fluctuations appear to be correlated in both space and velocity and, according to Falgarone and Pérault, are related to the turbulent velocity field of the cloud.

A recent study of the high–latitude translucent cloud toward HD 210121 (Gredel *et al.* 1991) also reveals considerable structure on both small and large scales, as shown in Figure 1. This contour map of peak antenna temperature in CO $J =1$-0 displays several prominent peaks and small–scale brightness fluctuations within them, very sharp boundaries (notably along the southwestern edge), and large contrasts in brightness between the peaks and the large interior regions of weak

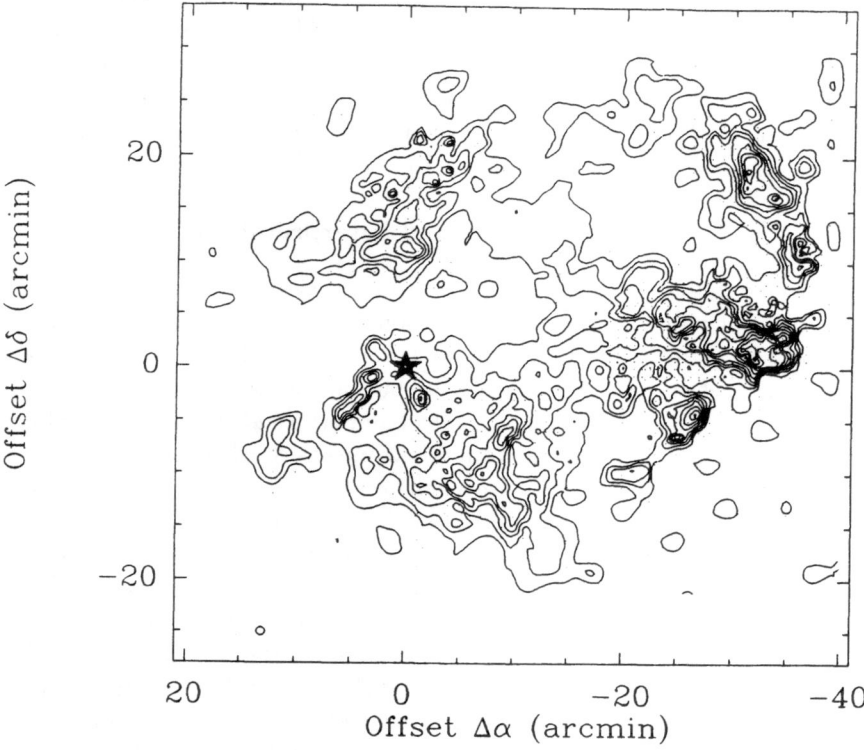

Fig. 1.   Map of the peak T_A^* in ^{12}CO J =1–0 in the high–latitude translucent cloud toward the star HD 210121. The lowest contour is T_A^*=1 K, and the increment between subsequent contours is 0.5 K. Map coordinates are offsets in arcmin from the star (from Gredel *et al.* 1991).

emission. Measurements of intensities of different lines have been used to estimate densities, temperatures, and abundances in parts of this cloud. The cloud appears to occupy an interesting regime of parameter space in which the abundance of CO has begun to be significant, but in which most of the carbon is still in atomic form. As discussed above, this is just the circumstance in which small variations in local conditions (e.g. density or total hydrogen column density) can produce very large fluctations in line–of–sight CO column density (and hence in observable CO line intensity). Therefore, Gredel *et al.* suggest that much of the apparent small–scale structure in this cloud reflects fluctuations in line–of–sight column density of CO, rather than large variations in density, since no significant differences in molecular excitation are found between various positions in the cloud. On the other hand, weak evidence is found for a small density gradient across one clump from observed variations in the CO excitation. As a general conclusion, it is not trivial to distinguish density variations from variations in other properties, since in the diffuse and translucent clouds, large fluctuations in observable intensities can result from small changes in physical properties.

Another method of finding small–scale structure involves the use of background stars in resolvable binaries (Meyer 1990) or in globular clusters (Bates *et al.* 1990, Langer *et al.* 1990) to map variations in interstellar absorption lines on angular scales of the order of arcseconds. Toward one nearby binary system, Meyer (1990) finds evidence of interstellar structure on scales less than 2000 AU. It is perhaps even more significant that Meyer finds no evidence of such structure in 5 other lines of sight. Some observational evidence exists for structures as small as 10^{14}–10^{15} cm in the general interstellar medium (Fiedler *et al.* 1987, Diamond *et al.* 1989) based on interferometric measurements of radio sources at centimeter wavelengths.

Complementary optical and radio observations of molecular clouds have been discussed by Crutcher (1985), Crutcher and Chu (1985), Jannuzi *et al.* (1988), van Dishoeck and Black (1989), van Dishoeck *et al.* (1991), and Gredel *et al.* (1991) among others. In simplest terms, the optical absorption line measurements provide one–dimensional information through a narrow cylinder of matter along the line of sight to a well placed background star. Species with weak lines, e.g. C_2, show little or no effect of saturation; therefore, the column densities are directly proportional to measured equivalent widths (integrated intensities). A more complicated curve-of–growth analysis is required to interpret strong, saturated absorption lines like those of CN. On the other hand, emission line measurements typically refer to $30''$–$60''$ antenna beams that average over all smaller–scale structure. In the case of CO lines, the beam–averaging is often non–linear in the sense that the lines are saturated and the radiative transfer within the cloud is, in principle, non–local. Recent observations of C_2 absorption and CO emission in the same translucent clouds yield somewhat discordant results for the densities that partly control the rotational excitation in both species (van Dishoeck *et al.* 1991). Intensity ratios for 3 transitions of ^{12}CO and 1 transition of ^{13}CO suggest hydrogen densities of the order of 10^3 cm^{-3} that are systematically higher than those, 200–500 cm^{-3}, indicated by the measured rotational populations of C_2. There are several possible explanations. One is chemical stratification : the CO and C_2 may be distributed differently with respect to gradients of density and temperature. A second lies in the distribution of density fluctuations with respect to the quite different effective "beam sizes" of the absorption and emission measurements. A third possibility is that the treatment of the CO line formation has not been sophisticated enough to represent correctly the nature of the velocity field and small–scale density fluctuations. In order to characterize the possible effects of chemical stratification, chemical models were constructed in which density and temperature were allowed to vary in discrete steps subject only to observational constraints on total column densities (van Dishoeck *et al.* 1991). It was found that the hydrogen density, weighted by the distribution of CO could be several times larger than the density weighted by the distribution of C_2 along a line of sight through the model, but only when ultraviolet starlight is incident on both sides of the plane parallel model. For model clouds illuminated on one side only, the situation can be reversed : C_2 is preferentially concentrated in the denser regions.

The CN molecule has been observed in diffuse and translucent clouds both by optical absorption line and millimeter emission line methods. Analysis of the excitation of CN excludes densities higher than $n(H_2) \approx 3 \times 10^3$ cm^{-3} on scales ranging from 10^{-3} to 60 arcsec in a sample of these clouds (Black and van Dishoeck 1990), and generally yields values consistent with the C_2 excitation. High quality CN absorption line data also provide measurements of the Doppler parameter b characterizing the saturation of the strongest lines. This value can be compared with that derived from CO or ^{13}CO mm observations, thus providing further information on the internal structure of the cloud (Gredel, van Dishoeck and Black 1991).

Another observed feature of translucent and high–latitude molecular clouds is that they often show asymmetrical, flat–topped, or possibly even centrally reversed profiles in CO emission lines. Theoretical calculations of line profiles formed in model clouds that have gradients of density, temperature, and CO abundance can reproduce the qualitative appearance of some of the complex profiles. Indeed, small central reversals can appear in such models whether the radiative transfer is treated according to the Sobolev approximation (large velocity gradients) or in a micro-turbulent description (van Dishoeck et al. 1991). Sometimes it may be difficult to distinguish such effects from multiple velocity components or profiles resulting from colliding clumps (Keto and Lattanzio 1989). The ominous implication of such calculations is that even in small clouds the observable line profiles are related to internal structure and chemical stratification in such a complicated way that it may be difficult—if not impossible—to infer much about the structure from the profiles!

4. Summary

Rapid advances in observational techniques and sensitivities have made it possible to observe very small–scale structure in nearby diffuse and translucent molecular clouds. Although these clouds are usually not directly implicated in the formation of stars, they provide important test cases for understanding the relationships among cloud structure (variations of density and temperature), chemistry, and turbulence. Every improvement in observation and in theoretical interpretation seems to increase the degree of apparent complexity! Studies of the smallest structures in the smallest clouds raise fundamental questions about chemical processes and about distinctions between microscopic and macroscopic phenomena.

Acknowledgements

J.H.B. is grateful to the Swedish National Science Research Council and NORDITA for support.

References

Adams, N.G., Smith, D., and Millar, T.J. 1984. *M. N. R. A. S.*, **211**, 857.

Bally, J., and Langer, W.D. 1982, *Ap. J.*, **255**, 143; **261**, 747.

Bates, B., Catney, M.G., and Keenan, F.P. 1990, *M. N. R. A. S.*, **245**, 238.

Black, J.H., and van Dishoeck, E.F. 1988, *Ap. J.*, **331**, 986.

Black, J.H., and van Dishoeck, E.F. 1990, *Ap. J. (Letters)*, submitted.

Boland, W., and de Jong, T. 1982, *Ap. J.*, **261**, 110.

Brown, R.D., Cragg, D.M., and Bettens, R.P.A. 1990, *M. N. R. A. S.*, **245**, 623.

Chièze, J.P., and Pineau des Forêts, G. 1989, *Astr. Ap.*, **221**, 89.

Crutcher, R.M. 1985, *Ap. J.*, **288**, 604.

Crutcher, R.M., and Chu, Y.-H. 1985, *Ap. J.*, **290**, 251.

Crutcher, R.M., and Federman, S.R. 1987, *Ap. J. (Letters)*, **316**, L71.

Dalgarno, A. 1988, *Ap. Lett. Commun.*, **26**, 153.

de Vries, C.P., and van Dishoeck, E.F. 1988, *Astr. Ap.*, **203**, L23.

Diamond, P.J., Goss, W.M., Romney, J.D., Booth, R.S., Kalberla, P.M.W., and Mebold, U. 1989, *Ap. J.*, **347**, 302.

Draine, B.T. 1987, in *Physical Processes in Interstellar Clouds*, (G.E. Morfill and M. Scholer, Eds.), 423–428, Reidel.

Falgarone, E., and Pérault, M. 1988, *Astr. Ap.*, **205**, L1.

Falgarone, E., and Phillips, T.G. 1990, *Ap. J.*, **359**, 344.

Federman, S.R., Glassgold, A.E., and Kwan. J. 1979, *Ap. J.*, **227**, 466.

Fiedler, R.L., Dennison, B., Johnston, K.J., and Hewish, A. 1987, *Nature*, **326**, 675.

Field, G.B., Somerville, W.B., and Dressler, K. 1966, *Ann. Rev. Astr. Astrophys.*, **4**, 207.

Flower, D.R., Heck, L., and Pineau des Forêts, G. 1989, *M. N. R. A. S.* **239**, 741.

Glassgold, A.E., Huggins, P.J., and Langer, W.D. 1985, *Ap. J.*, **290**, 615.

Gredel, R., van Dishoeck, E.F., de Vries, C.P., and Black, J.H. 1991, *Astr. Ap.*, submitted.

Gredel, R., van Dishoeck, E.F., and Black, J.H. 1991, in preparation.

Hartquist, T.W., Flower, D.R., and Pineau des Forêts, G. 1990, in *Molecular Astrophysics*, (T.W. Hartquist, Ed.), 99–112, Cambridge University Press.

Hobbs, L.M., and Welty, D.E. 1990, *Ap. J.*, submitted.

Jannuzi, B.T., Black, J.H., Lada, C.J., and van Dishoeck, E.F. 1988, *Ap. J.*, **332**, 995.

Jenkins, E.B., Lees, J.F., van Dishoeck, E.F., and Wilcots, E.M. 1989, *Ap. J.*, **343**, 785.

Joseph, C.L., and Jenkins, E.B. 1990, *Ap. J.*, submitted.

Keto, E.R., and Lattanzio, J.C. 1989, *Ap. J.*, **346**, 184.

Keto, E.R., and Myers, P.C. 1986, *Ap. J.*, **304**, 466.

Knapp, G.R., and Bowers, P.F. 1988, *Ap. J.*, **331**, 974.

Lambert, D.L., Sheffer, V., and Crane, P. 1990, *Ap. J. (Letters)*, **359**, L19.

Langer, G.E., Prosser, C.F., and Sneden, C. 1990, *Astron. J.*, **100**, 216.

Langer, W.D. 1990, in *Molecular Astrophysics*, (T.W. Hartquist, Ed.), 84–98, Cambridge University Press.

Langer, W.D., Glassgold, A.E., and Wilson, R.W. 1987, *Ap. J.*, **322**, 450.

Le Bourlot, J., Gérin, M., and Pérault, M. 1989, *Astr. Ap.*, **219**, 279.

Lepp, S., and Dalgarno, A. 1988, *Ap. J.*, **324**, 553.

Lepp, S., Dalgarno, A., van Dishoeck, E.F., and Black, J.H. 1988, *Ap. J.*, **329**, 418.

Magnani, L., and Siskind, L. 1990, *Ap. J.*, **359**, 355.

Magnani, L., Blitz, L., and Mundy, L. 1985, *Ap. J.*, **295**, 402.

Magnani, L., Blitz, L., and Wouterloot, J.G.A. 1988, *Ap. J.*, **326**, 909.

Magnani, L., Carpenter, J.M., Blitz, L., Kassim, N.E., and Nath, B.B. 1990, *Ap. J. Suppl.*, **73**, 747.

Meyer, D.M. 1990, *Ap. J. (Letters)*, **364**, in press.

Omont, A. 1986, *Astr. Ap.*, **164**, 159.

Penprase, B.E., Blades, J.C., Danks, A.C., and Crane, P. 1990, *Ap. J.*, Dec. 10 issue.

Pineau des Forêts, G., Flower, D.R., and Dalgarno, A. 1988, *M. N. R. A. S.*, **235**, 621.

150

Stecher, T.P., and Williams, D.A. 1967, *Ap. J. (Letters)*, **149**, L29.

Spitzer, L., Jr., and Jenkins, E.B. 1975. *Ann. Rev. Astr. Astrophys.*, **13**, 133.

Tarafdar, S.P., Prasad, S.S., Huntress, W.T., Jr., Villere, K.R., and Black, D.C. 1985, *Ap. J.*, **289**, 220.

Tielens, A.G.G.M., and Hollenbach, D. 1985, *Ap. J.*, **291**, 722.

van Dishoeck, E.F. 1990a, in *Molecular Astrophysics*, (T.W. Hartquist, Ed.), 55–83, Cambridge University Press.

van Dishoeck, E.F. 1990b, to appear in *The Evolution of the Interstellar Medium*, (L. Blitz, Ed.), Astronomical Society of the Pacific, in press.

van Dishoeck, E.F., and Black, J.H. 1986. *Ap. J. Suppl.*, **62**, 109.

van Dishoeck, E.F., and Black, J.H. 1988a, in *Rate Coefficients in Astrochemistry*, (T.J. Millar and D.A. Williams, Eds.), 209–237, Kluwer.

van Dishoeck, E... and Black, J.H. 1988b, *Ap. J.*, **334**, 771.

van Dishoeck, E.F., and Black, J.H. 1989, *Ap. J.*, **340**, 273.

van Dishoeck, E.F., Black, J.H., Phillips, T.G., and Gredel, R. 1991, *Ap. J.*, Jan. 1 issue.

Viala, Y. 1986, *Astr. Ap. Suppl.*, **64**, 391.

Wagenblast, R., and Hartquist, T.W. 1988. *M. N. R. A. S.*, **230**, 363.

Wagenblast, R., and Hartquist, T.W. 1989, *M. N. R. A. S.*, **237**, 1019.

Yee, J.H., Lepp, S., and Dalgarno, A. 1987, *M. N. R. A. S.*, **227**, 461.

VARIATIONS IN THE ABUNDANCE OF SMALL PARTICLES

F. BOULANGER

Radioastronomie, Ecole Normale Supérieure,
24 rue Lhomond, 75005 Paris, France

Abstract. IRAS images of nearby molecular clouds show that the mid-IR emission from small particles in the size range 10^2 to 10^5 atoms is distributed very differently from the 100 μm emission from large dust grains. Variations in color ratios by as much as one order of magnitude are seen on all angular scales. We summarize observational properties of the color variations and argue that neither their large amplitude nor their morphology can be explained by changes of the excitation by the UV radiation field only. The color variations reflect considerable inhomogeneities in the abundance of small particles. We suggest that the abundance variations are related to the cycling of interstellar matter between the gas phase and dust grains. This interpretation entails that clouds with distinct IR colors differ in their density and velocity structure and that cycling of matter between gas phase and dust grains is more ubiquitous and rapid that generally thought.

Keywords : Molecular Clouds, Grains : Size Distribution, IR Emission, IRAS Colors

1. Introduction

IRAS images of nearby molecular clouds at the four wavelengths 12, 25, 60, and $100\mu m$ provide a wealth of information on the composition of interstellar matter which has just begun to be analyzed. Particles contributing to the observed emission differ from one wavelength to the next. While the $100\mu m$ comes from large grains in thermal equilibrium with the radiation field, the 12 and $25\mu m$ emission, and, probably, part of the $60\mu m$ emission is radiated by particles with sizes in the range 0.5 to 10 nm for which the time between photon absorption is larger or comparable to the cooling time (Puget et al. 1985, Draine and Anderson 1985, Chlewicki and Laureijs 1988, Désert, Boulanger, and Puget 1990, hereafter DBP). These particles emit most of their emission at temperatures close to the peak temperature they reach after photon absorption. Since this temperature depends on the size of the particle, the spectral distribution of their IR emission is related to their size distribution (see Puget and Léger 1989). In their model DBP show that at least three dust components are needed to account for the IRAS data : (1) large molecules or cluster of molecules with one to a few hundred atoms (near-IR, 12 and part of $25\mu m$ emission), (2) very small dust grains with 10^3 to 10^5 atoms (25 and part of $60\mu m$ emission), and (3) dust grains with sizes larger than 10 nm in thermal equilibrium ($100\mu m$ and sub-mm emission). In practice the IRAS color ratios $R(12,100) = I_\nu(12\mu m)/I_\nu(100\mu m)$, $R(25,100) = I_\nu(25\mu m)/I_\nu(100\mu m)$ and $R(60,100) = I_\nu(60\mu m)/I_\nu(100\mu m)$ measure the relative abundances of these three families of particles.

Early in the analysis of IRAS data on nearby clouds heated by the general interstellar radiation field of the Galaxy (ISRF), it appeared that the ratio between mid-IR (12 and 25 μm) and 100 μm emission varies widely from cloud to cloud

(Boulanger et al. 1985, Leene 1985, de Vries and Le Poole 1985, Weiland et al. 1986, Boulanger and Pérault 1988, Heiles, Reach, and Koo 1988, Laureijs, Mattila, and Schnur 1987, Laureijs, Chlewicki, and Clark 1988, Beichman et al. 1988). Later, Boulanger (1989), Boulanger et al. (1989) and Puget (1989) stressed the existence of considerable variations (up to one order of magnitude) in IR colors on scales as small as the IRAS resolution (a few tenths of a parsec) within clouds in the complexes of Chamaeleon, Taurus, Ursa Major and Ophiuchus. Color variations within the first three complexes have been described in detail by Boulanger, Falgarone, Puget and Helou (1990, hereafter BFPH). Based on this work and preliminary comparison of IR images with molecular observations, we summarize observed properties of the color variations. Following BFPH we argue that the color variations reflect inhomogeneities in the abundance of small particles within and among clouds. Through the paper the dust model of DBP is used to discuss the excitation of small particles and quantify the abundance variations.

2. Observational Properties

2. 1. COLOR-BRIGHTNESS DIAGRAMS

Much can be learned on the variations of IR colors in molecular clouds by simply looking at the color images and cuts presented by BFPH, Boulanger (1989), Boulanger et al. (1989) and Puget (1989). To quantify these data, BFPH measured the IR brightnesses at the four IRAS wavelengths for several tens of positions within Chamaeleon, Taurus, and Ursa Major. Along most of the selected lines of sight $A_v < 2mag$. The colors computed at these positions are presented in figure 1 in three diagrams, R(12,25), R(12,100) and R(60,100) versus $I_\nu(100\mu m)$. In these diagrams $I_\nu(100\mu m)$ may be considered as a measure of the total column density along the line of sight. Direct comparisons of $I_\nu(100\mu m)$ with A_v in Taurus and Chamaeleon show a good correlation for $A_v < 3mag$ with a slope of 5 - 10 (MJy/sr)/mag, independent of the IR colors of the cloud.

The main feature of the diagrams is the considerable scatter of the R(12,100) and R(60,100) colors at all $I_\nu(100\mu m)$. Values of R(12,100) range from ~ 0.008 to 0.2, 1/5 to 5 times the average value for atomic gas in the Solar Neighborhood (Boulanger and Pérault 1988). Matter with a high $12\mu m$ emissivity, defined by a R(12,100) color higher than the average Solar Neighborhood value is observed only in the external parts of clouds. The $100\mu m$ brightness associated with this matter is at most 10 MJy/sr which corresponds to an A_v of 1 mag. This important property of the color variations translates in the diagram of figure 1 by a general decrease of R(12,100) with increasing $I_\nu(100\mu m)$. The median value of R(12,100) decreases from 0.14 for $I_\nu(100\mu m) \sim 2MJy/sr$, to 0.07 at $I_\nu(100\mu m) \sim 5MJy/sr$, and 0.03 for $I_\nu(100\mu m)$ between 10 and 20 MJy/sr. Values of R(12,100) larger than 0.05 for $I_\nu(100\mu m) > 10\,MJy/sr$ can all be ascribed to bright envelopes of $12\mu m$ emission. In each of these cases no $12\mu m$ emission is seen from the inner parts of the cloud, $I_\nu(12\mu m)$ is limb-brightened and is minimal at the position where $I_\nu(100\mu m)$ peaks.

R(60,100) is observed to vary from 0.05 to 0.3, the average Solar Neighborhood value being 0.2. This color appears roughly independent of $I_\nu(100\mu m)$. Within a cloud, variations in R(60,100) and R(12,100) are often correlated (Beichman et al. 1988) but this is not a systematic feature (Laureijs, Chlewicki and Clark 1988). The correspondence between these two colors is clearly not universal : it varies from cloud to cloud.

By contrast to the two other colors in figure 1, R(12,25) is remarkably constant. Taking into account uncertainties on the measurements the real scatter of R(12,25) is about a factor of 2, one order of magnitude smaller than that observed for R(12,100). This result suggests a close correlation between the populations of particles generating the 12 and $25\mu m$ emission. This is not a trivial result since dust models predict that only half of the $25\mu m$ emission is coming from particles giving rise to the $12\mu m$ emission (Chlewicki and Laureijs 1988, DBP).

2. 2. MORPHOLOGY AND PHYSICAL PROPERTIES

In view of the IRAS images, without oversimplifying the data we may distinguish three kinds of clouds. (1) Clouds with uniformly little or no 12 and $25\mu m$ emission (e.g. Cha I and III in Chamaeleon, L1495 and 1506 in Taurus), (2) clouds with an envelope of bright 12 and $25\mu m$ emission around more opaque pieces with no 12 nor $25\mu m$ emission (e.g. G300-15 in Chamaeleon, Heiles 2, B5), (3) clouds with randomly distributed IR colors (e.g. Ursa Major, L1497 and 1541). In this last type of clouds the spread of IR colors is generally smaller than the full range of colors observed between clouds of types (1) and (2). In Chamaeleon, matter with R(12,100) larger than the average Solar Neighborhood value accounts for 1/3 of the total $100\mu m$ emission of the complex. This emission fraction translates into a comparable (little smaller) fraction of the total mass of the complex. In the two other complexes the different types of clouds are not separated enough to make a similar estimate.

Clouds of type (1) have no noticeable envelope with warmer IR colors. Preliminary comparisons of CO and $100\mu m$ emission in a few of these clouds show that CO is detected all the way to the edge of the $100\mu m$ emission (see figure 2).

The peak column density through the envelopes with high R(12,100) characteristic of clouds of type (2), as measured from $I_\nu(100\mu m)$ or directly from star counts, corresponds to $A_v < 1 mag$. Images and cuts (BFPH, Boulanger 1989, and Puget 1989) illustrate the variety of morphologies of regions of high R(12,100). Their linear projected size varies from 0.3 pc to 10 pc. Based on the $100\mu m$ brightness and the cloud geometry the density of the gas emitting strongly at $12\mu m$ is estimated to be one to a few $100\, Hcm^{-3}$. CO observations for several clouds in different complexes indicate that this gas is not seen in emission in CO (see example in figure 2). For a few stars in Chamaeleon, absorption spectra were taken through this gas leading to the detection of CH (Boulanger et al. in preparation) which indicates the presence of H_2. Within many clouds of type (2) the color transition is unresolved by IRAS so that the size of the transition region is $< 0.2pc$. One example of such a

154

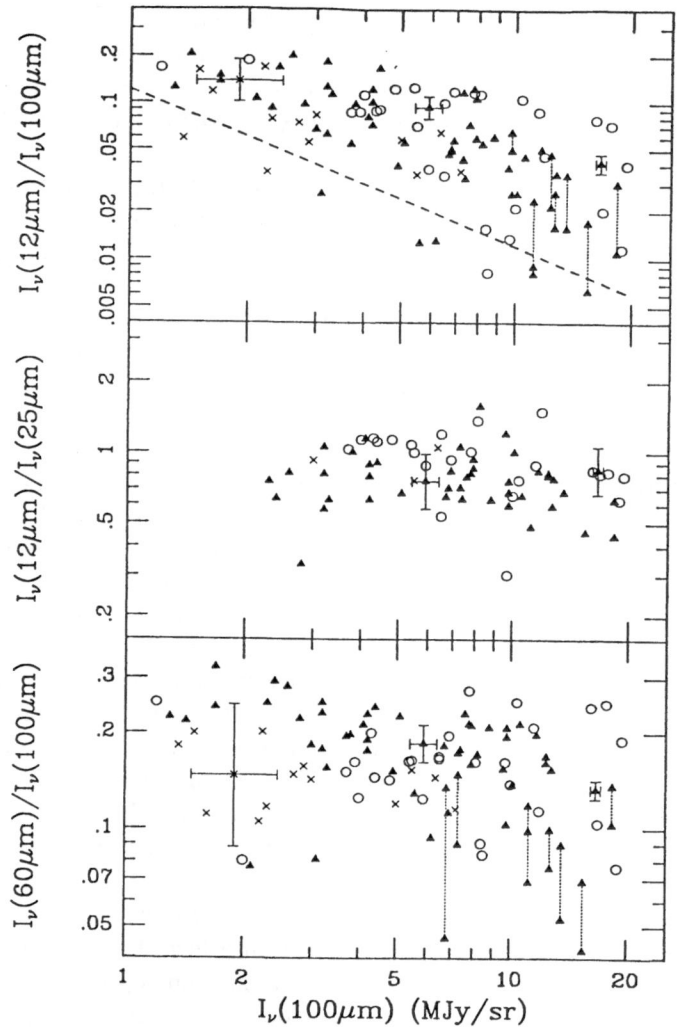

Fig. 1: (a) Plot of the color ratio $R(12,100) = I_\nu(12\mu m)/I_\nu(100\mu m)$ versus $I_\nu(100\mu m)$. The symbols are circles for Chamaeleon, solid triangles for Taurus, and crosses for Ursa Major. The dashed line shows the sensitivity limit for measurements of the $12\mu m$ brightness. All data points near this line are upper limits on $R(12,100)$. For a few positions we measured the colors for two baselines: a global one defined for the whole cut and one defined locally from local minima of emission near the position of the measurements. The corresponding pairs of data points are connected by a dotted line (shown also on the $R(60,100)$ diagram). The smaller value $R(12,100)$ and $R(60,100)$ always correspond to the local baseline. Error bars are only given for three data points selected as representative of points with low, average, and high $I_\nu(100\mu m)$. (b)Plot of $R(12,25)$ versus $I_\nu(100\mu m)$. Only data points with $I_\nu(25\mu m) > 0.25 MJy/sr$ are included.

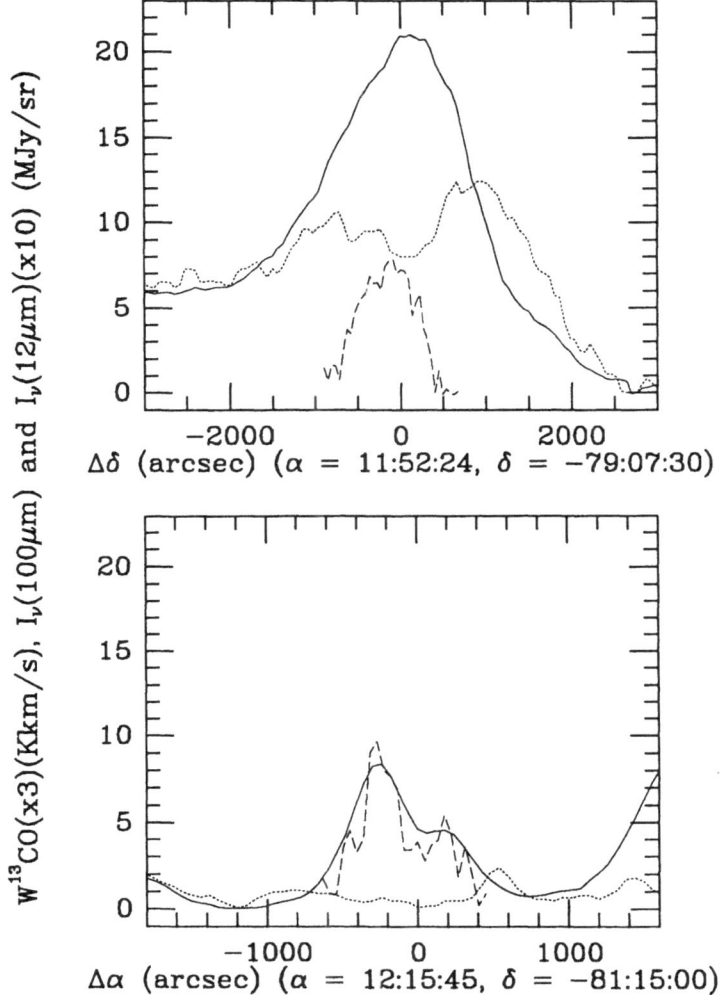

Fig. 2: $I_\nu(12\mu m)$ (dotted line), $I_\nu(100\mu m)$ (solid line), and W(^{13}CO) (J=1-0) (dashed line) cuts across two clouds in Chamæleon. Relative scaling factors between the three brightnesses are the same in the two figures. The molecular data was obtained with the SEST (Boulanger et al. in preparation). The first cloud (left figure) shows strong $12\mu m$ emission from the outer parts of the cloud with a minimum at the peak of the 100μ emission. The molecular and IR data are consistent with all of the ^{13}CO emission coming from a central core of the cloud with no $12\mu m$ emission and a $100\mu m/^{13}$CO emission ratio similar to that observed for the second cloud in the right figure. This cloud shows no $12\mu m$ emission and ^{13}CO emission all the way to the edge of the $100\mu m$ emission.

sharp transition is shown in figure 1 (cut 2) of Boulanger (1989). Other examples are L134 and L1780 studied by Laureijs, Clark and Prusti (1990). This sharpness may be a general feature only perceivable in clouds with a favorable geometry and orientation with respect to the observer. Existing molecular data and recent observations made with the SEST show that the color variations correlate with strong changes in gas density. Based on ^{13}CO J=2-1 and 1-0, and CS J=2-1 observations Boulanger et al. (in preparation) estimate the density of the gas not seen at 12 and $25\mu m$ to be $\sim 5 10^3 cm^{-3}$, a factor 50 larger than that of the gas with high R(12,100).

3. Excitation by UV Light

On average over a whole molecular complex, the emission from small particles at $\lambda < 25\mu m$ is $\sim 10^{-31} W$ per hydrogen atoms. This represents roughly 25% of the total power radiated by the clouds in the infrared and sub-millimeter. It is difficult to imagine a source of energy other than star light which could contribute for any significant fraction of this emission. For example the energy per nucleon associated with turbulent motions is for a cloud of size D, using the line width - size relation $\Delta V(FWHM) = (D/1pc)^{0.5} km/s$, $E_c = \frac{3}{2}\sigma_v^2 = 3(D/10pc)10^{-21} J/H$. If shocks associated with turbulent motions in clouds were at the origin of the emission from small particles the turbulent energy will be dissipated in $\sim 10^3 yrs$. No known mechanism is able to re-supply the turbulent energy of clouds at this rate. The emission from small particles is also higher than the heating rate of gas by photo-electrons by two orders of magnitude. Hence collisional excitation cannot be significant unless some unknown heating mechanism of the gas is active. Stellar light appears thus as the only possible source of excitation.

The small particles which have been proposed to account for the near and mid-IR emission from the interstellar medium absorb mostly in the ultraviolet (UV) (see Puget and Léger 1989). The variations of the mid-IR to far-IR emission ratio could thus result from changes in the rate of excitation of the small particles related to anisotropy and intensity variations in the UV radiation field. Beichman et al. (1988), Chlewicki and Laureijs (1988), and Laureijs et al. (1989) suggested that UV attenuation could account for the limb brightening of the 12 μm emission seen across some clouds. Models to test quantitatively this idea have been presented by these authors and DBP. Even in the most favorable case where large grains absorb a large fraction of their energy in the red due to the formation of mantles the decrease of R(12,100) with increasing $I_\nu(100\mu m)$ is only one tenth of the systematic decrease seen in figure 1 (BFPH). The model predicts small changes in R(12,100) because the attenuation of the UV radiation field reduces the 12 and $100\mu m$ emission at roughly the same rate. Since large grains absorb over the whole ISRF spectrum, attenuation of the UV part of the radiation field reduces their bolometric emission and temperature. The temperature decrease makes the reduction in the $100\mu m$ emission larger than that of the dust heating. This reduction of the $100\mu m$ emission predicted by the model and which is important to dismiss excitation changes as the cause of color variations is verified by $I_\nu(100\mu m) - A_v$ comparisons. It corresponds to the reduction by a factor 2 to 3 observed in the $I_\nu(100\mu m)/A_v$ ratio between atomic gas and molecular clouds (Boulanger 1989). Thus, models which fail to explain the limb brightening of the 12μ emission, do reproduce the observed decrease in the slope of the $100\mu m - A_v$ relation from atomic gas to molecular clouds (Chlewicki and Laureijs 1988, Bernard 1990).

Changes in the rate of UV excitation of the small particles are also inadequate in explaining the scatter of R(12,100) observed in figure 1 at any given $I_\nu(100\mu m)$. Again, to increase or decrease the R(12,100) color of a cloud of a given opacity by some factor, the UV radiation field at its surface must be increased or decreased by a much larger factor because large grains emitting at 100 μm also absorb in the UV. For example, in the calculations of Laureijs (1989) and DBP it is necessary to scale the UV part of the ISRF by a factor of ~ 5 to increase R(12,100) by 50%. Such a change in the radiation field implies an increase of the $100\mu m$ emissivity per dust grain measured by the ratio $I_\nu(100\mu m)/A_v$ of 3. The interpretation of the scatter in R(12,100) at a given $I_\nu(100\mu m)$ in Fig. 1 along this line would imply that the radiation field at the surface of the clouds varies by more than one order of magnitude from cloud to cloud. Such large inhomogeneities in the UV radiation field are completely excluded for the Chamaeleon, Taurus, and Ursa Major complexes which are located far-away from any OB association. While shadow effects between clouds can create regions of attenuated radiation field, UV radiation fields an order of magnitude more intense than the average ISRF, necessary to explain the high values of R(12,100), cannot be obtained without nearby O or early B stars. Further, such an interpretation would imply variations of $I_\nu(100\mu m)/A_v$ by more than an order of magnitude, correlated with those of R(12,100), which are excluded by comparisons between A_v and $I_\nu(100\mu m)$ (BFPH).

As for the R(12,100), variations in the spectrum and strength of the radiation field are inadequate in explaining the variations in the R(60,100) colors. For grains in thermal equilibrium the total amplitude of the R(60,100) variations corresponds to a change in the intensity of the radiation field by a factor of 20 (DBP), larger than what is believable for complexes located far away from OB associations and for clouds with moderate opacity to star light. The scatter in R(60,100) could be due to the contribution of small particles to the $60\mu m$ emission. Laureijs, Clark and Prusti (1990) have suggested that the formation of mantles on small particles could make them big enough to reduce their temperature fluctuation. For large grains the formation of mantles implies a change in optical properties and consequently in equilibrium temperature which could play a role in the R(60,100) color variations.

4. Abundance Variations

Since changes in the UV field are ineffective in affecting IR colors the R(12,100) and R(25,100) color variations must reflect changes in the abundance of small particles. The importance of this conclusion comes from the large amount of material necessary to account for the mid-IR emission measured by IRAS. On average in the atomic interstellar medium ~15% of the cosmic carbon is in small particles, PAH's or others, with less than a few hundred heavy atoms (Puget and Léger 1989). Based on this estimate, observed color variations imply that this fraction varies from place to place within molecular clouds from less than ~5 % to more than half of the cosmic abundance.

It is not clear how one can explain such strong abundance variations. BFPH have proposed that they are related to the cycling of interstellar matter between the gas phase and grain surfaces. In their scenario small particles are a product of chemistry on grain surfaces. Regions of low R(12,100) correspond to pieces of clouds where small particles are condensed onto grains, while high R(12,100) are found where small particles condensed on grains and those formed out of photo-processed molecular ices are released to the gas phase through photo-desorption by UV photons (Draine and Salpeter 1979) or as the result of localized heating (Duley 1989). From the strength of the $3.4\mu m$ feature in the direction of the Galactic center Tielens and Allamandola (1987) estimate that 25 to 50% of the carbon along this line of sight is in organic mantles. The quantity of carbon observed to be in mantles is thus comparable to the amplitude of the cycling necessary to account for the color variations.

The existence of abundance variations on small scales imply that the evolution time-scale regarding the abundance of small particles must be comparable to the mixing time-scale of matter by turbulent motions within clouds. This correspondence is readily understandable in clouds where the density structure is controlled by transient compressions associated with turbulent motions. Recent observations at high angular resolution support such a picture for clouds with moderate gas column density ($A_v <$ a few mag) (see Falgarone 1991 for a review). These observations have shown the existence, throughout clouds, of density peaks with $n_H \sim 510^3 cm^{-3}$, not massive enough to be bound by self-gravity. This density is also the one derived from molecular observations of moderate column density gas with no $12\mu m$ emission. For such a density, assuming a sticking probability of one, the condensation time scale of light molecules like H_2O is 3×10^5yr. For PAHs the condensation time scale depends on their relative velocity with respect to dust grains. Assuming a relative velocity of a few $10^3 cms^{-1}$ (see Omont 1986) the time scale for condensation of PAHs is about an order of magnitude higher than that of small molecules. For a sticking probability of one, it is comparable to the mutual coagulation time scale among the smallest particles. The question which remains to be answered is to know whether the time spent by the gas in dense regions is long enough to permit significant condensation of gas species on grains. The answer may differ from cloud to cloud which would explain observed differences in colors between clouds.

If the IR color variations are related to the formation and destruction of mantles one expects that the abundance of small particles will depend on the frequency at which matter cycles between dense and diffuse regions compared with the timescales for condensation and detachment. In a cloud where matter cycles from diffuse to dense regions quickly relative to the detachment time-scale but slowly compared to the condensation time-scale large mantles will build up on grains and the abundance of small particles in the gas phase will drop. This would correspond to clouds of type (1) described in the morphology section. In the case where cycling time-scale is large compared to both condensation and detachment time-scales the abundance of small particles will be low in dense regions and normal or high in diffuse parts leading to the limb-brightening of colors observed in clouds of type (2). Finally a random distribution of colors as in clouds of type (3) should indicate that all time-scales are comparable. This interpretation of the data predicts that the color morphology of clouds is related to the internal velocity and density structure which may be tested by molecular line observations. Preliminary comparisons between IR and molecular observations do suggest such a relation (see Observational Properties). Other observations which would confirm the existence of abundance variations and test the connection between color variations and the cycling and processing of interstellar matter on grains have been suggested by BFPH. If these observations demonstrate this hypothesis it will imply that cycling of matter between gas phase and dust grains is more ubiquitous and rapid that previously thought.

Whichever it is, the physical process at the origin of the color variations needs to involve a large amount of material and to take effect on short time-scales to account for the observed properties. It is thus necessarily a key mechanism in the physical and chemical evolution of molecular clouds. IRAS images have revealed us a conspicuous sign of the evolution of interstellar matter in clouds which is very worth investigating further.

160

References

Beichman, C., Wilson, R.W., Langer, W., and Goldsmith, P. 1988, Ap. J., 332, L81.

Bernard, J.P. 1990, Private Communication.

Boulanger, F., Baud, B., and van Albada, G.D. 1985, Astr. Ap. 144, L9

Boulanger, F., and Pérault, M., 1988, Ap. J. 330, 964.

Boulanger, F., Falgarone, E., Helou, G., and Puget, J.L., 1989, in Interstellar Dust Contributed Papers, eds. A.G.G.M. Tielens and L. J. Allamandola, NASA CP-3036.

Boulanger, F. 1989, The Physics and Chemistry of Interstellar Molecular Clouds, eds. G. Winnewisser and J. T. Armstrong, Springer Verlag, p. 30

Boulanger, F., Falgarone, E., Puget, J.L., and Helou, G. 1990, Ap. J. 364, 136

Chlewicki, G., and Laureijs, R.J., 1988, Astr. Ap. 207, L11.

Désert, F.X., Boulanger, F., and Puget, J.L. 1990, Astr. Ap. 237, 215.

de Vries, C.P., and Le Poole, R.S. 1985, Astr. Ap., 145, L7.

Draine, B.T., and Salpeter E.E. 1979, Ap. J. 231, 438.

Draine, B.T., and Anderson, N. 1985, Ap. J., 292, 494.

Duley, W.W. 1989, The Physics and Chemistry of Interstellar Molecular Clouds, eds. G. Winnewisser and J.T. Armstrong, Springer Verlag, p. 353.

Falgarone, E. 1991, From Ground-based to Space-borne Submillimeter Astronomy, eds. N. Longdon and B. Kaldeich, ESA Publ.

Heiles, C., Reach, W.T., and Koo, B.C., 1988, Ap. J. 332, 313.

Laureijs, R.J., Mattila, K., and Schnur, G. 1987, Astr. Ap. 184, 269.

Laureijs, R.J., Chlewicki, G., Clark, F.O., 1988. Astr. Ap. 192, L13.

Laureijs, R.J., Chlewicki, G., Clark, F.O., and Wesselius, P.R., 1989, Astr. Ap. 220, 226.

Laureijs, R.J. 1989, Ph. D. University of Groningen.

Laureijs, R.J., Clark, F.O., and Prusti, T. 1990, Ap. J. in press.

Leene, A. 1985, Astr. Ap., 154, 295.

Omont, A. 1986, Astr. Ap., 169, 159.

Puget, J.L., Léger, A., and Boulanger, F. 1985, Astr. Ap., 142, L19.

Puget, J.L. 1989, Interstellar Dust, eds. L. J. Allamandola and A.G.G.M. Tielens, Kluwer, p. 119.

Puget, J.L., and Léger, A., 1989, Ann. Rev. Astr. Ap., 27, 161.

Tielens, A.G.G.M., and Allamandola, L.J. 1987, Interstellar Processes, eds. D. Hollenbach and R. Thronson, Reidel, p. 397.

Weiland, J.L., Blitz, L., Dwek, E., Hauser, M.G., Magnani, L., and Rickard, L.J. 1986, Ap. J., 306, L101.

PHYSICAL AND CHEMICAL PARAMETERS IN DENSE CORES

C.M.WALMSLEY

Max-Planck-Institut für Radioastronomie, Auf dem Hügel 69, D-5300 Bonn 1, FRG

Abstract. The current state of our knowledge of the physical parameters and the chemical composition of dense cores in molecular clouds is discussed. In particular, I summarize what is known about the rate at which molecules condense out on grain surfaces. I discuss in turn : a) dense cores in nearby dust complexes such as Taurus, b) clumps in regions of massive star formation such as Orion and M17, c) hot dense cores near to newly formed O stars such as the Orion-KL hot core and d) the high density condensations which give rise to interstellar masers. Recent work on each of these categories is reviewed with emphasis on the chemical abundance determinations and estimates of the local density and temperature. Particular attention is given to recent work on OH, methanol, and ammonia masers.

Keywords : Molecular Clouds, Star Formation, Interstellar Chemistry, Masers, Clumps

1. Introduction

The physics of dense cores in nearby molecular clouds has been the subject of many reviews and it is doubtful whether it is useful to add to their number (but see e.g. Fuller and Myers 1987, Walmsley 1987, Shu et al. 1987, Wilson and Walmsley 1989). There is much less clarity about the chemical conditions and I therefore focus on this. Potentially, the observed molecular abundance distribution in molecular clouds can tell us a lot about local conditions and I discuss to what extent one can realise that potential. One of the questions which comes up in this regard is what fraction of the heavy elements are depleted out onto dust grain surfaces. I review briefly in section 2 what one expects theoretically and what is known from infrared studies. Section 3 deals with cold cloud cores and section 4 with the denser hotter clumps found in the neighbourhood of HII regions. In section 5, I give a very short discussion of hot core type regions and in section 6 I mention some new results for OH, methanol, and ammonia masers. Finally, in section 7, I give a very brief overview showing the range in temperature–density space covered by current observations.

2. Depletion onto dust grain surfaces

It has been realised for a long time that the timescale for a heavy atom or molecule hitting a dust grain and being removed from the gas phase is less than or of the same order as the timescale both for ion–molecule chemistry and for dynamical evolution of typical high density molecular cloud cores. Recent discussions of this problem are given by Walmsley (1985,1989) and by Williams(1990). One way of looking at the problem is to compare the free–fall timescale, which is an estimate of the time needed for dynamical evolution, with the time required for a large fraction of the molecules to deplete out on dust grains. The latter can be written as :

$$t_d(yr) = 1.5\,10^5 \,(10^4/n)\,(1/S)\,(2.1\,10^{-21}/\overline{\sigma_g})$$

1

where n is the hydrogen density (cm^{-3}) , S is the sticking coefficient (i. e. the fraction of molecule collisions with grains which lead to adhesion of the molecule to the surface) and $\overline{\sigma_g}$ is the mean grain cross–section per hydrogen atom for which one normally assumes the value ($2.1\,10^{-21}cm^2$) which is derived from interstellar extinction measurements. Figure 1 shows the values of t_d derived from this formula for sticking coefficients of 0.1 and 1 plotted as a function of the hydrogen density. The free–fall time ($(3\pi/(32G\rho))^{0.5}$) is shown in figure 1 for comparison. For virialized clumps, this should in a statistical sense be a lower limit on the clump age. One sees that even for a sticking coefficient of 0.1, one expects a large fraction of the molecules to deplete out in one free–fall time at a density of 10^5 cm^{-3}. At higher densities, the "freeze–out" should occur in a fraction of a free–fall time. There have been a considerable number of theoretical investigations of sticking coefficients (e.g. Leitch–Devlin and Williams (1984)) which suggest that for realistic grain materials, a sticking coefficient of below 0.1 is unlikely. Hence, depletion of molecules onto dust grain surfaces seems probable at densities higher than 10^5 cm^{-3}. This poses at first sight a problem since most of the high density clumps in or near which star formation goes on appear to have densities of the order of 10^5 or greater.

Part of the answer to this question is that infrared observations tell us unequivocally that depletion does occur in high density clumps. One observes directly in absorption towards embedded infrared sources bands of solid CO, water ice, and other species (see Tielens 1989,1990). However, the data presently available do not suggest that a large fraction (more than 50 percent) of the available heavy elements have been lost from the gas phase. In the first place there are lower limits of $5\,10^{-5}$ on the $[CO]/[H_2]$ abundance ratio towards deeply embedded IR sources in the NGC 2024 and NGC 2264 cloud cores (Black and Willner 1984, Black et al. 1990). These are high density lines of sight with CO column densities of order 10^{19} cm^{-2} and hydrogen densities at least 10^5 cm^{-3}. The data are much less equivocal than most radio observations in that one observes (in the 2-0 vibration–rotation band of CO) all rotation states of importance and in that one directly measures (or in this case puts limits upon) molecular hydrogen. These limits on $[CO]/[H_2]$ are very close to the values estimated on the basis of radio studies of nearby dust clouds (e.g. Dickman(1978)).

However, in regions such as the NGC 2024 molecular ridge which abuts upon an HII region, the dust temperatures may be sufficiently high that the more volatile components of the mantle such as CO can evaporate. Typical condensation temperatures for a sample of interstellar molecules are given by Nakagawa (1980) and a discussion of mantle evaporation is given by Léger et al. (1985). Above a dust temperature of 20 K, pure CO ice is likely to be evaporated but the reality is likely to be that the CO is in a matrix (see e.g. Tielens 1990) and in this case the fate of the mantle will probably be determined by less volatile species such as H_2O

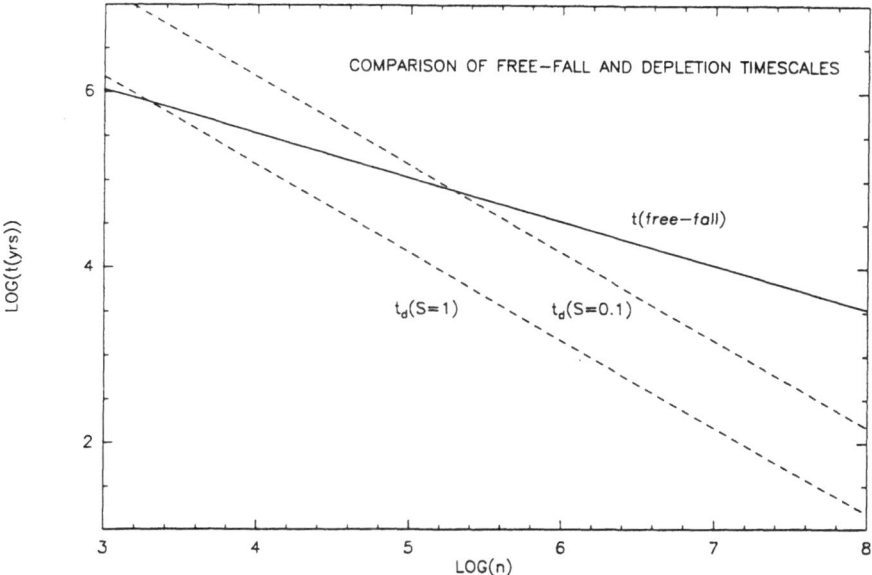

Fig. 1. The timescale for molecules condensing out on grain surfaces is plotted (dashed lines) as a function of the hydrogen density for two values of the sticking coefficient S. Equation 1 has been used with the normal diffuse gas value for grain surface area per hydrogen atom. The free fall time (full line) is shown for comparison

or CO_2 for which the corresponding condensation temperatures are 100 and 50 K respectively.

Direct evidence on the depletion question comes from the observations of the solid–state features of CO and H_2O towards infrared sources. In the sources examined by Lacy et al. (1984), the ratio of solid to gas phase CO varied between 0.02 and 0.3. Whittet et al. (1989) have observed 8 stars in the direction of the Taurus cloud and conclude that 30 percent of the available CO is in the solid form. On this basis, one might think that depletion is important but not overwhelming. However, it is conceivable that CO is processed upon grain surfaces and indeed d'Hendecourt and de Muizon have recently found solid CO_2 to have an abundance similiar to that of solid CO towards AFGL 961 and AFGL 989. This is in contrast to the gas phase where CO_2 is thought to be at least two orders of magnitude less abundant than CO (see Minh et al. 1988). It seems likely that the CO_2 has been made on the grain surface. Moreover, water ice is typically an order of magnitude more abundant on grain surfaces than CO (see Tielens 1989 for a summary of grain mantle abundances) and H_2O is probably formed on the grain surface. In fact, it seems reasonable to guess that water ice is the main repository of oxygen on the grain surface as is the case in comets. If this is so however, it implies that in the lines of sight examined to date, the column density of oxygen in the solid form is comparable to but not orders of magnitude greater than the column density of O in the gas phase (in all forms including CO).

The radio data relevant to the depletion problem are rather less clearcut than the infrared observations discussed above. I will mention some relevant measurements in the following sections and Mezger (this conference) will discuss the comparison of millimeter dust emission and molecular line maps. It is important to realise however that *it is not a valid assumption to suppose that molecular abundances are simply proportional to the depletion fraction*. With the exception of CO which is thought to be a major repository of gas phase carbon, the observed molecules (SiO is a special case also) contain trace amounts of the total gas phase heavy element content. As depletion progresses, their abundance may decrease more or less rapidly than the general loss of molecules. Ions such as HCO^+ , which are indirectly produced by cosmic ray ionization, may at first "not notice" the decrease in the CO abundance. Both the hydrocarbon radicals and deuterium fractionation can increase as a result of depletion (see the model calculations of Brown et al. 1988 ,Brown and Millar 1989, Brown and Charnley 1990). It is thus not possible to deduce that because, say, NH_3 decreases in abundance by a factor of a few close to a dust emission peak, that the molecules are being frozen out. The gas phase chemistry is just too complicated and too poorly understood for that and, incidentally, one knows of variations of the ammonia abundance in lower density regions where freeze–out is probably not the explanation (Olano et al. 1988). CO is different in this respect and its abundance should simply reflect the carbon depletion. In this sense, direct measurements of $[CO]/[H_2]$ such as those of Black et al. (1990) mentioned earlier are very important.

Having said all this, one sees from Figure 1 that depletion should occur on timescales short compared to free–fall for densities of 10^6 cm^{-3} or more. Moreover, the free–fall time is a minimum estimate for the lifetime of high density clumps and many scenarios have been envisioned where clumps are stabilised against collapse and exist for many free–fall times. Unfortunately, the abundance estimates discussed above refer mainly to regions between 10^4 and 10^6 cm^{-3}. In those regions which have been studied with densities above this limit, the dust temperature is typically of the order of 100 K or more and grain mantles can evaporate. These hot dust or "hot core" regions are very important in that they give us insight into the mantle composition. Nevertheless, the question remains of what happens in high density regions which are sufficiently far from a luminous infrared source that the dust is cold. One possibility of course is that the molecules do indeed *all* condense out. If this happens, it is interesting to note (Hartquist and Williams 1990) that a small fraction ($[CO]/[H_2]$ of order 10^{-7}) of molecules will remain in the gas phase due to photodesorbtion by Lyman photons secondary to cosmic ray ionization. This small quantity of CO may nevertheless be an important coolant for the gas and it is perhaps observable although confusion with surrounding CO makes this in practise difficult.

It is useful therefore to summarize (see Williams(1990) for a more extensive discussion) the processes which might prevent rapid depletion in high density regions. One such process is grain coagulation leading to a reduction in the total grain surface area or, in terms of equation 1, to a reduction in $\overline{\sigma_g}$. Mathis (1990) has summarized evidence for a change in the grain size distribution between "diffuse" and "outer cloud" dust. Cardelli et al. (1989) find for example a reduction by a factor of 2 or more in the integrated UV extinction per hydrogen atom towards two nearby stars and attribute this to grain coagulation. These data refer to material at much lower densities than in the cores where star formation occurs and a large decrease in $\overline{\sigma_g}$ in high density regions is a distinct possibility. It is worth noting also that the reduction in UV extinction per hydrogen atom leads to an increase in the UV penetration of molecular clouds. This should cause greater photodesorption and photo–processing in the exterior regions of clouds and raises the question of whether exchange of material between cloud interior and exterior can lead to depletion being kept at moderate levels even in high density clumps (see Boland and de Jong 1982, Chièze and Pineau des Forêts 1989). This type of model can probably explain observed depletion levels at densities of 10^4 cm^{-3} and below but not at higher values. Another mechanism which seems to fail at high densities is desorption due to the passage of cosmic rays. Léger et al. (1985) have studied this and conclude that spot heating by cosmic rays can lead to thermal desorption at densities below 10^4 cm^{-3}. Perhaps the most promising possibility is that shocks in molecular clouds can be sufficiently violent to desorb the more volatile components in grain mantles. Williams and Hartquist (1984) proposed a mechanism of this type to explain the apparent presence of atomic carbon deep in molecular clouds where theory predicts that most gas phase carbon should be in the form of CO. Charnley et al. (1988,1990) have extended this model and studied a situation where matter is continually cycled between dense clumps and interclump material. In this picture, clumps are ablated by the stellar winds due to newly formed pre–main sequence stars. Shocks in the interclump material cause mantle–desorption and one finds that for clump lifetimes of $1 - 5\,10^6$ years that a reasonable fit can be found to observed abundances. Given the large number of outflow sources which have been recently found to be embedded in dense regions, this mechanism seems plausible. A consequence would seem to be that there should be abundance differences between clumps with and clumps without embedded infrared sources.

3. Dark Cloud Cores and their Chemistry

Rightly or wrongly, it is generally accepted that dense clumps embedded in molecular clouds are the starting points for star formation. In nearby clouds such as Taurus, there is evidence for this in the form of the spatial correlations which have been found between infrared (IRAS) sources embedded in molecular clouds and high density molecular condensations observed in tracers such as NH_3 and CS. (e.g. Myers 1987, Fuller and Myers 1987). The masses of such clumps are typically only a few solar masses and it is natural to guess that they are in some sense "proto–protostars".

In order to verify such statements, one needs to estimate the physical parameters in the high density clumps. The ammonia data itself shows that 10 K is a reasonable estimate for the kinetic temperature in the nearby clouds seen in Taurus and similiar complexes. It is interesting however that in the clumps so far examined in the nearest GMC, Orion, temperatures are higher than in Taurus and linewidths also (Wouterloot et al. 1988). A very recent study (Harju et al. 1990 and this conference) suggests that the diameters of the Orion clumps are a factor of two larger than in Taurus. The local density is rather more debatable and studies to date have used tracers such as CS , HC_3N , NH_3 and C_3H_2. (see Cox et al. 1989, Walmsley and Wilson (1985), E.Lada (this conference), Zhou et al. 1989 and references therein). There is some disagreement between these different probes and also evidence that the spatial distribution of different species is not identical. For instance, clumps mapped in CS have a larger angular extent than in NH_3 (Zhou et al. 1989) suggesting that NH_3 is relatively less abundant in the lower density extended component. This is consistent with the fact that $NH_3(1,1)$ shows no signs of "self–absorbed" profiles. Other differences in the spatial distributions have been discussed by Swade (1989) and by Olano et al. (1988).

One potentially useful tracer of cloud history is the abundance of carbon rich molecules such as HC_3N in particular and the cyanopolyynes in general. There appears to be no obvious *physical* difference between the clumps where the cyanopolyynes are abundant and those where they are absent. It is possible that the important parameter is age and that the observed abundance distribution contains coded information about the dynamical history of the cloud. For example, given that current chemical models (e.g. Herbst and Leung 1989) predict that carbon–rich species should be abundant at "early times", it seems reasonable to conclude that the clumps where HC_3N and other cyanopolyynes are abundant are younger than the majority of "ammonia" clumps. Thus TMC1 should be younger than L183 and the majority of NH_3 cores (although why then the variations within TMC1 ?). If this is the case, one can use the relative abundance of the different members of the cyanopolyyne family as a "clock" (for a variant on this idea see Stahler 1984). Of course, older clumps should also show higher degrees of depletion onto dust grain surfaces and hence one might expect that high cyanopolyyne concentrations should correlate with low depletion. It is worth noting here that depletion of CO, H_2O and other species is expected to lead to increased deuterium fractionation (see e.g. Dalgarno and Lepp 1984). Hence, a comparison of cyanopolyyne abundances and deuterated ratios in several dense core sources might be useful. One can already compare the DCO^+/HCO^+ ratios found by Guélin et al.(1982) for L183 (relatively low cyanopolyyne abundance) with those for TMC1 (high cyanopolyyne abundance). These authors find an increase by a factor of 2 in the integrated intensity ratio of DCO^+ relative to $H^{13}CO^+$ between the TMC1 cyanopolyyne peak and L183. This is qualitatively what one expects but the effect is sufficiently small that it may not be significant. Unfortunately, there are considerable uncertainties in the detailed models for deuterium fractionation. (Brown and Millar 1989).

4. Cores in regions of massive star formation

In regions of massive star formation, densities and temperatures are typically higher than in the relatively low mass cores discussed in the previous section. Ammonia has been used as a "thermometer" by several groups (e.g. Batrla et al. 1983, Güsten and Fiebig 1988). The dense gas clumps bordering HII regions typically have temperatures in the range 20-50 K. Multi-transition observations of CS and H_2CO towards three HII regions have been carried out by Snell et al. (1984) and by Mundy et al. (1986,1987). These studies are relatively consistent with one another and suggest that the regions are clumped with local densities in the range $3\,10^5$ to 10^6 cm^{-3}. What is between the clumps remains something of a mystery. However, VLA maps of the clumps north of Orion-KL in the 2 cm K-doublet transition of formaldehyde (Wilson and Johnston 1989) seem to confirm the general picture by directly imaging the clumps. One of the observations, which any model of these regions must explain, is the presence of considerable amounts of warm CO (e.g. Schmid–Burgk et al. 1989, Stutzki (this proceedings)). Tauber and Goldsmith(1990) have developed a radiative transport model for such a clumpy cloud which gives a qualitative agreement with observation.

The "fragments" seen north of Orion-KL with the VLA (Wilson and Johnston 1989) have densities varying from $8\,10^5$ to $3\,10^7$ cm^{-3}, masses between 1 and 150 $M_\odot$ and temperatures of order 30 K. Thus the thermal pressure is of order 10^8 cm^{-3}K or roughly 10^4 times the general interstellar medium value. The pressures are not however greatly different from those pertaining in the HII region and this suggests that the ionized gas contributes directly or indirectly to the confinement of the clumps.

The molecular abundances in such high density clumps is a matter of interest for the same reasons as discussed above for the cold cores. At densities of 10^6 cm^{-3} , the freeze out timescale in such regions is of order 1000 years and thus considerably less than the dynamical timescale of order $r_c/\Delta v$ or roughly 10^4 years where r_c is the clump radius and Δv the line width. There have not been many abundance determinations in this type of high density core. However, in the clumps north of Orion-KL discussed above, both the ammonia abundance (see Batrla et al. 1983) and the methanol abundance (Menten et al. 1988a) are of order 10^{-8} relative to H_2 which is approximately the value observed in cold dust clouds (see Irvine et al. 1987). Moreover, in the high density region of M17, where the conditions seem to be quite similiar to those found in the Orion clumps, Stutzki and Güsten (1990) find that their CS data is comparable with an abundance of $2.5\,10^{-9}$. This is a "normal" dark cloud value and the M17 ammonia abundance (Güsten and Fiebig 1988) seems to be within a factor of 2-3 of that in Orion. Hence, while detailed abundance studies would certainly be useful, there is no evidence for generalised depletion of molecules onto grain surfaces in this type of high density region (Stutzki and Güsten find that the temperature is of order 50 K and the density 10^6 cm^{-3} in the M17 clumps).

5. Hot Core regions

Still hotter and denser are the so–called "hot core" regions which are found close to very luminous infrared sources. These clumps are observationally characterised by the fact that they show up on VLA ammonia maps (see e.g. Migenes et al. 1989, Genzel et al. 1982, Pauls et al. 1983). The temperature in such regions can be deduced both from the observed line brightness temperatures (the metastable ammonia lines are highly optically thick and thermalized) and from multi–line analyses. In the Orion-KL hot core, Hermsen et al. (1988) have derived a temperature of 160 K from their ammonia measurements. On the other hand, Loren and Mundy (1984) derive 275 K from their methyl cyanide measurements. It seems likely that this reflects different spatial distributions of the two species. The Migenes et al. maps show that there is considerable structure within the hot core region. Presumably, the clump is being battered by the wind blowing from the nearby luminous infrared object Irc2. In view of this structure, one can expect a considerable range in local densities. From the millimeter dust emission, the average Orion hot core density has been estimated to be $3\,10^7$ cm^{-3} and the mass of the region to be 45 M$_\odot$ (Mundy et al. 1988).

Close to newly formed massive stars, dust particles become sufficiently hot that icy dust mantles can evaporate. Because of the long timescales for gas phase chemistry, the newly ejected mantle material is not immediately destroyed and one can use observations of the abundance distribution in hot core regions to draw conclusions about the composition of the mantle. In this context, an interesting characteristic of the Orion hot core is the relatively high abundance of deuterated molecules in general and HDO in particular (Plambeck and Wright 1987, Jacq et al. 1990). This can be shown to be quite consistent with the idea that the hot core gas is to a large degree evaporated mantle material (Brown and Millar 1989). Also relevant is the general trend that saturated molecules are present in the hot core while radicals and ions are not (see Irvine et al. 1987 for a summary of abundance estimates). Some questions are however left open by all these studies. It is in particular frustrating that one still does not know in what form most of the hot core oxygen and nitrogen is present. The Jacq et al. study suggests that only a few percent of the oxygen is in the form of gas phase water (see also Knacke et al. 1988). One concludes then that either oxygen condenses out in some other form than water ice (e.g. O_2) or else complete heavy element depletion did *not* occur prior to the switch on of the luminous IR source IRc2 (which is presumed to heat the dust). In any case, one of the fundamental questions left in this field is "where is the O and N ?".

Another open question relates to methanol whose abundance also appears to jump in hot regions close to where massive stars are forming. CH$_3$OH is estimated to have an abundance of 10^{-6} in Orion-KL as compared to $3\,10^{-9}$ in the more extended ridge gas (Menten et al. 1988a). The latter value is similiar to that found in dark clouds (Friberg et al. 1988). It is tempting to relate the rise in the methanol abundance in high temperature regions to grain evaporation also. There is evidence (Tielens 1989,1990) that methanol is very abundant in mantle material but how this CH_3OH is produced is not clear. Also the CH_3OH spatial distribution in Orion-KL differs from the classical hot core distribution as seen say in NH_3 (Wilson et al. 1989). Partly as a consequence of this, Blake et al. (1987) have proposed that the methanol gets produced as a consequence of the mixing in of water–rich material from the Orion-IRc2 wind with surrounding ion–rich gas. Then, radiative association of H_2O and CH_3^+ can give rise to protonated methanol which in turn can recombine to CH_3OH. A recent study by Millar and Herbst (1990) suggests however that this scheme will not work in the original form.

6. Maser Regions

Interstellar masers allow us to study structures with sizes as low as a few astronomical units. They are important for a variety of reasons but in particular because they allow us to study detailed kinematics of star–forming regions. Good reviews of the subject are available from Reid and Moran (1981), Elitzur (1982), Genzel (1986) and Cohen (1989) among others. In this discussion, I confine myself to a few recent results concerning OH, methanol and ammonia masers.

6. 1. OH MASERS

The regions giving rise to OH masers may in fact be sub–structures of clumps which from the point of view of their physical parameters are similiar to the Orion hot core. OH masers tend to be found in the close vicinity of ultra–compact HII (UCHII) regions and it is natural to ask why this should be. Elitzur and de Jong (1978) for example put forward the idea that the OH abundance was enhanced due to photodissociation of water molecules which in turn were produced in the shock running ahead of the HII region ionization front. Andresen (1986) has suggested that the OH pump is basically chemical and also caused by water photo–dissociation which directly inverts the OH Λ-doublets. In principle, these two mechanisms could support one another although the chemical pump appears to be too inefficient (at most one maser photon per OH formation) to supply the observed flux. One of the difficulties in judging the Elitzur-de Jong proposal is that the OH abundance in "normal" gas clumps is poorly known and hence it is unclear whether an abundance enhancement is necessary. An interesting new suggestion (Hartquist and Sternberg 1990) is that the UV photons from the HII region can heat the dense gas close to the ionization front sufficiently (above 1000 K) that gas phase OH production mechanisms ($O+H_2$) can proceed efficiently.

The OH maser regions have also the characteristic that the masers in low lying states are often associated with absorption lines in highly excited states (see Baudry et al. 1981, Guilloteau et al. 1985, Wilson et al. 1990). The excitation requirements for the masers (typical densities 10^7 cm^{-3} and temperatures 100 K) are not greatly different from those for the absorption lines and indeed similiar to the hot core characteristics discussed in the previous section. Hence, the question is raised of whether the absorption lines and the masers form in essentially the same regions. In a recent study of OH maser emission towards W3(OH) for example, Cesaroni and Walmsley conclude that one can fit both ground and excited state data with a model which has temperature 150 K and density of order 10^7 cm^{-3}. In other words, the OH lines form in a hot core which is 3 kpc distant and which happens to contain an embedded compact HII region. The masing transitions can amplify the background continuum radiation and one observes emission or absorption lines depending upon the relative populations in the upper and lower states of the relevant transitions. The OH abundance in these models is $2\,10^{-7}$ which is comparable to that estimated in TMC1 (Irvine et al. 1987).

6. 2. METHANOL MASERS

Methanol masers in a series of transitions at 25 GHz were originally reported in Orion A by Barrett et al. (1971). More than ten years later, it was found that maser action can be observed in several other methanol transitions as well (Wilson et al. 1984,1985, Morimoto et al. 1984). It was also found that the strongest methanol masers are those found in the 12.1 GHz 2_0-3_{-1} E-type lines towards many compact HII regions (Batrla et al. 1988). Interferometric images of the 12 GHz sources show that the line brightness temperature can reach 10^{10} K (Menten et al. 1988b) and that individual maser spots have sizes of a few astronomical units.

It has also become apparent that there are in fact two "families" of methanol masers. Type 1 (or A) masers are found in the general neighbourhood of HII regions and other tracers of high mass star formation but, rather surprisingly, they are not' coincident with known infrared objects, radio sources, or other types of masers. This poses a puzzle because, irrespective of the precise pump mechanism, one expects observable radiation to emerge at other wavelengths from these regions. There is some evidence that the type 1 methanol masers are found at or close to regions which emit strongly in the vibrationally excited lines of molecular hydrogen (Plambeck and Menten 1990, Johnston et al. 1990). If so, this would suggest that the methanol is produced in shocks caused when stellar winds emanating from young stellar objects impact upon surrounding gas clumps. The mechanism for population inversion appears to be collisional (Menten et al. 1990) and theoretical studies suggest that the type 1 masers are quenched at densities much higher than 10^7 cm^{-3} .

The 12.1 GHz masers as well as several other transitions with similiar characteristics (Type 2 or B methanol masers) are found , as are OH masers, close to ultra–compact HII regions from which they presumably derive their power. Collisions anti–invert the 12.1 GHz line and hence the pump mechanism is quite different from that operating in the class 1 methanol masers discussed above. It seems likely that a far infrared pump is operating but a detailed model is lacking. An interesting feature is that the available maps (e.g. Menten et al. 1988b) of the 12.1 GHz line show the masers to be lined up as if they delineated a front.

6. 3. AMMONIA MASERS

Several high gain masers have been found in the (9,6) transition of NH_3 which is 1090 K above ground (Madden et al. 1986). The excitation of this maser is a puzzle. Neighbouring ammonia transitions (e.g. (8,6)) appear to be "normal". While the (9,6) maser is found towards W51 and other "hot core–type" regions, it behaves "normally" in Orion-KL. Interferometric measurements would be useful for the understanding of this phenomenon. The (6,3) line is also an "occasional" maser (Madden et al.) and various other transitions may be inverted (Wilson and Henkel 1988) towards the northern compact source in W51. Hot core type conditions seem to be a necessary but not sufficient condition for maser action in these transitions.

Another region which may be a hot core "silhouetted" against a background continuum source is the region seen in ammonia towards the ultra–compact HII region NGC7538. The source is unusual partly because of the $^{15}NH_3(3,3)$ masers which are observed in a region approximately 10^{16} cm in size towards the ultra–compact HII region NGC7538-IRS1(Johnston et al. 1989). Also, Schilke et al. (1990) have observed what appears to be a line of vibrationally excited ammonia in absorption towards the same source. Finally, there is a formaldehyde maser coincident within the errors with the $^{15}NH_3(3,3)$ (Rots et al. 1981). The interpretation of this is unclear but at least one can say that the bulk of the ammonia is at a temperature of around 150 K and the column density of NH_3 is estimated to be $5\,10^{18}$ cm^{-2}. An educated guess at the hydrogen column density is 10^{24} cm^{-2} which would put the [NH$_3$]/[H$_2$] ratio at $3\,10^{-6}$. This certainly is very remniscent of the Orion hot core. More recent 100-m data (Schilke, this conference) suggest that the $^{15}NH_3(4,3)$ and (4,4) lines are also masers and can be interpreted in a model where the maser inversion occurs due to pumping via the excited vibrational state. This makes the coincidence of the vibrationally excited absorption line and the $^{15}NH_3$ masers comprehensible. However, as far as the $^{15}NH_3(3,3)$ maser is concerned, there is another interpretation (Flower et al. 1990) which requires an overabundance of para-H_2 in the region where the maser forms. This implies relatively low temperatures and densities (say 100 K and 10^5 cm^{-3}) and an overabundance of para-H_2 relative to that expected in thermal equilibrium.

7. Overview

The regions discussed in this review span several orders of magnitude in density and pressure. In order to visualize this, I show in figures 2a and b the parameters derived for thermal pressure, density, and temperature in a variety of molecular cores. Median values for the derived local density have been taken from the studies of Cox et al. (1989), Harju et al. (1990), and Mundy et al.(1986). The choice of objects plotted in this diagram is arbitrary and the aim is merely to show what range in parameter space is covered. One sees interestingly that there is no evidence for cold (10 K) high density (10^6) regions although ρ Oph B is only a factor of 2 hotter than this. It is possible that the paucity of cold dense regions is due to selection but it may also be the case that high pressures (and by inference high densities) are caused by the proximity of newly formed O-B stars and HII regions which heat and compress their surroundings. Then, high pressure regions will only occur where O-B stars are forming and in such regions the temperature will also be high due to interaction with hot dust. On the other hand, one certainly cannot exclude the possibility that cold high density regions are not observed simply because molecules have been frozen out in such regions. As one sees in figure 2, the best approximation to a cold high density region seems to be the ρ Oph B core (Wadiak et al. 1985) which is unusual in that it shows emission in the 2 cm line of formaldehyde. This region stands out on figure 2 and is certainly worth further study. On the basis of presently available data, it seems to be exceptional.

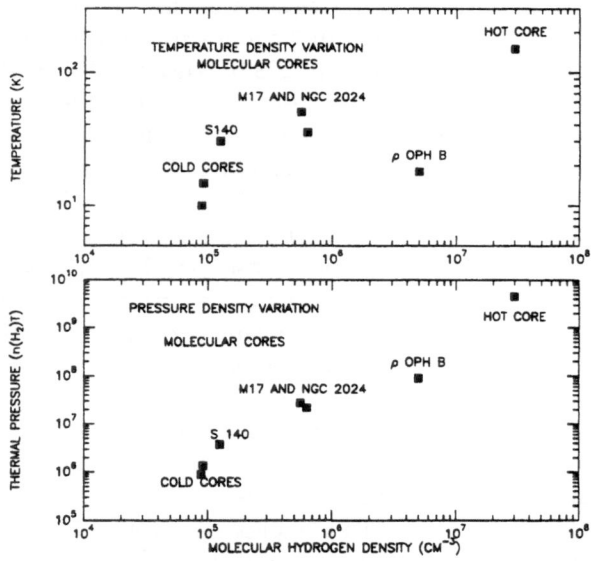

Fig. 2.   Thermal pressure and Temperature as a function of molecular hydrogen density for a sample of the molecular cloud cores discussed in the text. The cold core data is taken from Cox et al. (1989) and Harju et al. (1990, Orion cores). The data for M17, NGC 2024 and S140 is from Mundy et al. (1986) and for ρ Oph B from Wadiak et al. (1985).

On the basis of equation 1, one has a "freeze–out" time for ρ Oph B of 3000 years even assuming an "extreme" value for the sticking coefficient of 0.1. This compares with a free fall time of $2\,10^4$ years and hence one concludes that the grain surface area ($\overline{\sigma_g}$) must be an order of magnitude larger (at least) in this region than in the diffuse interstellar medium. Perhaps this is not as surprising as it might appear. It does seem worth carrying out in regions of this type extensive abundance analyses aimed at determining whether indications for partial depletion of the type discussed in section 2 (e.g. large D fractionation) can be observed. On the other hand, it might be useful to examine the consequences for the dust emission properties of a surface area reduction as discussed here.

Acknowledgements

Thanks are due to Rolf Güsten and Peter Schilke for their comments on the manuscript.

References

Andresen P. 1986 *Astron. Astrophys.* 154,42.
Barrett A.H., Schwarz P.R., Waters J.W. 1971 *Astrophys. J* 168,L101.
Batrla W., Wilson T.L., Bastien P., Ruf K. 1983 *Astron. Astrophys.* 128,279.
Batrla W., Matthews H.E., Menten K.M., Walmsley C.M. 1987 *Nature* 326,49.
Baudry A., Walmsley C.M., Winnberg A., Wilson T.L. 1981 *Astron. Astrophys.* 102,287.
Benson P.J., Myers P.C. 1989 *Astrophys. J. Suppl.* 71,89.
Black J.H., Willner S.P. 1984*Astrophys. J* 279,673.
Black J.H., van Dishoeck E.F., Willner S.P., Woods R.C. 1990 *Astrophys. J* 358,459.
Blake G.A., Sutton E.C., Masson C.R., Phillips T.G. 1987 *Astrophys. J* 315,621.
Boland W., de Jong T. 1982 *Astrophys. J* 261,110.
Brown P.D., Charnley S., Millar T.J. 1988 *Monthly Notices Roy. Astron. Soc.* 231,409.
Brown P.D., Millar T.J. 1989 *Monthly Notices Roy. Astron. Soc.* 237,661.
Brown P.D., Charnley S.B. 1990 *Monthly Notices Roy. Astron. Soc.* 244,432.
Cardelli J.A., Clayton G.C., Mathis J.S. 1989 *Astrophys. J* 345,245.
Cesaroni R., Walmsley C.M. 1990 *Astron. Astrophys.* (in press)
Charnley S.B., Dyson J.E., Hartquist T.W., Williams D.A. 1988 *Monthly Notices Roy. Astron. Soc.* 235,1257.
Charnley S.B., Dyson J.E., Hartquist T.W., Williams D.A. 1990 *Monthly Notices Roy. Astron. Soc.* 243,405.
Chièze J.P., Pineau des Forêts G. 1989 *Astron. Astrophys.* 221,189.
Cohen R.J. 1989 *Reports on Progress in Physics* 52,881.
Cox P., Walmsley C.M., Güsten R. 1989 *Astron. Astrophys.* 209,382.
Dalgarno A., Lepp S. 1984 *Astrophys. J* 287,L47.
Dickman R. 1978 *Astrophys. J. Suppl.* 37,407.
Elitzur M., de Jong T. 1978 *Astron. Astrophys.* 67,323.
Elitzur M. 1982 *Rev. Mod. Phys.* 54,1225.
Flower D.R., Offer A., Schilke P. 1990 *Monthly Notices Roy. Astron. Soc.* 244,4P.
Friberg P., Madden S.C., Hjalmarson A., Irvine W.M. 1990 *Astron. Astrophys.* 195,281.
Fuller G.A., Myers P.C. 1987 in *Physical Processes in Interstellar Clouds* (ed. Morfill G.E., Scholer M., publ.Reidel).
Genzel R., Downes D., Ho P.T.P., Bieging J.H. 1982 *Astrophys. J* 259,L103.
Genzel R. 1986 p233 in *Masers, Molecules, and Mass Outflows in Star Forming Regions*, (ed. A.D.Haschick, publ. Haystack Obs.).
Guélin M., Langer W.D., Wilson R.W. 1982 *Astron. Astrophys.* 107,107.

Guilloteau S., Baudry A., Walmsley C.M., Wilson T.L., Winnberg A. 1984 *Astron. Astrophys.* **131**,45.

Güsten R., Fiebig D. 1988 *Astron. Astrophys.* **204**,253.

Harju J., Walmsley C.M., Wouterloot J.G.A. 1990 *Astron. Astrophys.* (in press).

Hartquist T.W., Williams D.A. 1990 *Monthly Notices Roy. Astron. Soc.* (in press)

Hartquist T.W., Sternberg A. 1990 *Monthly Notices Roy. Astron. Soc.* (in press)

d'Hendecourt L.B., de Muizon M.J. 1989 *Astron. Astrophys.* **223**,L5.

Herbst E., Leung C.M. 1989 *Astrophys. J. Suppl.***69**,271.

Hermsen W., Wilson T.L., Walmsley C.M., Henkel C. 1988 *Astron. Astrophys.* **201**,285.

Irvine W.M., Goldsmith P.F., Hjalmarson A. 1987, p561 in *Interstellar Processes*, ed. Hollenbach D.J., Thronson H.A., publ. D.Reidel.

Jacq T. et al. 1990 *Astron. Astrophys.* **228**,447.

Johnston K.J., Stolovy S.R., Wilson T.L., Henkel C., Mauersberger R. 1989 *Astrophys. J* **343**,L41.

Johnston K.J., Gaume R., Stolovy S., Wilson T.L., Walmsley C.M., Menten K.M. 1990 *Astrophys. J* in press.

Knacke R.F., Larson H.P., Noll K.S. 1988 *Astrophys. J* **335**,L27.

Lacy J.H. et al. 1984 *Astrophys. J* **276**,533.

Léger A., Jura M., Omont A. 1985 *Astron. Astrophys.* **144**,147.

Leitch-Devlin M.A., Williams D.A. 1984 *Monthly Notices Roy. Astron. Soc.* **210**,577.

Loren R.B., Mundy L. 1984 *Astrophys. J* **286**,232.

Madden S.C., Irvine W.M., Matthews H.E., Brown R.D., Godfrey P.D. 1986 *Astrophys. J* , **300**,L79.

Mathis J.S. 1990 *Ann. Rev. Astron. Astrophys.* , (in press)

Menten K.M., Walmsley C.M., Henkel C., Wilson T.L. 1988a *Astron. Astrophys.* ,**198**,253.

Menten K.M., Reid M.J., Moran J.M., Wilson T.L., Johnston K.J., Batrla W. 1988b *Astrophys. J* , **333**,L83.

Migenes V., Johnston K.J., Pauls T.A., Wilson T.L. 1989 *Astrophys. J* , **347**,294.

Millar T.J, Herbst E. 1990 *Monthly Notices Roy. Astron. Soc.* , (in press)

Minh Y.C., Irvine W.M., Ziurys L.M. 1988 *Astrophys. J* , **334**,175.

Morimoto M., Ohishi M., Kanzawa T. 1984 *Astrophys. J* , **288**,L11.

Mundy L.G., Snell R.L., Evans N.J. II, Goldsmith P.F., Bally J. 1986 *Astrophys. J* , **306**,670.

Mundy L.G., Evans N.J. II, Snell R.L., Goldsmith P.F. 1987 *Astrophys. J* , **318**,392.

Mundy L.G. et al. 1988 *Astrophys. J* , **325**,382.

Myers P.C. 1987 in *Star Forming Regions*, IAU Symposium 115 (ed. Peimbert M., Jugaku J., publ. Reidel).

Nagakawa N. 1980 in *Interstellar Molecules*, IAU Symposium 87 (ed. Andrew B.H. , publ. Reidel).

Olano C.A., Walmsley C.M., Wilson T.L. 1988 *Astron. Astrophys.* , **196**,194.

Pauls T.A., Wilson T.L., Bieging J.H., Martin R.N. 1983 *Astron. Astrophys.* , **124**,23.

Plambeck R.L., Wright M.H.C. 1987 *Astrophys. J* , **317**,L101.

Plambeck R.L., Menten K.M. 1990 *Astrophys. J* ,(in press).

Reid M.J., Moran J.M. 1981 *Ann. Rev. Astron. Astrophys.* , **19**,231.

Rots A.H., Dickel H.R., Forster J.R., Goss W.M. 1981 *Astrophys. J* , **245**,L15.

Schilke P., Mauersberger R., Walmsley C.M., Wilson T.L. 1990 *Astron. Astrophys.* , **227**,220.

Schmid-Burgk J. et al. 1989 *Astron. Astrophys.* , **215**,150.

Shu F.H., Adams F.C., Lizano S. 1987 *Ann. Rev. Astron. Astrophys.* , **25**,23.

Snell R.L., Mundy L.G., Goldsmith P.F., Evans N.J. II, Erickson N.R. 1984 *Astrophys. J* , **276**,625.

Stahler S.W. 1984 *Astrophys. J* **281**,209.

Stutzki J, Güsten R. 1990 *Astrophys. J* **356**,513.

Swade D.A. 1989 *Astrophys. J* **345**,828.

Tauber J.A., Goldsmith P.F. 1990 *Astrophys. J* **356**,L63.

Tielens A.G.G.M 1989 in *Interstellar Dust*, IAU Symposium 135, ed. Allamandola L.J., Tielens A.G.G.M., publ. Kluwer.

Tielens A.G.G.M, 1990 in *The Chemistry and Spectroscopy of Interstellar Molecules* (ed. N.Kaifu), Proc. Of Symp. 228 of International Conference of Pacific Basin Societies.

Wadiak E.J., Wilson T.L., Rood R.T., Johnston K.J. 1985 *Astrophys. J* **295**,L43.

Walmsley C.M., Wilson T.L. 1985 in *Nearby Molecular Clouds*, (ed. Serra G., publ. Springer), Lecture Notes in Physics 237.

Walmsley C.M. 1985 in *Proceedings of ESO-IRAM-Onsala Workshop on "(Sub)-Millimeter Astronomy"* , edited by P.A.Shaver and K.Kjär, ESO Conference Proceedings 22,p 327.

Walmsley C.M. 1987 in *Physical Processes in Interstellar Clouds* (ed. Morfill G.E., Scholer M., publ. Reidel).

Walmsley C.M. 1989 p179 in *Evolution of interstellar dust and related topics*, edited Bonetti A., Greenberg J.M., Aiello S., Proceedings of the International School of Physics "Enrico Fermi", publ. North Holland.

Whittet D.C.B., Adamson A.J., Duley W.W., Geballe T.R., McFadzean A.D. 1989 *Monthly Notices Roy. Astron. Soc.* **241**,707.

Williams D.A., Hartquist T.W. 1984 *Monthly Notices Roy. Astron. Soc.* **243**,413.

Williams D.A 1990 In *Invited Review* at meeting on "Molecular Clouds" held in Manchester , March 1990.

Wilson T.L., Walmsley C.M., Snyder L.E., Jewell P.R. 1984 *Astron. Astrophys.* **134**,L7.

Wilson T.L., Walmsley C.M., Menten K.M., Hermsen W. 1985 *Astron. Astrophys.* **147**,L19.

Wilson T.L., Henkel C. 1988 *Astron. Astrophys.* **206**,L26.

Wilson T.L., Johnston K.J. 1989 *Astrophys. J* **340**,894.

Wilson T.L., Johnston K.J., Henkel C., Menten K.M. 1989 *Astron. Astrophys.* **214**,321.

Wilson T.L., Walmsley C.M. 1989 *The Astron. and Astrophys. Review* 1,141.

Wilson T.L., Walmsley C.M., Baudry A. 1990 *Astron. Astrophys.* **231**,159.

Wouterloot J.G.A., Walmsley C.M., Henkel C. 1988 *Astron. Astrophys.* **203**,367.

Zhou S., Wu Y, Evans N.J. II, Fuller G.A., Myers P.C. 1989 *Astrophys. J* **346**,168.

ABUNDANCE VARIATIONS OF TRACERS AND THEIR EFFECTS ON OUR DETERMINATION OF MOLECULAR CLOUD STRUCTURE

Paul F. Goldsmith
Five College Radio Astronomy Observatory
Department of Physics and Astronomy
University of Massachusetts, Amherst MA 01003

Our understanding of the molecular phase of the interstellar medium is critically dependent on use of various lines from different molecular species to trace this dense material. As our knowledge of molecular clouds becomes more refined, and we pursue in detail issues of molecular cloud structure, stability, and how star formation depends on and affects the molecular gas, it is appropriate to examine the basis by which we determine the morphology of clouds, their density, and other key parameters. This is obviously a major undertaking, well beyond the scope of the short presentation at this conference, so I will concentrate on one very basic, but critical issue, which is that of *abundance variations of tracers of density and molecular column density* which are widely used to delineate the denser portions of all types of molecular clouds. In this summary, I will first highlight some of the apparent indications of significant variations of abundance within individual clouds, as a way of indicating some potential dangers and the importance of the molecular tracer selected. I will also briefly suggest how such variations may be themselves important diagnostics of cloud structure and evolution.

The observation of abundance variations of different molecular species is certainly not new; while there is reasonable consistency for the fractional abundance of different species in various clouds, there are also clear differences. This issue is discussed in greater detail by Irvine, Goldsmith, and Hjalmarson (1987), Federman, Huntress, and Prasad 1990, and Irvine (1990). Comparison of the abundance of various molecular species requires properly analyzing molecular excitation and radiative transfer, a task which is made even more difficult when studying different clouds where the antenna beam is subtending different linear dimensions and thus likely averaging over regions with greatly varying conditions. Determining abundances within an individual cloud should be somewhat less susceptible to this type of error than is comparison of different clouds. Even when restricting observations to a single cloud, however, the variations of conditions from one line of sight to another make unambiguous determination of even such a basic quantity as column density of a particular species quite difficult.

We might expect the abundance of a particular molecular species within a given cloud or cloud complex to exhibit significant variations as a result of radiation field, density, temperature, and other even less well—understood conditions such as the passage of shock waves. This type of dependence is borne out in a number of cases where the structure of the cloud permits isolating variations in (hopefully) a single parameter. For example, many studies have been made of the correlation of

^{13}CO column density with total column density traced by visual extinction (e.g. Dickman 1978). For moderate extinctions it is clear that the fractional abundances of ^{13}CO and C^{18}O depend on the extinction characterizing the line of sight (cf. Goldsmith et al. 1980; Frerking, Langer, and Wilson 1982). Nevertheless, these same isotopes are widely used as tracers of total line of sight column density, with a *single* constant of proportionality employed for an entire cloud, or region being studied (e.g. Arquilla and Goldsmith 1983; Langer et al. 1987). While this is understandable in terms of ease of analyzing a map, and since it is difficult to accurately correct for abundance variations predicted in terms of a model, we must keep in mind that the assumption of a single fractional abundance for a specific molecular species even in a single cloud may be seriously misleading. From the perspective of theoretical models of interstellar cloud chemistry, these variations should be larger for less abundant species, such as the "high dipole–moment" molecules used to trace dense regions in clouds (Graedel, Langer, and Frerking 1982). Thus, **cloud structure may appear to be very different depending on which tracer is used, even when effects of excitation and radiative transfer are correctly accounted for.**

Major differences in cloud structure indicative of variations in fractional abundance are seen in the quiescent dark cloud L134N (L183). This region has been mapped by Swade (1989) in six molecules. No unambiguous explanation is available, but significant variations, particularly between the distributions of CS, H^{13}CO$^+$, SO, and NH$_3$ are apparent. A similar effect has been recognized in the TMC1 cloud, which is discussed in detail by Olano, Walmsley, and Wilson (1988). Again, in a region apparently devoid of star formation, significant abundance gradients in the case of NH$_3$ and cyanopolyynes are found.

In the case of clouds with **massive** star formation, very large variations in the abundance of many species have been found. Although it is not easy to directly determine the total H$_2$ column density, relative abundance variations can be determined with some confidence. For example, in the region surrounding the HII region complex Sgr B2(N) having a luminosity of $\approx 10^7$ L$_o$, Vogel Genzel and Palmer (1987) find a fractional abundance of NH$_3 \cong 10^{-5}$, **an enhancement of a factor of 100 to 1000** relative to the general cloud material. These authors, and others, have suggested that this may be the result of evaporation of molecules frozen on grain surfaces in regions warmed by newly–formed stars. On the other hand, Zhou Evans, and Mundy (1990) have found evidence for an **order of magnitude reduction** in the fractional abundance of NH$_3$ in the region immediately surrounding the young stars in NGC 2071. Given these highly disparate results, it is difficult to have unbridled confidence in the ability of ammonia to trace regions of high mass star formation in molecular clouds.

In Figure 1 I show maps of the Sgr B2 region in SO and HNCO (taken from Lis and Goldsmith, 1990) which highlight the difference found on relatively large scales in the relative abundance of various molecular species. These species have been found by Goldsmith et al. (1987) to exhibit significant abundance differences on a much smaller scale, between the Middle and Northern cores in Sgr B2.

While it is difficult to eliminate completely the issue of nonuniform excitation, it seems likely that the evolutionary state of a dense cloud core with an embedded massive protostar has a major effect on the molecular content of the surrounding material, as suggested by Vogel, Genzel and Palmer and by Goldsmith et al.

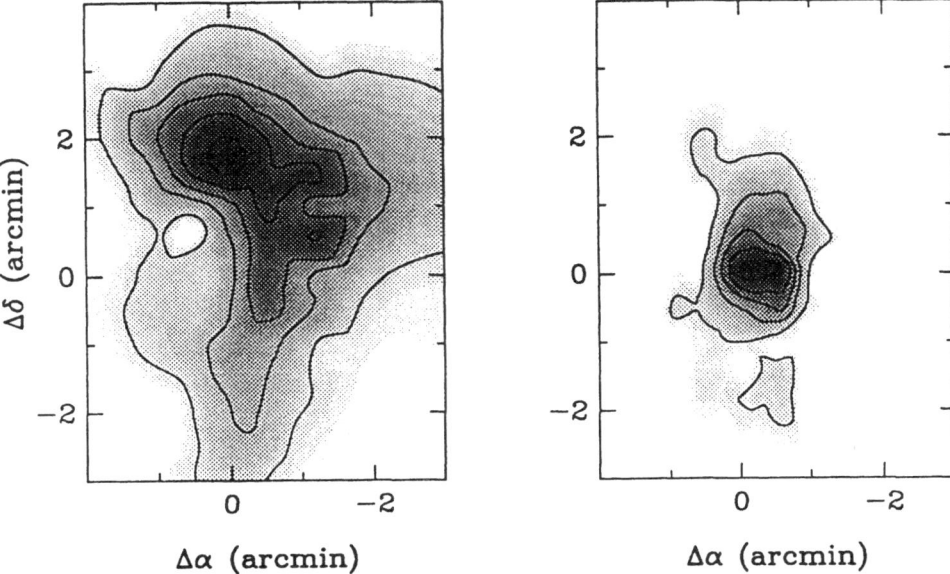

FIG. 1 — Maps of integrated intensity of HNCO 5_{05}–4_{04} (left) and SO 2_3–1_2 (right) in Sgr B2 made with FCRAO antenna. The coordinates of the central position are $\alpha = 17^h44^m10^s.5$, $\delta = -28°22'05''$.

Within the central portion of the Orion molecular cloud are found components having different spatial and kinematic structure (extended ridge, compact ridge, hot core, and plateau; cf. Johansson et al. 1984; Blake et al. 1987), which also have significantly different chemical composition (Plambeck and Wright 1988).

Detailed, multi–transition studies of HC$_3$N have yielded the H$_2$ density distribution in Sgr B2, which does not closely resemble either of the above maps in appearance, but rather follows the general distribution of C^{18}O found by Lis and Goldsmith (1989). Thus, there appear to be highly significant parsec–scale abundance variations which can seriously affect our perception of the overall cloud structure. The origin of these variations is quite uncertain at the present time. The enhanced abundance of HOCO$^+$ relative to predictions of standard ion–molecule astrochemical models led Pineau des Forets, Flower and Roueff (1989) to postulate its production via destruction of HCO$^+$ in MHD shocks. The primary HNCO peak seen in Figure 1, and also a secondary maximum in HC$_3$N J =12–11, coincide with the strongly peaked HOCO$^+$ emission mapped by Minh, Irvine, and Ziurys (1988). While it is desirable to carry out more complete modeling of the effects of shocks on cloud chemistry, the lack of any apparent driving source (HII region or massive young star) as well as of any effect on the line profiles makes this explanation problematic for the "2' North" peak in Sgr B2.

Another cloud whose structure seems to be highly dependent on the molecular tracer used is NGC 2071 N. This region was studied by Iwata, Fukui, and Ogawa (1988) who found very different appearance as traced by NH$_3$ and C^{18}O. In a recently–completed project, Goldsmith et al. (1990) have obtained maps with the

FCRAO antenna of three carbon monoxide isotopes, as well as of CS J = 2–1, and HCO⁺ J = 1–0. The region contains a fairly well–defined bipolar outflow, and a 40 L$_0$ infrared source IRAS 0541+0037. The C¹⁸O emission shows four well–defined condensations, while the ¹³CO emission is relatively featureless, presumably as a result of its high opacity. The CS and the HCO⁺ emission each shows a single clump. Impressively enough, as shown in Figure 2, **the CS condensation does not coincide with any of the C¹⁸O peaks** or with the HCO⁺ maximum. Also, it is the CS peak which is coincident with the IRAS source, which is also near the center of the outflow—**none of the C¹⁸O peaks is close to this position!** The ammonia emission studied by Iwata, Fukui and Ogawa (1988) is also highly clumped, with one condensation overlapping, but extending to the east of the CS maximum, and another coincident with the C¹⁸O peak at (−75", 0"). Two additional NH$_3$ peaks do not correspond to regions of enhanced emission in any of the species we have observed.

Again, I must emphasize that we have observed only a single transition of each species, and cannot eliminate the effect of variations in excitation, and of optical depth, especially for the CS. Nevertheless, it is difficult to believe that these are responsible for all of the difference in the structures that we see. Rather, it seems necessary to confront the fact that CS and C¹⁸O—two widely used tracers of cloud structure—give very different pictures of the cloud. The magnitude of the relative abundance variations indicated is on the order of a factor of 10. If we observed only C¹⁸O, we would simply not be able to make sense of the bipolar flow, while CS alone would give a poor idea of the general cloud structure.

What can be responsible for this **differential clumpiness?** One possibility is that the chemical evolution in different clumps is in different phases: the time dependence of the abundance of different species (cf. Graedel, Langer, and Frerking 1982) results in an apparent enhancement of different molecules in different condensations at any moment of time. To be verified, this concept would require a great deal more observational data and analysis as well as detailed modeling, but it would be a significant change in our concept cloud structure.

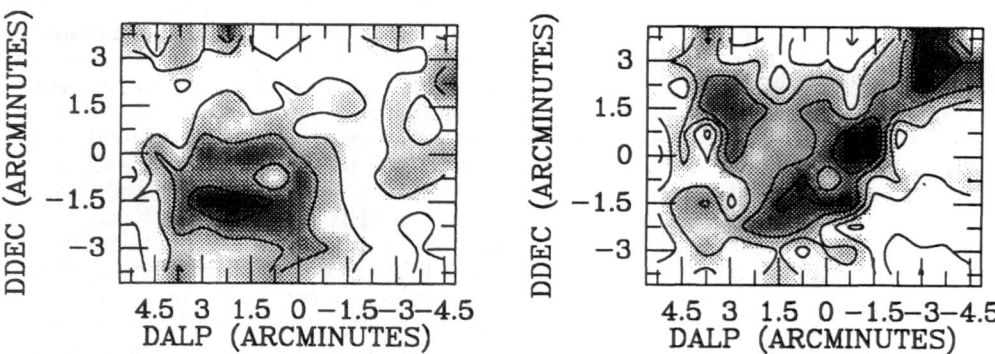

FIG 2. – Maps of CS integrated intensity of J = 2–1 CS (left) and J = 1–0 C¹⁸O (right) in NGC 2071 N. The (0,0) position is $\alpha = 05^h45^m$, $\delta = 0^o39'$. The coordinates of IRAS 05451+0037 are +120",−90" relative to the map center.

Understanding the extent of density inhomogeneities in a cloud and the role that they play in their structure is critically dependent on unraveling the issue of molecular abundance variations so that we can properly analyze density structure rather than just nonuniformities in the abundance of trace molecules. Dust continuum emission is subject to variations in physical conditions as well as to systematic uncertainties, but these problems are quite different from those which affect molecular emission. It thus appears highly desirable to use dust emission as a complement to observations of molecules.

Limiting studies to a single, or a few molecular species, even those widely accepted as "good" tracers of molecular material, can lead to serious errors in our perception of the structure of molecular clouds. This is largely the result of significant variations in the abundance of the various tracers within individual clouds. This effect is found in quiescent dark clouds as well as giant molecular clouds, but is particularly prominent in regions where star formation has occurred. In this sense, the apparent complication of molecular abundance variations may prove to be a valuable tracer of the structure and evolutionary state of molecular clouds.

I would like to acknowledge the considerable help of my colleagues Darek Lis, Mike Margulis, Ron Snell, and Bill Irvine in preparing this talk. Astronomy research at FCRAO is supported by the NSF under grant AST 88–15406.

REFERENCES

Arquilla, R., and Goldsmith, P.F. 1985, **Ap.J.**, 297, 436.
Blake, G.A., Sutton, E.C., Masson, C.R., and Phillips, T.G. 1987, **Ap.J.**, 315, 621.
Dickman, R.L. 1978, **Ap.J. Suppl.**, 37, 407.
Federman, S.R., Huntress, W.T. Jr., and Prasad, S.S. 1990, **Ap.J.**, 354, 504.
Frerking, M.A., Langer, W.D., and Wilson, R.W. 1982, **Ap.J.**, 262, 590.
Goldsmith, P.F., Langer, W.D., Carlson, E.R., and Wilson, R.W. 1980, in **Proc. IAU Symp. 87**, B. Andrew, ed. Dordrecht: Reidel, p. 417.
Goldsmith, P.F., Snell, R.L., Hasegawa, T., and Ukita, N. 1987, **Ap.J.**, 314, 525.
Goldsmith, P.F., Margulis, M., Snell, R.L., and Fukui, Y. 1990, in preparation.
Graedel, T.E., Langer, W.D., and Frerking, M.A. 1982, **Ap.J.Suppl.**, 48, 321.
Irvine, W.M., Goldsmith, P.F., and Hjalmarson, A. 1987, in **Interstellar Processes**, H. Thronson and D. Hollenbach, eds. Dordrecht: Reidel, 561.
Irvine, W.M. 1990, in **Chemistry and Spectroscopy of Interstellar Molecules**, N. Kaifu, ed. Tokyo: University of Tokyo, in press.
Iwata, T., Fukui, Y., and Ogawa, H. 1988, **Ap.J.**, 325, 372.
Johansson, L.E.B. et al. 1984, **Astr. Astrophys.**, 130, 227.
Langer, W.D., Wilson, R.W., Goldsmith, P.F., and Beichman, C.A. 1989, **Ap.J.**, 337, 355.
Lis, D.C., and Goldsmith, P.F. 1989, Ap.J., 337, 704.
Lis, D.C., and Goldsmith, P.F. to appear in **Ap.J.**, 1991.
Minh, Y.C., Irvine, W.M., and Ziurys, L.M. 1988, **Ap.J.**, 334, 175.
Olano, C.A., Walmsley, C.M., and Wilson, T.L. 1988, **Astr. Astrophys.**, 196, 194.
Pineau des Forets, G., Roueff, E., and Flower, D.R. 1989, **J. Chem. Soc. Faraday Trans.**, 2, 85(10), 1665.
Plambeck, R. and Wright, M. 1988, in **Molecular Clouds in the Milky Way and External Galaxies**, R.Dickman, R.Snell, J.Young, eds. Berlin: Springer, 182.
Swade, D.A. 1989, **Ap.J. Suppl.**, 71, 219; **Ap.J.**, 345, 828.
Vogel, S.N., Genzel, R., and Palmer, P. 1987, **Ap.J.**, 316, 243.
Zhou, S., Evans, N.J. II, and Mundy, L. 1990, **Ap.J.**, 355, 159.

IV - SHOCKS AND INSTABILITIES

MAGNETOHYDRODYNAMIC SHOCK WAVES IN MOLECULAR CLOUDS

B. T. Draine
Princeton University Observatory
Peyton Hall
Princeton NJ 08544
U.S.A.

ABSTRACT. The fluid dynamics of MHD shock waves in magnetized molecular gas is reviewed. The different types of shock solutions, and the circumstances under which the different types occur, are delineated. Current theoretical work on C*- and J-type shocks, and on the stability of C-type shocks, is briefly described. Observations of the line emission from MHD shocks in different regions appear to be in conflict with theoretical expectations for single, plane-parallel shocks. Replacement of plane-parallel shocks by bow shocks may help reconcile theory and observation, but it is also possible that the observed shocks may not be "steady", or that theoretical models have omitted some important physics.

1. Introduction

Shock waves are a relatively common phenomenon in molecular clouds, particularly in star-forming regions. A shock wave may be described as a "hydrodynamic surprise" – a pressure-driven disturbance propagating into the ambient medium with a speed v_s larger than the "signal speed" for compressive waves in the unperturbed gas.

The structure of a shock wave depends upon the shock speed v_s and on the properties of the ambient medium. The density $n_H \equiv n(H) + 2n(H_2)$ extends from $n_H \approx 10^2$ cm^{-3} in relatively diffuse regions to $\gtrsim 10^6$ cm^{-3} in dense cores; the fractional ionization x_e ranges from $\sim 10^{-4}$ in diffuse clouds to $\sim 10^{-8}$ in regions of density $\sim 10^6$ cm^{-3}. Our limited knowledge of magnetic fields in interstellar clouds appears to be roughly consistent with $B_0 \approx b_0(n_H/\text{cm}^{-3})^{1/2}\mu G$, with $b_0 \approx 1$(see the review by Heiles [1991]).

In the absence of the magnetic field, the molecular gas would be able to transmit ordinary sound waves with a sound speed $c_n \approx 0.49(T_n/40\,\text{K})^{1/2}\,\text{km s}^{-1}$. However, for $b_0 \approx 1$ the Alfven velocity $v_A = 1.84b_0\,\text{km s}^{-1}$ is much larger than the sound speed, and the medium is able to transmit compressive waves perpendicular to $\vec{B}_0$ at the "magnetosonic" speed $c_{in} = (v_A^2 + c_n^2)^{1/2} \approx v_A$.

When fluid flows in the ambient cloud have velocities large compared to c_{in}, shock waves will occur. In OMC-1, H_2 line emission is observed with line wings extending at least $\pm 100 \, \mathrm{km \, s^{-1}}$ from line center (Brand *et al.* 1989 *a*; Moorhouse *et al.* 1990), so there is little doubt that strong shock waves must be present. Hypersonic flows are also observed in other star-forming regions.

The present review will touch on some current topics in the fluid dynamics of shock waves in molecular clouds. Shock chemistry, though important, will not be addressed; see the recent review of chemistry in diffuse cloud shocks by Hartquist, Flower, and Pineau des Forets (1990).

2. Fluid Dynamics of MHD Shock Waves

The fluid dynamics of MHD shocks have been recently reviewed (Shull and Draine 1987). It is important to use a two-fluid treatment (Mullan 1971; Draine 1980): the neutral gas (containing most of the inertia) and the magnetized plasma (the bearer of the magnetic stresses) are treated as distinct fluids, with coupling between the two fluids provided explicitly through ion-neutral collisions, ionization, and recombination; a simple derivation of the fluid equations is given by Draine (1986). In a typical quiescent cloud, the coupling between the two fluids is sufficient to ensure that they are essentially comoving. In regions where large gradients are present (e.g., large amplitude, short wavelength disturbances, with shock waves as the extreme case), appreciable "slip" velocities can develop between the neutral and ionized fluids. "Frictional coupling" is provided by ion-neutral collisions, so that slippage produces momentum transfer between the fluids and energy dissipation.

The nature of a steady fluid flow depends critically on the "Mach number" of the flow, where the generalized "Mach number" is defined as the ratio of the flow velocity (in the reference frame where the flow is stationary) to the signal speed for compressive waves. We discuss here the simple case where the preshock magnetic field $\vec{B}_0$ is perpendicular to the shock velocity $\vec{v}_s$; the case of arbitrary orientation of the magnetic field is qualitatively similar (Wardle and Draine 1987). First of all, in order to have a shock we must have $v_s > c_{in}$, where $c_{in} \approx 1.84 b_0 \, \mathrm{km \, s^{-1}}$ is the signal speed (in the preshock medium) for long wavelength waves (in which the neutral and ionized fluids move together).

The local signal speed for short wavelength waves in the magnetized plasma is initially $c_i \approx 70 b_0 (x_e/10^{-4})^{-1/2} \, \mathrm{km \, s^{-1}}$, and it increases as the plasma is compressed [here x_e is the fractional ionization and we assume a mean ionic mass of $\sim 10 m_H$]. Therefore for $v_s \lesssim 50 \, \mathrm{km \, s^{-1}}$ shock waves in molecular clouds we may safely assume that the flow velocity of the magnetized plasma will everywhere be "subsonic" (in the frame of reference where the shock is stationary). It is a familiar result of steady flow theory that discontinuities in the flow can only occur at a transonic point where the flow makes a transition from supersonic to subsonic. Therefore the flow of the magnetized plasma must be everywhere *continuous*. One way of interpreting this result is that the large signal speed in the magnetized plasma allows it to "communicate" upstream and "inform" the upstream plasma of the approaching compression; the plasma therefore is not "surprised".

3. Types of MHD Shock Waves

3.1. C-TYPE SHOCKS

The local signal speed (for short wavelength waves) in the neutral fluid is just the thermal sound speed $c_n \approx 0.49(T_n/40 \text{ K})^{1/2} \text{ km s}^{-1}$. If the neutral fluid remains cold (either because the shock is extremely weak or because radiative cooling is very effective) then the neutral fluid may remain everywhere supersonic, with no supersonic$\rightarrow$subsonic transition and therefore no discontinuity. Such shocks – in which both the ion and neutral flow variables are everywhere continuous – are termed "C-type" (Draine 1980). In C-type shocks the energy dissipation (and entropy generation) is due entirely to the ion-neutral collisions in the region of ion-neutral "slip". Fluid velocities in a C-type shock are illustrated in Fig. 1.

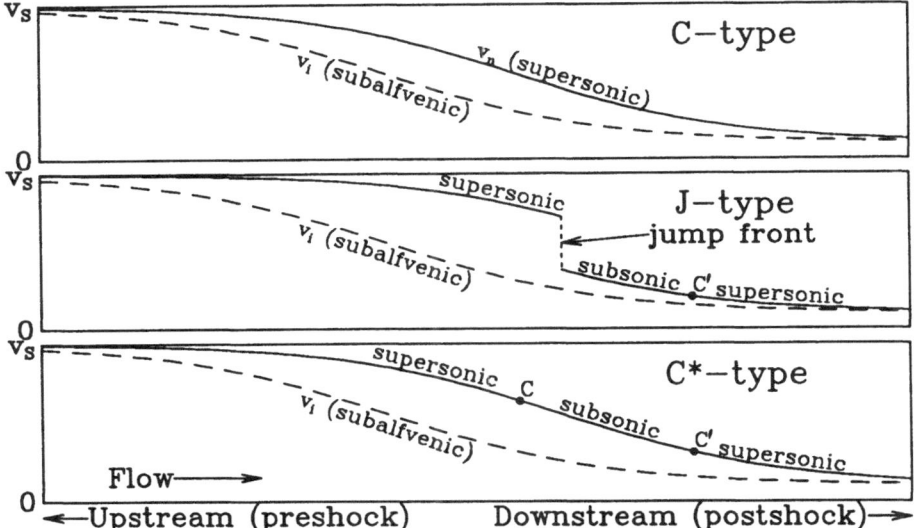

Figure 1. v_n (solid) and v_i (broken) for C-type, C*-type, and J-type shock waves.

3.2. J-TYPE AND C*-TYPE SHOCKS

If cooling is unable to hold down the neutral temperature, then the neutral sound speed will increase as the neutral flow velocity decrease; the Mach number will decrease and a supersonic$\rightarrow$subsonic transition may occur in the flow. It turns out there are two very different ways in which this can occur. Under some circumstances the supersonic$\rightarrow$subsonic transition takes place at a "viscous subshock", where the neutral fluid undergoes a "jump" effected by ordinary molecular viscosity (just as for classical single fluid shocks): this is termed a "J-type" shock. Fig. 1 shows the velocity structure in a J-type shock; the sonic point (C') at which the neutral gas makes the subsonic$\rightarrow$supersonic transition is indicated.

Under other conditions, however, it is possible for the supersonic$\rightarrow$subsonic transition to take place *smoothly* as the neutral gas is heated and decelerated (in the "shock frame") by collisions with streaming ions. Such solutions – termed "C*-type" shocks – have been discussed by Chernoff (1987) and Roberge and Draine (1990). The velocity structure in a C*-type shock is shown in Fig. 1, including the two sonic points (C and C')

3.3. SHOCK TRAJECTORIES IN THE PHASE PLANE

Chernoff (1987) invented a "phase plane" analysis which proves very useful in understanding the different shock types. Once the preshock conditions and shock speed are specified, then, to an excellent approximation, the laws of conservation of mass and momentum permit the physical conditions at a given point in a shock to be fully determined by just two flow variables: the neutral and ion flow velocities v_n and v_i. In particular, given these two flow variables, the neutral temperature T_n can be determined. The flow of the fluid through the shock can therefore be represented by a "trajectory" on a "phase plane" with coordinates $q = v_i/v_s$ and $r = v_n/v_s$. The trajectory begins at the point $U = (1, 1)$. The local derivative of the trajectory is an explicit function of q and r, which can be written $dr/dq = R(q, r)/(M^2 - 1)$, where $M = v_n/c_n$ is just the neutral Mach number, and R is an explicit function of q and r. Chernoff showed that there were three important lines on this phase plane: the line $q = r$ (the trajectory begins and ends at points U and D on this line); the line $M = 1$ (where dr/dq is singular); and the line where the function $R(q, r) = 0$. These three lines are shown qualitatively in Fig. 2. Simple arguments can be used to show that certain regions bounded by these curves are forbidden: a trajectory which begins at U either cannot enter these regions or, if in the region, cannot reach the downstream solution D. These forbidden regions have been shaded in Figs. 2 – 4. Note in particular the two potential "sonic points" C and C' where the $R = 0$ and $M = 1$ curves intersect. These are the only places where it is possible for continuous trajectories to cross the $M = 1$ or $R = 0$ curves. Note that the region *above* the $M = 1$ line (which includes the initial point U and, we will assume, the downstream point D) is the locus of *super*sonic flow – if the cooling is able to keep the gas cool (and the sound speed low), then the entire trajectory will remain above the $M = 1$ line. Now if a trajectory which begins at $U = (1, 1)$ always remains above the $M = 1$ line, then it is everywhere continuous and it must inevitably terminate at the downstream solution D: such a "C-type" trajectory is illustrated in Fig. 2.

If, on the other hand, the neutral temperature rises so that the trajectory beginning at U will intersect the $M = 1$ curve, there are two possibilities. If the trajectory beginning at U collides with the $M = 1$ curve at a point F, as shown in Fig. 3, then it is necessary to abandon this trajectory at some earlier point J, and invoke a "jump" in the neutral flow variables (i.e., change r while $q = constant$, using the Rankine-Hugoniot jump conditions) to a point J' (below the $R = 0$ curve), and begin integrating a new trajectory beginning at J'. The "jump" point J must be chosen so that J' lies on the (unique) trajectory which passes through the sonic point C' – otherwise there is no way to reach the (unique) downstream steady state D. Iterative numerical techniques to accomplish this have been implemented (Roberge and Draine 1990). It is also worth noting that integration of the trajectory in the region below the $M = 1$ line is numerically delicate: small errors grow

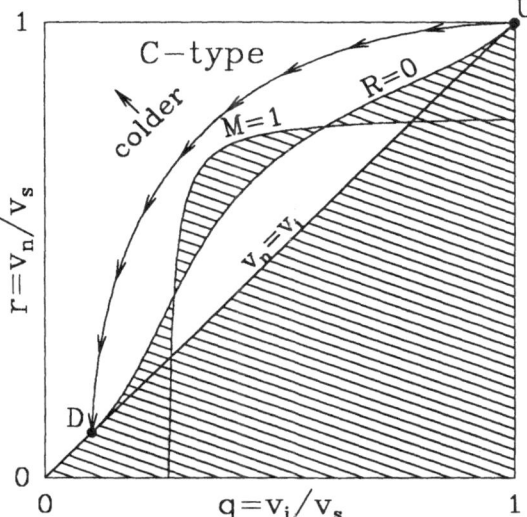

Figure 2. Phase space trajectory for C-type shock. Shaded areas are forbidden.

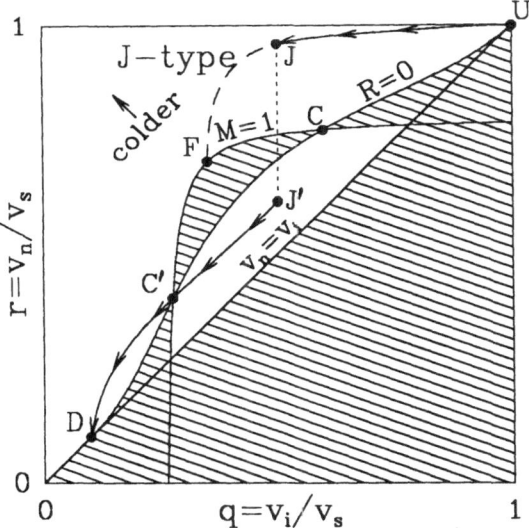

Figure 3. Phase-space trajectory for J-type shock.

exponentially. "Shooting" techniques are therefore employed to successfully integrate the $J' - C'$ portion of the trajectory (Roberge and Draine 1990).

At first sight it might appear that only a fortuitous choice of initial conditions could produce a trajectory beginning at U which reached the sonic point C – it turns out, however, that the sonic point C is an "attractor" in this phase plane: trajectories in the supersonic

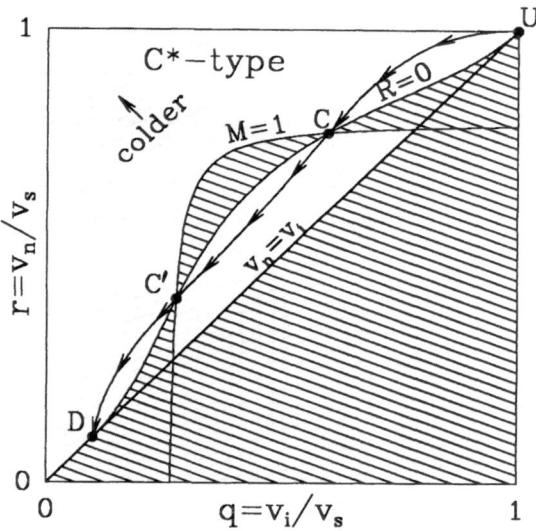

Figure 4. Phase-space trajectory for C*-type shock.

region which come sufficiently close to C will be "sucked" into it, thereby permitting a smooth transition from supersonic to subsonic flow. On the subsonic side of C there are an infinite number of trajectories which originate from C – one chooses the one which reaches the second sonic point C'. Such a solution is referred to as a "C*-type" shock. From the standpoint of the fluid dynamics these solutions are fundamentally different from the C-type shock solutions. However, from the standpoint of the thermal structure in the flow, there is a fundamental similarity in that, just as for C-type shocks, the heating is a smooth, continuous function of space – the "impulsive" heating which is associated with a "jump" is absent – and therefore the cooling processes may be able to keep the temperature from rising to high levels. (Of course, the very fact that the C*-type solutions include a subsonic portion implies that cooling is unable to hold the temperature down to very low values.)

3.4. BREAKDOWN

It is important to recognize that the multifluid character of the flow depends on the weak coupling between the ion and neutral fluids. Appreciable slip velocities can exist over appreciable distances only if the fractional ionization remains low. Furthermore, if the shocks are to account for strong molecular line emission (e.g., in OMC-1) then H_2 must survive in the shock; survival of H_2 in fast shocks requires that the ion density remain low so that molecular line cooling is able to radiate away the heat dissipated in ion-neutral collisions. Therefore, for given preshock conditions (n_H, $\vec{B}_0$, x_e) there is a "breakdown" shock speed above which the shock destroys essentially all of the H_2 (Draine, Roberge, and Dalgarno 1983; Hollenbach, Chernoff, and McKee 1989; Smith and Brand 1990a). There are two ways this can happen: (1) High-velocity ion-neutral collisions may increase the ionization, a process which tends to "run away" as the ion density and heating rate increase. (2) Even if the ionization remains approximately constant, if the neutral temperature rises too high the H_2 will be collisionally-dissociated.

It should be noted that even above the "breakdown" velocity there may still be appreciable molecular emission from the "magnetic precursor" at the leading edge of the shock, unless the shock speed is so high that UV radiation from the shocked gas is able to raise the preshock ionization and thereby reduce the extent of this precursor.

4. Instabilities

Nearly all of the theoretical work on MHD shocks in molecular clouds has been restricted to steady, plane-parallel flows; with these assumptions, numerical solutions to the fluid equations can be found. It is, however, crucially important to know whether or not such flows are *stable*.

Wardle (1990 a,b, 1991) has analyzed the stability of steady, plane-parallel, C-type shocks. Wardle pointed out that one could intuitively anticipate the possibility of an instability which would involve the "buckling" of field lines, in a manner reminiscent of (but fundamentally different from) the Parker instability for magnetized gas in a gravitational field.

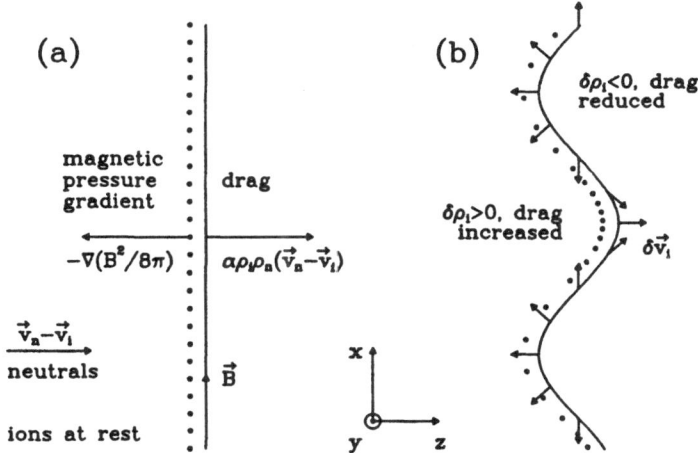

Figure 5. Possible instability mode described in text (from Wardle 1990a). The dots represent ions along a magnetic field line. In the unperturbed case (a), the ion density is uniform along a field line. In the perturbed case (b), the ions are driven into the "valleys" by the "neutral wind", resulting in an increase in the local force density which may further deform the magnetic field.

Consider the dynamics of the ionized fluid at a point in the flow, and adopt a reference frame moving with the velocity of the ionized fluid. The plane-parallel, steady solution is illustrated in Fig. 5a, for the special case where the magnetic field $\vec{B}$ is perpendicular to the direction of propagation of the shock. In the adopted reference frame the ions are (locally) stationary, as are the magnetic field lines (frozen into the ionized fluid). Consider the dynamics of the ions: they are subject to a force density $\alpha \rho_n \rho_i (\vec{v}_n - \vec{v}_i)$ due to collisions with the atoms in the "neutral wind" which blows from left to right in Fig. 5a (here α is just a constant proportional to the rate coefficient for ion-neutral scattering), and they are also subject to a force density $-\nabla(B^2/8\pi) = (1/c)\vec{J} \times \vec{B}$. Because the ions have essentially no inertia, these two opposing force densities must almost exactly balance! Note that so long as the field lines remain straight, the force on the ions due to the "neutral wind" is perpendicular to $\vec{B}$.

Now consider what may happen if the magnetic field $\vec{B}$ for some reason has a bend in it, as shown in Fig. 5b. The force due to the neutral wind is still directed from left to right, but at most points this force now has a component parallel to $\vec{B}$, which cannot possibly be balanced by the $\vec{J} \times \vec{B}$ force! This means that the "neutral wind" will tend to drive the ions *along* the field lines, resulting in an increase in ρ_i in the "valleys" defined by $\vec{B}$. This in turn implies that the force density $\alpha \rho_n \rho_i (\vec{v}_n - \vec{v}_i)$ will increase in these valleys, in which case it may now exceed $-\nabla(B^2/8\pi)$, in which case the bends in the field lines will tend to grow! This is the basic mechanism of the Wardle instability, where the "neutral wind" plays the role analogous to that of gravity in the Parker instability.

Wardle has performed a linear stability analysis of the fluid equations for the special case where $\vec{B} \cdot \vec{v}_s = 0$, both for an isothermal equation of state (Wardle 1990a) and for realistic radiative cooling (Wardle 1990b). The stability analysis has been extended to general case of arbitrary orientation of preshock $\vec{B}$ (Wardle 1991). The stability depends upon a number of parameters, but Wardle's results may be approximately summarized as follows: the stability of C-type shock waves depends primarily on the "Alfven Mach number" $M_A = v_s/v_A$, where $v_A = 1.84 b_0$ km s^{-1} is the Alfven speed in the ambient gas. Wardle found C-type shocks to be stable when $M_A \lesssim 5$, and unstable for $M_A \gtrsim 5$. This implies that MHD shocks will only be stable for $v_s \lesssim 9 b_0$ km s^{-1}! Steady, plane parallel multifluid MHD shock models are therefore of questionable validity for $v_s \gtrsim 10 b_0$ km s^{-1}.

Wardle's investigation was limited to a linear stability analysis; the nonlinear development of the instability remains unknown. In particular, it is not clear whether the instability will grow to a large amplitude or whether nonlinear terms may limit the growth of the instability before it becomes large. It seems likely that the Wardle instability mechanism (based upon the delicate balance of momentum transfer from the "neutral wind" against the gradient of magnetic pressure) will also apply in J-type and C*-type shocks, but this has not yet been demonstrated. These are important topics for future investigation.

It therefore appears that plane-parallel steady models of MHD shocks may be regarded as accurate approximations for C-type shocks in uniform media only for $v_s \lesssim 10 b_0$ km s^{-1}. For higher shock speeds, we can at present only hope that the plane-parallel, steady shock models provide a good estimate for the average emission from and chemistry in the shock; if, however, the instability grows to large amplitudes this will not be the case.

5. Puzzles

5.1. H_2 Line Emission from OMC-1

Powerful line emission is observed from H_2 and high$-J$ CO in the BNKL region of OMC-1. In an attempt to explain this emission, two different groups (Draine and Roberge 1982; Chernoff, Hollenbach and McKee 1982) proposed C-type shock models. The models attempted to account for all of the observed emission using a single shock (approximated as spherically-symmetric). The C-type shock models were reasonably successful at accounting for the observations, and it appeared at the time that the shock models – while clearly an oversimplification in aspects such as spherical symmetry – were probably a good first approximation.

The C-type models did not, of course, provide a perfect fit to observed line ratios; as additional emission lines were observed, and the reddening by dust was more accurately determined, it became clear that the C-type models predicted insufficient emission in both the lowest excitation $[v = 0 \rightarrow 0S(2)]$ and highest excitation [e.g., $v = 0 \rightarrow 0S(17)$ and $v = 4 \rightarrow 3S(3)$] transitions of H_2 (Brand et al. 1988).

In an attempt to understand the observed emission line ratios, Brand et al. (1988) proposed a very simple model: a *nonmagnetic* single-fluid radiative shock wave, in which dissociation of and emission from H_2 dominate the cooling. The Brand et al. model provides a good fit to the observed H_2 line ratios, and further predicts that the line ratios should be quite insensitive to the shock speed for v_s in the range $10 - 25 \, \mathrm{km \, s^{-1}}$.

The Brand et al. model ostensibly has no "adjustable parameters", and yet achieves impressive agreement with the observations of H_2 emission, whereas the complex C-type models, with a number of adjustable parameters, failed. This is even more remarkable when one considers the fact that we have fundamental theoretical reasons for rejecting the Brand et al. model: (1) The Brand et al. model assumes H_2 to be the only coolant, whereas all theoretical studies of the chemistry in such shocks conclude that a large fraction of the O is converted to OH and H_2O, which, at the densities $n_H \gtrsim 10^6 \, \mathrm{cm^{-3}}$ in question, should dominate the cooling. (2) In order to obtain the observed H_2 line intensities, high preshock densities ($n_H \gtrsim 10^6 \, \mathrm{cm^{-3}}$) must be assumed; the resulting densities ($n_H \gtrsim 5 \times 10^6 \, \mathrm{cm^{-3}}$) in the shock-heated gas are so high that excessive amounts of emission from very high rotational levels ($J \gtrsim 35$) of CO would be produced, in conflict with observations. Note that good agreement with the observed CO emission spectrum is obtained in the C shock models of Chernoff, Hollenbach, and McKee (1982), where the density in the emitting region is only $n_H \approx 5 \times 10^5 \, \mathrm{cm^{-3}}$. (3) We have independent reasons for believing that dynamically-important magnetic fields must be present in molecular clouds; if our estimates of Alfven velocities $v_A \approx 2 \, \mathrm{km \, s^{-1}}$ are even approximately correct, then the magnetic fields *must* have an important effect on the fluid dynamics, and multifluid flow models must be used.

5.2. H_2 Line Ratios in OMC-1 and Other Objects

The emission spectrum computed for C-type shock models (Draine and Roberge 1982;

Chernoff, Hollenbach, and McKee 1982; Draine, Roberge and Dalgarno 1983) was a sensitive function of the shock parameters: shock speed v_s, preshock density n_H, magnetic field B_0, and fractional ionization x_e. Since it seemed certain that physical conditions (particularly the density n_H and the orientation of local $\vec{B}$ relative to $\vec{v}_s$) would vary from point-to-point along the shock front, it is fair to say that one of the firm predictions of the C-type shock models was that the actual H_2 line ratios would be found to vary with position in OMC-1.

Brand $et\,al.$ (1989b) cleverly chose to study the $v = 1 \to 0 O(7)$ and $v = 0 \to 0 S(13)$ transitions – two lines with nearly identical wavelengths (3.807 and 3.847 μm, respectively, so that differential extinction may be neglected) but originating from levels with energies differing by $\Delta E/k = 9093$ K. The intensity ratio for this line pair would be expected to be a quite sensitive function of shock parameters for C-type shock models: the levels in question have sufficiently high energies that the emission from a C-type shock would be dominated by the region of peak temperature, and the peak temperature varies as the shock parameters are varied. Brand $et\,al.$ mapped the line ratio over a significant area in OMC-1. Amazingly, the line $ratio$ remained $constant$ to within obserational uncertainties, even as the line $intensities$ varied by an order of magnitude from one position to another! Such a result is very difficult to understand if the emission originates in a plane-parallel C-type shock. The observed line ratio is, however, close to that "predicted" by the nonmagnetic shock models (Brand $et\,al.$ 1988), but this appears to be fortuitous since, as discussed above, there are strong reasons for believing such nonmagnetic shock models to be inapplicable.

To further complicate the observational constraints, Burton $et\,al.$ (1989) measured the $1 \to 0 O(7)/0 \to 0 S(13)$ line ratio in CRL618, HH7, and two positions in IC443. Even though the line ratio was found to be constant within OMC-1, it $does$ vary from object to object, although only by a factor of ~ 2 among the objects and positions observed.

6. Discussion and Future Directions

What are these observations telling us? It seems clear that simple, planar C-type shock models are not consistent with the observations, but it is not entirely clear how the models need to be changed.

There seem to be two distinct avenues to explore. Smith, Brand, and Moorhouse (1990) argue that the observed H_2 line intensities from OMC-1 and IC443 (the two best-studied outflows) can be understood in terms of $bow\ shocks$ in which the shock transition at a given point on the bow shock is computed as for a planar multifluid C-type shock, but with the emission from a single bow shock structure involving a sum over shock parameters. For high-velocity bow flows the inner portions of the bow shock are dissociative; the bulk of the H_2 line emission comes from the outer portions of the bow shock where the component of the bow velocity $normal$ to the shock surface is below the "breakdown" velocity. Using approximate treatments of the shock structure, Smith $et\,al.$ show that good agreement with observations can be obtained. Recent H_2 line images of OMC-1 (Bally 1990; Beichman 1990) are indeed suggestive of "interstellar bullets", in which case this bow shock interpretation may indeed be called for. However, Smith and Brand (1990b) concluded that the H_2 line *profiles* in OMC-1 were *not* consistent with the predictions of their bow shock model.

It should further be noted that a C-shock in OMC-1 will have a hot transition zone of thickness $\sim 10^{15}$ cm (Draine and Roberge 1982) – corresponding to $0.13''$ at the 500 pc distance of Orion. The "bow shock" modelling of Smith *et al.* presumes the radius of the "bullet", and hence the radius of curvature of the shock, to be large compared to this shock thickness. We would therefore expect observations with a small aperture (say $\sim 1''$) to sample only a portion of the bow shock structure, in which case line ratios such as $I[0 - 0S(13)]/I[1 - 0O(7)]$ would be expected to vary with position. The observational study by Brand *et al.* (1989*b*) employed a $5''$ aperture; future observations with a smaller aperture will be of value to test the "bow shock" interpretation.

It is important to keep in mind that the existing multifluid shock calculations make a number of approximations, some of which are of questionable validity. Perhaps the flows are unstable, and perhaps somehow the nonlinear development of the instabilities can account for the observed line ratios in a "natural" way. The existing shock models have used a highly simplified and approximate treatment of the grain dynamics; perhaps a more refined and exact treatment of the grain dynamics will lead to a more satisfactory H_2 emission spectrum. Perhaps we are overlooking some other important physical or chemical processes. Only further theoretical investigation will tell!

Acknowledgements

This work was supported in part by NSF grant AST-8612013 and NASA grant NAGW-1973.

References

Bally, J. 1990, private communication.

Beichman, C. A. 1990, private communication.

Brand, P. W. J. L., Moorhouse, A., Burton, M. G., Geballe, T. R., Bird, M., and Wade, R. 1988, "Ratios of Molecular Hydrogen Line Intensities in Shocked Gas: Evidence for Cooling Zones", *Ap. J. (Letters)*, **334**, L103-L106.

Brand, P. W. J. L., Toner, M. P., Geballe, T. R., and Webster, A. S. 1989*a*, "The velocity profile of the 1-0S(1) line of molecular hydrogen at Peak 1 in Orion", *M.N.R.A.S.*, **237**, 1009-1018.

Brand, P. W. J. L., Toner, M. P., Geballe, T. R., Webster, A. S., Williams, P. M., and Burton, M. G. 1989*b*, "The constancy of the ratio of the molecular hydrogen lines at 3.8 μm in Orion", *M.N.R.A.S.*, **236**, 929-934.

Burton, M. G., Brand, P. W. J. L., Geballe, T. R., and Webster, A. S. 1989, "Molecular hydrogen line ratios in four regions of shock-excited gas", *M.N.R.A.S.*, **236**, 409-423.

Chernoff, D. F. 1987, "Magnetohydrodynamic Shock Waves in Molecular Clouds", *Ap. J.*, **312**, 143-169.

Chernoff, D. F., Hollenbach, D. J., and McKee, C. F. 1982, "Molecular Shock Waves in the

BN-KL Region of Orion", *Ap. J. (Letters)*, **259**, L97-L102.

Chernoff, D. F., and McKee, C. F. 1990, "Shocks in dense molecular clouds", in *Molecular Astrophysics*, ed. T. R. Hartquist, (Cambridge: Cambridge Univ. Press), pp. 360-373.

Draine, B. T. 1980, "Interstellar Shock Waves with Magnetic Precursors", *Ap. J.*, **241**, 1021-1038 (erratum: *Ap. J.*, **246**, 1045).

Draine, B. T. 1986, "Multicomponent, reacting MHD flows", *M.N.R.A.S.*, **220**, 130-148.

Draine, B. T., and Roberge, W. G. 1982, "Origin of the Intense Molecular Line Emission from OMC-1", *Ap. J. (Letters)*, **259**, L91-96.

Draine, B. T., Roberge, W. G., and Dalgarno, A. 1983, "Magnetohydrodynamic Shock Waves in Molecular Clouds", *Ap. J.*, **264**, 485-507.

Hartquist, T. W., Flower, D. R., and Pineau des Forets, G. 1990, "Shock chemistry in diffuse clouds", in *Molecular Astrophysics*, ed. T. R. Hartquist, (Cambridge: Cambridge Univ. Press), pp. 99-112.

Heiles, C. E. 1991, this volume.

Hollenbach, D. J., Chernoff, D. F., and McKee, C. F. 1989, "Infrared diagnostics of interstellar shocks", in *Infrared Spectroscopy in Astronomy*, ed. B. H. Kaldeich (Noordwijk: ESA Publications Division), pp. 245-258.

Moorhouse, A., Brand, P. W. J. L., Geballe, T. R., and Burton, M. G., "Velocity profiles of high-excitation molecular hydrogen lines", *M.N.R.A.S.*, **242**, 88-91.

Mullan, D. J. 1971, "The structure of transverse hydromagnetic shocks in regions of low ionization", *M.N.R.A.S.*, **153**, 145-170.

Roberge, W. G., and Draine, B. T. 1990, "A New Class of Solutions for Interstellar MHD Shock Waves", *Ap. J.*, **350**, 700-721.

Shull, J. M., and Draine, B. T. 1987, "The Physics of Interstellar Shock Waves", in *Interstellar Processes*, ed. D. Hollenbach and H. Thronson (Dordrecht: Reidel), 283-319.

Smith, M. D., and Brand, P. W. J. L. 1990a, "Cool C-shocks and high-velocity flows in molecular clouds", *M.N.R.A.S.*, **242**, 495-504.

Smith, M. D., and Brand, P. W. J. L. 1990b, "H_2 profiles of C-type bow shocks", *M.N.R.A.S.*, **245**, 108-118.

Smith, M. D., Brand, P. W. J. L., and Moorhouse, A. 1990, "Bow shocks in molecular clouds: H_2 line strengths", *M.N.R.A.S.*, submitted.

Wardle, M. 1990a, "The stability of magnetohydrodynamic shock waves in molecular clouds", *M.N.R.A.S.*, in press.

Wardle, M. 1990b, "The instability of radiative C-type shock waves", *M.N.R.A.S.*, submitted.

Wardle, M. 1991, in preparation.

Wardle, M., and Draine, B. T. 1987, "Oblique Magnetohydrodynamic Shock Waves in Molecular Clouds", *Ap. J.*, **321**, 321-333.

THERMAL PROCESSES IN MOLECULAR GAS

J.P. Chièze and C. de Boisanger
Commissariat à l'Energie Atomique
Centre d'Etudes de Bruyères-le-Châtel
Service PTN
91680 Bruyères-le-Châtel, France

ABSTRACT. The dynamics of the cold atomic and molecular gas, on which we focus here, is strongly affected by non equilibrium heating and cooling processes. We give two different examples, in which the breaking of the thermal balance is due respectively to variations of the incident ultraviolet radiation flux, and non equilibrium abundances of H_2 molecules in molecular clouds envelopes. Fluctuations of the ultraviolet radiation flux in clumpy molecular cloud envelopes result in the formation or the destruction of dense regions. Large density contrasts, greater than one order of magnitude, are easily achieved in cloud regions of moderate visual extinction. Condensation or expansion develop on quite short time scales, of the order of a few tenth of million year, and induce collective motions which can feed turbulence.

Another example of the importance of out of equilibrium thermochemical processes is furnished by the study of the $H - H_2$ transition layers in molecular clouds envelopes. They turn out to be unstable against convection–like motions, driven by the energy released by H_2 photodestruction. The gas velocities involved in these motions are, again, typical of the observed turbulent velocity in clouds envelopes.

Keywords: Interstellar medium: clouds – Interstellar medium: kinematics and dynamics of – Interstellar medium: thermal processes.

1. Introduction

Roughly speaking, the interstellar gas has the unusual property to get cooler with increasing density. However, this is only true if one assumes *thermal balance* between the various heating and cooling processes. Large departures from thermal equilibrium are easily achieved in (hydro)dynamical processes, which can result from an instability – such as thermal instability – or can be triggered from the environment. Since the work of Field, Goldsmith and Habing (1969), thermal instability has been extensively studied, in particular in the context of HI cloud formation and cooling flows. Recently, Balbus (1986,

1988) and Balbus and Soker (1989) have established a close connection between thermal instability and convection. Hattori and Habe (1990) have performed two–dimensional hydrodynamical calculations of thermal instability in cooling flows.

The heating or cooling times in the envelopes of molecular clouds are of the order of a few tenth of million years. By no means the thermodynamic evolution of the gas can be properly mimicked by a definite polytropic transformation, since the polytropic exponent Γ (such that locally $P \propto \rho^\Gamma$) depends explicitely on the rate of variation $\omega = 1/\rho \, d\rho/dt$ of the density. This is due to the fact that the heating and cooling terms do have finite characteristic times which have to be compared to the hydrodynamical time scale ω. Accordingly, reasonable approximations to the heating and cooling term, when coupled to hydrodynamics, reveal a new class of phenomena closely related to clump formation and turbulence in molecular cloud envelopes.

The thermal balance of the gas can be affected in three different ways. First, variations of the density influence markedly the collisional cooling terms (per atomic hydrogen nucleus), since they are proportional to the density. Second, variations of the incident ultraviolet radiation flux alter the dust grain photoelectric heating rate, including or not heating by PAHs, which are dominant in regions with visual extinction $A_v \leq 3 \ mag$. Large ultraviolet intensity fluctuations are suggested by the analysis of the tranfer in clumpy media (Boissé 1990). Finally, various species which play an important role in thermal processes, such as CO, C^+ or H_2, can be brought far from their chemical equilibrium abundances by the hydrodynamical evolution of the gas. In each case, the actual thermal pressure of the gas can be very different from the estimates based on the assumption of thermal balance. The resulting non uniform pressure or temperature variations may dominate the dynamics of the interstellar gas.

In the followings, we illustrate these points by two examples, in which we analyze some consequences of local variations of the UV incident flux, due to density inhomogeneity, and the effect of turbulence in the transition region of a molecular cloud where the gas turns from atomic (HI) to molecular (H_2). In the first case, the extinction variations result essentially in the condensation of the shielded regions. In the second case, a strong instability is shown up, which is closely akin to convection.

2. The gas response to fluctuations of the UV radiation field

Many clumps in clouds envelopes have masses far below the Jeans mass, and exhibit clear cut edges, hardly reconciliable with a strict gravitational origin. Up to extinctions $A_v \sim 3 \ mag$, the interstellar ultraviolet radiation flux regulates the physical state of large amounts of molecular gas. Thus, it should play an important role in the formation of such clumps, through the ionization of carbon atoms and dust grain photoelectric heating. Very small particles or PAHs may be a major source of heating of the low density gas (d'Hendecourt and Léger 1987, Lepp and Dalgarno 1988, Puget and Léger 1989). However, charge effects both on the dust grain or PAH photoelectric heating rates push the threshold of thermal instability down to a density lower than $n_H \sim 1 \ cm^{-3}$.

Even in the absence of thermal instability, spatial fluctuations of the local UV intensity generate a non uniform thermal balance and substantial density variations are achieved when the gas tends to recover local thermal equilibrium. This dynamical process sets matter in motion, for at least two reasons: condensation (expansion) flows are expected in cooling (heating) regions, and the formation of density inhomogeneities upsets the local hydrostatic equilibrium, giving rise to bulk motions driven by the Archimedes force.

The description of the actual dynamics requires a careful treatment of the hydrodynamical interactions between regions with different thermal evolutions (see Sect.3). However, useful informations can be easily obtained if one assumes that the perturbed regions roughly remain in pressure equilibrium with the unperturbed surroundings.

We first present the *isobaric* gas response to an external attenuation of the UV field, which, due to the strong decrease of the dust grains photoelectric heating rate, leads to the condensation of the cooling gas. The adopted unpertubed UV intensity, at zero visual extinction, is (Mathis *et al.* 1983): $\phi_0 = 2.5 \ 10^{-8} - 4.6 \ 10^{-8} \ photon \ cm^{-2} \ s^{-1} \ Hz^{-1}$ at 1000 and 1101 Å. Indirectly, the abundances of the major cooling and heating agents (C^+, CI, OI, CO, H_2) are altered during the evolution. The time dependent abundances are calculated using a realistic chemical network (Pineau des Forêts *et al.* 1988) and the method described by Heck *et al.* (1990).

Regarding H_2 photodestruction, self-shielding is treated following de Jong *et al.* (1980), with the rate:

$$R_D(H_2) \ = \ 4. \ 10^{-11}\beta(\tau_{H_2})exp(-2.5A_v) \quad s^{-1} \tag{1}$$

where A_v is the visual extinction, $\beta(\tau_{H_2})$ is the probability that UV photons penetrate in the considered region of a cloud with optical depth τ_{H_2}, related to the H_2 column density by (de Jong *et al.* 1980):

$$\tau_{H_2} = 1.2 \ 10^6 \ \left(\frac{\mathcal{N}(H_2)}{10^{20} \ cm^{-2}} \right) \left(\frac{\delta V_D}{1 \ km \ s^{-1}} \right)^{-1} \tag{2}$$

where δV_D is the gas velocity dispersion. A difficulty arises in the evaluation of this latter expression, since the total shielding H_2 column density, $\mathcal{N}(H_2)$, is *a priori* unknown. It could be calculated self-consistently at each point of a cloud model, at the cost, however, of a loss of generality. We choose to estimate the value of the optical depth τ_{H_2} as a function of the visual extinction A_v, using the observed correlation between $\mathcal{N}(H_2)$ and the color excess E(B-V) (Savage *et al.* 1977), and assuming a constant color ratio R=A_v/E(B-V)=3.2. The resulting function $\tau_{H_2}(A_v)$ exhibits a sharp discontinuity of the H_2 self-shielding at the visual extinction $A_v^* = 0.25 \ mag$.

The gas density evolution (at constant pressure P_0) which follows the extinction perturbation ΔA_v is described by the equation:

$$\frac{dn}{dt} = \frac{2}{5} \frac{n\mathcal{L}}{P_0} + \sum_i \left(\frac{dn_i}{dt} \right)_{ch} \tag{3}$$

where $\mathcal{L}$ is the net cooling rate per unit volume (cooling minus heating), n is the total number density of particles and the index "ch" denotes purely chemical transformations. The temperature T of the gas obeys the equation:

$$\frac{dT}{dt} = -\frac{2}{5} \frac{T\mathcal{L}}{P_0} - \frac{T}{n} \sum_i \left(\frac{dn_i}{dt} \right)_{ch} \tag{4}$$

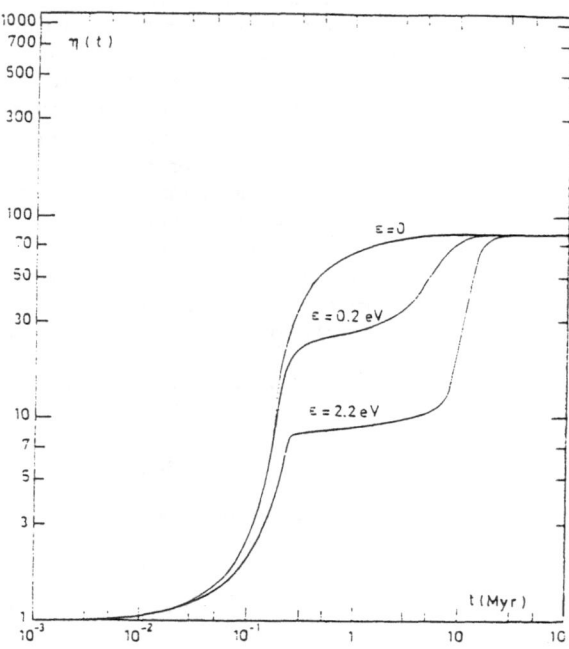

Figure 1: *Time dependent isobaric condensation in low density gas following an extinction variation $\Delta A_v = 1$ mag. Curves are labeled with three distinct values of the kinetic energy released per H_2 formation.*

The condensation factor $\eta(t) = n_H(t)/n_H(t = 0)$ is presented in Fig.1 for the cloud envelope initial conditions $n_{H_0} = 20$ cm^{-3}, $A_{v_0} = 0.25$ mag. The visual extinction is decreased by two magnitudes. The condensation factors are calculated for three different values of the uncertain kinetic energy input due to H_2 formation on dust grains, $\epsilon(H_2)$ (Williams 1987). The condensation factor increases with larger incident UV flux. A typical value $\eta = 10$ is achieved in denser gas, $n_{H_0} = 100$ cm^{-3} with extinction $A_{v_0} = 0.5$ mag (de Boisanger and Chièze 1991).

The process discussed so forth is efficient in producing non gravitationally driven gas condensation by more than one order of magnitude. The amplitude of the condensation is higher in low density gas, in which it exceeds two orders of magnitude, than in $n_H \sim$ 100 cm^{-3} gas for which density contrasts of about 10 are achieved. Quite naturally, a high UV intensity strengthens condensation. Furthermore, condensation factors are quite sensitive – through the heating rate – to the grain size, the photoelectric efficiency, the presence or not of PAHs, and – through the cooling rate – the depletion of carbon. The condensation growth time is always quite short: density enhancements by a factor of 10 are commonly achieved by $t \approx 0.1 - 0.2\ Myr$, which is about two orders of magnitude shorter than the free fall time. The linear size of the condensed regions depends only on the coverage of the perturbation.

3. Two dimensional hydrodynamical simulations

Hydrodynamical simulations, including the relevant heating and cooling terms, confirm the values of the condensation factors, and present further the advantage of a correct estimation of the *velocity field* induced by the extinction perturbation. As an example of such calculations, Fig. 2 represents the gas response to a (gaussian) perturbation of the extinction of 1 *mag*, moving with the velocity $v = 0.2\ kms^{-1}$. The gas is initially thermally balanced and in hydrostatic equilibrium with a density of 30 cm^{-3} at the free surface, and $10^6 cm^{-3}$ at the base of the structure, while the (unperturbed) extinction varies from 0.1 *mag* to about 100 *mag*. Lateral periodic conditions are assumed, together with a fixed boundary at the base, and a free surface maintained at constant pressure throughout the calculation.

The condensation lags slightly behind the position of the perturbation, and induces a convection–like velocity pattern with a maximum velocity $v_{max} = 0.7\ km\ s^{-1}$ after 1 Myr, which steadily increases with time up to 0.9 $km\ s^{-1}$ for $t \geq 3$ Myr. Downwards velocity in the condensed material are set up by the Archimedes force. Two rarefaction waves are naturally driven on both sides of the condensation, which result in gas heating and upwards motions.

The nature of this mechanism favors the growth of "elephant-trunk" shaped condensations, and can be a source for turbulence in cloud envelopes.

4. Instability of the $H - H_2$ transition layer

Molecular clouds are generally closely associated with large amounts of atomic hydrogen. Due to H_2 self–shielding, the transition to molecular hydrogen is very sharp, and occurs in a cloud at an extinction level of $Av \sim 0.25 - 0.5\ mag$. Loosely speaking, the gas located below this level is mostly molecular, while it is mostly atomic above. In other words, large amounts of H_2 molecules are present in the vicinity of a region where they can be quickly photodissociated.

This is not without sequel if one consider that cloud envelopes are turbulent. Since the gas velocity dispersion in cloud envelopes is at least of the order of $\sigma_v = 0.5\ km\ s^{-1}$, mixing should occur accross the transition layer. As a consequence, H_2 molecules advected in the essentially unshielded regions are rapidly destroyed, releasing an amount $\Delta\epsilon = 0.5\ eV$ of kinetic energy per photodestruction. This is a large extra heating term for still H_2-rich rising parcels of gas, which, in most cases, overcomes adiabatic cooling due to their expansion in the decreasing pressure gradient of the cloud envelope.

The problem is closely related to convection, at least in its simplest acceptation. Consider a gas bubble perturbed from its original position in the shielded region. We follow its motion in the $HI - H_2$ transition layer, that is the region where the UV optical depth τ_{H_2} varies abruptly, by calculating the thermal balance between the various cooling

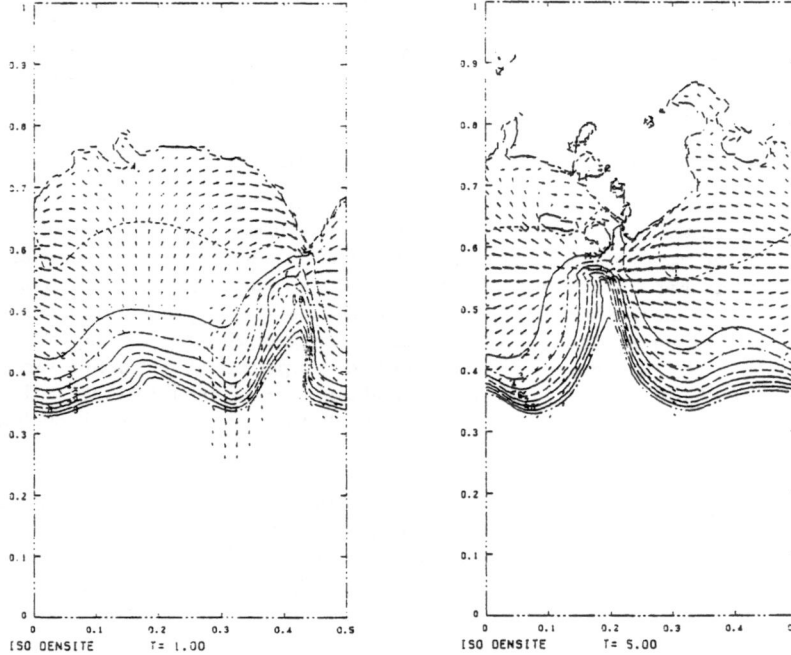

Figure 2: *Two dimensional simulation of a cloud envelope perturbed by an extinction variation $\Delta A_v = 1$ mag moving with a velocity of 0.2 km s^{-1}. Density contours are equally spaced from 50 to 850 cm^{-3}. Low lying dense and heavily obscured material is unaffected by the perturbation. The maximum gas velocity is 0.72 km s^{-1} after 1 Myr (left) and 0.90 km s^{-1} at $t = 5$ Myrs (right). Horizontal and vertical scales are in parsecs.*

and heating processes, including notably cooling due to the expansion of the bubble and heating due to H_2 photodissociation. As in the standard analysis of convection, it is assumed that the rising gas parcell maintains pressure equilibrium with its unperturbed surroundings. If the temperature in the rising bubble turns out to be greater than the ambient temperature, the Archimedes buoyant force will reinforce its motion.

The result is that, despite the fact that the outward entropy gradient in a molecular cloud envelope is markedly positive – which means convective stability according to the classical *adiabatic* Schwarzschild criterion – the energy deposition by H_2 photodissociation is by far sufficient to induce large buoyant forces. By contrast with convection as generally treated in the context of stellar models, the motion of moving fluid elements in a molecular cloud are strongly non adiabatic, due to the presence of many heating and cooling processes, always unbalanced in a time dependent thermodynamical evolution. Calculations show that the Archimedes acceleration is in most cases greater than the gravitational acceleration by more than a factor of 10. Neglecting viscosity (and hydrodynamics) it would result in bubbles rising soon at supersonic velocities. Thus, if some amount of turbulence is required to trigger the instability, it is likely to be amplified in these regions of cloud envelopes.

5. Conclusions

The examples we have presented stress the role of the heating and cooling processes both on non gravitational condensation of molecular gas and the generation of a complex velocity field, comparable in magnitude to the observed turbulent velocities. By itself, the UV process does not produce very high density clumps, but since the Jeans mass varies as η^{-2} during an isobaric condensation, local decrease of the Jeans mass by two or three orders of magnitude are expected to occur in regions shielded from UV exposure by preexisting clumps. The condensation time scales are short, of the order of a few tenth of million years. Since the driving perturbations are not stationary, this means that molecular gas may experience large density and extinction variations with such short turn over time scales. From a complementary point of view, this points to non–equilibrium molecular chemistry. It should be noticed that calculations of chemical abundances taking into account short time scale mixing and density variations are effective in producing much larger amounts of complex molecules, together with high C/CO ratios, than equilibrium calculations (Chièze and Pineau des Forêts 1989, Chièze et al. 1991). Finally, thermal processes are able to induce large gas velocities, and thus should be considered as a natural driving term for cloud turbulence.

Acknowledgements

The 2–D hydrodynamical simulations have been performed in collaboration with J. Gambard and B. Meltz, at Département de Mathématiques Appliquées, Centre d'Etudes de Limeil (Commissariat à l'Energie Atomique).

References

Balbus, S.A.: 1986, *Astrophys. J.*, **303**, L79
Balbus, S.A.: 1988, *Astrophys. J.*, **328**, 395
Balbus, S.A. and Soker, N.: 1989, *Astrophys. J.*, **341**, 611
Black, J.: 1987, in *Interstellar Processes*, eds Hollenbach and Thronson, Reidel Publishing Company, p. 731.
de Boisanger,C. and Chièze, J.P.: 1991, *Astron. Astrophys. (in press)*
Boissé, P.: 1990, *Astron. Astrophys.* **228**,483.
Chièze, J.P., Pineau des Forêts, G.: 1987, *Astron. Astrophys.* **183**, 98.
Chièze, J.P., Pineau des Forêts, G.: 1989, *Astron. Astrophys.* **221**, 89.
Chièze, J.P., Pineau des Forêts, G., Herbst, E.: 1991, *Astrophys. J. (in press)*
Elitzur, M., Watson, W.D.: 1978, *Astron. Astrophys.* **70**, 443.
Field, G.B., Goldsmith, D.W., Habing, H.J.: 1969, *Astrophys. J. Letters* **155**, 149.
Graff, M.M.: 1989, *Astrophys. J.* **339**, 289.
Hattori, M., and Habe, A.: 1990, *Monthly Notices Roy. Astron. Soc.* **242**, 399.
Heck, L., Flower, D.R., Pineau des Forêts, G.: 1990, *Comp. Phys. Comm.* **58**, 169.

204

d'Hendecourt, L.B., and Léger, A.: 1987, *Astron. Astrophys.* **180**, L9.

de Jong, T., Dalgarno, A., Boland, W.: 1980, *Astron. Astrophys.* **91**, 68.

Lepp, S. and Dalgarno, A.: 1988, *Astrophys. J.* **324**, 553.

Mathis, J.S., Mezger, P.G., Panagia, N.: 1983, *Astron. Astrophys.* **128**, 212.

Meyer, J.P.: 1989, in *Cosmic Abundance of Matter* , ed C.J. Waddington, AIP Conference Proceedings 183, The American Institute of Physics, New York, p. 245.

Pineau des Forêts, G., Flower, D.R., Dalgarno, A.: 1988, *Monthly Notices Roy. Astron. Soc.* **235**, 621.

Puget, J.L. and Léger A.: 1989, *Ann. Rev. Astron. Astrophys.* **27**, 161.

Savage, B.D., Bohlin, R.C., Drake, J.F., Budich, W.: 1977, *Astrophys. J.* **216**, 291.

Stutzki, J., Stacey, G.J., Genzel, R., Harris, A.I., Jaffe, D.T., Lugten, J.B.: 1988, *Astrophys. J.* **332**, 379.

Williams, D.,A.: 1987, in *Physical Processes in Interstellar Clouds*, eds Morfill and Scholer, Reidel Publishing Company, p. 377.

ON THE ORIGIN OF BROAD LINE WINGS IN MOLECULAR CLOUD SPECTRA

Bruce G. Elmegreen
IBM Thomas J. Watson Research Center
P.O. Box 218, Yorktown Heights, N.Y. 10598 USA

ABSTRACT. The broad line wings in molecular cloud spectra are proposed to result from strong magnetic waves on the periphery of dense cores and in the intercore regions where the Alfvén velocity should be larger than average. The observed line profiles are reproduced by a simple but realistic model, and the ratio of the broad to the narrow line components is found to equal approximately three, independent of cloud parameters, as long as the core/intercore contrast in the local average density is sufficiently large. Interactions between the magnetic waves should produce dense clumps in the non-linear splash regions between converging flows.

1. Line Profiles in Magnetic Clouds with Density Gradients

The broad line wings found in molecular cloud spectra by Blitz and Stark (1986) and Falgarone and Phillips (1990; see also Phillips, this conference) are proposed to result from non-linear magnetic wave motions in regions with gradients in the average density of gas. This result follows from a cloud model in which most of the mass is in the form of dense clumps that are clustered together into regions of higher than average clump number density. The regions with high average densities will be called cores here. Many of the cores in molecular clouds appear to be virialized. The regions with lower-than-average densities will be called intercore regions. Because the actual radiating matter is presumed to be at the uniformly high density of the clumps everywhere in the cloud, while the average density, i.e., averaged over many clumps, varies from place to place, this model agrees with the observation that the ratios of the 2-1 to the 1-0 line strengths are somewhat uniform throughout a cloud, even though the brightness temperature varies from the cores to the intercore regions (see Solomon or Falgarone, this conference).

Molecular clouds are considered to be magnetic, with average magnetic energies comparable to the gravitational and kinematic energy densities (Myers and Goodman 1988). This near equality implies that the "average" Alfvén speed in a cloud is comparable to the virial velocity. But for a cloud with a core/intercore structure, the Alfvén speed is rarely equal to its average value; it varies instead from a low value in the cores to a high value in the intercore regions. If the core/intercore average density contrast is sufficiently large, then the Alfvén speed in the intercore region can significantly exceed the virial velocity. Gas motions at such high velocities will produce excess radiation in the line wings of molecular cloud spectra, giving the impression that material is moving at greater than escape speeds, although in fact this motion should not lead to any real loss of matter from the cloud.

The molecular line profiles from magnetic clouds with average density gradients can be determined approximately by assuming that the volume emissivity is proportional to the average density and that the normalized line profile is Gaussian. This assumption of *locally* Gaussian motions is consistent with expectations from energy equipartition for an ensemble of clumps that collisionally interact (probably via non-touching magnetic linkages). Then the contribution to the line profile from the gradient region is given by

$$I(\upsilon) \propto \int_{r_1}^{r_2} \frac{n(r)}{\sigma(r)} \, e^{-\upsilon^2/2\sigma(r)^2} dr \tag{1}$$

where r_1 and r_2 span the distance over which the density and Alfvén speed vary.

Now we identify $\sigma(r)$ with the Alfvén speed at r and consider that $\sigma \propto n^{-\beta}$ where β equals 0.5 for a uniform field strength and $\beta = 0.25$ for a slight field divergence along the path from the core to the intercore region. (If $\beta = 0$, then there would be no broad line wings in this model.) There is a critical assumption at this step, that the magnetic wave motions contributing to σ are *non-linear*; otherwise the Alfvén speed would not correspond to a material speed that can be observed by Doppler shifts in a spectrum. Non-linear waves contain mass motions both parallel and perpendicular to the mean field direction, and they also change the average density as a result of these motions. Linear waves produce only slow motions perpendicular to the field, and extremely slow motions and small density changes parallel to the field. Thus *our identification of broad line wings with magnetic motions in the low density regions of clouds requires that the perturbed field strengths in these regions be a large fraction (0.3 – 1) of the mean field strength there.*

The large-scale, core/intercore density variations in a cloud are assumed to result from local equilibria between pressure gradients and self-gravitational forces. This gives $n(r) \propto r^{-\alpha}$ for $\alpha \sim$ 1 to 2. We identify n_2 with the density in the core and n_1 with the density in the intercore region; σ_2 is the dispersion in the core.

For the sake of generality, we also add a contribution to the line profile from a purely Gaussian velocity distribution in the core region. This cloud core emission can be strong or weak compared to the gradient emission written in equation (1), depending on the relative emissivity of the core and intercore gas. This relative emissivity cannot be obtained from first principles, but presumably depends on the evolutionary state of the cloud if, for example, the cores become denser and more massive with time from accretion, or if the intercore region becomes denser with time from core ablation.

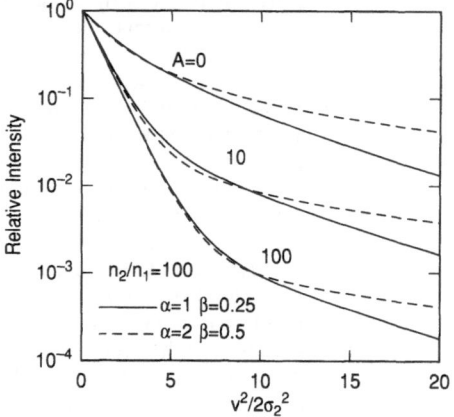

Figure 1. Sample line profiles from equation (2).

The resultant line profile, normalized to the peak value, is

$$\frac{I(\upsilon)}{I(0)} = \frac{\displaystyle\int_C^1 x^\gamma e^{-x\upsilon^2/2\sigma_2^2}dx + Ae^{-\upsilon^2/2\sigma_2^2}}{\displaystyle\frac{1}{\gamma+1}\left[1 - C^{(\gamma+1)}\right] + A}$$

(2)

where $\gamma = (\alpha - \alpha\beta - 1)/2\alpha\beta$ and $C = (n_1/n_2)^{2\beta}$. This profile is shown in Figure 1 for various α, β, and A, as indicated. The coordinates for the figure are such that a Gaussian line would appear as a straight line. The excess above the extrapolated emission on the right is the line wing. These theoretical spectra resemble the observed spectra in Falgarone and Phillips (1990). Variations in the relative strength of the line core and the line wing are interpreted in terms of variations in the core and intercore masses or emissions.

The theoretical line profiles were fitted to two Gaussians for comparison with the results of Falgarone and Phillips (1990), who approximated the observed spectra in this way. The ratios of the broad to the narrow dispersions in these fits are shown in Figure 2 as functions of the core to intercore density contrast, and for various α, β, and A. The ratio is approximately 3, as observed for molecular clouds, for a reasonable range of parameter values as long as the core-to-intercore average density contrast is not too small.

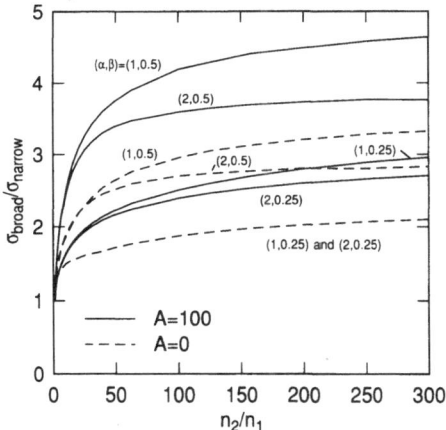

Figure 2. The ratio of Gaussian fit velocity dispersions for the broad and narrow components of the line profile generated by equation (2).

Evidently the velocity distribution functions in magnetic clouds containing strong wavelike motions and gravitationally relaxed density gradients can produce spectral line profiles from Doppler shifts that have core-wing structures similar to those observed by Falgarone and Phillips (1990).

2. Splashy Collisions Between Non-Linear Alfvén Waves

The previous section considered the effect of non-linear waves on the velocity distribution function in clouds with density gradients. These same waves also have the interesting property that interactions between them can produce sharp density peaks which may be visible in molecular clouds as clumps, filaments, or sheets, depending on the wave geometry and viewing perspective. The density peaks form when the convergence of material flowing along the field lines in the non-linear wave pushes the neutrals through the ion fluid, which is attached to the field lines. The neutrals are then squeezed into a clump, filament or sheet, depending on the shape of the waves (the ion fraction continuously adjusts as this squeezing occurs, so ions should always be present in the neutral fluid.) This result follows from the non-linear equations for Alfvén waves including ion-neutral diffusion. If diffusion were not included, there would still be a density peak at the interface between the colliding wavetrains, but it would not be as large because the magnetic pressure would generally dominate the thermal pressure for typical turbulent Mach numbers in molecular clouds, i.e., for low temperatures.

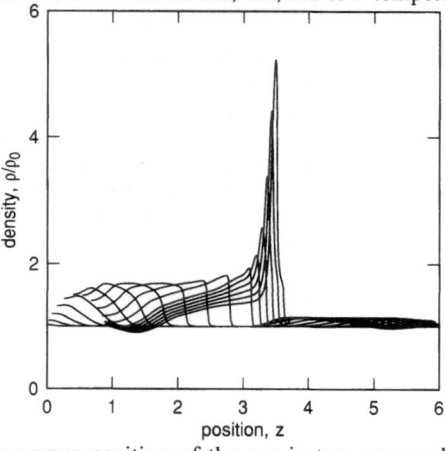

Figure 3. Density versus position of the gas in two approaching waves, with different times indicated by lines, for times 0.25 to 3 in steps of 0.25 and 3.1 to 3.5 in steps of 0.1 time units.

An example of this wave-wave interaction is shown in Figure 3, which plots the density as a function of position for different times (each line is a different time). Two waves approach each other from each side of the grid. The wave from the left is strong, driven by a transverse sinusoidal perturbation at the left hand end of the grid with a velocity equal to the unperturbed Alfvén speed. The wave from the right is weaker, driven by a transverse motion at half the Alfvén speed. Both waves contain streaming motions parallel to the field, with amplitudes equal to ~0.6 and ~0.15 of the unperturbed Alfvén speed (not shown). The plotted quantity is the gas density, which increases by only a factor of ~1.5 in the strong wave until the two waves interact. Then the density increases by at least a factor of ~5, at which point the calculation had to be stopped because it could not follow the sharp gradients.

The number of grid spacings in the calculation is 600, and the grid points move with the parallel flow in a Lagrangian coordinate system. The computer program was tested for accuracy by comparing the speed and amplitude of the crest of a very weak wave with the Alfvén speed and expected constant amplitude from the linear theory. The program was also tested by comparing the quantities on either side of a shock front with the theoretical jump conditions for a shock. Shock fronts readily form when the waves are strong. A shock can be seen in Figure 3 when the wave from the left reaches a grid position of about 2.5. The accuracy of the computer code is estimated to be better than 3%.

The formation of density peaks between colliding non-linear waves should be a common event in molecular clouds, where strong waves should abound because of virialized motions of the cores in the potential well of the cloud complex. The dissipation of energy that accompanies such shocks will not necessarily halt their formation because the loss of energy corresponds to a loss of pressure, which should be followed by a contraction of the whole cloud. This contraction then pumps more energy into the cloud (in the form of compressional energy, PdV), and the cloud remains in approximate virial equilibrium. The source of the energy is gravity. Non-linear magnetic waves should be present in clouds even in the absence of stars or other direct stirring sources. They can account for the observed line profile and for some of the small scale clumpy structure.

A more complete description of this work is in Elmegreen (1990).

References

Blitz, L., and Stark, A.A. 1986, *Astrophys.J.(Letters)*, **300**, L89.
Elmegreen, B.G. 1990, *Astrophys.J.(Letters).*, in press.
Falgarone, E., and Phillips, T. 1990, *Astrophys.J.*, **359**, 399.
Myers, P.C., and Goodman, A.A. 1988, *Astrophys.J.(Letters)*, **326**, L27.

THE DEVELOPMENT OF MODES OF THERMAL INSTABILITY IN A NON-STATIONARY MEDIUM (PRODUCTION OF OH MASER CONDENSATIONS)

V. V. BURDYUZHA
Astro-Space Centre Lebedev's Physical
Institute Academy of Sciences of the USSR
Profsoyuznaya,84/32, 117810, Moscow USSR

ABSTRACT.   In a nonstationary medium (behind the shock front, for instance) the development of isobaric and adiabatic modes of the thermal instability are more preferable. Some examples of the fragmentation of the medium on clouds in case of OH masers are given.

In a nonstationary medium the development of modes of thermal instability is more stable than in a stationary that and the increments of increase of perturbation may be rather high (Hunter, 1970; Burdyuzha, Ruzmaikina, 1974; Schekinov, 1978). In real conditions the situation with nonperturbation nonstationary state is realized more often. That kind the state of gas is after passage of shock waves. The production of maser condensations OH in envelopes of young massive stars and in envelopes of IR stars by means of the fragmentation of the medium in result of thermal instability behind the shock front was considered in papers (Burdyuzha, Ruzmaikina, 1974; Burdyuzha, Ruzmaikina, 1975). It seems that the fragmentation of the medium on clouds in envelopes of young massive stars (the cloudy structure is observational fact Reid (1990)) takes place probably because of realization of condensational (isobaric) mode of thermal instability since in the case the perturbation for the cooling from

$$T_2 \sim 10^4 \text{ K to } T \sim 10^2 \text{ K} \quad \text{can increase from } 10^2 \text{ to } 10^4$$

times if the law of them the increase is

$$\frac{\delta n}{n} = -\frac{\delta T}{T} \sim (\frac{T_2}{T})^{2-\alpha} \qquad (1)$$

here $0 \leqslant \alpha \leqslant 1$ (see in detail Burdyuzha,Ruzmaikina,1974).

Let's consider and compare the role of condensational and adiabatic mode of thermal instability forming behind the shock wave front; for instance in IR stars envelope. The criterion of condensation mode development of thermal instability, that is formed behind the shock wave front, is fulfilled in the following inequality

$$(\frac{\partial L}{\partial T})_p - \frac{L}{T} < 0 , \qquad (2)$$

where L - is the function of cooling. The criterion of adiabatic mode development is fulfilled in the following inequality

$$\frac{1}{\gamma-1} \frac{\rho}{T} (\frac{\partial Z}{\partial\rho})_T + (\frac{\partial Z}{\partial T})_\rho - \frac{Z}{T} < \frac{\gamma}{(\gamma-1)^2} \frac{R}{\mu} (4 \frac{d\ln\rho}{dt} - \frac{1}{Z} \frac{d\ln T}{dt}),$$

$$(3)$$

where γ is adiabatic index; R is equal to universal gas constant; $Z = (L-\Gamma)$ is the generalized loss function; L and Γ is the cooling and heating velocity per mass unit.

Because of the critical scale defined by thermal conductivity the additional condition of thermal instability is the fulfillment of the following inequality $\lambda > l_{cr}$

$$l_{cr} \sim (\frac{\varkappa T}{\mu m_p n L})^{1/2} , \qquad (4)$$

where $\varkappa$ is the coefficient of thermal conductivity, m_p- is mass of proton.

In case of adiabatic mode

$$l_{cr} < \lambda_{adiab} \ll \lambda_{cold} \qquad (5)$$

and in case of isobaric mode

$$l_{cr} < \lambda_{isob} \overset{\sim}{=} \lambda_{cold} \qquad (6)$$

where $\lambda_{cold} \overset{\sim}{=} c\,\tau_{cold}$ (c - is sound velocity).

We have $\lambda_{adiab} \overset{\sim}{=} \beta c \tau_{cold}$; $\beta \ll 1$ $\qquad (7)$

for adiabatic perturbations.
    If $\lambda_{adiab} > l_{cr}$, then

$$\beta > (\frac{\varkappa T}{\mu m_p nL})^{1/2} \frac{1}{c \, \tau_{cold}} > \frac{1}{c} (\frac{\varkappa}{kn\tau_{cold}})^{1/2}, \quad (8)$$

where k is Boltzman constant.
The velocities of shock waves in giant envelopes are unlikely to be more than $V_{sh} < 20$ km/s (the t^0 behind the
front is estimated as follows $T_2 \approx 3500 \, [V_{sh}/10 \text{ km/s}]^2$
Shull, Hollenbach (1978)). At the degree of ionization
$x \geqslant 10^{-2}$ since $T_2 \geqslant 10^4$K the cooling time may be estimated using our formula (Burdyuzha, Ruzmaikina, 1974)

$$\tau_{cold} \approx \frac{5 \, 10^7 T_2}{x \, n} \qquad s \qquad . \qquad (9)$$

By substituting the value of electron thermal conductivi-
ty $\varkappa \sim 2 \, 10^{-6} T^{5/2}$ since the general role is due to electron excitation of $n < 10^5$ cm^{-3}, and τ_{cold} and n   in
(8) we'll have $\beta > 0.017$ what is compatible to the inequ-
ality $\beta \ll 1$. Therefore $\lambda_{cold} \sim 5 \, 10^{13}$ cm, $\lambda_{adiab} \sim 10^{12}$
cm, $l_{cr} \sim 7 \, 10^{11}$ cm.
    The fulfillment of the criterion (3) appeares to be
more favorable than that of (2) for isobaric perturbations increase. It is explained by the fact that gas temperature behind shock front is decreased because of radiation. The density increases with the characteristic
time close to cooling time because of the pressure being
stable. Therefore the positive value is placed in the
right part of inequality (3). Besides adiabatic mode of
thermal instability can easily develop in stationary
medium with heating source. By using the formula of heating rate in gas due to collision with dust (Oppenheimer, 1977) we have

$$\Gamma = 1/2 \, n_{H_2} n_g \sigma_g v \, k \, (T_d - T_{kin}) \qquad (10)$$

have $\sigma_g v \sim 3 \, 10^{-6} T^{1/2}$; $n_g \sim 3 \, 10^{-12} n_{H_2}$.

At $n_{H_2} \sim 10^5$ cm^{-3} , $T_d \sim 200$ K, $T_{kin} \sim 100$ K

we have $\Gamma \sim 6 \ 10^{-21}$ erg/cm^{3}s .

At this value of heating rate the isobaric mode of thermal instability may develope under very specific conditions only. Therefore in case shock waves in giant and supergiant envelopes are not avar lable the adiabatic mode will cause medium fragmentation. The presence of condensed and rare gas regions is the characteristic feature of adiabatic mode development. The variability of some OH sources in IR stars may be understood in this way. The characteristic time of adiabatic mode development is much less than cooling time i.e. $t_h = (kc)^{-1} \ll \tau_{cold}$. At sound velocity $c \sim 10^5$ cm/s and $k = 1/\lambda_{adiab} \sim 10^{-12}$cm^{-1} t_h is equal to 10^7 s . Cooling rate $\tau_{cold} \sim$ $5 \ 10^8$ s if $n \sim 10^5$ cm^{-3} , $x \sim 10^{-2}$, $T \sim 10^4$ K.
    Therefore the maser condensation OH in envelopes of young massive stars and in envelopes a giants and supergiants can be produced for the fragmentation in result of the development of isobaric or adiabatic modes of thermal instability both in nonstationary and in stationary medium.

References

Burdyuzha V.V., Ruzmaikina T.V. 1974, Soviet. Astron. 18, 201.
Burdyuzha V.V., Ruzmaikina T.V. 1975, Astron. Astrophys. 40, 233.
Hunter J.H. 1970, Astrophys. J. 161, 451.
Oppenheimer M. 1977, Astrophys. J., 211, 400.
Reid M.J. The Paper at the Thirg Haystack conference,1990
Schekinov Yu.A. 1979, Astrophysica, 15, 347.
Shull J.H., Hollenbach D.J. 1978,Astrophys. J., 220, 525.

LARGE-SCALE FRAGMENTATION OF GAS CLOUD ROTATING AT THE GALACTIC CENTER

M. FUJIMOTO and Y. TATEMATSU
Department of Physics, Nagoya University
Furo-cho, Chikusa-ku, Nagoya, 464-01

ABSTRACT. A rotating and slowly-contracting gas cloud is followed in the deep gravitational potential of the galactic center. When the gas density increases as high as $(10^2 - 10^3)H_2$ cm^{-3}, which is more than twenty times as large as the background matter density, the self-gravity of the cloud becomes dominant to govern its dynamical structure. The cloud elongates and then splits into two separated objects, as observed at the centers of IC342, NGC6946, and Maffei 2 where we have two symmetric peaks on the major axis of the ^{12}CO (J=1-0) cloud.

1. Elongation of Gaseous Ellipsoid Rotating at the Galactic Center

A self-gravitating uniform gaseous ellipsoid rotates and contracts slowly in the potential $\Phi = AX^2 + BY^2 + CZ^2$, where X, Y, and Z denote the coordinates referred to the galactic center, and A, B, and $C > 0$. When $A < B$, Φ represents a bar-like potential elongated in the X-direction in the galactic plane. When Φ rotates slowly, it would mimic a central gravitational potential of the barred spiral. By applying a method developed by Tatematsu and Fujimoto (1990), we can follow the flattening and elongation of a uniform gaseous ellipsoid as in figure 1, where a_1, a_2 are the principal axes of the ellipsoid on the equatorial plane and a_3 is the one coincident with the galactic

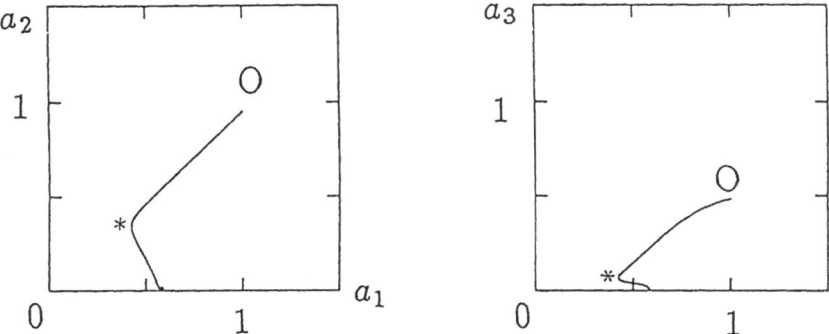

Fig. 1. Elongation and flattening of a contracting homogeneous gaseous ellipsoid.

axis. The gaseous ellipsoid, which was circular initially, elongates quickly at $*$s where $\rho \geq 20\rho_b$, with ρ and ρ_b the gas density of the ellipsoid and the background matter density $(A + B + C)/2\pi G$. The direction of the major axis of the ellipsoid (a_1 in figure 1) is not parallel to that of the background bar potential of $(B - A)/A = 0.1$: The ellipsoidal configuration rotates freely against the background bar-potential.

2. Large-Scale Fragmentation of Elongated Gas Cloud at the Galactic Center

When the nonaxisymmetric contraction proceeds beyond the point $*$ in figure 1 where $\rho > 20\rho_b$, the gaseous ellipsoid spins up and is ever more dynamically subject its gravity and rotation. Then the third-order (pear-shaped) perturbation grows when $a_2/a_1 < 0.33$ (Chandrasekhar 1969), leading to a large-scale fragmentation (fission) of the elongated cloud. Since the density ratio ρ/ρ_b and the axis ratios of the molecular clouds at the centers of IC342 (Lo et al. 1984), NGC6946 (Ball et al. 1985) and Maffei 2 (Ishiguro et al. 1989) exceed these theoretical values, the dumbbell-like distribution of ^{12}CO (J=1-0) line intensity would be due to the fission of high-density, $(10^2 - 10^3)H_2$ cm^{-3}, gas clouds accreted onto and elongated at the galactic center.

3. Numerical Simulation for the Fission of Elongated Gaseous Cloud

In order to see the fission of the elongated gas cloud in figure 1, we use a fluid-particle simulation for following the motion of gaseous component (Fujimoto, Tatematsu and

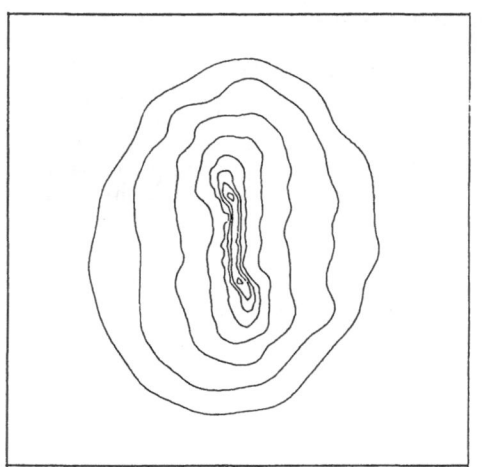 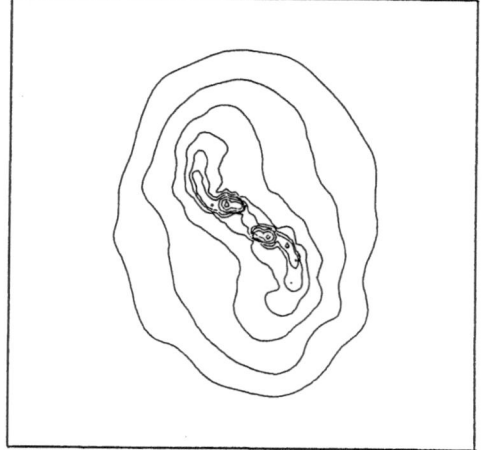

Figure 2   Figure 3

Fig.2. Shocked condensation of gas along the bar-like potential. Note that this potential is due to the self-gravity of the gas.

Fig.3. Fission of the elongated gas cloud.

Miyama 1990). Figure 2 shows that when gas crosses over the major axis of the elongated cloud, it hits the bar-like potential and is compressed as a shock wave along the major axis. Although the background gravitational force stabilizes the gas cloud, the self-gravity due to the strongly compressed gas becomes dominant to divide it into two objects (figure 3). As the contraction proceeds in each fragment, it becomes a gravitationally more bound object and spins up due to the local conservation of angular momentum.

4. Inflow of Gas Clouds into the Galactic Nucleus

When the fission proceeds and each fragment condenses into a more-bound object, it moves as a ballistic particle in the background potential Φ. Since the fragment mass, M_f, is $(10^7 - 10^8)M_\odot$, the dynamical friction due to the background stars, η, is so large that it spirals in towards the center. The decay time, τ, of the orbital motion or the dynamical lifetime of the fragment is estimated as

$$\tau \sim \frac{1}{\eta}$$

$$\sim 2.5 \times 10^7 \left(\frac{r}{1\ kpc}\right)^3 \left(\frac{M_f}{10^8\ M_\odot}\right)^{-1} \left(\frac{\Omega}{100\ km\ s^{-1}\ kpc^{-1}}\right)\ yr$$

where $\Omega \sim \sqrt{2A}$ is the angular velocity of the rigid rotation curve of the galactic center. Thus the supply of mass to the center is estimated as $M_f/\tau \sim 1 M_\odot$ yr^{-1}.

References

Ball, R., Sargent, A.I., Scoville, N.Z., Lo, K.Y., and Scott, S.L. 1985, *Astrophys. J. Letters*, **298**, L21.

Chandrasekhar, S. 1969, *Ellipsoidal Figures of Equilibrium* (Yale University Press, New Haven).

Fujimoto, M., Tatematsu, Y., and Miyama, S.M. 1990, *Publ. Astron. Soc. Japan*, 42, No.3 in press.

Ishiguro, M., Kawabe, R., Morita, K.-I., Okumura, S.K., Chikada, Y., Kasuga, T., Kanzawa, T., Iwashita, H., Handa, K., Takahashi, T., Kobayashi, H., Murata, Y., Ishizuki, S., and Nakai, N. 1989, *Astrophys. J.*, **344** 763.

Lo, K.Y., Berge, G.L., Claussen, M.J., Heiligman, G.M., Leighton, R.B., Masson, C.R., Moffet, A.T., Phillips, T.G., Sargent, A.I., Scott, S.L., Wannier, P.G., and Woody, D.P. 1984, *Astrophys. J. Letters*, **282**, L59.

Tatematsu, Y., and Fujimoto, M. 1990, *Publ. Astron. Soc. Japan*, **42** 217.

V - SMALL SCALE STRUCTURE

CLOUDS, CORES, AND STARS
IN THE NEAREST MOLECULAR COMPLEXES

P. C. MYERS

Harvard-Smithsonian Center for Astrophysics,
60 Garden Street, Cambridge, MA 02138, USA
E-mail MYERS@CFA.BITNET

(Received : 1 October, 1990; accepted : 1 November, 1990)

Abstract. The properties and structure of six molecular complexes within 500 pc of the Sun are described and compared. They are generally organized into elongated filaments which appear connected to less elongated, more massive clouds. Their prominent star clusters tend to be located in the massive clouds rather than in the filaments. The complexes have similar structure, but big differences in scale, from a few pc to some 30 pc. They show a pattern of regional virial equilibrium, where the massive, centrally located clouds are close to virial equilibrium, while the less massive filaments and other small clouds have too little mass to bind their observed internal motions. Complexes can be ranked according to increasing size, mass, core mass, and the mass and number of the associated stars : they range from Lupus to Taurus to Ophiuchus to Perseus to Orion B to Orion A. The cores in nearby complexes tend to have maps which are elongated, rather than round. The core size, velocity dispersion, and column density of most cores are consistent with virial equilibrium. Cores in Orion tend to exceed cores in Taurus in their line width, size, temperature, mass, and in the mass of the associated star, if any. Stars in Orion tend to be more numerous and more massive than in Taurus, while those in Taurus tend to be more numerous and more massive than in Lupus. The mass of a core tends to increase with the mass of the cloud where it is found, with the mass of the star cluster with which it is associated, and with its proximity to a star cluster. These properties suggest that complexes and their constituent cores and clusters develop together over time, perhaps according to the depth of the gravitational well of the complex.

Keywords : star formation, molecular clouds

1. Introduction

This article summarizes and compares properties of six molecular cloud complexes within 500 pc of the Sun : Lupus, Taurus-Auriga, Ophiuchus, Perseus, Orion B, and Orion A. The dimensions, density, structure, internal motions, and prevalence of equilibrium are discussed on the scale of the complex and on the scale of individual dense cores. The stellar content in individual stars and in clusters is described and compared. Interrelations among the complexes, and their cores and stars are discussed. This paper focuses on a cloud complex as a system, and touches only briefly on formation of individual stars. For more detailed discussion of star formation, see the papers in these proceedings by Evans, Lada, and Larson.

2. Complexes

The complexes in Lupus (Murphy, Cohen, and May 1986; Krautter 1990), Taurus (Ungerechts and Thaddeus 1987), Ophiuchus (Loren 1989a,b), Perseus (Bachiller, Cernicharo, and Duvert 1985), Orion B (Lada 1990), and Orion A (Maddalena and Thaddeus 1986) appear similar in their overall structure, in that they are generally organized into elongated filaments. The filaments terminate in rounder, more massive structures in Taurus, Ophiuchus, and Perseus. These more massive clouds contain the embedded star clusters "Group I" in Taurus (Jones and Herbig 1979), the cluster associated with HD147889 in Ophiuchus (Lada, Wilking, and Young 1989), and the NGC1333 cluster in Perseus (Strom, Vrba and Strom 1976). Also, the Orion A filament terminates in the Trapezium and Kleinman-Low cluster, whose molecular gas environment is much denser than, although not distinct in shape from, that in the elongated cloud L1641 to the South. In Lupus, the small cluster which contains HR 5999 and some 20 lower-mass stars appears at the end of a filament of high extinction (Schwartz 1977). Thus most of the embedded star clusters in these six complexes lie at the ends of filaments, where in three cases they are surrounded by unusually massive and less elongated concentrations of gas.

These similarities of structure and stellar concentration in complexes are accompanied by marked differences in the "scale" of several complex properties : the size, column density, mass, and velocity dispersion increases significantly from Lupus to Perseus to Orion. For example, the linear extent of the CO emission map in Lupus is smaller than in Orion by a factor ~ 5. Furthermore the properties of associated stars also increases in the same sense : from Lupus to Perseus to Orion, the number of stars, the number of massive stars, and the degree of clustering of stars also increase. For example, there are some 50 T Tauri stars known in Lupus, but some 1000 stars known in Orion A. Taking these factors into account, we can rank the "scale" of the six complexes under consideration from Lupus, where it is smallest, to Taurus, to Ophiuchus, to Perseus, to Orion B, to Orion A.

The brightest and most massive parts of many molecular cloud complexes appear close to virial equilibrium (Larson 1981), but detailed mapping of complexes shows that in many cases the less massive and filamentary clouds farther from the center of a complex tend to be subvirial : they have too little mass to bind the internal motions in their line widths. This property is evident in Taurus from the CO maps of Ungerechts and Thaddeus (1987), and in Ophiuchus from the ^{13}CO maps of Loren (1989a, b). This interesting property is not fully understood. It is also of interest because it is related to the presence or absence of the well-known correlation between the line width Δv and the size R. For example, in Ophiuchus Loren (1989a) found that 13 of the 89 clouds he catalogued have line widths within a factor of 2 of the value expected in virial equilibrium, and that the 89 clouds show no correlation between Δv and R . But examination of the 13 clouds close to virial equilibrium shows that they exhibit such a correlation, approximately as $\Delta v \propto R^{0.5}$. This property is easily understood, since one can write the identity

$$\Delta v^2 \propto N R \left(\frac{\mathbf{K}}{\mathbf{G}} \right)$$

where N is the mean column density over the map of size R, and where $\mathbf{K}$ and $\mathbf{G}$ are respectively the kinetic and gravitational energy density in the cloud. Thus a cloud close to virial equilibrium has $\Delta v \propto N R$. When the clouds under consideration have a relative range of N smaller than the relative range of R, it follows that $\Delta v \propto R^{0.5}$ as observed in Ophiuchus. This result follows regardless of the reason for the small range of N, which could arise from selection, sensitivity limits, or from equipartition between magnetic and kinetic energy and a relatively small range of magnetic field strength.

3. Dense Cores

Comparison of properties of dense cores in the nearby complexes reveals significant differences in core size, line width, column density, temperature, and mass, depending on which complex one considers, and also on position within the complex.

Before comparing core properties, it is important to recognize that a comparison requires a consistent definition of core identity. The spectral line emission which is used to define cores can have a complex spatial distribution, and any definition is somewhat idealized. Thus a statistic as basic as the number of cores in a cloud can vary greatly from definition to definition. For example, an emission region with two peaks separated by a "valley" higher than half the height of the higher peak would have one core, if the core is defined by the contour of half-maximum intensity. It would have two cores, if a core is defined by a distinct local maximum of intensity. It would have three or more cores if a core is defined by application of a "clean" iteration procedure. In such a procedure, a two-dimensional Gaussian is fit to each local maximum, subtracted from the observed map, and then a new fit is applied to the residual profile. All three of these definitions are in current use. In the following discussion we use the half-maximum contour of line intensity.

The shape of a core map is generally elongated, with aspect ratio about $2 : 1$, in Taurus, Ophiuchus, and other smaller dark cloud regions, according to a study of 16 cores, each mapped in lines of NH_3, CS, and $C^{18}O$ (Fuller 1989; Benson and Myers 1989; Myers et al 1991). In at least six cores, this elongation probably corresponds to a prolate, rather than oblate, shape in three dimensions. Prolate shape is especially interesting because it implies that the core cannot be modelled by an isolated self-gravitating region in stable equilibrium with a simple combination of isotropic motions, rotation, and magnetic fields. Models which include the effects of the core environment are needed.

Despite the uncertainty about core geometry, most cores appear near to virial equilibrium, independent of which geometrical model one adopts. Equilibrium models of the relation between core line width, size, and column density were compared with observed values of these quantities for spheres with various power laws of density with radius; for rotationally supported oblate spheroids; for magnetically supported oblate spheroids; and for segments of an infinitely long, magnetically supported, prolate cylinder. In all cases the observed quantities are consistent, within their scatter, with the range of possible values for each equilibrium geometry (Myers *et al* 1991).

The NH_3 line mapping studies of cores in Taurus by Benson and Myers (1989) and in L1641 in the Orion A region by Harju, Walmsley, and Wouterlout (1990) offer a good comparison of core properties between regions with substantially different size scale and stellar content, especially because the studies were made with nearly the same linear resolution, FWHM about 0.08 pc. These observations indicate that cores in Orion have significantly broader lines (median line width 0.75 km s^{-1}) and larger sizes (median FWHM 0.17 pc) than do cores in Taurus (median line width 0.29 km s^{-1}, median FWHM 0.08 pc). Figure 1, adapted from Harju, Walmsley, and Wouterlout (1990), illustrates these properties. Furthermore, the Orion cores are substantially hotter and more massive than their Taurus counterparts, and are associated with IRAS sources substantially more luminous, and thus presumably with stars substantially more massive, than those in Taurus. Thus the cores in Orion and Taurus show a ranking of properties similar to those already evident in the larger-scale properties of the complexes.

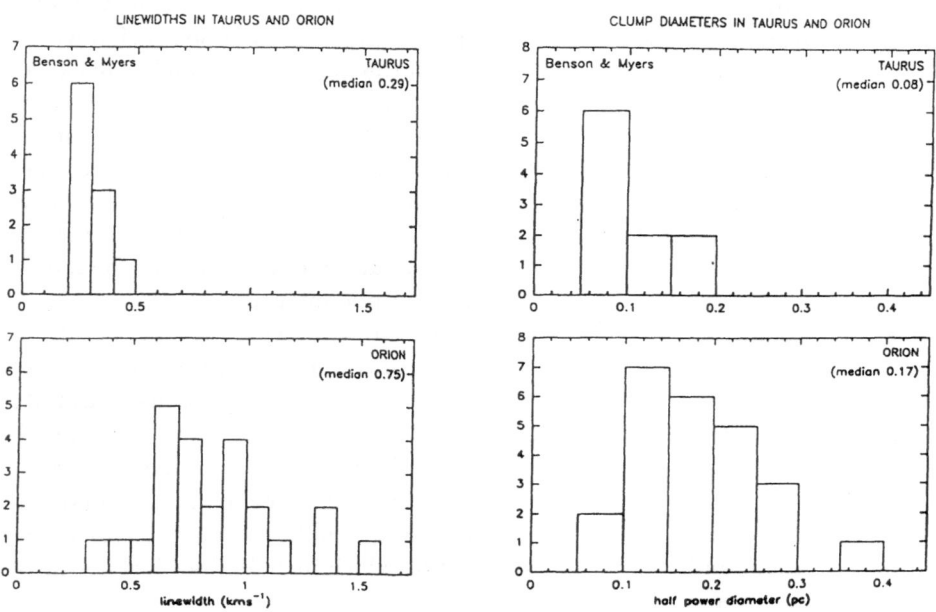

Fig. 1.   Distributions of NH_3 Line Width and Map Size in Taurus and Orion

In Taurus, Ophiuchus, and other relatively small, nearby complexes, surveys for cores based on criteria independent of the presence or absence of an embedded star found that about half of the cores have of the cores have associated stars, and half are "starless" (Myers and Benson 1983; Beichman *et al* 1986). To date, most of the surveys for cores in larger, more distant complexes have been based on the presence of an embedded star. A notable exception is that of E. Lada (1990) in the Orion B region. In order to compare "initial conditions" for star formation from region to region it will be necessary to carry out more unbiased surveys, to allow more starless cores to be identified and compared with their counterparts in other regions.

4. Stars

The T Tauri stars in Orion, Taurus, and Lupus differ in their distributions of spectral type, in the sense that Orion exceeds Taurus, and Taurus exceeds Lupus, in the number of stars, and in the mass of the typical star. Comparison of the number distributions in Orion and Taurus-Auriga, from Cohen and Kuhi (1979), and in Lupus, from Krautter (1990), shows that Orion, Taurus, and Lupus have respectively 112, 81, and 58 T Tauri stars with known spectral types, and have median spectral type respectively K4, K7-M0, and M1.5.

Furthermore, comparison of the circumstellar reddening of *IRAS* sources in L1641 in Orion and in Taurus indicates that the Orion sources are considerably redder than those in Taurus, suggesting that the Orion sources have significantly higher column density of circumstellar dust than the Taurus sources (Strom, Margulis, and Strom 1989).

These differences among young stars in Orion, Taurus, and Lupus appear consistent with the differences in properties of the complex gas, and core gas, in the sense that the larger, more massive complexes tend to have larger, more massive cores, and to have a greater number of young stars, and on average stars of greater mass.

5. Interrelations Among Clouds, Clusters, and Cores

The core size, velocity dispersion, extinction, and mass appear to increase with increasing complex size and mass, as described above. Furthermore, these same core properties also increase from one part of a complex to another, in two ways related to the mass of an associated star cluster. The core mass and its correlated properties appear to increase with the mass of its associated star cluster. Also, a core associated with a star cluster tends to be more massive than cores in the same complex, located in the less massive filaments extending away from the cluster. Table I presents examples of these two kinds of core mass variation.

In Table I, the cluster mass increases from top to bottom, roughly in proportion of the number of associated stars. Similarly, the mass, size, velocity dispersion, and extinction of the associated core also increase from top to bottom. Further, the core in each nearby cloud is distinctly smaller and less massive than the core associated with the cluster.

Table I suggests that in addition to the variation in core mass and other properties from complex to complex, the core mass increases with the mass of the associated star cluster, and with proximity to the cluster. These variations can be understood if the cores, stars, clusters, and complexes all scale together in response to a common cause. If so, the simplest explanation is that the gravitational well of a complex sets the scale of the core and star formation in the complex. Within a complex, the biggest concentration of mass occurs where the well is deepest. There, the most massive cores tend to form, and thence the most massive stars. Of course this picture is extremely idealized, and many other factors may also be at work. Nonetheless, this picture has the virtues of simplicity and the ability to be checked by observational data.

TABLE I

Cores Associated With Clusters, and With Nearby Clouds

Complex	Cluster	Stars	Ref.	Core	Ref.	Nearby Core	Ref.
Tau-Aur	AB Aur	5	1	L1517	2	—	—
Tau-Aur	HK Tau	10	3	TMC2	4	L1506	5
Oph	HD147889	80	6	L1688	7	L1709	7
Per	N1333	30	8	L1450	9	B1	10
Orion	Trapezium	500	11	OMC-1	12	OMC-2	12

References–1, Herbig and Bell (1988); 2, Benson and Myers (1989); 3, Jones and Herbig (1979); 4, Ho *et al* (1977); 5, Greenberg (1988); 6, Lada, Wilking, and Young (1989); 7, Loren, Wootten, and Wilking (1990); 8, Strom, Vrba, and Strom (1976); 9, Ho and Barrett (1980); 10, Bachiller, Menten, and del Rio (1990); 11, McCaughrean (1989); 12, Batrla *et al* (1983).
Note–Each cluster is named by its NGC name, or the name of its most prominent star.

6. Conclusion

The six nearby molecular cloud complexes discussed here appear to have similar rank in their size, column density, mass, and velocity dispersion on scales from ~10 pc to 0.1 pc. Their stellar content appears also to follow this ranking in their number of stars, in their typical stellar mass, in the degree of stellar clustering, and in their circumstellar column density. The core properties also increase with increasing mass of their associated star cluster, and with proximity to the star cluster. These tendencies suggest that over a wide range of spatial scales, molecular gas in a complex is densest where the gravitational potential well is deepest, and embedded star clusters tend to lie close to these concentrations of gas.

Many of these properties can be understood if the mass of a molecular cloud increases throughout nearly all of its lifetime, in contrast to the mass of a star. If so, relatively small complexes like Lupus and Taurus start out with relatively shallow gravitational wells, which favor the formation of relatively small, low-density cores and relatively low-mass stars, which appear singly or in sparse clusters. As the complex grows, its mean density increases. Its regions of denser gas, which have already begun to form small groups of stars, increase in mass faster than do their neighboring regions. These denser clouds tend to form larger and denser cores, perhaps by accretion and mergers of smaller, preexisting cores, and perhaps by dispersal of cores too small to survive tidal forces, now stronger than they were in the past. As long as low-density peripheral gas is available to feed the increasing gravitational pull of the complex, the growth of the complex and its constituent clouds, cores, stars and clusters should continue until the effects of stellar winds and luminosity disperse the gas. This picture is similar to that proposed by Herbig (1962), but with two new ingredients : the continual growth of the molecular cloud complex and the role of dense cores as intermediaries in forming stars.

Acknowledgements

I thank Edith Falgarone, Jean-Loup Puget, Michel Guelin, Alain Castets, and other members of the Organizing Committee for their support of my attendance at this meeting, their kind hospitality, and their invitation to speak. I thank my collaborators Priscilla Benson, Gary Fuller, Ned Ladd, and Alyssa Goodman.

References

Bachiller, R., Cernicharo, J., and Duvert, G. 1985, "The Taurus-Auriga-Perseus Complex of Dark Clouds." *Astron. Astrophys. (Suppl.)*, 149, 273–282.

Bachiller, R., Menten, K. M., and del Rio-Alvarez, S. 1990, "Anatomy of a Dark Cloud : a Multimolecular Study of Barnard 1." *Astron. Astrophys.*, in press.

Batrla, W., Wilson, T. L., Bastien, P., and Ruf, K. 1983, "Clumping in Molecular Clouds. The Region Between OMC1 and 2." *Astron. Astrophys.*, 128, 279–290.

Beichman, C. A., Myers, P. C., Emerson, J., Harris, S., Mathieu, R.D., Benson, P. J., and Jennings, R. 1986, "Candidate Solar-Type Protostars in Nearby Molecular Cloud Cores." *Ap. J.*, 307, 337–349.

Benson, P. J., and Myers, P. C. 1989, "A Survey for Dense Cores in Dark Clouds," *Ap. J. (Suppl.)*, 71, 89–108.

Cohen, M., and Kuhi, L. V. 1979, "Observational Studies of Pre-Main Sequence Evolution." *Ap. J. (Suppl.)*, 41, 743–843.

Fuller, G. A. 1989, *Molecular Studies of Dense Cores*, Ph. D. thesis, Astronomy Department, U. of California, Berkeley.

Greenberg, M. A. 1988, *Dense Cores and Star Formation in the Molecular Clouds L1506 and L1529*, Senior Thesis, Department of Astronomy, Harvard University.

Harju, J., Walmsley, C. M., and Wouterlout, J. G. A. 1990, "Young Ammonia Clumps in the Orion Molecular Cloud." *Astron. Astrophys.*, in press.

Herbig, G. H. 1962, "The Properties and Problems of T Tauri Stars and Related Objects." *Adv. Astr. Ap.*, **1**, 47–101.

Herbig, G. H., and Bell, K. R. 1988, "Third Catalog of Emission-Line Stars of the Orion Population." *Lick Obs. Bull.*, No. 1111.

Ho, P. T. P., and Barrett, A. H. 1980, "Observations of Herbig-Haro Objects and Their Surrounding Dark Clouds." *Ap. J.*, **237**, 38–54.

Ho, P. T. P., Martin, R. M., Myers, P. C., and Barrett, A. H. 1977, "Gas Temperatures and Motion in the Taurus Dark Cloud." *Ap. J. (Letters)*, **215**, L29–L33.

Jones, B. F., and Herbig, G. H. 1979, "Proper Motions of T Tauri Variables and Other Stars Associated with The Taurus-Auriga Dark Clouds." *A.J.*, **84**, 1872–1889.

Krautter, J. 1990, "The Star Forming Region in Lupus." preprint of paper presented at ESO Workshop on Low Mass Star Formation and Pre-Main-Sequence Objects, Garching, 1989.

Lada, E. 1990, *Global Star Formation in the L1630 Molecular Cloud*, Ph. D. thesis, Astronomy Department, University of Texas at Austin.

Larson, R. B. 1981, "Turbulence and Star Formation in Molecular Clouds." *MNRAS*, **194**, 809–826.

Loren, R. B. 1989a, "The Cobwebs of Ophiuchus I. Strands of ^{13}CO : The Mass Distribution." *Ap. J.*, **338**, 902–924.

Loren, R. B. 1989b, "The Cobwebs of Ophiuchus II. ^{13}CO Filament Kinematics." *Ap. J.*, **338**, 925–944.

Loren, R. B., Wootten, H. A., and Wilking, B. A. 1990, "Cold DCO^+ Cores and Protostar-Like Inclusions in the Warm ρ Ophiuchi Clouds." *Ap. J.*, in press.

Maddalena, R. J., and Thaddeus, P. 1986, "The Large System of Molecular Clouds in Orion and Monoceros." *Ap. J.*, **303**, 375–391.

McCaughrean, M. J. 1989, "Multicolour Near Infrared Imaging of the Orion Nebula and Trapezium Cluster." *B.A.A.S.*, **21**, 712.

Murphy, D. C., Cohen, R., and May, J. 1986, "CO Observations of Dark Clouds in Lupus." *Astron. Astrophys.*, **167**, 234–238.

Myers, P. C., and Benson, P. J. 1983, "Dense Cores in Dark Clouds. II. NH_3 Observations and Star Formation." *Ap. J.*, **266**, 309–320.

Myers, P. C., Fuller, G. A., Goodman, A. A., and Benson, P. J. 1991, "Dense Cores in Dark Clouds. VI. Shapes." submitted to *Ap. J.*

Schwartz, R. D. 1977, "A Survey of Southern Dark Clouds for Herbig-Haro Objects and H-Alpha Emission Stars." *Ap. J. (Suppl.)*, **35**, 161–170.

Strom, K. M., Margulis, M., and Strom, S. E. 1989, "A Study of The Stellar Popolation in the Lynds 1641 Dark Cloud : Deep Near-Infrared Imaging." *Ap. J. (Letters)*, **345**, L79–L82.

Strom, K. M., Vrba, F. J., and Strom, S. E. 1976, "Infrared Surveys of Dark Cloud Complexes. II. The NGC 1333 Region." *A. J.*, **81**, 314–316.

Ungerechts, H., and Thaddeus, P. 1987, "A CO Survey of the Dark Nebulae in Perseus, Taurus, and Auriga." *Ap. J. (Suppl.)*, **63**, 645–690.

Wilking, B. A., Lada, C. J., and Young, E. T. 1989, "*IRAS* Observations of the ρ Ophiuchi Infrared Cluster : Spectral Energy Distributions and Luminosity Function." *Ap. J.*, **340**, 823–852.

DENSE CORE STRUCTURE AND FRAGMENTATION
IN THE RHO OPHIUCHI MOLECULAR CLOUD

ALWYN WOOTTEN

Observatoire de Paris–Meudon, France and
N. R. A. O., Edgemont Road, Charlottesville, VA 22903, USA
E–mail AWOOTTEN@NRAO.BITNET

Abstract. About a dozen distinct dense cores have been identified in the Rho Ophiuchi molecular cloud. The properties of these cores are summarized and compared to the properties of cores in the Taurus molecular cloud, a less efficient region of star formation, and in DR21(OH), a more massive region of star formation. The data are consistent with a picture in which more massive clouds have a higher surface density of cores, which in turn are more massive. The adjacent cores in L1689N have been studied with very high resolution; one has formed stars and one never has. The structure of these cores shows a tendency for duplicity of structures from the largest scales (1 pc) to the smallest (50 AU).

Keywords : Star Formation, Dense Molecular Cores, Ophiuchus Molecular Complex

1. Cores in Molecular Clouds

Locally, stars form in molecular clouds. These clouds encompass structures of many densities, but certainly the densest large objects within them are the stars which have formed there. Less dense, though spanning the density range of twenty orders of magnitude between stellar densities and 10^4 cm^{-3} are the molecular cloud cores. Understanding the connection between these dense cores and the stars we suppose to form within them poses a difficult challenge. How does the character and extent of star formation in a region depend upon the number and physical properties of these dense cores? What is the structure of the cores–are they clustered and do they fragment? How many stars form within each core–do more massive cores produce more massive stars, less massive stars, or do they reproduce the initial mass function? Do all cores produce stars, or are there processes which inhibit star formation in some cores?

As a first step toward the answer to these questions we have begun a census of dense cores and their properties in the ρ Ophiuchi and DR21(CH) molecular clouds. One survey (Loren, Wootten and Wilking 1990) focussed on identification of cold cores which have not yet created stars. This was accomplished through mapping of DCO$^+$ emission in the ρ Oph complex. A second survey, in progress, has mapped the central region of the complex in ammonia emission. A third, recently begun, concentrates on CS emission. In DR21(OH) Mangum (1990) has completed an interferometric survey in C^{18}O and NH$_3$ emission, and in the millimeter dust continuum.

2. The ρ Oph Cores : General Properties

2. 1. CLUSTERING

Twelve well-defined cores lie in the central star-forming region of the ρ Oph molecular complex. These twelve dense cores lie within the boundaries of four dense clumps of material defined by Loren (1989) from maps of ^{13}CO emission. Each ^{13}CO clump which contains a denser core encompasses at least two denser cores. In turn, each of the denser cores which has been mapped at higher resolution shows evidence for further fragmentation, usually into pairs of similar structures. On scales of 10,000 AU (0.05 pc), the mean separation between closest pairs of dense cores is 0.3 ± 0.13 pc. In the Taurus molecular cloud, with the exception of two close cores in L1495, separations exceed 0.6 pc (Benson and Myers 1989). For the DR21(OH) cores listed by Mangum (1990), the mean separation is 0.25 $\pm$ 0.14 pc. The surface density of cores measured by Loren et al. (1990) is 3-4 cores pc^{-2} in the ρ Oph complex, in contrast to the 1.3 cores pc^{-2} in Taurus and the 11.3 cores pc^{-2} in DR21(OH) found by Mangum (1990). Mangum (1990) also notes that in the OMC1-OMC2 complex, the core density is 9.2 cores pc^{-2}. The sequence of increasing core surface density Taurus$\rightarrow$Ophiuchus$\rightarrow$DR21(OH) is also a sequence of mass and luminosity in the star-forming complexes. The conclusion which emerges is that *cores are more tightly clustered in more massive star-forming regions.*

2. 2. MASS

Loren et al. (1990) determined masses for their cores both from their study of the excitation of DCO$^+$, and from the virial theorem. The mean core mass lay near 25 $M_\odot$ with a range $8 < M_{core} < 44$ $M_\odot$. In contrast, the mass of the ammonia cores in Taurus tabulated by Benson and Myers (1989) spans a range $0.3 < M_{core} < 33$ $M_\odot$ with a mean of 6 $M_\odot$. Mangum (1990) finds that the DR21(OH) cores are substantially more massive, with an average mass of 500 $M_\odot$. Individual cores appear to be more massive in more massive star-forming regions. Further study of cores and their masses will be necessary to determine if less massive cores also exist in the massive regions.

2. 3. STELLAR CONTENT

Loren et al. (1990) noted the strong tendency of the infrared objects with the coldest spectra to lie within cores. In fact, there are several young stars within the boundary of the typical core in the ρ Oph cloud, while there is seldom more than a single young star near the Taurus cores. The locations of young stars in DR21(OH) are unknown as no sensitive survey of the region exists. The data suggests the existence of a trend for more massive cores to produce several stars, but this trend must be confirmed.

2. 4. STERILE CORES

Several cores, or core fragments, in the ρ Oph cloud have no luminous infrared sources within their boundaries although sensitive searches have been made. As these cores have apparently not produced stars, we refer to them as sterile. Such cores are particularly hard to locate. However, study of their physical conditions may uncover differences between these and star-forming cores, differences which may lead to a better understanding of how cores form stars.

The B1 core and its more massive neighbor B2 (see Loren *et al.* 1990 for locator maps of the cores and infrared sources) each have several solar masses of material. There are no luminous sources within the B1 core, although Rieke and Rieke (1990) have located a very faint, possibly substellar, object there. No sources have yet been located within the B2 core (see Mezger, Sievers and Zylka, 1990). However, the infrared sources WL 3, 4, 5, 6, and YLW12A (Wilking, Lada and Young 1989) lie in a cavity between B1 and B2. This geometry suggests that these cores are fragments remaining from a larger core which once produced stars industriously. The cores may appear sterile as the fragments reorganize under the influence of, among other things, gravity and the nearby infrared sources.

The two cores of the L1689S cloud center to the south and west of two near-IR sources corresponding to IRAS16288, the easternmost of which is a 6 $L_\odot$ star (Wilking 1990) accompanied by a bipolar CO outflow (Loren *et al.* 1990, Wootten *et al.* 1991). There is no evidence for an infrared source at 2.2 μm (Wilking 1990) within the very dense SE component where a water maser is found (Wootten *et al.* 1991). Since no water masers are known to exist very far from infrared sources, any star which is present must lie embedded deeply enough to escape detection. Presumably, a star so deeply located within the cloud is quite young. This core may appear sterile owing to the lack of a survey at the long infrared wavelengths necessary to probe to the core interior.

The L1689N cloud contains two cores, one of which contains the young IRAS16293-2422 binary star. The other, discussed by Wootten and Loren (1987), contains no infrared sources, although its dense core does contain compact features. This core, cold but gravitationally evolved, is the best candidate for a sterile core.

Centrally condensed cores which have not formed stars, "sterile" cores, appear to be scarce in the ρ Oph complex, accounting for three, at most, of the twelve cores. Cores may appear to be sterile for several reasons : they may be fragments of larger star-forming cores, they may have formed substellar objects, or they may just not have been probed deeply enough in the infrared to detect luminosity sources.

Because the L1689N cloud has been well-mapped and structures defined over a wide range of scales, and because the adjacent cores offer a contrast in their stellar content, we examine them more closely in the next section.

3. Structure of Cores within the L1689N Cloud

Loren designated a region of enhanced ^{13}CO emission in the northern part of the L1689 cloud as clump R55, containing 200 $M_\odot$ of material. Toward the western edge of the clump lies the dense double core, one of which has apparently not produced a star, mentioned in the previous paragraph. In this section we review the heirarchy of structure in these cores, noting the remarkable pairing of structures from the 1 pc scale down to the 50 AU scale.

The sterile core contains about 10 $M_\odot$ of material. Interferometric ammonia observations have shown it to contain cold (13K) substructures of size $\sim$ 3000 AU (Wootten and Loren 1987), elongated perpendicularly to the magnetic field in the region, with densities not exceeding $5 \times 10^5 cm^{-3}$.

In contrast, Mundy, Wootten and Wilking (1990) found the ammonia emission about IR16293, 0.7 pc to the west, to be oriented in a smaller disk-like structure of similar density and orientation but warmer (15-20K). Over much of the 6000 AU extent this structure is quiescent, with narrow lines stationary in velocity. Toward the northwest end, however, ammonia emission broadens and weakens, and the weak line wings hint at the presence of a rotating component. Mundy, Wilking and Myers (1986) had found a disk-like dust distribution here, encompassing the infrared source and mimicing the geometry of the outer disk on smaller ($\sim$1500 AU) scales. Mundy et al. (1990) showed that $C^{18}O$ emission from this structure was strong, and apparently centered on one of two radio continuum sources discovered by Wootten (1987). The $C^{18}O$ structure rotates quite rapidly, contains $\sim$ 2 $M_\odot$ of material, and appears to be warmed by the young star it envelops. Dynamically, the inner disk is distinct from the outer disk–the transition region between the non-rotating outer ammonia disk and the rapidly-rotating inner disk was unresolved in a 6" beam.

On smaller scales, the inner disk has recently been resolved (Mundy et al. 1991) at OVRO ($\lambda 2.75 mm$) into smaller dust condensations centered on the 2cm radio sources. The IR16293A source, brighter at frequencies below 20 GHz and centered on the $C^{18}O$ emission, is marginally resolved (640 AU). The IR16293B source, 830 AU to the northwest remains unresolved (<400 AU at 2.75mm). IR16293B shows higher *surface* brightness at frequencies higher than 20 GHz; at 22 GHz the region of free-free emission is less than 16 AU in extent.

IR16293A, the southeastern component of this protobinary system, has been resolved into two 2cm sources separated by 50 AU along a line perpendicular to the major axis of the disk (Wootten 1987). These sources lie 50 AU northeast of the center of a cluster of twelve water masers spread about their position centroid with an rms offset of 45 AU. The centroid position of the positive velocity masers (with respect to ambient cloud velocity) lie southwest of the centroid of negative velocity masers, a pattern mimicking the spatial structure the continuum sources, and the dynamical structure of the CO outflow mapped on larger scales. Hence, the source of the outflow is undoubtedly IR16293A, and the flow is collimated on scales less than that of Saturn's orbit within our solar system.

4. Summary

A census of the cores in the ρ Oph cloud has shown that the surface density of cores lies between that in the Taurus clouds, which have been relatively inefficient in star production, and that in the Orion or DR21(OH) molecular clouds, which have apparently been more productive. The masses of the cores lie in the intermediate region between the Taurus and DR21(OH) cores also.

The core structure becomes more complex on finer scales. In the L1689N cores the structures are generally similar and show a tendency to be paired at successively finer levels down to scales smaller than the solar system.

Acknowledgements

Co-workers on most of the projects whose results are reported here have been Bob Loren, Jeff Mangum, Lee Mundy and Bruce Wilking. I am grateful for the hospitality of P. Encrenaz at Meudon.

References

Benson, P. J., and Myers, P. C. : 1989, *Ap. J. Suppl.*, **71**, 89.
Loren, R. B. : 1989, *Ap. J.*, **338**, 902.
Loren, R. B., and Wootten, A. : 1986, *Ap. J.*, **306**, 142.
Loren, R. B., Wootten, A. and Wilking, B. : 1990 *Ap. J.*, **365**, in press.
Mangum, J. : 1990, Ph. D. Dissertation, The University of Virginia.
Mezger, P., Sievers, A. and Zylka, R. : 1990, this volume.
Mundy, L. G., Wilking, B. A., and Myers, S. T. : 1986, *Ap. J. (Letters)*, **311**, L75.
Mundy, L. G., Wootten, A., and Wilking, B. A. : 1990, *Ap. J.*, **352**, 159.
Mundy, L. G., Wilking, B. A., Blake, G. A., Sargent, A. I. and Wootten, A. : 1991, *Ap. J. (Letters)*, in preparation.
Rieke, G. H. and Rieke, M. J. : 1990 *Ap. J. (Letters)*, **362**, L21.
Wilking, B. : 1990, private communication.
Wilking, B., Lada, C. and Young, E. : 1989, *Ap. J.*, **340**, 823.
Wootten, A. : 1989, *Ap. J.*, **337**, 858.
Wootten, A., Loren, R. B., Mangum, J. and Butner, H. M. : 1991, in preparation.
Wootten, A., and Loren, R. B. : 1987, *Ap. J.*, **317**, 220.

UV PENETRATED CLUMPY MOLECULAR CLOUD CORES

JÜRGEN STUTZKI

I. Physikalisches Institut der Universität zu Köln,
Zülpicher Straße 77, D-5000 Köln 41, Fed. Rep. of Germany

REINHARD GENZEL, URS GRAF, ANDREW I. HARRIS, AMIEL STERNBERG

Max-Planck-Institut für Physik und Astrophysik,
Institut für extraterrestrische Physik,
D-8046 Garching b. München, Fed. Rep. of Germany

ROLF GÜSTEN

Max-Planck-Institut für Radioastronomie,
Auf dem Hügel 69, D-5300 Bonn 1, Fed. Rep. of Germany

Abstract. Recent observations of sub-mm and far-IR atomic fine structure and molecular rotational lines give evidence that due to the clumpiness of the molecular cloud cores the UV radiation from newly formed stars affects a very large fraction of the cloud material. Direct observations of the clumpy structure in M17 SW allow to derive several parameters of the clump distribution, in particular the clump mass spectrum and the volume filling factor. Implications of these results in regard to star formation are shortly discussed.

1. UV Radiation and Molecular Clouds

Far-UV radiation longward of the Lyman continuum edge (912 Å) can escape from the HII region around newly formed stars and affect the surface of the parent molecular cloud. It dissociates most molecules, notably the most abundant ones, H_2 and CO, and ionizes several atomic species, in particular carbon (C^+). In addition the UV radiation directly heats the gas via two processes : first, fast photoelectrons ejected from dust grains collide with molecules, thus transferring the energy from the UV photons to the thermal energy of the gas. Second, and dominant at high densities, UV-pumped vibrational levels in H_2 molecules collisionally deexcite to transfer their energy to the kinetic energy of the gas.

The UV-radiation thus creates a transition region, named photodissociation region (PDR), with a thickness of a few A_v on the molecular cloud surface, where the chemical composition progressively changes from HI to H_2, and in parallel, from C^+ to C^0 to CO. The temperature drops from about 1000 K at the cloud surface to the ambient cloud temperature of the UV shielded gas deep inside. The fine structure transitions of [CII] and [OI], as well as the mid- and high-J rotational transitions of CO, all observable in the sub-mm and far-IR spectral range, are the dominant coolants of the gas.

Several groups have studied theoretically the structure of PDR's, starting with the early work by de Jong, Dalgarno and Boland (1980), through the extensive modelling of the Orion region by Tielens and Hollenbach (1985a,b), and the more recent work by Sternberg and Dalgarno (1989), Burton et al. (1988,1990) and Sternberg (1990). The intensities of the different fine structure transitions are generally reproduced rather nicely. However, all theoretical models fail to explain the large amount of warm molecular gas inferred from mid- and high-J CO and isotopic CO observations. This discrepancy is discussed further in the article by Graf et al. (this volume). Very recently Parmar, Lacy and Achtermann (priv. comm.) suceeded in the first observations of the J=4→2 (12 μm) and J=3→1 (17 μm) rotational transitions of molecular hydrogen in a strip map across the Orion Bar PDR. Their preliminary analysis gives hydrogen column densities and temperatures in apparently good agreement with the theoretical models discussed above.

2. Clumpiness of Molecular Clouds and UV Radiation

The different fine structure transitions of ionized carbon [CII] at 158 μm and neutral atomic carbon [CI] at 370 μm and 609 μm, and the mid- and high-J rotational transitions of CO (e.g. J=6→5, 7→6, 14→13) are ideally suited to study the $C^+/C^0/CO$ transition structure of PDR's. They all show the emission regions of ionized and neutral atomic carbon, as well as warm molecular material traced by CO, to be rather extended and to spatially coincide (Figure 1).

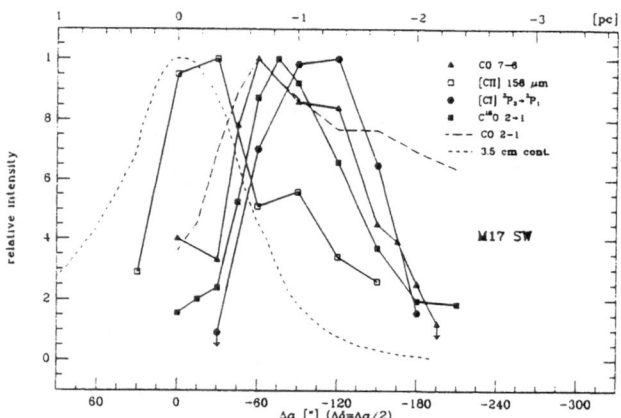

Fig. 1. Cut across the M17 SW HII region/molecular cloud interface showing the distribution of ionized, neutral atomic, and molecular carbon.

Moreover, the [CII] emission typically extends over regions where the average column density corresponds to A_v of 10 to 100 in all regions observed, typical examples being Orion (Fig. 2), M17 SW (Fig. 3), the circumnuclear in the Galactic Center, W3 and NGC 1977. No UV photons able to ionize carbon can reach that far in, unless the cloud material is distributed very clumpy or filamentary, with a

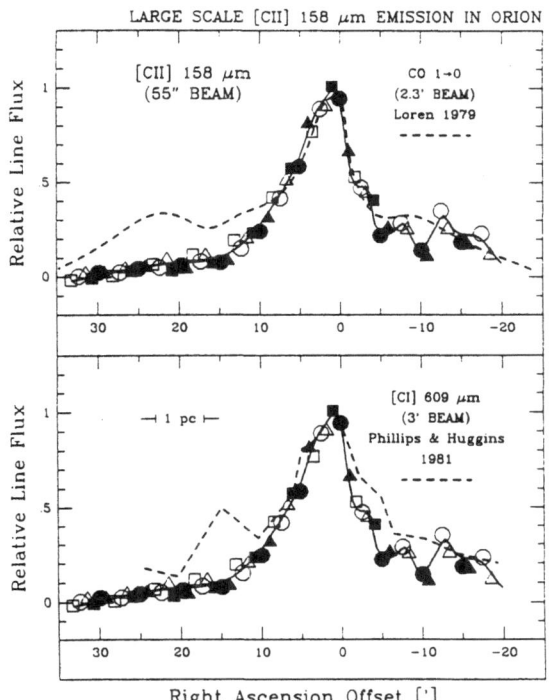

Fig. 2. Large scale distribution of [CII], [CI] and CO emission in Orion (from Stutzki et al. 1990a). The different symbols denote the sets of strip scans with different chopper throw from which the [CII] data have been constructed.

high clump/interclump (column) density contrast. The observed intensity is then an average over many PDR's on the surfaces of individual, unresolved clumps.

This picture has several very attractive features. Detailed modeling, taking into account the attenuation of the UV-radiation with depth into the cloud by the blocking and scattering on clumps (Boissé 1990 presents a very detailed study of continuum radiative transfer in a clumpy medium), and incorporating the dependence of the emergent [CII] on the exciting UV intensity are very successful in reproducing not only the spatial distribution, but also the absolute intensity of the observed [CII] emission (Stutzki et al. 1988, Howe, Jaffe and Genzel 1990; Figure 4). The picture also naturally explains the coexistence of [CI] emission with the bulk of the molecular material (Keene et al. 1985, Genzel et al. 1988), without having to invoke special chemistry or untypically high initial carbon abundances.

Due to the clumpiness, the density in individual PDR's on clump surfaces is higher than estimated from the average density of the emission region. Higher densities help to explain the large amounts of warm molecular CO observed. At

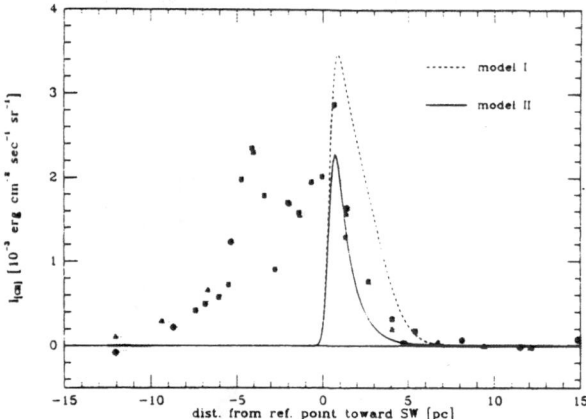

Fig. 3. The measured [CII] distribution compared with two simple models for the emission expected for a clumpy, UV penetrated cloud core (see Stutzki et al. 1988 for details).

higher densities, the molecular reformation rate is much faster and the molecular layer in the PDR is pushed out towards the higher temperature regime (Burton et al. 1990; Sternberg 1990). However, the extremely large column densities of warm molecular material implied by the recent ^{13}CO J=6→5 detection (Graf et al. 1990, and this volume) still are not understood in the framework of theoretical PDR models.

The model of a clumpy, UV-penetrated cloud implies that the interpretation of CO lines in particular needs great care : the optical depth of CO rotational lines in the mm and sub-mm spectral regime is typically close to unity for column densities corresponding to A_v of a few, that is for the thickness of individual PDR's. The temperature and the chemical composition of the emitting region thus varies strongly across the line formation region. Simple attempts to interpret the observed intensity ratios of more than two lines in a single component radiative transfer model (LVG or escape probablity codes) in consequence lead to intrinsically contradicting results (Dutrey et al. this volume). A more sophisticated approach, incorporating the temperature, density and chemical gradients on the clump surface (PDR) in the radiative transfer (Gierens, this volume), and comparing the observed intensity ratios with the approriate average over many clumps embedded in a (spatially varying) UV-field (Tauber and Goldsmith 1990) gives a straightforward explanation of the observed line intensities.

3. Direct Observations of Clumpiness

Several indirect, but nevertheless compelling arguments for the clumpy structure of molecular clouds are now commonly accepted. In some cases the observed bright-

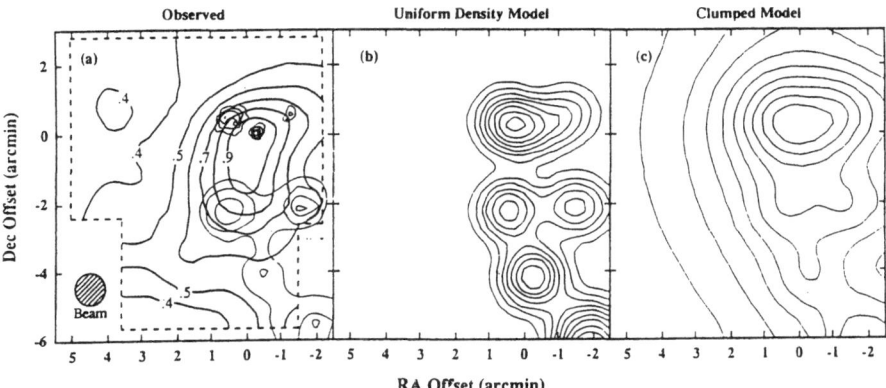

Fig. 4. The observed [CII] emission *(left)* of W3 compared with the one predicted assuming a homogeneous *(mid)*, and a clumpy *(right)* cloud (from Howe, Jaffe and Genzel 1990).

ness temperature of a particular molecular transition is much lower than the excitation temperature independently derived from the comparison of other line ratios. Ammonia is a classical example (see Ho and Townes 1983). The missing intensity is then ascribed to a beam filling factor (typically around 20%) of the emitting material, implying that the emitting material has structure on scales much smaller than the resolution of the observing telescope.

In other cases the density necessary for excitation of a particular line observed is much higher than the average density inferred from the ratio of column density and (assumed) source extent along the line of sight, again implying that the emission region is clumpy (this argument typically applies to multiline studies in CS, e.g. Snell et al. 1984, 1986, Evans et al. 1987, Mundy et al. 1986).

In an independent argument, the line profiles of different CO isotopes, in particular their width, which are observed to grow much less with optical depth than indicated by a simple curve of growth analysis, are successfully interpreted in a macroturbulent cloud model where the cloud consists of many individual, unresolved clumps of rather narrow intrinsic line width, but with a rather wider interclump velocity dispersion (Martin, Sanders and Hills 1984).

Only few direct observations of the clumpy structure of molecular clouds are published in the literature. Many of them refer to observations of very bright central regions, often in some molecular species for which it is not clear to what extent it traces the structure of the bulk molecular gas. CO observations or, due to the large optical depth of ^{12}CO, rare isotopic CO observations with very high angular resolution should overcome this problem. Pérault, Falgarone and Puget (1985) published a study of nearby, cold clouds showing hierachical structure down to scales well below the Jeans length.

Stutzki and Güsten (1990) used the $C^{18}O$ J=2→1 transition at the IRAM-30m telescope to map the molecular cloud core next to the edge on HII-region/molecular cloud interface M17 SW. The same region was simultaneously mapped in $C^{34}S$ J=3→2 and 2→1 with somewhat poorer angular resolution. These data show small scale spatial and velocity structure, best seen in the comparison of two 0.5 km/s wide velocity channel maps of the region (Figure 5). Clumps are identifiable down to the spatial resolution of the telescope (13" at 218 GHz). Individual clumps typically have line widths of 0.5 to 2 km/s, compared to an overall width of the position averaged line profile of 5 km/s. Note the very high peak brightness temperatures of even the rare isotopic $C^{18}O$ J=2→1 emission, implying column densities of up to several times 10^{23} cm^{-2} in individual clumps.

Together with the typical clump dimensions of a few tenths of a pc, these high column densities immediately imply average densities in the $C^{18}O$ emission region of at least 10^5 cm^{-3}, much higher than the critical density necessary for excitation of the J=2→1 transition and rather similar to the densities inferred from CS multi-line studies. The clump decomposition procedure discussed below confirms these high densities. The total amount of gas traced by $C^{18}O$ is similar to that traced by $C^{34}S$. In addition, the $C^{34}S$ J=3→2 and $C^{18}O$ J=2→1 maps, once the latter is smoothed to the lower angular resolution of the former, look very similar. These results confirm that the bulk of the gas in M17 SW is in high density clumps.

Note that lower angular resolution observations in main isotopic CO lines can be very misleading under these circumstances. High angular resolution ^{12}CO spectra show very high peak brightness temperatures (around 100 K) and deep absorption notches (Stutzki and Güsten 1990). Their shape varies rapidly from position to position, resulting in a rather smooth, almost Gaussian line profile when observed at an angular resolution of about 1'. The intensity observed with low angular resolution, typically 50 K, which is usually interpreted as a measure for the gas kinetic temperature being in good agreement with the dust temperature, obviously has no direct relation to the global properties of the gas, but happens to result from a positional average over the much more complicated and intrinsically much brighter spectra described above.

With an appropriate clump decomposition procedure Stutzki and Güsten (1990) were able to separate the emission from M17 SW into about 170 individual clumps, assuming that the individual clumps have a Gaussian density and velocity distribution. The clump sizes and velocity widths closely follow the virial equilibrium relation. The volume filling factor is around 20 to 30%, with slightly higher values towards the more central regions.

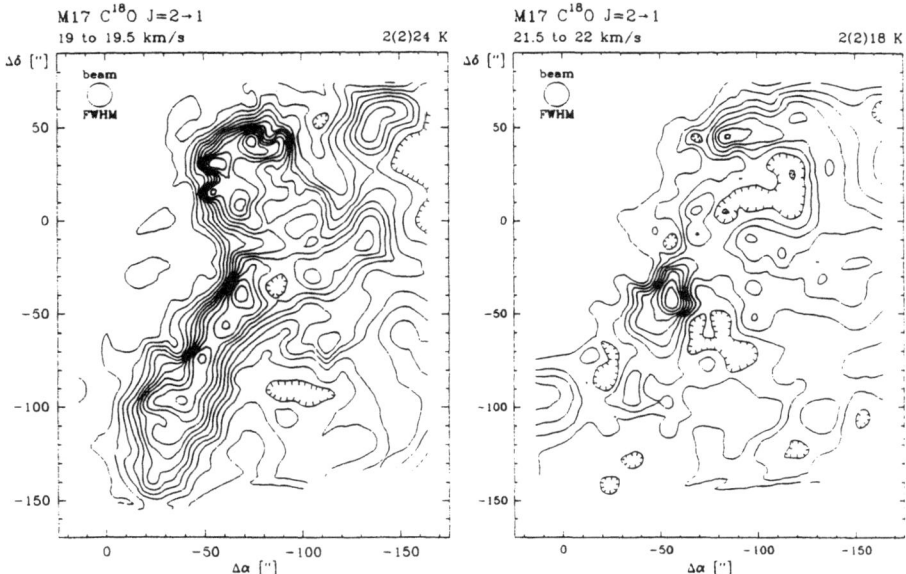

Fig. 5. Representative channel maps of the M17 SW $C^{18}O$ $J=2\rightarrow1$ emission, delineating the clumpy nature of the cloud in the spatial and the velocity domaine. The contours are in steps of 2 K, from 2 to 24 K, in units of main-beam brightness temperature, averaged over 0.5 km/s wide channels, centered at $v_{LSR} = 19.25$ *(left)* and 21.75 km/s *(right)*.

This volume filling factor allows an estimate of the clump/interclump density contrast of >23, assuming that confusion with emission from lower density inter-clump gas starts at the 10% level of the peak emission. The extent of the observed [CII] emission, and hence the UV penetration depth into the cloud needs a density contrast of >40 and probably more like 100.

The clumpiness of the molecular cloud structure does not seem to be related to the proximity of the M17 SW core to its exciting UV sources. A similar $C^{18}O$ map of a remote area of the M17 cloud, close to peak C in the notation of Elmegreen, Lada and Dickinson (1979), which is far off any detectable radio continuum emission or IR point source, shows the same clumpy structure (Stutzki et al. 1990b). Only the total (i.e. clump to clump) velocity dispersion is about a factor 2-3 lower, and hence the clump contrast is somewhat smeared out in individual velocity channel maps.

4. Clump Mass Spectrum

The derived clump masses show a spectrum $dN/dM \propto M^{-\alpha}$ with $\alpha = 1.7$ over two orders of magnitude (Figure 6). A mass spectrum similar to the one found in the M17 SW study of Stutzi and Güsten (1990) is derived in many other cloud cores (Blitz 1987, Loren 1989, Lada 1990) and over a wide range of clump masses, from very large scale structures of neutral, atomic hydrogen clouds (Perry and Helfer 1972, Dickey and Garwood 1989), over giant molecular clouds (Casoli, Combes and Gérin 1984), down to fragments of order one solar mass and less. The derived values of the power law index α range between 1.1 and 1.8. A power law index of $\alpha = 1.5$ is predicted in a simple coagulation and fragmentation scenario (Field and Saslaw 1965, Spitzer 1982, Nakano 1984) and depends only weakly on the details of the interclump interaction processes (Field and Hutchings 1968; Taff and Savedoff 1972, 1973; Arny and Weissman 1973; Silk and Takahashi 1979).

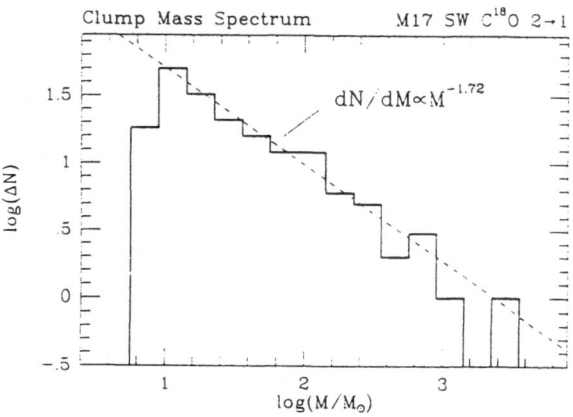

Fig. 6. The mass spectrum of the clumps identified in M17 SW is well fitted by a power law with an index $\alpha = 1.7$.

Note that a power law with $\alpha = 1.7$ together with an efficiency to convert clump mass into stellar mass $M_* = \epsilon M_{clump}^{p}$ translates into an initial mass function $dN/dM \propto M^{-q}$ where $q = (\alpha + p - 1)/p$. For $\alpha = 1.7$ as observed, and $p = 0.5$, a plausible estimate based on present theoretical models of the protostellar evolution (Zinnecker 1989, and references therein), the resultant initial mass function has $q = 2.4$, very close to the Salpeter initial mass function. One may thus speculate that the shape of the initial mass function is determined by the dynamical processes governing the fragmented structure of molecular clouds, combined with the efficiencies involved in forming individual stars, binaries and multiple systems with few members out of individual clumps.

5. Stability of the clumpy structure

The picture emerging from the observational results discussed above shows that the molecular cloud core consists of individual, high density clumps with a volume filling factor of 20 to 30% and a very high density contrast between the clumps and the interclump medium. It is not clear at the moment what stabilizes the clumps against evaporating into the interclump phase. Notably, the pressure in the adjacent HII region, in the PDR envelopes and in the dense clump cores is quite comparable, about a few times 10^6 K cm^{-3}. If we postulate a neutral atomic interclump medium of comparable pressure to stabilize the clumps, the density contrast derived above predicts its density and temperature typically to be around $n = 500 - 1000$ cm^{-3} at $T = 3000 - 6000K$. Such a medium would be very difficult to detect with any presently known tool.

Alternatively, the clumps may all be marginally unstable against gravitational collapse and turn into a cluster of young stars within a few free fall times. This may eventually happen to all clumpy cores. The recent IR-camera observations (Lada 1990) showing dense stellar clusters around most OB groups observed, support the idea that the clumps discussed above eventually turn more or less at once into a cluster of young stars, and that this mode may be the dominant mode for star formation. The resultant picture is thus a more refined version of the old ideas of triggered star formation, where the first generation of a stellar cluster, due to the UV radiation from its brightest members, increases the pressure in the clumpy parent cloud (possibly by ablating more and more material from the clumps into the interclump medium), thus compresses the clumps until they become gravitationally unstable and turn into the next generation stellar cluster.

6. Open Problems

The present observations indicate that we have no satifactory understanding of the heating processes relevant in PDR's, missing some heating mechanism that is able to create rather large column densitites of warm molecular material. Also, in order for higher angular resolution observations to be useful, we need more versatile radiative transfer codes to study the radiation from a clumpy, turbulent medium.

The discussion above is largely based on observational data of only a few sources. Expansion of the observational data base, covering a larger variety of sources with different physical parameters, and equally important, observations with higher angular resolution, sufficient to resolve individual PDR surfaces on clumps, are necessary to understand the physical processes controlling the evolution of a clumpy molecular cloud core pentrated by UV radiation.

References

Arny, T., and Weissman, P., 1973, *Astron. J.* **78**, 309.

Blitz, L., 1987, in *Physical Processes in Interstellar Clouds*, eds. G.E. Morfill and M. Scholer, (Dordrecht : Reidel), p. 35.

Boissé, P., 1990, *Astron. Astrophys.* **228**, 502.

Burton, M., Hollenbach, D., and Tielens, A., 1989, *Proceedings of 22nd Eslab Symposium on Infrared Spectroscopy in Astronomy*, Salamanca, 7-9 Dec. 1988 (ESA SP series).

Burton, M., Hollenbach, D., and Tielens, A., 1990, preprint.

Casoli, F., Combes, F., and Gérin, M., 1984, *Astron. Astrophys.* **113**, 99.

de Jong, T., Dalgarno, A., and Boland, W., 1980, *Astron. Astrophys.* **91**, 68.

Dickey, J.M., and Garwood, R.W., 1989, *Astrophys. J.* **341**, 201.

Elmegreen, B.G., Lada, C.J., and Dickinson, D.F., 1979, *Astrophys. J.* **230**, 415.

Evans, N.J., Mundy, L.G., Davis, J.H., and Vanden Bout, P., 1987, *Astrophys. J.* **312**, 344.

Field, G.B., and Hutchins, J., 1968, *Astrophys. J.* **153**, 737.

Field, G.B., and Saslaw, W.C., 1965, *Astrophys. J.* **142**, 568.

Genzel, R., Harris, A.I., Jaffe, D.T., and Stutzki, J., 1988, *Astrophys. J.* **332**, 1043.

Graf, U.U., Genzel, R., Harris, A.I., Hills, R.E., Russell, A.P.G., and Stutzki, J., 1990, *Astrophys. J. Letters*, in press.

Ho, T.P.T., and Townes, C.H., 1985, *Ann. Rev. Astron. Astrophys.* **21**, 239.

Howe, J., Jaffe, D.T., and Genzel, R., 1990, preprint.

Keene, J., Blake, G.A., Phillips, T.G., Huggins, P.J., and Beichman, C.A., 1985, *Astrophys. J.* **299**, 967.

Lada, E., 1990, Ph.D. Thesis, University of Texas at Austin.

Loren, R.B., 1989, *Astrophys. J.* **338**, 902.

Martin, H.M., Sanders, D.B., and Hills, R.E., 1984, *Monthly Notices Roy. Astron. Soc.* **208**, 35.

Mundy, L.G., Snell, R.L., Evans, N.J., Goldsmith, P.F., and Bally, J., 1986, *Astrophys. J.* **306**, 670.

Nakano, T., 1984, *Fund. Phys.* **9**, 139.

Pérault, M., Falgarone, E., and Puget, J.L., 1985, *Astron. Astrophys.* **152**, 371.

Perry, J.F., and Helfer, H.L., 1972, *Astrophys. J.* **174**, 341.

Sanders, D.B., Scoville, N.Z., and Solomon, P.M., 1985, *Astrophys. J.* **289**, 373.

Silk, J., and Takahashi, T., 1979, *Astrophys. J.* **229**, 242.

Snell, R.L., Mundy, L.G., Goldsmith, P.F., Evans, N.J., and Erickson, N.R., 1984, *Astrophys. J.* **276**, 625.

Snell, R.L., Goldsmith, P.F., Ulich, B.L., Lada, C.J., Martin, R., and Schulz, A., 1986, *Astrophys. J.* **304**, 780.

Spitzer, L., 1982, *Searching between the Stars* (New Haven : Yale University Press), pp. 148-151.

Sternberg, A., and Dalgarno, A., 1989, *Astrophys. J.* **338**, 197.

Sternberg, A., 1990, in prep.

Stutzki, J., Stacey, G.J., Genzel, R., Harris, A.I., Jaffe, D.T., and Lugten, J.B., 1988, *Astrophys. J.* **332**, 279.

Stutzki, J. et al., 1990a, in *Submillimetre Astronomy*. eds. G.D.Watt and A.S.Webster (Dordrecht : Kluwer), p. 269.

Stutzki, J., et al. 1990b, in preparation.

Stutzki, J., and Güsten, R., 1990, *Astrophys. J.* **356**, 513.

Taff, L.G., and Savedoff. M.P., 1972, *Monthly Notices Roy. Astron. Soc.* **160**, 89.

Taff, L.G., and Savedoff. M.P., 1973, *Monthly Notices Roy. Astron. Soc.* **164**, 357.

Tauber, J.A., and Goldsmith, P.F., 1990, *Astrophys. J. Letters* **356**, L61.

Tielens, A., and Hollenbach, D., 1985a, *Astrophys. J.* **291**, 722.

Tielens, A., and Hollenbach, D., 1985b, *Astrophys. J.* **291**, 747.

Zinnecker, H., 1989, in *Evolutionary Phenomena in Galaxies*, ed. J. Beckman (Cambridge : Cambridge University Press), p.115.

DUST EMISSION FROM STAR FORMING CLOUDS:
A PROGRESS REPORT

P.G. MEZGER, A. SIEVERS, R. ZYLKA

MPIfR bolometer group*
Max-Planck-Institut für Radioastronomie,
Auf dem Hügel 69, 5300 Bonn 1

* Other members of the MPIfR bolometer group are: R. Chini, H.P. Gemünd, G. Haslam, E. Kreysa, R. Lemke and J. Wink.

1. Dust emission and star formation: The working plan

Model computations of protostellar evolution depend very strongly on the initial conditions: Fragmentation of massive cloud cores or coagulation of substellar condensations, the physical state of gas and dust (e.g. the formation of ice-mantles and grain coagulation), the presence of magnetic fields and its effect on gas and dust, and the formation of accretion disks as a consequence of an initial angular momentum of the protostellar condensation. The MPIfR bolometer group together with the molecular spectroscopists R. Mauersberger and T.L. Wilson have embarked on a program aimed at the exploration of the earliest evolutionary stages of high- and low-mass star formation. Here follows a brief progress report.

1.1 DUST EMISSION AS A PROBE OF DENSE REGIONS OF INTERSTELLAR MATTER.

In principle both isotopic molecular lines and dust emission can be used as probes of the ISM. In a recent review Mezger (1990) has discussed the merits and limitations of both methods. Measurements of dust emission yield the beam averaged surface brightness

$$I_\nu = S_\nu/\Omega_A = B_\nu(T_d)(1-e^{-\tau_\nu}) \tag{1}$$

Here S_ν is the flux density per beam solid angle Ω_A, T_d an appropriate average dust temperature, $\tau_\nu = N_H \sigma_\nu^H$ the optical depth of the emitting dust layer, σ_ν^H the dust absorption cross section per H-atom and $N_H = N(H)+2N(H_2)$ the total hydrogen column density. If $\tau_\nu \ll 1$ and both σ_ν^H and T_d are known the hydrogen column density can be derived from the relation

$$N_H = \frac{1}{\sigma_\nu^H} \frac{I_\nu}{B_\nu(T_d)} \tag{2}$$

A parametrized representation of $\sigma_H^H(\lambda)$ and formulae to estimate dust temperatures, hydrogen column densities and masses are given in Appendix A of Mezger, Wink and Zylka (1990) and in the above mentioned review. We refer to these papers and summarize here briefly those results needed in the following discussion.

i) The dust absorption cross section in molecular clouds (MCs) and cloud cores can be approximated by

$$\sigma_H^H = \tau_\nu/N_H = (Z/Z_\odot) \; b \begin{cases} 7 \; 10^{-21} \lambda_{\mu m}^{-2} & \lambda_{\mu m} \geq 100 \\ 7 \; 10^{-22} \lambda_{\mu m}^{-1.5} & 40 \leq \lambda_{\mu m} \leq 100 \end{cases} \tag{3}$$

Here $Z/Z_\odot$ is the relative metallicity and $\lambda_{\mu m}$ is the wavelength in μm. b=1 corresponds to the Draine and Lee (1984) dust model. For dust clouds of medium density ($n_H \leq 10^5 cm^{-3}$) and high density ($n_H \geq 10^6 cm^{-3}$ we recommend the use of b=1.9 and 3.4, respectively.

ii) For clouds of medium density isotopic CO transitions and dust emission yield values of N_H and M_H which agree within a factor of ~2. However, in dense ($n_H > 10^6 cm^{-3}$) and cold ($T_d \sim T_g < 20-30K$) regions molecular transitions become unreliable tracers of column densities, probably because molecules freeze out and form ice mantles around refractory grain cores. Due to the formation of ice mantles the parameter in the above relation appears to increase to b≤3.4.

iii) Line contamination of broadband dust emission becomes important only at relatively high gas temperatures.

iv) Grain coagulation has been predicted and observed in circumstellar disks but may also occur in protostellar condensations. If grains coagulate and form compact particles of average size $\bar{a} \ll 100\mu$m but large compared to NIR wavelengths, one expects little effects on the FIR/submm absorption cross section but a strong reduction of the extinction cross section in the wavelength range from the UV to the NIR. If, on the other hand, coagulated grains comprise a large volume fraction of empty space the dust cross section can be affected just in the opposite direction (Mathis and Whiffen, 1989) i.e. no change at optical/NIR wavelengths but an increase of the absorption cross section at FIR/mm wavelengths.

v) When observing dust emission from dense condensations the telescope beam usually samples both cold (~10-20K) and warm(~30-100K) dust. Since warm dust is in most cases the principal contributor to the total IR luminosity cold dust contributes in essence only to the mm/submm part of the combined spectrum. Second to a model fit a spectral decomposition of the observed dust emission spectrum of the form

$$S_\nu = \sum_i B_\nu(T_{d,i})[1 - e^{-\tau_\nu, i}]\Omega_{s,i} \tag{4}$$

will usually yield meaningful (dust) mass averaged grain temperatures if a minimum of i=2 for $\lambda > 30\mu$m and i=3 for $\lambda > 10\mu$m components are used. In many cases high angular resolution observations are needed to determine the fit parameters $\tau_{\nu,i}$ and $\Omega_{s,i}$.

1.2 MAPPING OF DUST EMISSION

For the observations discussed in the following sections we used the MPIfR ^{3}He cooled bolometer (Kreysa, 1985) with the 3m IRTF and the 2.2m U. of Hawaii telescope (UHT) on Mauna Kea, Hawaii, ($\lambda=1300\mu m$, $\theta_A=90$"; $\lambda=350\mu m$, $\theta_A=30$" for the IRTF and $\lambda 1300\mu m$, $\theta_A=128$" for the UHT) and with the 30m MRT on Pico Veleta, southern Spain ($\lambda=1300\mu m$, $\theta_A=11$"; $\lambda=870\mu m$, $\theta_A=7$").

1.3 IS THERE A CHANCE TO DETECT INDIVIDUAL PROTOSTARS?

Stellar masses in the galaxy range from ~100m$_\odot$ to ~0.1m$_\odot$. The present star formation rate is ~2-3m$_\odot$yr^{-1}. What triggers star formation on the one hand, and what prevents gravitationally unstable clouds to collapse on the other hand is not yet well understood. Bimodal star formation, as originally formulated by Güsten and Mezger (1983) to explain galactic abundance gradients, suggests induced formation of massive stars (m $\gtrsim$3m$_\odot$) in GMCs in spiral arms and the central region of the Galaxy, and spontaneous formation of stars in the whole stellar mass range, occuring primarily in clouds in the interarm region.

In any case cold (T$_d$~10-20K) condensations of gas and dust are expected to represent the first evolutionary stages of star formation. Combination of the point source sensitivity of S$_{1300}\gtrsim$150mJy (~3 times the rms noise) of the 30m MRT operated in the mapping mode with a scan velocity of 8'/min, with an earlier estimate of the number of isothermal protostars within 1Kpc (Mezger et al. 1988) leads us to conclude that most of the nearby MCs (D$_{Kpc}$~0.1-0.2) should contain protostars of low and medium mass. Most of these protostars together with about hundred high mass protostars should be detectable with the MRT in the mapping mode.

1.4 THE WORKING PLAN

Most star forming MCs have been mapped in isotopic transitions of CO and CS with low and intermediate angular resolution. Molecular transitions and mm/submm dust emission yield comparable results for medium density clouds (n$_H\lesssim$10^{5}cm^{-3}), easily detect gas outflow from PMS objects, but fail to detect high-density protostellar condensations. Our observing strategy is based on these considerations and the fact that focal plane chopping limits high angular resolution MRT and SEST maps to a maximum size of ~8'x6'. We use molecular emission as a guide line for the dust emission surveys.

2. Regions of low-mass star formation

2.1 MYERS CLOUD CORES

Myers and collaborators have – by means of molecular spectroscopy – surveyed nearby dust clouds and have catalogued a number of condensations. (For a recent review see Benson and Myers, 1989). It has been suggested that "Myers cores" might represent the earliest stages of low-mass star formation. This suggestion appeared to be strengthened by the fact that a number of low-luminosity IRAS sources were observed close to the center of some of the cores. We have begun a survey of Myers cores with the MPIfR bolometer used at $\lambda 1300$ and $350\mu m$ with the IRTF and at $\lambda 1300\mu m$ with the UHT. Results of this survey can be summarized as follows (Chini et al., priv.comm.).

Table 1
Statistical summary of Mauna Kea Observations
of Myers cores without associated IRAS sources

$\lambda_{\mu m}$	θ_A/arcsec	N_{obs}	N_{det}	S_{min}/Jy	S_{det}/Jy
350	30	22	6	0.5	1.8-8.5
1300	90(128)	39	32	0.2	0.4-4.4

A survey (i.e. maps of size ~3'x2') of the cores with positive detections and of a number of additional cores with associated IRAS sources was made at $\lambda 1300\mu m$. Compiled in Table 2 are the observed and derived characteristics of 10 of these cores which are located in the Taurus cloud at a distance of 140pc. For a constant column density the hydrogen mass within the telescope aperture (expressed in Table 2 as linear size d) should decrease $\propto d^2$. This is in rough agreement with the hydrogen masses derived for the Myers cores with different telescope apertures. There are especially no signs of a dense central core as would be expected to develop in the case of free-fall contraction of a protostar. The only exception is TMC-1A, which is the position of a low-luminosity IRAS source about ~2.5 west of TMC1 and which therefore is probably unrelated with TMC1. In the MRT map shown in Fig.1 TMC-1A appears as a slightly extended source of integrated flux density S_{1300}~330mJy.

Table 2
Observed and derived characteristics of some Myers cores in the Taurus cloud which are not associated with IRAS sources

Name	α_{1950}	σ_{1950}	S_ν/Jy $\lambda_{\mu m}$ = 1300 θ_A=128"	90"	11"	350 30"	$M_H/M_\odot$ T_d = 10K d=2.7E17	1.9E17	6.3E16	2.3E16
L1489	04^{h}01^{m}45^{s}5	26°10'33"0	2.7		<0.15	2.4	3.4		0.1	<0.2
L1498	07 50.0	25 02 13.0	1.2		"	1.8	1.5		0.1	<0.2
L1495C	10 25.0	28 08 20.0	1.4		"	<0.5	1.7		<0.03	<0.2
L1495	10 57.5	28 03 53.0	1.3		"	<0.5	1.6		<0.03	<0.2
L1495B	12 30.0	28 38 39.0	2.1		"	<0.5	2.6		<0.03	<0.2
L1506	15 29.0	25 12 10.0	1.5		"	<0.5	1.9		<0.03	<0.2
TMC-2	29 43.0	24 18 54.0		1.4	"	2.9		1.7	0.2	<0.2
TMC-1A	36 31.2	25 35 56.0		0.5	0.33	<0.5		0.6	<0.03	0.4
TMC-1	38 42.0	25 35 45.0		1.0	<0.15	2.2		1.2	0.1	<0.2
L1521D	24 42.5	26 12 13.0		1.2	"	-		1.5	-	<0.2

Mean volume and column densities of the observed Myers cores are of order n_H~10^5cm^{-3} and N_H ~ some 10^{23}cm^{-2}. The cloud cores TMC1 and TMC2, although well investigated by molecular spectroscopy, could not be detected in $\lambda 1300\mu m$ dust emission.

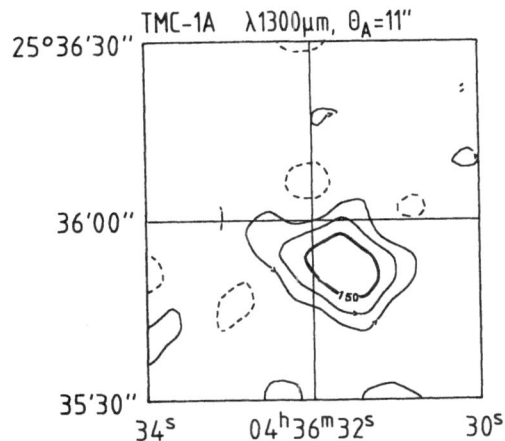

TMC-1A $\lambda 1300\mu m$, $\theta_A = 11''$

Fig.1.: MRT $\lambda 1300\mu m$ map of TMC-1A. Contour levels are in mJy $(11''-beam)^{-1}$.

2.2 LOW-MASS PMS OBJECTS

Here we discuss three PMS objects which have been amply investigated and discussed in the literature. Dust emission from all three sources was observed at $\lambda 1.3mm$ by Walker, Adams and Lada (1990, in the following referred to as WAL). We compare our average dust temperatures and associated gas masses with corresponding results by WAL to emphasize the importance of a multi-component analysis of the dust emission spectrum. WAL used a single-temperature, variable m fit to their data. All these sources have been investigated in various molecular transitions but a detailed comparison with dust emission is outside the scope of this paper. IRAS 16293-2422 is a PMS object with gas outflow located in the Rho Ophiuchus cloud (D_{Kpc}~0.16). WAL derive a flux density of S_{1300}=7.0Jy. Although they do not resolve this source they observe emission out to ~60". Their spectral fit yields T_d=32K, m=1.3, an optical depth of τ_{1300}=3.0E-3 and a gas mass of M_H~2.0m$_\odot$. Mundy, Wilking and Myers (1986) resolved this source at $\lambda 2.7mm$ into an elongated feature of size 11"x<5" when convolved with the 4"x6" beam.

Fig.2a shows the $\lambda 1300\mu m$ MRT image of this source. It consists of a central point source of FWHP~3" (not resolved with the MRT but derived from the spectral fit) with a flux density of 6.7Jy and an extended envelope of size ~40"x35" and flux density 7.6Jy. Fig.2b shows the spectrum fitted with a two component model. Note that the $\lambda 350\mu m$ flux density was obtained with a 30"-beam in the ON-OFF mode so that part of the contribution from the envelope is suppressed and that the interferometer observations at 2.7mm refer primarily to the central point source. In this fit the point source with T_d=45K and Ω_s~10(arcsec)^2 becomes opaque at $\lambda 1100\mu m$ and has a gas mass of M_H~1.5m$_\odot$. The cold (T_d~20K) envelope has an optical depth of τ_{1300}~1.2E-2 and a gas mass of M_H~4.5m$_\odot$. (In both cases b=3.4 has been adopted). Assuming a spherical density distribution one derives hydrogen volume and column densities for core and envelope of $n_H = 5 \ 10^9 cm^{-3}$, $N_H = 5.1 \ 10^{25} cm^{-2}$ and $10^7 cm^{-3}$ and $8.5 \ 10^{23} cm^2$, respectively. Obviously one needs to know the core–halo structure for a correct interpretation of the observations.

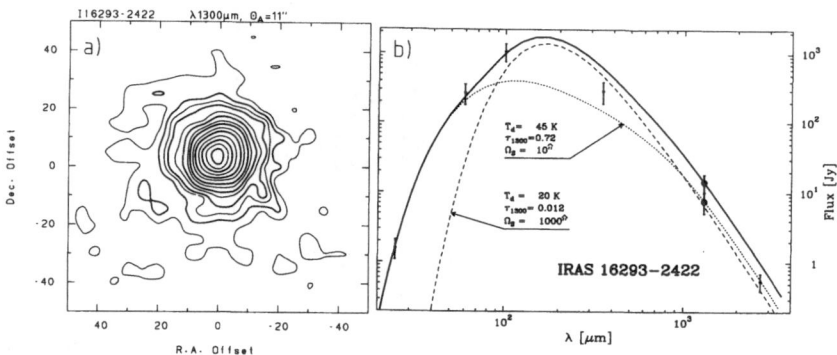

Fig.2.: IRAS 16293-2422. a) MRT $\lambda 1300\mu m$ map. Coordinates are offset relative to the position $\alpha_{1950} = 16^h 29^m 20.9$, $\delta_{1950} = -24° 22'13''$. Contour levels are 100, 200, 300, 600, 800, 1100, 1400, 1900, 2400, 3100, 3800, 4500..mJy (11''-beam)^{-1} b) Decomposed FIR/mm spectrum. Flux densities are from WAL and this paper ($\lambda 1300$ and $350\mu m$).

L1551 IR5 is another outflow source at a distance $D_{Kpc} \sim 0.16$ for which WAL derive $T_d = 48K$, $m=1.2$, $\tau_{1300} = 4.8E-4$, $L_{IR} = 19.4 L_\odot$ and – with a flux density of $S_{1300} = 4.3Jy$ – a total gas mass of $\sim 0.6 m_\odot$. Their contour map shows a source of $\sim 1'$ extent with a fairly flat EW–plateau of emission at its center.

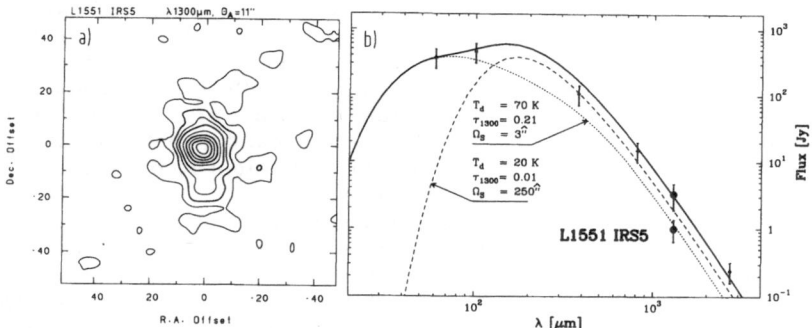

Fig.3.: L1551 IRS5. a) MRT $\lambda 1300\mu m$ map. Coordinates are offset relative to the position $\alpha_{1950} = 4^h 28^m 40.5$, $\delta_{1950} = 18° 1'42''$. Contour levels are 100, 200, 300, 400, 550, 700...mJy (11''-beam)^{-1}. b) Decomposed FIR/mm spectrum. Flux densities are from WAL and this paper ($\lambda 1300\mu m$).

Fig.3a shows our MRT map. The structure is resolved into a point source of $S_{1300} \sim 0.8Jy$ and an extended envelope of size $\sim 37'' x 20''$ and flux density $\sim 2.5Jy$. The spectrum shown in Fig.3b requires again a two-component fit and yields source sizes of ~2" for the point source (the MRT observations yield only an upper limit of the source diameter) and ~16" for the envelope. Corresponding gas masses of core (T_d=70K) and envelope (T_d=20K) are (with b=3.4) M_H~0.1 and 1.5$m_\odot$, respectively. B335 is a large isolated dust globule at D_{Kpc}~0.4 with mass outflow and appears to have a stellar core of low luminosity (Keene et al., 1983). WAL derive T_d=25K, m=1.1, τ_{1300}=1.4E-3, L=8$L_\odot$ and with S_{1300}=0.8Jy a total gas mass of 1.1$m_\odot$. Actually, the spectrum of this source can be most simply interpreted in terms of a thin outer layer of warm dust (~19K) surrounding the interior cold (~14K) dust (Mezger, Mathis and Panagia, 1982). This is the typical dust temperature structure of a source which is mainly heated from outside by the general ISRF. A more recent spectral analysis confirms that the bulk of the emission comes from dust with ~14K (Gee et al., 1985). Fig.4 shows the λ1300μm MRT map where some spatial structure is indicated. We confirm the low 1300μm flux density of 0.8Jy observed by WAL. The corresponding mass is M_H~5$m_\odot$.

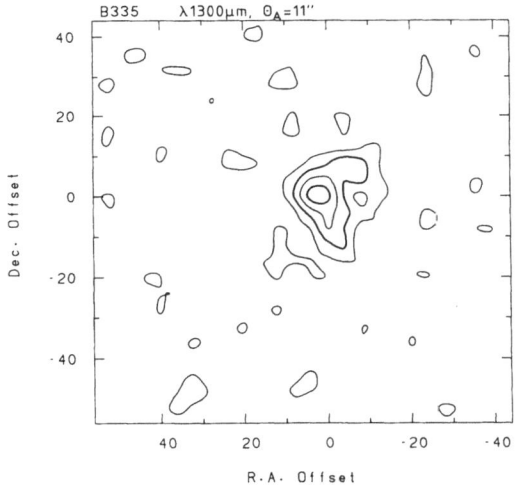

Fig.4.: *B335. MRT λ1300μm map. Coordinates are offset relative to the position α_{1950}= 19^{h}34^{m}34.7, δ_{1950}= 7°27'20". Contour levels are in mJy (15"-beam)$^{-1}$.*

2.3 THE RO OPHIUCHUS CLOUD

This cloud south of the star ρ Oph is located at a distance of 160pc from the sun. It is considered to be the prototype of a low- to medium-mass star forming cloud. Wilking and Lada (1983) based on a $C^{18}O$ survey detected a 1x2pc ridge with hydrogen column densities as high as 10^{23}cm^{-2}, which extends from SE to NW. It contains a total mass of M_H~400-500$m_\odot$. The internal structure of this cloud has been thoroughly investigated by means of molecular spectroscopy. Loren (1989) has described 89 dense clumps of which 12 very dense cores are claimed to be active sites of star formation. Wooten (1989) in a recent survey gives a very vivid but also somewhat gloomy description of this work, from which we quote: *"that indeed the perception of the structure of dense cores depends critically upon the molecular transition in whose light it is imaged."*

In an attempt to detect low-mass isothermal protostars we began a survey of the λ1300μm dust emission from this cloud using both the IRTF and MRT. With the IRTF we observed at λ1300μm a cross section in α, δ through the point $\alpha_{1950}=16^h23^m25^s87$, $\delta_{1950}=-24°17'20''7$ and found extended emission ranging from ~1 to 10Jy(90"-beam)$^{-1}$. This corresponds for T_d~15K to hydrogen column densities ranging from N_H~2 to 20 10^{22}cm^{-2} and tends to confirm the existence of the above mentioned ridge structure. A number of MRT maps ranging in size from 3'x2' to 6'x5' were observed around peak positions of molecular line emission and some complex dust emission structure was indeed detected. One such structure is a submm source recently investigated at λ1100, 800 and 350μm with the UKIRT 3.8m telescope by Ward-Thompson et al. (1989) and referred to as SM1. This source lies $\Delta\alpha$~4^{s}6 west and 33" south of a previsouly known FIR source observed by both Harvey, Campbell and Hoffmann (1979) and by IRAS which is referred to as IRS1. Ward-Thompson et al. derive from fits to the dust emission spectra of SM1 and IRS1 longward of λ25μm dust characteristics of T_d=15K, m~2.2 and T_d=36K, m~1.0. Their luminosities are L_{IR}~60 and 110L$_\odot$.

Fig.5 shows the MRT map of SM1. It contains three condensations in an extended NS ridge. Flux densities are: S_{1300}~40Jy within the 150mJy/beam contour and ~16Jy within the ridge structure defined by the 450mJy/beam contour. The condensations have flux densities of (from N to S) 4.5, 6.5 and 3.9Jy. For T_d=15K and b=1.9 the total mass contained in the map is ~60m$_\odot$; masses of the condensations range (for b=3.4) from ~3 to 5.5m$_\odot$. Based on both the low dust temperature (T_d~15K) and luminosity (L_{IR}~60L$_\odot$) we conclude that the three condensations represent isothermal low-mass protostars. The FIR source IRS1, on the other hand, appears to be a self-luminous PMS star embedded in a dust shell. The relatively low luminosity of L_{IR}~110m$_\odot$ suggests a medium mass star as heating source.

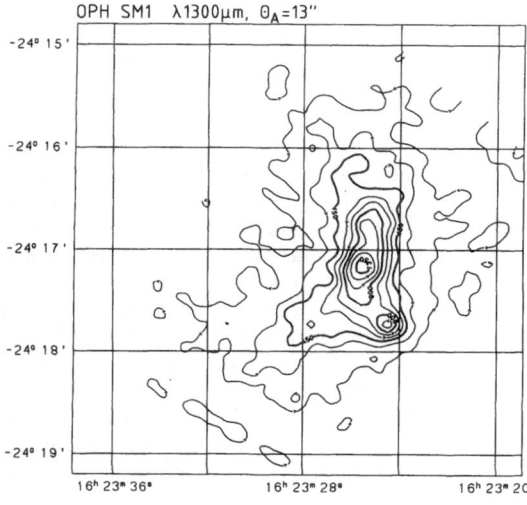

Fig.5.: MRT λ1300μm map of the submm source SM1 in the ρOph cloud. Contours are in mJy(13"-beam)$^{-1}$.

We also surveyed regions in ρ Oph B where Loren et al. (1980) and Martin-Pintado et al. (1983) had detected strong $\lambda 2cm$ H_2CO emission which indicates the presence of cloud cores and condensations with volume and column densities of $\geq 10^6 cm^{-3}$ and $\sim 10^{23} cm^{-2}$, respectively. Fig.6 shows a contour map of $\lambda 1300 \mu m$ dust emission from this region. An elongated structure with a surface brightness of $\sim 200mJy$ in a 15" beam is visible, which corresponds for $T_d = 15K$ and $b = 3.4$ to a hydrogen column density of $N_H \sim 1.5\ 10^{23} cm^{-2}$, in good agreement with the H_2CO results. The total mass within the 200mJy per beam contour is $\sim 5m_\odot$. Within this area there are no indications of condensations in the mass range $M_H \geq 0.3 m_\odot$.

3. Regions of spontaneous high-mass star formation

The Orion GMC is the region closest to the sun with (probably) spontaneous high-mass star formation. Two of its most prominent features are the Orion A and B clouds each of which contain some $10^5 m_\odot$ of gas. Observations of dust emission from two cores in these clouds which contain $\sim 1000 m_\odot$ of gas have been published previously (NGC2024: Mezger et al., 1988; Mezger, 1988. OMC1/2, Mezger, Wink and Zylka, 1990). Embedded in these cores are massive isothermal protostars (NGC2024) and PMS objects (OMC1). Derived characteristics of cloud cores and dense condensations are given in Table 3. Fig.7a and b show unpublished MRT $\lambda 870 \mu m$ maps of the dense protostellar condensations − where free-free contamination is negligible − which fully confirm the morphology of the published $\lambda 1300 \mu m$ maps.

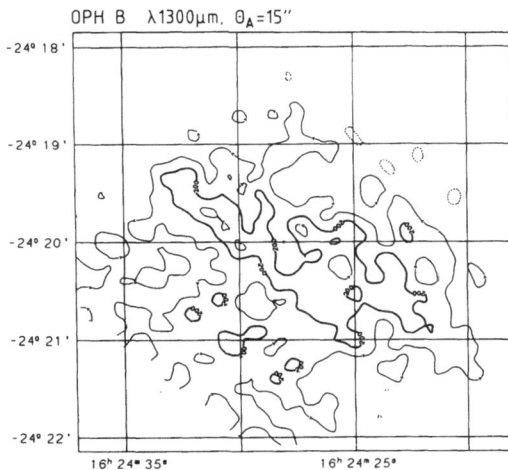

Fig.6.: MRT $\lambda 1300 \mu m$ map of the cloud core ρOph B. Contours are in $mJy(15"-beam)^{-1}$.

A review of the wealth of more recent observations of the NGC2024 cloud core is beyond the scope of this paper. Here we just want to point out some problems in interpreting limited data of a very complex region of the ISM.

Dust and gas temperatures: Moore et al.(1989) made a single fit to the dust emission spectrum with a modified Planck function $B_\nu(T_d)\nu^m$ and $T_d=47K$, $m=1.6$. Since opaque dust with 16K emits in essence only at wavelengths $\lambda \gtrsim 400\mu m$ it is clear that the composite spectrum can be represented by dust with an opacity spectrum less steep than m^{-2}. Such a fit ignores, however, the fact that warm dust, represented by $\lambda 60\mu m$ emission, outlines the HII region (which is extended in RA) while cold dust, represented by $\lambda 1300\mu m$ mission, outlines the molecular line emission (which is extended in DEC).

Table 3

Physical conditions in the cloud cores NGC2024 and OMC1.

Source	n_H/cm^{-3}	N_H/cm^{-2}	$M_H/m_\odot$	T_d/K	Ref
NGC2024 core	$2\ 10^5$	$3\ 10^{23}$	800	16	
isoth. protostars	$4\text{-}14\ 10^6$	$2\text{-}6\ 10^{25}$	10-60	16	1)
OMC1	10^5	$2\ 10^{23}$	1700	25	
hot core=FIR2	$3\text{-}9\ 10^7$	10^{25}	50	$1\text{-}2\ 10^2$	2)

1) Mezger et al., 1988
2) Mezger, Wink and Zylka, 1990

Mezger et al. (1988; see their Fig.8) noted that the cold and dense condensations which are clearly discernible in the high-resolution dust emission maps are not well traced by various molecular transitions. Wilson, Mauersberger and Mezger (in prep.) using the MRT have recently mapped the NGC2024 cloud core in rare CO and CS isotopes ($C^{18}O$ J=2-1 and 1-0; $C^{17}O$ J=2-1; $C^{34}S$ J=3-2). Molecular emission follows in general the dust emission but maxima in the line emission tend to coincide with dust emission minima. Such an anticorrelation of dust and molecular emission has been also found in the case of OMC1 (Mezger, Wink and Zylka, 1990).

4. W49 as an example of induced high-mass star formation

We have mapped a number of dust clouds associated with giant HII regions which are probably located in main spiral arms and thus should be typical for induced high-mass star formation. Here we will discuss some preliminary observations of W49 which is located at a distance of ~14Kpc (Sievers et al., in prep.).

Fig.8a shows a $\lambda 1300\mu m$ map obtained with the IRTF. This map is not corrected for free-free emission which accounts for ~36% of the integrated flux density of ~113Jy. A map obtained at the same wavelength with the MRT containes ~26Jy after correction for free-free emission using the $\lambda 3mm$ maps of Salter et al. (1989). This is about one third of the dust emission contained in the IRTF map. Fig.8b shows a MRT map at $\lambda 870\mu m$, where contamination by free-free emission is no longer a problem. Fig.8c shows the spectral index $r(870/1300\mu m)$ of the surface brightness of dust emission. This spectral index has been derived by comparing 870/1300μm MRT maps convolved to the angular resolution of

29''5, with which free-free emission has been measured (Salter et al., 1989) and which has been subtracted from the shorter wavelength maps.

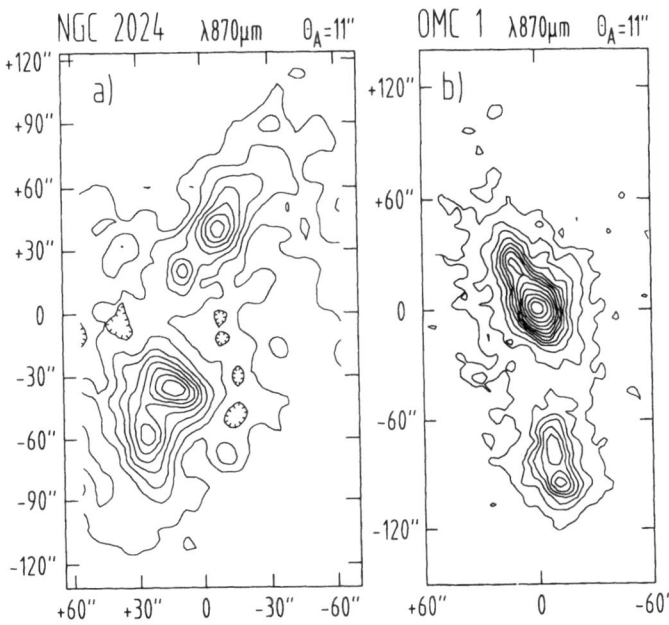

Fig.7.: *MRT* $\lambda 870\mu m$ *dust emission maps of NGC2024 and OMC1 Gauss-convolved to 11 arcsec (unpublished). Coordinates are offset relative to the position* $\alpha = 05^h 39^m 12.0$, $\sigma = -01°56'30''$ *(NGC2024) and* $05^h 32^m 46.9$, $\sigma = -05°24'26''$ *(OMC1).*

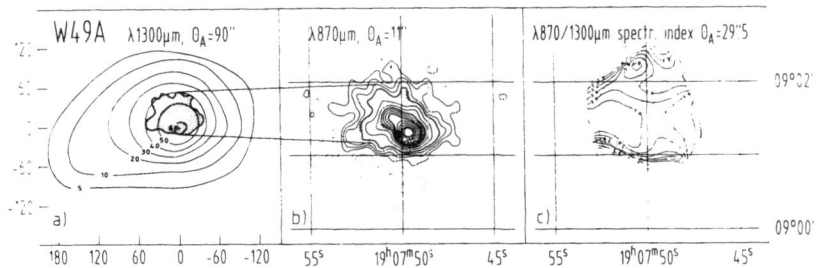

Fig.8.: *The GMC associated with the Giant HII region W49 a) IRTF* $\lambda 1300\mu m$ *map not corrected for free-free emission. The third-lowest contour of Fig.b) comprises the stipled area. Contours are in Jy* $(90''-beam)^{-1}$ *b) Gauss-convolved MRT* $\lambda 870\mu m$ *map. Contours are in mJy* $(11''-beam)^{-1}$ *c) Contour map representation of the spectral index* $r(870/1300\mu m)$ *derived from MRT maps at* $\lambda 1300$ *and* $870\mu m$, *which are Gauss convolved to a FWHP of* $\theta_A = 29''5$ *and from which free-free emission is subtracted.*

Over most of the central region of the map the surface brightness has a spectral index of r^{-4}, as is expected for an opacity law $\sigma\slashed{b}(\lambda)\propto\lambda^{-2}$ and temperatures $T_d \gtrsim 50K$. Towards the edge of the cloud the spectral index appears to decrease indicating the presence of dust with lower temperatures. This is in agreement with a decomposition of the submm/FIR/MIR spectrum shown in Fig.9 into three components with $T\sim 30$, 53 and 140K, respectively. Characteristics of these three components are given in Table 4. Note that the hydrogen masses derived from the luminosities of the three components are systematically lower than those derived from $\lambda 1300\mu m$ flux densities. This is due to opacity of the sources in the wavelength range of maximum emission. The extent of 30K and 53K dust emission is approximately outlined in Fig.8a by the 0.2 contour of the IRTF map and the second lowest contour of the MRT map, respectively.

Table 4

Observed and derived characteristics of "warm" GMC associated with W49.

$\frac{T_d}{K}$	$\frac{S_{1300}}{Jy}$	$\frac{M_H}{m_\odot}$ a)	$\frac{n_H}{cm^{-3}}$	$\frac{size}{pc}$	$\frac{N_H}{cm^{-2}}$	$\frac{L_{IR}}{L_\odot}$	$in\%$	$\frac{M_H}{m_\odot}$ b)
30	46	2.2E5	$\sim 10^4$	~ 11	$\sim$3E23		~ 20	1.0E5
53	26	6.6E4	$\sim 10^5$	~ 3.7	$\sim$1E24	$3\ 10^7$	~ 60	1.3E4
140	~ 0.2	~ 1.7E2	~ 20	-	-		~ 20	2.8E1

a) derived from the flux density at $\lambda 1300\mu m$ and b=1.9
b) derived from the luminosity of the individual components and the dust luminosity-to-hydrogen mass ration computed for T_d.

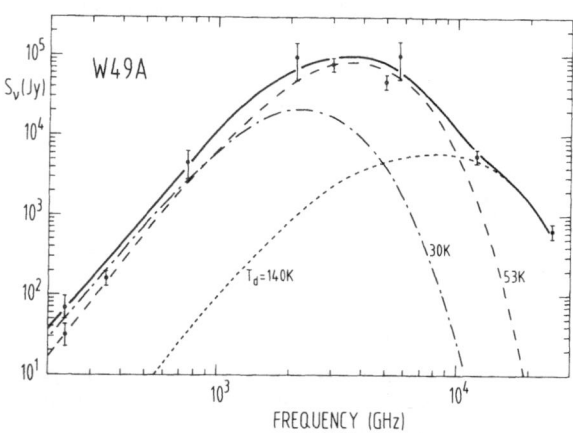

Fig.9.: Decomposed FIR/submm spectrum of W49. Characteristics of the components are given in Table 4.

References

Benson, D.J., Myers, P.C.: 1989, ApJ. Suppl. **71**, 89

Draine. B.T., Lee, H.M.: 1984, ApJ. **285**, 89

Gee, G., Griffin, M.J., Cunningham, Ch.T., Emerson, J.P., Ade, P.A.R.:
    1985, MNRAS **215**, 15p

Güsten, R., Mezger, P.G.: 1983, Vistas in Astronomy **26**, 159

Harvey, P.M., Campbell, M.F., Hoffmann, W.F.: 1979, ApJ. **228**, 445

Keene, J. Davidson, J.A., Harper, D.A., Hildebrand, R.H., Loewenstein, R.F.,
    Low, F.J., Pernie, R.: 1983, ApJ. **274**, L43

Kreysa, E.: 1985 in Int. Symp. on "mm- and submm wave Astronomy" URSI,
    Granada, p.153

Loren, R.B., Wooten, A., Sandquist, A., Bernes, C.: 1980, ApJ. **240**, L169

Loren, R.: 1989, ApJ. **338**, 902

Martin-Pintado, J., Wilson, T.L., Gardner, F.F., Henkel, C.: 1983,
    A&A **117**, 145

Mathis, J.S., Whiffen, G.: 1989, ApJ. **341**, 808

Mezger, P.G., Mathis, J.S., Panagia, N.: 1982, A&A **105**, 372

Mezger, P.G., Chini, R., Kreysa, E., Wink, J.E., Salter, C.J.: 1988,
    A&A **191**, 44

Mezger, P.G., Zylka, R., Salter, C.J., Wink, J.E., Chini, R., Kreysa, E.,
    Tuffs, R.: 1989, A&A **209**, 337

Mezger, P.G.: 1988 in "Galactic and Extragalactic Star Formation"
    NATO-ASI Series **232**, Pudritz and Fich, eds. p.227

Mezger, P.G., Wink, J.E., Zylka, R.: 1990, A&A **228**, 95

Mezger, P.G.: 1990 in Conf.Proc. "Physics and Composition of the
    Interstellar matter", Torun, Poland (in press)

Moore, T.J.T., Chandler, C.J., Gear, W.K., Mountain, C.M.: 1989,
    MNRAS **237**, 1p

Mundy, L.G., Wilking, B.A., Myers, P.C.: 1986, ApJ **311**, L75

Salter, C.J., Emerson, D.T., Steppe, H., Thum, C.: 1989, A&A **225**, 167

Walker, C.K., Adams, F.C., Lada, C.J.: 1990, ApJ. **349**, 515 (WAL)

Ward-Tomson, P., Robson, E.I., Whittet, D.G.B., Gordon, M.A., Walther, D.M.,
    Duncan, W.D.: 1989, MNRAS **241**, 119

Wilking, B.A., Lada, Ch.J.: 1983, ApJ. **274**, 698

Wooten, A.: 1989, Proc. IAU Coll.No.120, eds. Tenorro-Tagle, Mole,
    Melznick, Lect.Notes in Phys. **350**, p.210

VI - STAR FORMATION

SOME PROCESSES INFLUENCING THE STELLAR INITIAL MASS FUNCTION

RICHARD B. LARSON
Yale Astronomy Department
Box 6666
New Haven, Connecticut 06511
U.S.A.

ABSTRACT. Current evidence suggests that the stellar initial mass function has the same basic form everywhere, and that its fundamental features are (1) the existence of a characteristic stellar mass of order one solar mass, and (2) the existence of an apparently universal power-law form for the mass spectrum of the more massive stars. The characteristic stellar mass may be determined in part by the typical mass scale for the fragmentation of star forming clouds, which is predicted to be of the order of one solar mass. The power-law extension of the mass spectrum toward higher masses may result from the continuing accretional growth of some stars to much larger masses; the fact that the most massive stars appear to form preferentially in cluster cores suggests that such continuing accretion may be particularly important at the centers of clusters. Numerical simulations suggest that forming systems of stars may tend to develop a hierarchical structure, possibly self-similar in nature. If most stars form in such hierarchically structured systems, and if the mass of the most massive star that forms in each subcluster increases as a power of the mass of the subcluster, then a mass spectrum of power-law form is predicted. Some possible physical effects that could lead to such a relation are briefly discussed, and some observational tests of the ideas discussed here are proposed.

1. Introduction

A major goal of studies of star formation is to understand the spectrum of masses with which stars are formed, since this is of fundamental importance in many areas of astronomy. It has, however, become increasingly clear that the initial mass function (IMF) of stars depends on many complex and still poorly understood processes, and it may even be that processes not yet recognized will turn out to be important. Therefore it is not yet possible to enumerate or discuss with any confidence all of the processes that may play a role in determining the IMF. Many theories have of course been proposed; these have been reviewed a number of times, for example by Miller and Scalo (1979), Elmegreen (1985), Larson (1986b, 1989), and Zinnecker (1987, 1989), and for the most part they will not be reviewed again here. Instead, I shall first summarize what seem to be the essential features of the IMF that need to be understood, and then I shall briefly mention some possible approaches to understanding these properties, with emphasis on the relevant evidence and on further observations that may help to clarify the processes involved.

2. Some Basic Properties of the IMF

The observational evidence on the stellar IMF has been summarized in extensive reviews by Miller and Scalo (1979) and by Scalo (1986), and further reviews of some of the evidence have been given by Larson (1986b), Rana (1987), Scalo (1987, 1990), and Zinnecker (1987). The most familiar and well-established property of the IMF is that, at least for stars more massive than about one solar mass, it can be approximated by a power law; within the uncertainties, the slope of this power law appears to be much the same everywhere, and it is similar to but probably somewhat steeper than the slope originally found by Salpeter (1955). If we define the IMF as the number of stars formed per unit logarithmic mass interval, and if for masses above one solar mass we approximate this function by

$$dN/d \log m \; \propto \; m^{-x}$$

where m is the stellar mass, then the value of the exponent x, according to Scalo (1986), is 1.7 ± 0.5. This is somewhat larger than Salpeter's value of 1.35, although the uncertainties remain large and Salpeter's slope is not yet excluded. All of the systems for which extensive direct star counts exist, including the solar neighborhood (Scalo 1986), nearby galaxies (Freedman 1985), and star clusters in our Galaxy and the Magellanic Clouds (Mateo 1988, 1990), show a very similar form for the upper IMF, and there is at present no convincing evidence for any variability in its slope.

A second basic and apparently general feature of the IMF is that it does not keep rising indefinitely with the same slope toward smaller masses, but instead falls well below an extrapolation of the above power law at masses less than about 0.5 solar masses. According to Miller and Scalo (1979), the IMF for solar neighborhood field stars becomes approximately flat ($x \sim 0$) at the smallest masses, while according to Scalo (1986), it has a peak at a mass near 0.25 solar masses and drops steeply toward lower masses. Although the IMF for the least massive stars remains particularly uncertain because of the poorly determined mass-luminosity relation for faint stars, and although some continuing increase of the IMF toward smaller masses is possible (Zinnecker 1987, 1989), it nevertheless appears that the IMF does not rise much, if at all, in the region from several tenths of a solar mass down to the hydrogen-burning limit at 0.08 solar masses. Unless it turns sharply upward again at yet lower masses, for which significant data do not yet exist, there cannot be much mass in stars of very small mass or "brown dwarfs". In fact, no evidence has yet been found for the existence of large numbers of brown dwarfs, despite extensive searches. If there is indeed very little mass in such objects, this would have the important implication that the typical mass with which stars form is not much smaller than one solar mass.

The upper and lower parts of the IMF for solar neighborhood field stars can be properly matched together only if an assumption is made about the past history of the star formation rate. It has been conventional to assume a constant star formation rate, but models of galactic evolution have also been proposed in which the star formation rate decreases strongly with time; in the latter case, the inferred IMF does not decline monotonically with mass near one solar mass but has a second peak just above one solar mass. Models of this type with a double-peaked or "bimodal" IMF might account for the unseen mass that has been thought to exist in the solar neighborhood, since they predict a substantial amount of mass to be in the form of stellar remnants (Quirk and Tinsley 1973; Larson 1986a). However, recent data and analyses imply a smaller surface density of mass in the local Galactic disk than had been found in earlier studies, and together with increased estimates of the amount of mass in visible stars and gas, these results no longer leave much room for any unseen mass (Kuijken and Gilmore 1989; Gould 1990). This is a very important finding, if correct, since it implies that neither brown dwarfs nor stellar remnants can constitute a large fraction of the mass in the local Galactic disk, and consequently that neither the lower nor the upper end of the IMF can ever have been greatly enhanced above the standard form discussed above. Any bimodality of the IMF is therefore strongly constrained by these results, and in particular the model of Larson (1986a) now seems excluded by them; instead, the current data seem consistent with a relatively simple and conventional model of galactic evolution in which there has been only a modest decline of the star formation rate with time (Larson 1990b).

It is worth noting that similar constraints are also placed on the IMF in other stellar systems by recent studies of their dynamics, which have generally yielded relatively small mass-to-light ratios. For example, studies of the mass distributions in the disks of other spiral galaxies have found mass-to-light ratios of the order of 2.5 or less which are similar to that of the local Galactic disk, and which again leave little room for unseen mass (Kent 1986,1987; Athanassoula, Bosma, and Papaioannou 1987). Also, the best-studied globular clusters have mass-to-light ratios that are very similar to each other and that mostly lie between 1.0 and 1.5, consistent with the directly observed stellar content of these clusters and arguing against any large variations in at least the upper part of the IMF, which would influence the mass-to-light ratio through a varying content of stellar remnants (Pryor *et al.* 1989). Thus, the current evidence supports a considerable degree of universality in the basic form of the IMF, and it also demonstrates that the processes that determine this apparently universal form operate even on the scale of individual star clusters.

It remains possible that there is some variability in the lower part of the IMF, even if at a more modest level than in the various "bimodal star formation" hypotheses that have been considered. For example, there is some evidence for differences in the mass spectra of the T Tauri stars in different regions of star formation, as was discussed by Larson (1982); in the Taurus clouds the median mass of the known T Tauri stars is about 0.6 solar masses, while in the Orion cloud it is about 1.1 solar masses (Larson 1986b). A possibly related difference between these two regions is that in the Taurus clouds the young stars are distributed in scattered small groups, whereas in the Orion cloud they are more concentrated into a single large cluster. Additional evidence for differences in the mass spectra of T Tauri stars in different regions of star formation and for possible correlations with cloud properties has been presented at this meeting by P. C. Myers. There may be a correlation between the typical masses of newly formed stars and the typical masses of the clumps in the associated molecular clouds, and the typical clump mass may in turn correlate with cloud mass; this is suggested by the fact that the Orion cloud, which is by far the most massive nearby star forming cloud, contains the most massive clumps and also seems to be forming stars with the largest typical mass. It is not yet entirely clear whether the apparent deficiency of low mass stars in the Orion cloud compared with other regions is real or is just due to incomplete sampling of the least massive stars, owing to the relatively large distance and high opacity of the Orion cloud; in any case, a real turndown in the mass function at the smallest masses is suggested by the steep drop at faint magnitudes in the K-band luminosity function of the Trapezium cluster (McCaughrean et al. 1990; H. Zinnecker, private communication). A similar deficiency of faint stars may also be present in the young cluster NGC 2023 in the Orion B cloud (DePoy *et al.* 1990).

Another difference in the observed mass spectra of young stars in different star forming regions is that the mass of the most massive star in each cloud or complex increases systematically with the mass of the cloud, varying approximately as the 0.43 power of cloud mass (Larson 1982). This result is consistent with, and in fact adds support to, the existence of a universal power-law form for the upper part of the IMF; however, it also implies that the most massive stars do not form in random locations but only in the most massive aggregates of gas and young stars. The most massive stars also appear to form preferentially in the dense core regions of these massive star forming complexes; the Trapezium stars in Orion provide a well-known example. These properties of star forming regions suggest that the upper part of the IMF is built up systematically in a way that depends on the total amount of matter present as well as on the spatial structure of the system (see Section 4).

Summarizing the evidence that has been discussed, I conclude that there are two principal facts to be explained about the IMF: (1) There is a characteristic mass scale for star formation that is of the order of one solar mass; defined as the mass such that half of the matter condensing into stars goes into less massive stars and half into more massive ones, this characteristic mass is within a factor of 2 of one solar mass. (2) The IMF has an extended tail toward larger masses that is approximately of power-law form, with an apparently universal slope of $x = 1.7 \pm 0.5$. Both the characteristic stellar mass and the slope of the power law may vary to some extent with location, but such variations appear not to be large. In the following sections, I shall review briefly some possible ways of trying to understand these basic properties of the IMF, and also some relevant types of observations that could help to clarify the physical processes involved.

3. The Mass Scale for Star Formation

A classical and well-known mass scale that is presumably relevant to star formation is the Jeans mass, which is the minimum mass that a cloud fragment must have in order for its gravity to dominate over thermal pressure at a given temperature and density. The validity of the Jeans criterion as originally derived has always been somewhat questionable, since the Jeans analysis is inconsistent in neglecting the collapse of the postulated initially nearly uniform medium as density fluctuations begin to grow in it. A more rigorous and probably more relevant stability analysis which predicts a similar critical mass can be made if it is assumed that the initial state of a fragmenting cloud is, instead of a uniform medium, an equilibrium sheet, disk, or filament (Larson 1985). These latter configurations all have qualitatively similar dispersion relations, which imply in each case a well-defined characteristic length and mass scale for fragmentation; this is true even in the presence of rotation and magnetic fields, which act primarily to reduce the growth rate but do not change the basic length or mass scales involved. The predicted fragmentation scale depends on two readily measurable parameters, namely the temperature and the surface density of the fragmenting cloud, so it is easy to calculate a characteristic mass scale for fragmentation in star forming clouds, provided that these clouds are not far from being in hydrostatic equilibrium locally.

A variety of numerical simulations of the fragmentation of collapsing clouds have yielded results that are in general agreement with the above predictions. The relatively crude simulations by Larson (1978) of the fragmentation of rotating collapsing clouds showed that the number of objects formed was comparable to the number of Jeans masses in the initial cloud; however, fragmentation did not actually become pronounced in these simulations until the cloud had flattened to a disk, and the results were also roughly consistent with the predictions for the fragmentation of disks (Larson 1985). More detailed simulations of the fragmentation of rotating collapsing clouds by Miyama, Hayashi, and Narita (1984) again showed marked fragmentation following the collapse of the cloud to a disk, and in this case a larger number of fragments was found for similar initial conditions, in better agreement with the theory. Analytic and numerical studies of the fragmentation of non-rotating sheets and filaments by Miyama, Narita, and Hayashi (1987a,b) suggest that fragmenting sheets may first tend to break up into filaments, such as are often seen in star forming clouds, and that the filaments may then fragment into clumps. The most recent simulations of the fragmentation of rotating collapsing clouds by Monaghan and Lattanzio (1990) include a proper treatment of gas cooling, and they show the formation of a network of filaments and voids reminiscent of those obtained in cosmological simulations; they also show the formation of dense knots at the intersections of the filaments. Lattanzio, Keto, and Monaghan (1990) suggest that these results may be able to explain the ring-like distribution of massive young stars seen in W49A. In all of these numerical investigations, there appears to be a fairly well-defined characteristic mass scale for fragmentation, a result which is in at least qualitative agreement with the theoretical expectations.

Nevertheless, it may still be questioned, given the obvious complexity in the structure and dynamics of molecular clouds, whether stability analyses and collapse simulations such as those discussed above really provide an accurate description of how star formation is initiated and how a characteristic stellar mass comes about. A more general approach to estimating a mass scale for clump formation, which again leads to essentially equivalent results, is to suppose that clumps form, whether dynamically or quasi-statically, in approximate pressure balance with the ambient molecular cloud; if the internal pressure in the clumps is predominantly thermal, the minimum mass that a clump must have in order for gravity to dominate over pressure is then

$$M_{\text{crit}} \sim c^4 / G^{3/2} P^{1/2},$$

where c is the sound speed and P is the ambient pressure. Here the relevant ambient pressure is the total pressure acting to support the molecular cloud against gravity, and in general it includes thermal, turbulent, and magnetic contributions; on large scales, turbulent and magnetic pressures dominate, while on the smallest scales thermal pressure is dominant (Larson 1981; Myers 1983). The scaling relations that appear to hold between the sizes, densities, and internal turbulent velocities of molecular clouds imply that the supporting pressure is of similar magnitude in different clouds and in different parts of clouds, possibly reflecting a tendency toward general pressure balance; for example, if the supporting pressure is largely magnetic, then the data imply a nearly constant magnetic field strength (Myers and Goodman 1988). If the associated typical pressure in molecular clouds is substituted into the above relation along with a typical sound speed of 0.2 km/s, the minimum mass predicted for bound clumps is about one solar mass. The fact that this predicted mass scale for clump formation is essentially the same (within perhaps a factor of 3) as the observed characteristic stellar mass suggests that stellar masses may be determined, at least in part, by the mass scale for clump formation in molecular clouds.

A different view has been advocated by Shu and Terebey (1984), Shu, Adams, and Lizano (1987), and Shu et al. (1988). These authors argue that since the clumps or cloud cores in which stars form are not sharply bounded but merge smoothly with their surroundings, a forming star would continue to accrete matter and grow almost indefinitely in mass if the accretion process were not somehow shut off, for example by a stellar wind; they suggest, moreover, that the typical stellar mass is determined by the effect of winds in stopping accretion, and not by any property of star forming clouds. In this view, the onset of a stellar wind is assumed to be associated with the beginning of deuterium burning in a forming star, although it is not clear that deuterium burning is actually responsible for initiating winds, or even that a wind will necessarily stop accretion (Stahler 1988; Zinnecker 1989). Stellar winds probably do play a role in limiting the accretional growth of embryonic stars, but other effects may also act to prevent the accretion of the outer part of a cloud core by a forming star. Outside the thermally supported inner region of the core, which has the mass discussed above, magnetic fields and turbulence dominate the dynamics and may tend to inhibit further infall (e.g. Mouschovias 1987); tidal forces may also limit the amount of mass that can be accreted (Zinnecker 1989), and it even seems possible that in some star forming regions the outer parts of cloud cores may be ablated away by large-scale gas flows. Clearly, it is difficult to predict exactly how much of the core mass will end up in a star, since this depends on various intricacies of the dynamics of the accreting gas that are not yet well understood.

In any case, the question of whether stellar masses depend on cloud properties can be addressed observationally, given sufficiently good data. If typical stellar masses depend on the properties of star forming clouds, the lower part of the IMF might be expected to vary between different regions of star formation. For example, if the typical stellar mass depends on the mass scale for fragmentation, and if this in turn depends most importantly on cloud temperature, as suggested by Larson (1985), then warmer clouds might be expected to form stars with higher typical masses. Such a difference may exist between the Orion cloud and other nearby star forming clouds, as was noted in Section 2, although more data are needed to confirm this possibility. Continuing studies of the stellar content of star forming regions, particularly at infrared wavelengths, will help to provide better answers to such questions.

4. The Origin of the Power-Law Upper IMF

Unlike the lower part of the IMF, the upper part shows no evidence for any preferred mass. Instead, for the larger stellar masses there is approximately a constant *ratio* between the numbers of stars formed in adjacent logarithmic mass intervals: for example, the number of stars formed per unit logarithmic mass interval decreases by about a factor of three for each factor of two increase in stellar mass. This suggests that the upper part of the IMF may be generated by processes that operate similarly on different mass scales. Many ideas have been proposed about how to account for a scale-free upper IMF of the observed form, and a few of the possibilities are mentioned below.

Recent observations of the clump mass spectrum in several large molecular clouds have yielded results that are fairly similar and are approximately of power-law form; for example, in the M17 cloud the clump mass spectrum is approximately a power law with $x \sim 0.7$, which is similar to the form predicted by coagulation theories of clump growth (Stutzki and Güsten 1990). If each clump forms one star, and if there is a relation between clump mass and stellar mass such that the efficiency of star formation decreases with increasing clump mass, then the clump mass spectrum may translate directly into a stellar mass spectrum of the observed form (Zinnecker 1989). However, it is not clear that there is such a direct relation between the clump mass spectrum and the stellar mass spectrum, since it has not yet been shown that there is a one-to-one correspondence between clumps and stars, or that there is a relation of the required form between clump mass and stellar mass. In fact, reality is almost certainly more complicated than such a simple picture, since we have seen evidence that at least the most massive clumps may form not just single stars but entire groups or clusters of stars, and it is possible that most stars actually form in such groups or clusters. If this is the case, then in order to understand the upper IMF, it is necessary to understand the processes that control the way in which matter is apportioned among the stars in a forming system of stars.

A possible way to generate a scale-free upper IMF in a forming system of stars is via the continuing growth in mass of the stars by gas accretion at a rate varying as a power of the stellar mass. Accretion was found to play an important role in building up the masses of the most massive objects obtained in the fragmentation simulations of Larson (1978), and circumstantial evidence that accretion is important for the formation of massive stars is provided by the fact that these stars appear to form preferentially in dense cluster cores, where conditions for continuing accretion are likely to be especially favorable (Larson 1982). If the accretion rate is proportional to a power of the stellar mass greater than unity, then an extended power-law tail is built up on any initially peaked mass distribution. For example, if the accretion rate is proportional to the square of the stellar mass, as in classical accretion theory, then a power-law upper IMF with a slope of $x = 1$ is generated (Zinnecker 1982); this is similar to, but not quite as steep as, the observed upper IMF. A steeper IMF would result if the accretion rate were proportional to a higher power of the stellar mass. Since the accretion rate increases with increasing gas density and with decreasing relative velocity between the accreting object and the ambient gas, a stronger dependence of the accretion rate on stellar mass would result if the more massive objects were preferentially located in denser regions or were moving more slowly with respect to the ambient gas. Correlations of this sort between the masses of newly formed stars and their locations and motions in young clusters are in fact suggested by the observational evidence, and they are also to be expected theoretically as a consequence of gravitational drag effects, as discussed below.

Considerable evidence, much of it obtained only recently with infrared array cameras, suggests that many and perhaps most stars, even of low mass, actually form in compact groups or clusters deeply embedded in dense molecular cloud cores. Some of this evidence has been presented and discussed at this meeting by E. A. Lada and by K.-W. Hodapp, and a review of some of the early results obtained with infrared arrays has been given by Gatley, DePoy, and Fowler (1988). Some well-known examples of compact young clusters that are still partly embedded in molecular cloud cores are the Trapezium cluster in the Orion A cloud (Herbig and Terndrup 1986) and the ρ Ophiuchi cluster (Wilking, Lada, and Young 1989). An infrared photograph of the Trapezium cluster obtained by McCaughrean (1989) shows clearly the extremely high central density of this cluster and also its symmetrical and centrally concentrated structure, which is less apparent at visible wavelengths where part of the cluster is not seen. Particularly striking is the presence of the Trapezium multiple system itself right at the center of this cluster, since it contains several of the most massive stars and also the single most massive star in the entire Orion region of star formation. A central location for the most massive stars is not unique to the Trapezium cluster, since other examples are provided by the massive young clusters in NGC 3603 and the 30 Doradus nebula (Moffat, Seggewiss, and Shara 1985), each of which has at its center a multiple system containing some of the most massive stars in the cluster (Baier, Ladebeck, and Weigelt 1985; Weigelt and Baier 1985). Most, but not all, of the compact clusters in the Orion B cloud studied by E. A. Lada (1990) and described at this conference also appear to have their most luminous stars located near the center. This evidence suggests that accretion processes occurring at the centers of dense clusters may play an important role in the formation of the most massive stars.

An additional effect that might be important in organizing the way in which matter is accreted by the objects in a forming system of stars is a tendency toward the development of hierarchical clustering in the system. Some of the simulations of cloud fragmentation of Larson (1978) yielded hierarchical multiple systems containing subsystems similar to, but scaled down from, the larger system; some of these groupings of objects also contained a dominant centrally located object (see, for example, Fig. 6(a) of that paper). It was suggested in that paper that forming systems of stars might tend to develop a self-similar or fractal hierarchical structure, and it was noted that if the most massive star in each subsystem acquires a mass that is a constant fraction of the mass of the subsystem, this would lead to a power-law IMF. Hierarchical clustering may be of quite general importance in cosmogony, since it has been found or suggested to occur in other contexts as well; for example, the simulations by West and Richstone (1988) of the formation of clusters of galaxies show a similar phenomenon, even though the physics is simpler and involves only stellar dynamics. Hierarchical clustering may also be important for the formation of individual galaxies (Larson 1990c). In the simulations of West and Richstone (1988), many small subclusters initially form, and they subsequently merge into progressively larger ones; at each stage, the most massive objects tend to be centrally located in these subclusters. This tendency of the most massive objects to be centrally located is a result of the action of dynamical friction, which is most important for the most massive objects and causes them to sink rapidly toward the centers of local mass concentrations (Binney and Tremaine 1987; Richstone 1990). Similar gravitational drag effects will be even more important in young clusters of stars that still contain a significant amount of gas, since the gas produces a drag on all of the stars and not just the most massive ones; again, however, the effect is most important for the most massive stars, so the result is likely to be the formation of very dense clusters dominated by central subgroups of very massive stars (Larson 1990a), similar to the observed young clusters mentioned above.

If the most massive stars in a forming cluster or subcluster quickly settle to the center and then remain nearly at rest there because of gravitational drag effects, this will favor their continuing rapid growth by accretion. Because of this positive feedback effect, the amount of matter accreted by a star may be a strong function of its mass, just as would be required to generate an extended upper IMF of the observed form. If systems of stars are built up hierarchically from smaller systems, as suggested above, then a central most massive star may form in each subsystem, and central stars of progressively larger mass may form as larger and more centrally concentrated systems of stars are built up. In this way, a correlation may be introduced between the mass of the most massive star that can form in each system and the mass of the system. Negative feedback effects may also tend to introduce a coupling between the masses of accreting stars and the distribution of matter in their surroundings by limiting or regulating the accretion; for example, radiation pressure is a very important effect for luminous stars, and it may inhibit or prevent the formation of massive stars by conventional spherical accretion processes (Wolfire and Cassinelli 1987). The net result of such effects may be that the most massive stars can form only at the centers of exceptionally massive and condensed aggregates of gas and young stars, where the deep potential well produced by the surrounding matter, as well as possible interactions with neighboring objects (Larson 1982, 1990a), may allow accretion to continue even onto very luminous stars. Thus, there may be several reasons for the mass of the most massive star that can form in a cluster to depend on overall properties of the system such as its total mass and central density.

The mass of the most massive star that forms might then be expected to depend on at least two basic mass scales, the minimum fragment mass or Jeans mass and the total cluster mass. On very general grounds, the Jeans mass might be expected to be relevant because it determines the typical size of the clumps in which the matter is distributed, while the cluster mass should be relevant because the mass of the most massive object that can form should increase with the total mass available. In many problems in physics where two fundamental scales are involved, the scale of the phenomenon of interest is the geometric mean of these two fundamental scales; if we were to guess that similar behavior holds in the present problem, the maximum stellar mass might then be something like the geometric mean of the Jeans mass and the cluster mass. Such a simple guess would not obviously conflict with the observational evidence; for example, the maximum stellar mass in the Trapezium cluster is about 42 solar masses (Larson 1982), not far from the geometric mean of the Jeans mass and the cluster mass, which are of the order of one solar mass and a thousand solar masses, respectively.

More generally, one might consider the hypothesis that the mass of the most massive star that can form in a cluster or subcluster is related to the mass of the associated system by a scaling relation of the power-law form

$$M_{star,max} \propto M_{cluster}{}^n .$$

The existence of such a relation is already suggested by the fact that the mass of the most massive young star in each region of star formation is found to vary approximately with the 0.43 power of the mass of the associated molecular cloud (Larson 1982). It was noted by Larson (1982) that this dependence of maximum stellar mass on cloud mass can be understood if the IMF has a universal power-law form with a slope of $x = 1/0.43 = 2.3$, and if all clouds form stars with the same efficiency. Here, I note that the inverse can also be argued: if the processes of star formation result in a general power-law relation between the mass of the most massive star and the mass of the associated stellar system, and if stellar systems are built up from subsystems in all of which the same relation holds, then an IMF of power-law form is predicted. For example, if the maximum stellar mass is proportional to the square root of the cluster mass, as in the simple example of the previous paragraph, and if each system of stars contains or is built from two subsystems whose most massive objects are less massive by a factor of $2^{1/2}$ on the average, then the number of stars per unit logarithmic mass interval increases by a factor of 2 with each decrease by a factor of $2^{1/2}$ in stellar mass, which corresponds to a power-law IMF with a slope of $x = 2$. More generally, if the maximum stellar mass is proportional to a power n of the cluster mass, then the slope of the resulting IMF is $x = 1/n$. For example, the observed IMF slope of $x = 1.7 \pm 0.5$ would be reproduced if n were equal to 0.6 ± 0.2, close to the simple square-root dependence suggested above.

The above scaling law is suggested here only as a possibly useful working hypothesis to be tested observationally, without necessarily being based on any very specific theory or model. A variety of possible theoretical explanations for such a relation might of course be proposed, and some have already been suggested in the literature (for example, a possible explanation based on feedback effects was suggested by Larson 1989). However, it is likely that many complex processes are actually involved in determining the form of the upper IMF, and it is not yet clear which ones are most important, or even that all of the important ones have yet been recognized. Therefore any attempted theoretical explanation for such a scaling relation would at best be incomplete. Like the other scaling relations that have been discussed at some length at this meeting, a scaling law such as that suggested above might serve as a useful description of empirical trends even though it does not have a well-understood theoretical basis; nevertheless, all such scaling relations should be treated with some caution, because the processes that occur in star forming clouds are clearly so complex that any regularities that might exist are bound to show many exceptions and a wide dispersion about any mean trend.

5. Some Questions for Observers

All of the effects that have been discussed in this review, as well as others that have not been discussed, probably play a role in determining the form of the stellar IMF, so a complete understanding of the subject is still a distant goal. Therefore, I conclude not with any answers but with a list of questions for further observational research suggested by some of the issues that have been discussed. Much progress in this subject can be expected on the observational front, given the large amount of relevant data now being produced by various new techniques, but it may be useful to propose some specific questions or tests that can be addressed with these data. Some of these questions were already anticipated by P. C. Myers in his review at this meeting, in which some important first efforts at correlating and organizing the large amount of relevant information were presented. My suggested questions are:

(1) What is the relation between the clumpy structure of molecular clouds and newly formed stars? Do all dense clumps form stars? If not, which ones actually form stars?

(2) In the clumps that form stars, is there a simple relation between clump mass and stellar mass? Can the clump mass spectrum, as a result, be translated directly into a stellar mass spectrum?

(3) Do some clumps, for example the larger ones, form not just single stars but entire compact groups or clusters of stars?

(4) What fraction of all stars is formed in such compact groups and clusters?

(5) In clumps or cloud cores that form clusters of stars, is there a relation between the clump mass and the mass of the cluster?

(6) Is there a relation between the maximum mass of a star that can form in a group or cluster and the total mass or other properties of the cluster?

(7) What is the spatial structure of forming systems of stars? In particular, is any tendency toward hierarchical clustering observed?

(8) What is the spatial distribution of stars in forming clusters as a function of mass? Are the most massive stars centrally located, and are stars progressively more widely dispersed with decreasing mass?

(9) What are the properties of the residual gas in forming clusters? Is it concentrated around the stars, for example in disks? Can any evidence be found for accretion or gravitational drag effects?

Better answers to these and similar questions will lead to a better observational understanding of how the stellar IMF is built up, and will also provide a stimulus for further theoretical efforts to understand the physical processes involved.

REFERENCES

Athanassoula, E., Bosma, A., and Papaioannou, S. (1987) 'Halo parameters of spiral galaxies', *Astron. Astrophys.* **179**, 23-40.

Baier, G., Ladebeck, R., and Weigelt, G. (1985) 'Speckle interferometry of the central object in the giant HII region NGC 3603', *Astron. Astrophys.* **151**, 61-63.

Binney, J. and Tremaine, S. (1987) *Galactic Dynamics*, Princeton University Press, Princeton.

DePoy, D. L., Lada, E. A., Gatley, I., and Probst, R. (1990) 'The luminosity function in NGC 2023', *Astrophys. J. (Letters)* **356**, L55-L58.

Elmegreen, B. G. (1985) 'The initial mass function and implications for cluster formation', in R. Lucas, A. Omont, and R. Stora (eds.), *Birth and Infancy of Stars*, North-Holland, Amsterdam, pp. 257-277.

Freedman, W. L. (1985) 'The upper end of the stellar luminosity function for a sample of nearby resolved late-type galaxies', *Astrophys. J.* **299**, 74-84.

Gatley, I., DePoy, D. L., and Fowler, A. M. (1988) 'Astronomical imaging with infrared array cameras', *Science* **242**, 1264-1270 and cover photograph.

Gould, A. (1990) 'Galactic disc column density by maximum likelihood', *Mon. Not. Roy. Astron. Soc.* **244**, 25-28.

Herbig, G. H., and Terndrup, D. M. (1986) 'The Trapezium cluster of the Orion Nebula', *Astrophys. J.* **307**, 609-618.

Kent, S. M. (1986) 'Dark matter in spiral galaxies. I. Galaxies with optical rotation curves', *Astron. J.* **91**, 1301-1327.

Kent, S. M. (1987) 'Dark matter in spiral galaxies. II. Galaxies with HI rotation curves', *Astron. J.* **93**, 816-832.

Kuijken, K., and Gilmore, G. (1989) 'The mass distribution in the galactic disc - II. Determination of the surface mass density of the galactic disc near the Sun', *Mon. Not. Roy. Astron. Soc.* **239**, 605-649.

Lada, E. A. (1990) 'Global star formation in the L1630 molecular clouds', this volume; Ph.D. thesis, University of Texas.

Larson, R. B. (1978) 'Calculations of three-dimensional collapse and fragmentation', *Mon. Not. Roy. Astron. Soc.* **184**, 69-85.

Larson, R. B. (1981) 'Turbulence and star formation in molecular clouds', *Mon. Not. Roy. Astron. Soc.* **194**, 809-826.

Larson, R. B. (1982) 'Mass spectra of young stars', *Mon. Not. Roy. Astron. Soc.* **200**, 159-174.

Larson, R. B. (1985) 'Cloud fragmentation and stellar masses', *Mon. Not. Roy. Astron. Soc.* **214**, 379-398.

Larson, R. B. (1986a) 'Bimodal star formation and remnant-dominated galactic models', *Mon. Not. Roy. Astron. Soc.* **218**, 409-428.

Larson, R. B. (1986b) 'The initial mass function', in C. A. Norman, A. Renzini, and M. Tosi (eds.), *Stellar Populations*, Cambridge University Press, Cambridge, pp. 101-123.

Larson, R. B. (1989) 'Fragmentation and the initial mass function', in G. Tenorio-Tagle, M. Moles, and J. Melnick (eds.), *Structure and Dynamics of the Interstellar Medium*, IAU Colloquium No. 120, Springer-Verlag, Berlin, pp. 44-54.

Larson, R. B. (1990a) 'Formation of star clusters', in R. Capuzzo-Dolcetta, C. Chiosi, and A. Di Fazio (eds.), *Physical Processes in Fragmentation and Star Formation*, Kluwer Academic Publishers, Dordrecht, pp. 389-400.

Larson, R. B. (1990b) 'Galactic evolution', in D. L. Lambert (ed.), *Frontiers of Stellar Evolution*, Astronomical Society of the Pacific Conference Series, San Francisco, in press.

Larson, R. B. (1990c) 'Galaxy building', *Publ. Astron. Soc. Pacific* **102**, 709-722.

Lattanzio, J. C., Keto, E. R., and Monaghan, J. J. (1990) 'Hydrodynamical models of the W49A star-forming region', poster presented at this conference.

Mateo, M. (1988) 'Main-sequence luminosity and initial mass functions of six Magellanic Cloud star clusters ranging in age from 10 megayears to 2.5 gigayears', *Astrophys. J.* **331**, 261-293.

Mateo, M. (1990) 'The initial mass functions of Magellanic Cloud star clusters', in R. Capuzzo-Dolcetta, C. Chiosi, and A. Di Fazio (eds.), *Physical Processes in*

Fragmentation and Star Formation, Kluwer Academic Publishers, Dordrecht, pp. 401-414.

McCaughrean, M. J. (1989) 'Multicolour near infrared imaging of the Orion Nebula and Trapezium cluster', *Bull. Amer. Astron. Soc.* **21**, 712; photograph published in *Sky and Telescope* **77**, 352 (1989).

McCaughrean, M., Zinnecker, H., Aspin, C., and McLean, I. (1990) 'Low mass pre-main sequence clusters in regions of massive Galactic star formation', in R. Elston and C. A. Pilachowski (eds.), *Astrophysics with Infrared Arrays*, Astronomical Society of the Pacific Conference Series, San Francisco, in press.

Miller, G. E. and Scalo, J. M. (1979) 'The initial mass function and stellar birthrate in the solar neighborhood', *Astrophys. J. Suppl.* **41**, 513-547.

Miyama, S. M., Hayashi, C., and Narita, S. (1984) 'Criteria for collapse and fragmentation of rotating, isothermal clouds', *Astrophys. J.* **279**, 621-632.

Miyama, S. M., Narita, S., and Hayashi, C. (1987a) 'Fragmentation of isothermal sheet-like clouds. I. Solutions of linear and second-order perturbation equations', *Prog. Theor. Phys.* **78**, 1051-1064.

Miyama, S. M., Narita, S., and Hayashi, C. (1987b) 'Fragmentation of isothermal sheet-like clouds. II. Full nonlinear numerical simulations', *Prog. Theor. Phys.* **78**, 1273-1287.

Moffat, A. F. J., Seggewiss, W., and Shara, M. M. (1985) 'Probing the luminous stellar cores of the giant HII regions 30 Dor in the LMC and NGC 3603 in the Galaxy', *Astrophys. J.* **295**, 109-133.

Monaghan, J. J. and Lattanzio, J. C. (1990) 'A simulation of the collapse and fragmentation of cooling molecular clouds', *Astrophys. J.*, in press; poster presented at this conference.

Mouschovias, T. Ch. (1987) 'Star formation in magnetic interstellar clouds: I. Interplay between theory and observations', in G. E. Morfill and M. Scholer (eds.), *Physical Processes in Interstellar Clouds*, D. Reidel Publishing Company, Dordrecht, pp. 453-489.

Myers, P. C. (1983) 'Dense cores in dark clouds. III. Subsonic turbulence', *Astrophys. J.* **270**, 105-118.

Myers, P. C. and Goodman, A. A. (1988) 'Magnetic molecular clouds: Indirect evidence for magnetic support and ambipolar diffusion', *Astrophys. J.* **329**, 392-405.

Pryor, C., McClure, R. D., Fletcher, J. M., and Hesser, J. E. (1989) 'Mass-to-light ratios for globular clusters. I. The centrally concentrated clusters NGC 6624, M28 (NGC 6626), and M70 (NGC 6681)', *Astron. J.* **98**, 596-610.

Quirk, W. J. and Tinsley, B. M. (1973) 'Star formation and evolution in spiral galaxies', *Astrophys. J.* **179**, 69-83.

Rana, N. C. (1987) 'Mass function of stars in the solar neighbourhood', *Astron. Astrophys.* **184**, 104-118.

Richstone, D. (1990) 'Dynamical evolution of clusters of galaxies', in W. R. Oegerle, M. J. Fitchett, and L. Danly (eds.), *Clusters of Galaxies*, Cambridge University Press, Cambridge, PP. 231-255.

Salpeter, E. E. (1955) 'The luminosity function and stellar evolution', *Astrophys. J.* **121**, 161-167.

Scalo, J. M. (1986) 'The stellar initial mass function', *Fundam. Cosmic Phys.* **11**, 1-278.

Scalo, J. M. (1987) 'The initial mass function, starbursts, and the milky way', in T. X. Thuan, T. Montmerle, and J. Tran Thanh Van (eds.), *Starbursts and Galaxy Evolution*, Editions Frontières, Gif sur Yvette, pp. 445-465.

Scalo, J. M. (1990) 'Top-heavy IMFs in starburst galaxies', in G. Fabbiano, J. S. Gallagher, and A. Renzini (eds.), *Windows on Galaxies*, Kluwer Academic Publishers, Dordrecht, pp.125-140.

Shu, F. H., and Terebey, S. (1984) 'The formation of cool stars from cloud cores', in S. L. Baliunas and L. Hartmann (eds.), *Cool Stars, Stellar Systems, and the Sun*, Springer-Verlag, Berlin, pp. 78-89.

Shu, F. H., Adams, F. C., and Lizano, S. (1987) 'Star formation in molecular clouds: observation and theory', *Ann. Rev. Astron. Astrophys.* **25**, 23-81.

Shu, F. H., Lizano, S., Adams, F. C., and Ruden, S. P. (1988) 'Beginning and end of a low-mass protostar', in A. K. Dupree and M. T. V. T. Lago (eds.), *Formation and Evolution of Low Mass Stars*, Kluwer Academic Publishers, Dordrecht, pp. 123-137.

Stahler, S. W. (1988) 'Deuterium and the stellar birthline', *Astrophys. J.* **332**, 804-825.

Stutzki, J. and Güsten, R. (1990) 'High spatial resolution isotopic CO and CS observations of M17 SW: The clumpy structure of the molecular cloud core', *Astrophys. J.* **356**, 513-533.

Weigelt, G. and Baier, G. (1985) 'R136a in the 30 Doradus nebula resolved by holographic speckle interferometry', *Astron. Astrophys.* **150**, L18-L20.

West, M. J. and Richstone, D. O. (1988) 'The formation of clusters of galaxies containing dark matter', *Astrophys. J.* **335**, 532-541.

Wilking, B. A., Lada, C. J., and Young, E. T. (1989) 'IRAS observations of the ρ Ophiuchi infrared cluster: Spectral energy distributions and luminosity function', *Astrophys. J.* **340**, 823-852.

Wolfire, M. G. and Cassinelli, J. P. (1987) 'Conditions for the formation of massive stars', *Astrophys. J.* **319**, 850-867.

Zinnecker, H. (1982) 'Prediction of the protostellar mass spectrum in the Orion near-infrared cluster', in A. E. Glassgold, P. J. Huggins, and E. L. Schucking (eds.), *Symposium on the Orion Nebula to Honor Henry Draper*, *Ann. New York Acad. Sci.* **395**, 226-235.

Zinnecker, H. (1987) 'A review of the IMF', in J. Palous (ed.), *Evolution of Galaxies*, Proceedings of the 10th European Regional Meeting of the IAU, Vol. **4**, Czechoslovak Academy of Sciences, Prague, pp. 77-85.

Zinnecker, H. (1989) 'Towards a theory of star formation', in J. E. Beckman and B. E. J. Pagel (eds.), *Evolutionary Phenomena in Galaxies*, Cambridge University Press, Cambridge, pp. 113-127.

A Comparative Study of Star Formation Efficiencies
in Nearby Molecular Cloud Complexes

Yasuo Fukui and Akira Mizuno
Department of Astrophysics
Nagoya University

Abstract   We present new observational data of ^{13}CO J=1-0 emission in nearby
molecular cloud complexes in Taurus, Ophiuchus, and Orion, and compare star formation
efficiencies in the three regions.   The ^{13}CO data have been obtained with the 4m
millimeter telescope at Nagoya equipped with a SIS receiver having receiver temperature
of 23 K in double side band. The molecular distribution is often characterized by highly
filamentary, continuous distribution, rather than well defined clumps. Active star
formation is taking place almost exclusively in the densest regions in the molecular
distributions.   The filamentary distributions often show velocity gradients perpendicular
to their elongation, suggesting that they are spinning around their major axes. Roles of
photoionization and the spinning motion on formation of stars and clouds are discussed as
a step toward understanding the mechanism regulating star formation efficiencies.

1. Introduction

Stars are being formed in molecular clouds.   Nearby molecular clouds allow detailed
observations at various wavelengths with little contamination by background objects, and
are, thereby, most suitable to study star formation in molecular clouds.   The present
millimeter observations of molecular clouds are far from complete because of limited
spatial coverage in high angular resolution studies, or because of limited angular resolution
in large scale surveys.   This is particularly true for nearby dark cloud complexes
subtending huge angular extents, like Taurus and Ophiuchus. It is of vital importance to
have complete knowledge on molecular distributions of a few arc min resolution in these
regions in order to better understand low mass star formation.

We have been carrying out a systematic survey of star formation regions within 1 kpc
of the sun in the CO J=1-0 emission with the 4 m millimeter telescope at Nagoya
University.   The principal aim of the survey is to obtain an unbiased view of molecular
distributions in these regions.   The ^{12}CO emission is powerful to probe outflow
phenomenon from recently formed stars, whereas the ^{13}CO emission, or the $C^{18}O$
emission in regions with highest molecular column densities, serves as a tracer of
molecular cloud cores.   The beam size of this survey, ~ 3', is suitable for detecting these
key observational features of star formation in nearby cloud complexes.   The ^{12}CO
survey has resulted a discovery of 52 outflows, corresponding to about a third of the CO

outflows known at present (Fukui et al. 1986, Fukui 1989, Fukui et al. 1990a), and have revealed an evolutionary trend of molecular outflows indicating that molecular outflow corresponds to the main accretion phase of a solar type protostar (Fukui et al. 1989). The collaborators in this project are H. Ogawa, K. Kawabata, T. Iwata, K. Sugitani, S. Nozawa, J. Yang, Y. Minoshima, Y. Teshima, H. Dobashi, K. Imaoka, and T. Nagahama.

In this paper, we shall present new ^{13}CO data of three nearby molecular complexes in Taurus, Ophiuchus, and Orion, use them in order to derive star formation efficiencies (=M_{star}/M_{cloud}), and discuss their astrophysical implications.

2. Observations

Most of the molecular observations presented in this paper were made in ^{13}CO J=1-0 emission with the 4m telescope at Nagoya University equipped with a 4 K cooled niobium superconducting mixer receiver having a receiver temperature of 23 K at 110 GHz in double side band (Ogawa et al. 1990). The velocity resolution provided by an acousto-optical spectrometer was 0.1 km s^{-1}, sufficiently high for probing detailed velocity distributions in regions of low mass star formation. Most of the observations were carried out from December 1989 to May 1990, and 50000 ^{13}CO spectra with typical rms noise level of 0.2 K in 0.1 km s^{-1} resolution have been obtained in Taurus, Ophiuchus, and Orion (the Orion south giant molecular cloud including L1641). Areas of 50 square degrees were fully sampled every 2' with a 3' beam. The ^{13}CO data were used to calculate LTE molecular masses with an assumed ^{13}CO/H$_2$ ratio of 2 x 10^{-6}. H$_2$ column density is converted to Av by the following equation; N(H$_2$) = 1.25 x 10^{21} Av (cm^{-2}).

3. Taurus   (d = 140 pc)

A fully sampled ^{13}CO map of the Taurus complex, consisting of 20000 spectra, is shown as a false color map in Figure 1. In addition to several molecular condensations, filamentary distributions of molecular gas are clearly seen with a general trend of elongation nearly along the galactic plane with a tilt angle of ~ 20°. In Figure 2, individual clouds are identified such as Heiles cloud 2 (HCL2) including TMC1, B213, L1495, B18 including TMC2, and L1536. Two major condensations, HCL2 and L1495, have molecular masses of ~ 800 M$_\odot$, respectively, and the total molecular mass of the clouds shown in Figure 1 is ~ 3000 M$_\odot$. One of the most remarkable features of the ^{13}CO distribution is the filamentary cloud toward B213. This filament, characterized by an unusually large line width of 3 km s^{-1}, is associated with another filament that is extending normal to the B213 filament to the north at l ~ 170° (Figure 1). There is no OB association in Taurus.

A series of channel maps of ^{13}CO intensity are shown in Figure 3, illustrating the highly filamentary nature of the ^{13}CO distribution even more clearly. A general trend obvious in Figure 3 is that HCL2 and L1495 are linked by several filamentary clouds, showing a large scale velocity gradient in the sense that the east part is blue shifted relative to the west part. A detailed account of the data will be published elsewhere (Mizuno et al. 1990b, in preparation).

277

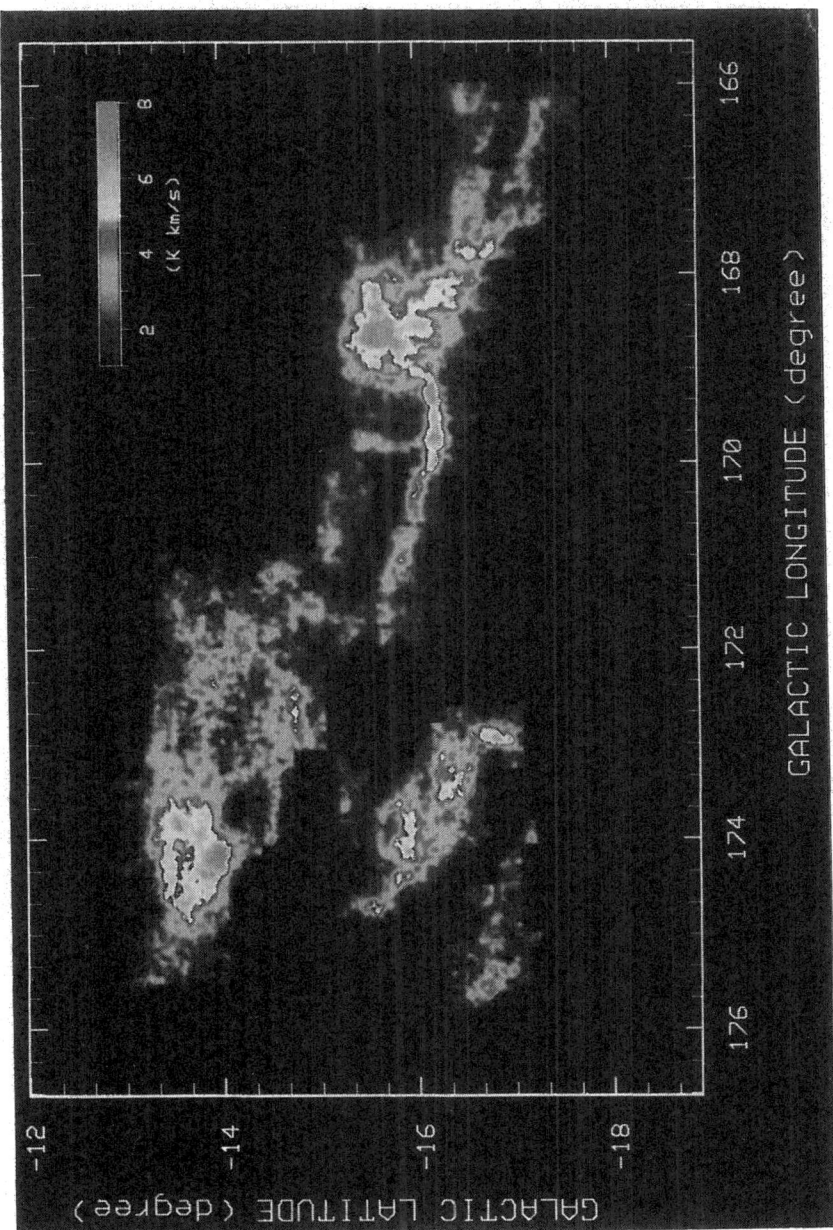

Figure 1 False-color image of total intensity in ${}^{13}CO$ (J=1-0) in the Taurus region.
(See Color-plates section)

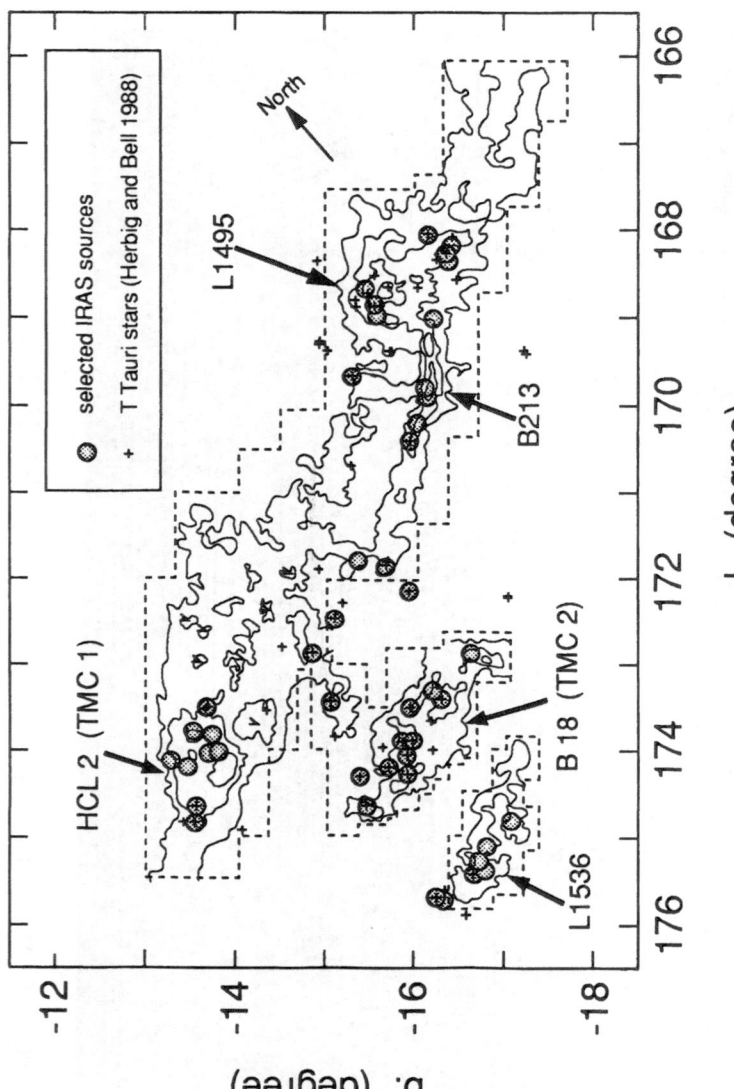

Figure 2 The selected IRAS sources and T Tauri stars (Herbig and Bell 1988) are superposed on the ^{13}CO (J=1-0) total intensity contour map. Contours are every 2K km s^{-1} from 2 K km s^{-1} in T$_A$*. The boundaries of the observed area are denoted by dashed lines. The selection criteria for the IRAS point sources are as follows; i) significantly detected at more than two wavelengths (quality flags are 2 or 3), and ii) flux densities more than 1 Jy at either 25 μm or 60 μm.

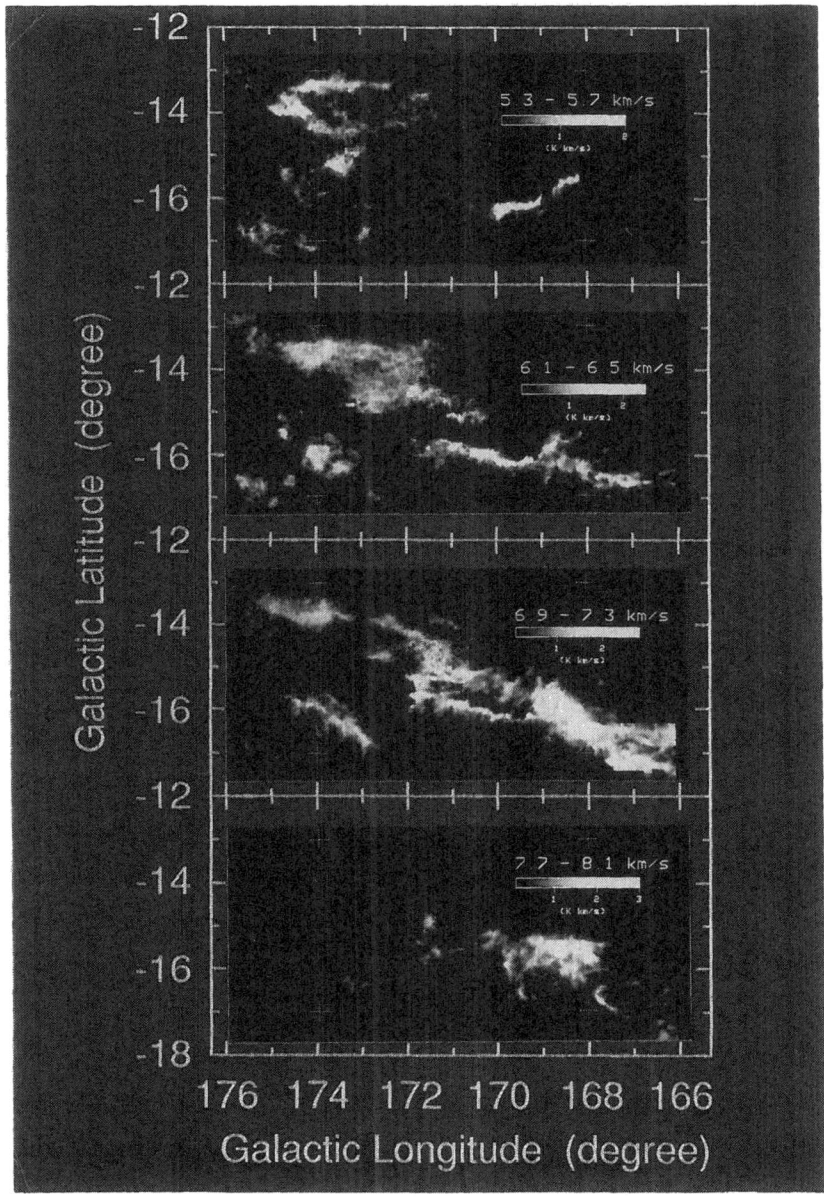

Figure 3   Grey-scale images of ^{13}CO (J=1-0) velocity channel maps in the Taurus region. The velocity intervals for the four panels are   5.3 - 5.7 km s^{-1}, 6.1 - 6.5 km s^{-1}, 6.9 - 7.3 km s^{-1}, and 7.7 - 8.1 km s^{-1} (from top to bottom), respectively.

We used data on T Tauri stars compiled by Herbig and Bell (1988) and on IRAS point sources (IRAS Catalogs and Atlases) as indications of recent star formation. These young stellar objects are certainly highly concentrated in the densest parts of the molecular gas as indicated in Figure 2, where the lowest contour corresponds to H_2 column density of 2-3 x 10^{21} cm^{-2}. Since the IRAS and ^{13}CO data sample the region almost fully, the remarkable coincidence between them strongly indicates that stars are being formed in the molecular gas having large column densities, i.e. large Av's. The outer parts of HCL2 and L1495, having low column densities, have no associated T Tauri stars or IRAS sources. On the other hand, L1536, showing also active star formation, has a smaller H_2 column density, and may be somewhat exceptional in the Taurus region. The luminosity range of the IRAS sources and T Tauri stars is from 0.1 to 10 $L_\odot$, indicating that only low mass stars are formed in Taurus. Searches for CO outflows in Taurus are not yet complete, and are to be made in a more systematic manner.

4. Ophiuchus   (d = 160 pc)

The Ophiuchus dark cloud complex is extended in an area from l ~ 350° to 10°, and b ~ 13° to 23°. A ^{13}CO map of this region is shown in Figure 4 (Nozawa et al. 1990). In the southern edge of the complex, the ρ Oph main cloud is located at l ~ 355°, and b ~ 16°, consisting of two dense condensations associated with long (~ 10 pc) filamentary clouds, so called "interstellar streamers" (see also Loren, 1989a, 1989b). On the other hand, the northern half of the complex is highly clumpy, consisting of ~ 50 ^{13}CO clumps (Nozawa et al. 1990). The molecular mass of the ρ Oph main cloud is estimated to be ~ 2500 $M_\odot$, while that of the northern complex is ~ 3000 $M_\odot$. Sco OB2 association on the southwest of the complex is probably located at the same distance with the molecular complex.

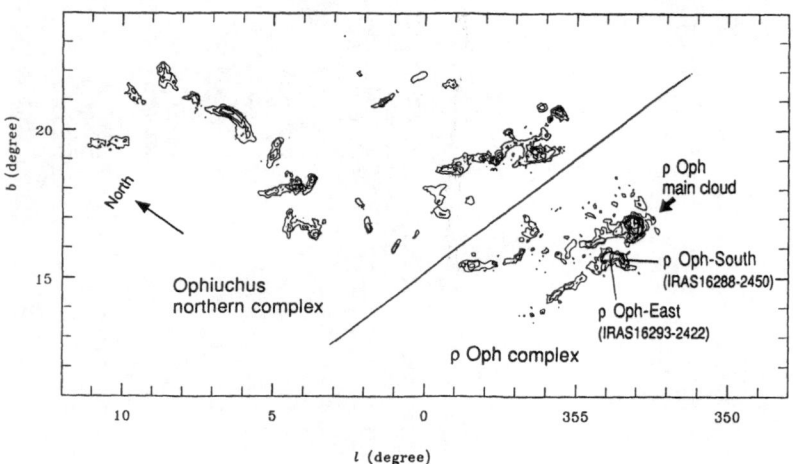

Figure 4   Total intensity map in ^{13}CO (J=1-0) in the Ophiuchus region. Contours in T_R^* are every 2.0 K km s^{-1} from 2.0 K km s^{-1} for the Ophiuchus northern complex and every 3.2 K km s^{-1} from 3.2 K km s^{-1} for the ρ Oph complex, respectively.

The high star formation efficiency in the ρ Oph main cloud is well recognized. Along the $C^{18}O$ ridge embedded in the main cloud, about hundred infrared sources are detected, suggesting SFE of ≳ 22% (Wilking, Lada, and Young 1989). On the contrary, the neighboring secondary core is associated with only a few infrared sources, IRAS16293-2422 (ρ Oph-East outflow; Mizuno et al. 1990a), IRAS16288-2450 (ρ Oph-South outflow; Fukui 1989) etc., and even lower star formation activity is found in the two streamers. The northern part of the complex is also found to show a low level of star formation with a star formation efficiency of ~ 0.3 % (Nozawa et al. 1990). The ρ Oph main cloud is therefore characterized by an unusually enhanced star formation activity in the whole Ophiuchus complex.

5. Orion (d = 480 pc)

The molecular cloud associated with Ori A, including L1641, is the Orion south giant molecular cloud. A new ^{13}CO map of this cloud is shown in Figure 5. This map shows several new features that are not recognized in previous ^{13}CO maps (e.g. Bally et al. 1987), particularly weak extended features around the most intense elongated ridge containing Ori KL object (l ~ 209°). The northern half of this cloud shows sign of massive star formation as indicated by the Trapezium stars (l ~ 209°) exciting the Orion Nebula and other luminous infrared sources. On the other hand, the luminosities of embedded infrared sources decrease toward the south, indicating that low mass stars are being formed in the southern half, consisting of the L1641 cloud, at l ~ 211°- 214°. We shall in the following focus on the L1641 cloud.

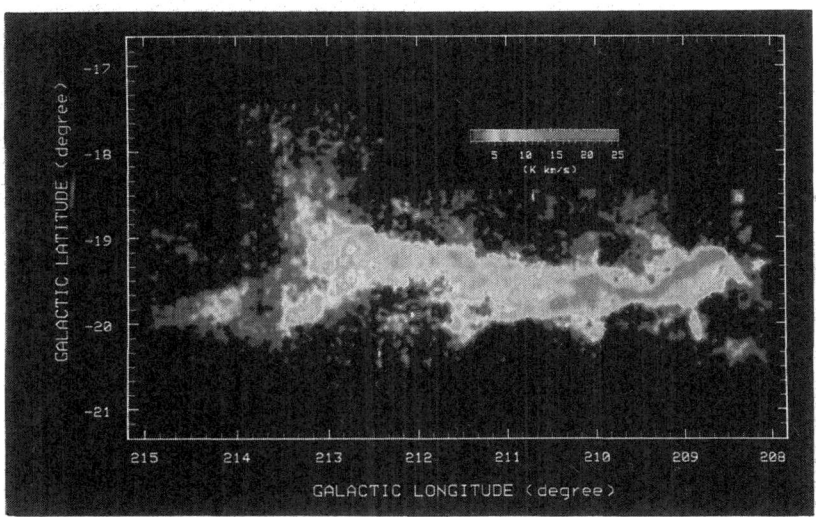

Figure 5   False-color image of the ^{13}CO (J=1-0) total intensity for the Orion south giant molecular cloud, involving the L1641 cloud. *(See Color-plates section)*

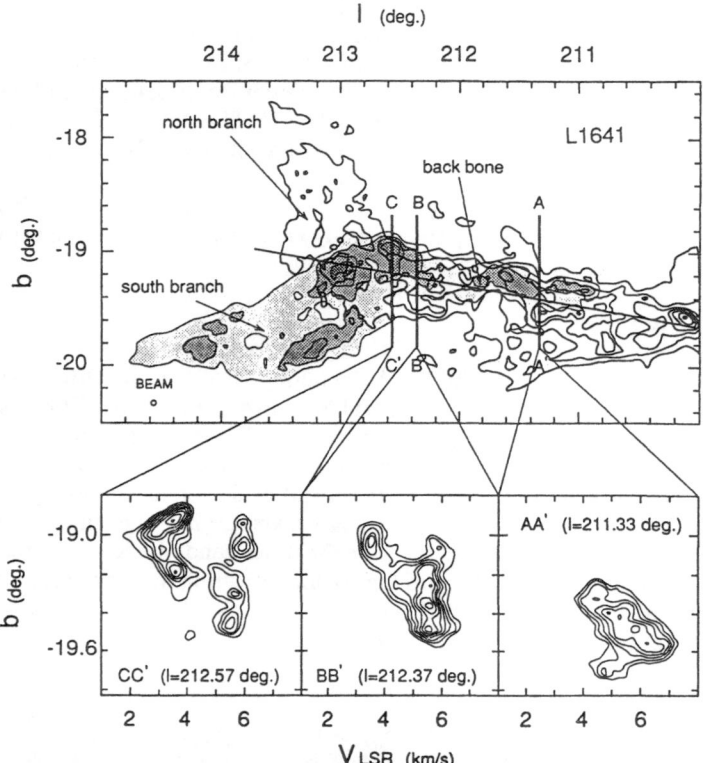

Figure 6   Distribution of the two velocity components in the L1641 cloud.   In the upper panel, the ^{13}CO intensity integrated over 1 to 5 km s^{-1} and 5 to 8 km s^{-1} are shown by hatchings and contours, respectively. Contours for each are every 2 K km s^{-1} from 2 K km s^{-1} as the lowest.   The lower panels show typical latitude-velocity diagrams at three longitudes.   Contours of the lower panels are every 0.5 K from 2.0 K as the lowest, with velocity resolution of 0.1 km s^{-1}.

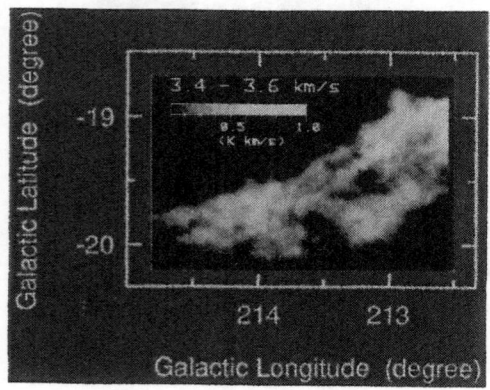

Figure 7   A grey-scale image of channel map (v$_{LSR}$ = 3.4 to 3.6 km s^{-1}) for the "fish tail" part of the Orion south giant molecular cloud.   The highly filamenrtary nature of the "fish tail" is seen.

The most interesting new result from our observations is the finding that the southern edge of the ^{13}CO cloud consists of a system of two branches of highly ordered structures. These two branches are seen as a V shape in Figure 5 at l ~ 212°.5 to 215° and b ~ -18° to -20°. We shall call this V shaped structure as the "fish tail" structure from its appearance. This "fish tail" exhibits a striking velocity structure; i.e. its north and south branches show clear velocity difference (Figure 6). The south branch is blueshifted relative to the main cloud, appearing at v_{LSR}=2-5 km s^{-1}, while the north branch is redshifted at v_{LSR}=5-7 km s^{-1}. The full width of the "fish tail" in latitude is ~ 15 pc at l ~ 213°, becoming systematically smaller in both the north and south branches to ~ 3 pc at l ~ 212°.5. The highly filamentary nature of the "fish tail" is demonstrated in a representative channel map of a small velocity range, v_{LSR}=3.4 - 3.6 km s^{-1}, in Figure 7.

According to a ^{12}CO map obtained with the Columbia telescope, two streams of filamentary ^{12}CO emission can be traced nearly along the galactic plane down to l ~ 216°.5 (Maddalena et al. 1986), suggesting that the "fish tail" is further extended as lower density molecular gas not detectable in the ^{13}CO emission with the current sensitivity. It is known that the molecular complex is embedded in a huge HI cloud of 10^5 M$_\odot$ (Chromey, Elmegreen, and Elmegreen 1989). The masses of the north and south branches of the "fish tail" structure are 2000 M$_\odot$ and 2700 M$_\odot$, respectively. The total mass of the "fish tail" is then ~ 5000 M$_\odot$, about one sixths of the total mass of the main cloud between l ~ 208° and 213°, ~ 30000 M$_\odot$.

The L1641 cloud (l ~ 211°-214°) is an active site of low mass star formation. Observational signatures for them are more than 50 IRAS point sources (e.g. Strom et al. 1989) and 6 molecular outflows (Fukui et al. 1989, Levereault 1988). The distribution of the IRAS sources and molecular outflows indicate that these young objects are remarkably well correlated with the ^{13}CO ridge of the L1641 cloud (Fukui et al. 1989). Most of the IRAS sources have luminosities in a range from 1 to 10 L$_\odot$, consistent with that of a solar-mass star in the pre-main sequence contraction phase. Although only about one fifth of them are identified as T Tauri stars (Herbig and Bell 1988), it seems very reasonable to suppose that they are heavily obscured T Tauri stars.

6. Discussion

6.1. Star formation efficiencies

It is difficult to estimate stellar masses precisely in the early stage of star formation, since the luminosities of protostars dramatically change in time (e.g. , Shu, Adams, and Lizano 1987, Stahler, Shu, and Taam 1980). Although masses of the new-born stars are not well determined, we can estimate a frequency of star formation from the infrared and optical data. As a first order approximation of SFE, we use the number of IRAS sources and visible T Tauri stars per unit solar mass cloud. Figure 8 shows SFE's thus derived as a function of stellar bolometric luminosities in the three regions, Taurus, the ρ Oph main cloud, and L1641 in Orion. The masses of these clouds are estimated based on the present ^{13}CO (J=1-0) data. The databases of the stellar luminosities are the IRAS point source catalog for the IRAS sources, and, as for T Tauri stars, Herbig and Bell (1988) and Cohen, Emerson, and Beichman (1989) for Taurus, Wilking, Lada, and Young (1989) for

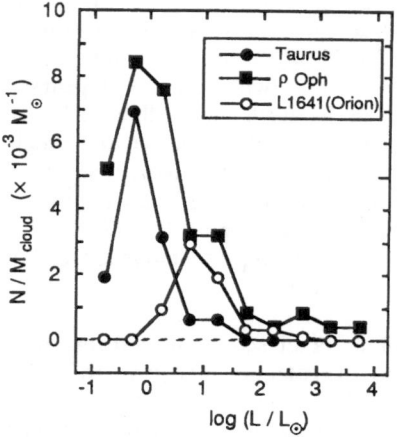

Figure 8   A plot of star formation efficiencies (S.F.E.) as a function of lunmionosity for Taurus, the ρ Ophiuchus main cloud, and the L1641 cloud in Orion.

ρ Oph, and Strom et al. (1989) for L1641.   In Figure 8, the peak position of SFE in L1641 is by one order of magnitude larger than those in Taurus and ρ Oph.   However, it is to be noted that the detection limit of IRAS data is ~ 3 $L_\odot$ in L1641, causing the decrease in the SFE below 3 $L_\odot$ in L1641.   The difference of peak positions does not mean that L1641 is a massive star forming region compared with Taurus and ρ Oph.   Indeed, objects more luminous than 1000 $L_\odot$ are not detected in L1641, whereas such objects are detected in ρ Oph.   Above the detection limit ($\geq$ 3 $L_\odot$), the number of new-born stars per unit mass in Taurus, ρ Oph, and L1641 are ~ 1 x 10^{-3} $M_\odot^{-1}$, ~ 9 x 10^{-3} $M_\odot^{-1}$, and ~ 6 x 10^{-3} $M_\odot^{-1}$, respectively.   In this luminosity range, the SFE of Taurus is much smaller than those of ρ Oph and L1641.   However, in a luminosity range of 0.3 - 3 $L_\odot$, the number of new-born stars per unit mass in Taurus ,~ 1.2 x 10^{-2} $M_\odot^{-1}$, is about a half of that in ρ Oph, ~ 2.1 x 10^{-2} $M_\odot^{-1}$.   These results indicate that the Taurus region is an active low mass star forming region.

6.2. Ionization degree

Molecular clouds are ionized mainly by cosmic ray protons and ultraviolet radiation of OB stars.   This ionization couples the molecular gas to the magnetic field, depending on the ionization degree, and results in the cloud collapse time scale an order of magnitude larger than the free fall time scale because of the magnetic pressure.   This implies that star formation is effectively regulated by the degree of ionization in molecular clouds.   In the deep interior of the molecular clouds, cosmic ray protons must be dominant as an ionization agent.   On the other hand, near the cloud surface contribution of ultraviolet photons may be important.   Photoionization therefore may well regulate star formation in molecular clouds (McKee 1989).   Recent theoretical estimates have shown that ultraviolet radiation can effectively ionize molecular gas to an ionization degree of $\gtrsim$ 10^{-8} up to Av of ~ 4 mag from the cloud surface for a typical CO cloud having a density of 10^3 cm^{-3} (Nakano and Fukui 1990, in preparation).

The Ophiuchus northern complex is characterized by clumped distribution, favorable for penetration of radiation, has a typical Av of ~ 7 mag, and shows a small star formation efficiency (Nozawa et al. 1990). In addition, Sco OB2 association raises the ultraviolet radiation flux by an order of magnitude above the general interstellar value. Therefore, it is probable that relatively high ionization degrees are realized in the Ophiuchus northern complex, leading to the small star formation efficiency. On the other hand, the ρ Oph main cloud, having an unusually large Av of ~ 100 mag, must have small ionization degrees, allowing efficient ambipolar diffusion of magnetic flux and active star formation as observed.

The Taurus complex has a typical Av of several magnitudes, implying that the general ultraviolet radiation field may be important in regulating ionization degrees. This might explain that the SFE in the densest parts of HCL2, L1495, and B18 in Taurus is significantly smaller than that in the ρ Oph main cloud. The rather large SFE in L1536 having a small Av may need some other explanation; this cloud may have a larger gas column density previously, and may be dispersing the placental cloud material at present. This possibility could be tested by a careful analysis of the ages of the IRAS sources in L1536.

L1641 in Orion has an average Av three times larger than that in Taurus, and active star formation is expected, if photoionization is playing a role. This is actually the case, and SFE is relatively high in L1641 (Figure 8). To summarize section 6.2, it seems that regulation of star formation by photoionization provides to a first approximation a viable explanation for the difference in star formation activity among the three regions.

6.3. Magnetic field

Optical polarization data provide information on projected direction of magnetic fields, and Zeeman measurements of atomic and molecular spectral lines may determine the field strength and orientation. The optical polarization measurements of Taurus indicate that the polarization vectors are generally normal to the molecular filaments, whereas in the ρ Oph region the optical polarization vectors are nearly parallel to the molecular filaments. Thus, there appears no simple relationship between magnetic fields and cloud elongation if the field configuration is simply assumed straight and parallel (see e.g. Goodman et al. 1990).

If the fields have a more complicated, but ordered configuration, we may find a possible solution for the problem. Such an example is found in L1641, where the optical polarization vectors are at an angle of ~ 30° to the major axis of the cloud, and the HI Zeeman splitting measurements indicate that the field line is reversed on the both sides of the cloud. On the basis of these data, it is suggested by Y. Uchida and J. Bally that the general magnetic field configuration is consistent with a large scale *helical* pattern (Bally 1989). This suggests that a more elaborate field configuration may be needed to explain the apparent diversity between optical polarization vectors and cloud elongation in other regions.

6.4. Origin of the filamentary distributions

The present study of three nearby molecular complexes indicates that the distribution of molecular clouds is highly filamentary. Since one of them, Taurus, is associated with no massive stars, dynamical or radiative effects of OB stars do not provide a general explanation for these filamentary distributions. Alternatively, magnetic fields may play an important role in producing the filamentary distributions, and an idea to explain formation of molecular filaments via magnetohydrodynamical mechanism is proposed by Uchida, Mizuno, Nozawa, and Fukui (1990a). We analysed the velocity field in the ρ Oph streamer, one of the most remarkable filamentary distributions, and found that the streamer shows a systematic velocity gradient of ~ 1.5 km s^{-1} pc^{-1} across its major axis. This velocity gradient is in the same sense with the main cloud, and we interpreted that the velocity gradient indicates a spinning motion of the molecular filament accelerated by the torsional Alfven waves emitted from the massive main cloud. This idea is essentially an interstellar version of the magnetohydrodynamical model for molecular outflow acceleration (Uchida and Shibata 1985, Pudritz and Norman 1986). Recent numerical simulations of a spinning interstellar cloud suggest that such a process really works to form filamentary clouds (Uchida, Shibata, and Rosner 1990, in preparation).

In the following, we shall show that the L1641 cloud also exhibits a remarkable spinning motion. In order to see the detailed velocity distribution, in Figure 6 are shown a series of position-velocity diagrams (b-v$_{LSR}$ diagrams) taken at three representative longitudes near the root of the "fish tail". As shown in the cut at l=211°.33, the main ridge of the L1641 cloud is characterized by a significant velocity gradient of ~ 1 km s^{-1} pc^{-1}, and this gradient is interpreted in terms of spinning motion around the major axis of the cloud by Uchida et al. (1990b). We shall call this main ridge as the "backbone" of the "fish tail". The radius of the "backbone", ~ 1 pc, may be determined as a result of establishment of a nearly dynamical equilibrium state among magnetic pressure, centrifugal force, and self gravity. The magnetic field strength is then estimated to be ~ 100 μG, corresponding to an Alfven speed of ~ 5 km s^{-1}.

We suggest that the two branches of the "fish tail" represent merging filamentary structure, being wound up by the spinning main cloud. The spinning motion will pull the two branches of the "fish tail" into a twisted rope, and this merging processes the west ward extension of the "backbone". The apparent merging of the two components of the "fish tail", and their funneled shape may be well explained by such a picture. The parallel filamentary distribution of molecular gas shown in Figure 7 is a natural result of the stretching due to this twisting up. To be quantitative, the total molecular mass of the L1641 "backbone", ~ 10000 M$_\odot$, is large enough to dynamically influence the "fish tail" structure via magnetic tension, and the whole Orion south cloud including the Ori KL region having ~ 30000 M$_\odot$ may contribute in magnetically winding up the "fish tail". Thus, the total angular momentum of the main cloud is sufficient to dynamically influence the "fish tail". It is noteworthy that the helical magnetic field suggested in section 6.3 is a natural consequence of such winding up by a massive spinning cloud.

In the present picture, the coalescence, accompanying compression due to the pinching effect is causing the increase of density from ~ 200 cm^{-3} in the "fish tail" to ~ 2000 cm^{-3} in the "back bone". It is shown in section 5 that IRAS point sources, good candidates of protostars, are distributed along the "backbone" in L1641, suggesting active low mass star formation. This star formation may be enhanced owing to the increase of molecular column density as well as the compression due to the coalescence of the two branches of the "fish tail".

The spinning picture is in a sense a very natural idea to collect matter from less dense raw material; we, human beings, have been spinning cotton into yawn since the prehistoric age. It is no wonder that the molecular clouds do the same in the interstellar space. It is however to be noted that many physical aspects remain to be clarified; the process should accompany MHD shocks, and, in some cases, reconnection of field lines. All these details are to be thoroughly investigated theoretically. It is also important to increase the number of such spinning clouds with filamentary distributions. The B213 filament in Taurus also shows a velocity gradient normal to its elongation, and a detailed analysis for it is in progress.

7. Summary

Star formation efficiencies in three nearby cloud complexes are described on the basis of new ^{13}CO data. It is shown that young stellar objects are preferentially located in the densest, heavily obscured regions in the molecular clouds. Differential photoionization may well regulate the trend in star formation efficiencies through affecting ionization degrees in molecular gas, and thereby controlling the ambipolar diffusion of the magnetic flux leading to star formation. The cloud morphology is often filamentary. Recent detailed analyses of the kinematics of these filamentary clouds suggest that the clouds are spinning along their major axes. A possible role of such spinning motion in cloud formation is discussed on the basis of the L1641 data.

We are grateful to K. Sugitani and Y. Minoshima for their invaluable help in preparing this manuscript. This research was in part financially supported by the Grant-in-Aid for Specially Promoted Research No. 01065002 in the Ministry of Education, Science, and Culture.

References

Bally, J. 1989, in *Low Mass Star Formation and Pre-Main Sequence Objects*, ed. B. Reipurth, (European Southern Observatory, Garching bei München), p. 1.
Bally, J., Langer, W. D., Stark, A. A., and Wilson, R. W. 1987, *Astrophys. J. Letters)*, 312, L45.
Chromey, F. R., Elmegreen, B. G., and Elmegreen, D. M. 1989, *Astron. J.*, 98, 2203.
Cohen, M., Emerson, J. P., and Beichman, C. A. 1989, *Astrophys. J.*, 339, 455.
Fukui, Y. 1989, in *Low Mass Star Formation and Pre-Main Sequence Objects*, ed. B. Reipurth, (European Southern Observatory, Garching bei München), p. 95.
Fukui, Y., Iwata, T., Mizuno, A., Bally, J., and Lane, P. A. 1990a, in *Protostars and Planets III*, (in press).
Fukui, Y., Iwata, T., Takaba, H., Mizuno, A., Ogawa, H., Kawabata, K., and Sugitani, K. 1989, *Nature*, 342, 161.
Fukui, Y., Minoshima, Y., Uchida, Y., Mizuno, A., Iwata, T., Ogawa, H., and Takaba, H. 1990b, (submitted to *Publ. Astron. Soc. Japan*).

288

Fukui, Y., Sugitani, K., Takaba, H., Iwata, T., Mizuno, A., Ogawa, H., and Kawabata, K. 1986, *Astrophys. J. (Letters)*, **311**, L85.
Goodman, A. A., Bastien, P., Myers, P. C., and Ménard, F. 1990, *Astrophys. J.*, **359**, 363.
Herbig, G. H., and Bell, K. R. 1988, University of California, Lick Observatory Bulletin No.1111.
IRAS Catalogs and Atlases, Explanatory Suppl. 1984, eds. C. A. Beichman, G. Neugebauer, H. J. Habing, P. E. Clegg, and T. J. Chester, (U.S. Government Printing Office, Washington D.C.).
Levereault, R. M. 1988, *Astrophys. J. Suppl.*, **67**, 283.
Loren, R. B. 1989a, *Astrophys. J.*, **338**, 902.
Loren, R. B. 1989b, *Astrophys. J.*, **338**, 925.
Maddalena, R. J., Morris, M., Moscowitz, J., and Thaddeus, P. 1986, *Astrophys. J.*, **303**, 375.
McKee, C. F. 1989, *Astrophys. J.*, **345**, 782.
Mizuno, A., Fukui, Y., Iwata, T., Nozawa, S., and Takano, T. 1990a, *Astrophys. J.*, **355**, 184.
Nozawa, S., Mizuno, A., Teshima, Y., Ogawa, H., and Fukui, Y. 1990, (submitted to *Astrophys. J. Suppl.*).
Ogawa, H., Mizuno, A., Hoko, H., Ishikawa, H., and Fukui, Y. 1990, *Int. J. of Infrared and Millimeter Waves*, **11**, 717.
Pudritz, R. E., and Norman, C. A. 1986, *Astrophys. J.*, **301**, 571.
Shu, F. H., Adams, F. C., and Lizano, S. 1987, *Ann. Rev. Astron. Astrophys.*, **25**, 23.
Stahler, S. W., Shu, F. H., and Taam, R. E. 1980, *Astrophys. J.*, **241**, 637.
Strom, K. M., Newton, G., Strom, S. E., Seaman, R. L., Carrasco, L., Cruz-Gonzalez, I., Serrano, A., and Grasdalen, G. L. 1989, *Astrophys. J. Suppl.*, **71**, 183.
Uchida, Y. 1989, in *Low Mass Star Formation and Pre-Main Sequence Objects*, ed. B. Reipurth, (European Southern Observatory, Garching bei München), p. 141.
Uchida, Y., and Shibata, K. 1985, *Publ. Astron. Soc. Japan*, **37**, 515.
Uchida, Y., Fukui, Y., Minoshima, Y., Mizuno, A., Iwata, T., and Takaba, H. 1990b, (submitted to *Nature*).
Uchida, Y., Mizuno, A., Nozawa, S., and Fukui, Y. 1990a, *Publ. Astron. Soc. Japan*, **42**, 69.
Wilking, B. A., Lada, C. J., and Young, E. T. 1989, *Astrophys. J.*, **340**, 823.

LUMINOSITY FUNCTION, STAR DENSITY, AND STAR FORMATION EFFICIENCY IN REGIONS OF STAR FORMATION

- NEAR INFRARED OBSERVATIONS -

Klaus-Werner Hodapp
John Rayner
Hua Chen

Institute for Astronomy, University of Hawaii
2680 Woodlawn Drive, Honolulu, HI 96822, U.S.A.

ABSTRACT

Clusters of young stars have been found near a number of compact HII regions. These clusters do not show a turnover in the K-band luminosity and are probably several million years old. In L 1641 only moderate clustering tendency has been observed and many sources show signs of extremely young age.

Introduction

Near infrared (NIR) observations, here meaning the wavelength range from 1.0-2.5µm, can penetrate about an order of magnitude deeper into molecular clouds than visible light studies. In typical starforming molecular clouds, most embedded stars become visible in the NIR. The most heavily obscured very young stars escape direct detection even at 2.2µm and are often only seen in scattered light. Obtaining luminosity information requires extinction correction which however are still unreliable since the NIR extinction law is poorly known in molecular clouds. NIR detector technology has matured rapidly in the last 5 years. We can expect that in the near future, all the basic observational techniques of optical astronomy (imaging, photometry, polarimetry, and spectroscopy) will also be available in the NIR. This paper reports initial results of NIR imaging observations of starforming regions, obtained during the last 2 years at the University of Hawaii.

1. S 106

The first object we want to discuss is S 106, for which we have the most complete results. S 106 is a bipolar HII region excited by a O9 or B0 star. The bipolar morphology of S 106 in the radio continuum is commonly explained by an anisotropic stellar wind blocking UV photons from ionizing the equatorial plane of the bipolar structure (Felli et al. 1984). In addition, optical and infrared images as well as molecular line studies show that the object is surrounded by a clumpy torus of molecular gas, whose dust absorption leads to the broad absorption band bisecting the HII region at visible and NIR wavelengths.

Our K band image of this region is a mosaic of 129 frames obtained with a 128x128 HgCdTe detector array. The image covers a 4'x7'field of view. In the extended emission east of the illuminating star, some linear features are extending away from the central source into a region of high extinction in the dust lane. We interpret these features as illumination (shadow and light) effects of a clumpy molecular gas and dust distribution around the central star, being projected into the dense molecular material east of this star and becoming visible by scattering.

Even considering the uncertainties in where exactly the clumpy material is located, this projection effect allows a "magnified" view of this clumpy material, confirming the existence of clumpiness on the scale of 1" = 600 AU in the molecular torus around S 106.

Fig. 1: K-band mosaic image of S 106, obtained with a 128x128 HgCdTe infrared array.

Another but maybe more important feature in Fig. 1 is a cluster of approx. 200 stars surrounding the bipolar nebula. The center of this cluster is located some 30" NW of the ionizing star, i.e. the latter is not located precisely in its center.

Quantitative photometric analysis of mosaic images like Fig. 1 is very difficult. Due to responsivity nonuniformities of the array, the noise is not constant in each flatfield-corrected frame. Sky background and to some extent the PSF vary from frame to frame. In addition, source crowding and the extended nebulosity make photometry very difficult and automated source searches prone to errors. To check the reliability of our search routines under these circumstances, we inserted fake stars with the same PSF and noise properties than real images into the frame at random locations. The probability of rediscovering these fake stars by the search routine was then measured. Fake stars fainter than K=12 show increasing photometric errors and at K=14, the rediscovery rate drops rapidly. Down to this completeness limit of K=14, the K-band magnitude distribution rises smoothly. Beyond the completeness limit, it drops rapidly to zero, as expected. We do not see any indication for a turnover in this brightness distribution.

Only some very tentative results on the evolutionary status of the stars in the S 106 cluster can be derived from a K-band image alone. The majority of stars in this cluster are not associated with localized reflection nebulosity, that is frequently found around very young stars in their outflow phase. Only 4 stars, including IRS 4 show localized reflection nebulosity around them (as opposed to the generally distributed nebulosity in this region). Polarization studies of stars in the S 106 area have only found one star showing strong intrinsic polarization, which is also an indicator of young age (HL Tau being a prototype for such a strongly polarized young T Tauri star). Taken together, only 5 stars out of 200 stars detected in total show signs of extreme youth associated with the formation of a dense disk and the outflow phase, commonly assumeed to last about 50000 years (Fukui, 1989). Assuming that the star formation rate in S 106 was constant in the past, this leads to a very rough estimate for the age of the oldest stars in S 106 of 2 million years. The assumption of constant star formation rate may of course be wrong and star formation may have proceeded in bursts, but the above estimate indicates that the S 106 cluster as a whole is not an extremely young structure, and that the individual stars in this cluster may have a substantial spread in their ages.

2. Clusters associated with other HII regions

S 106 is not a unique object. We are currently conducting a survey of a sample of compact HII regions in an attempt to study the frequency of cluster formation. We are concentrating on HII regions in the distance range from 2 to 5 Kpc, which can be imaged on a single frame of our infrared camera. In the majority of cases studied, we have found young embedded clusters, examples being S 255, S 269, W3(OH), and GL 437 (Rayner et al. 1990). The formation of groups or clusters of young stars therefore seems to be common in molecular clouds massive enough to form O or B stars sometimes during their star formation phase.

3. The L 1641 Cloud

In an attempt to study the star formation efficiency in a whole cloud, we obtained NIR images of outflow sources in L 1641. L 1641 is the southern part of Orion A and, following the data presented by Bally (this conference) and Fukui (this conference) the L 1641 cloud consists of two molecular filaments that are being twisted around each other, thereby creating a dense region where, according to the IRAS survey, most of the star formation occurs. The winding up of the filaments progresses along the axis of L 1641 from north to south. In addition, there is evidence that further to the north, the Orion A cloud has been compressed by the passage of a shock front. In summary, the available data suggest that star formation is spreading through the Orion A cloud from north to south.

NIR imaging of outflow sources in Orion A seems to confirm this result. In the north, there is the dense and rich cluster of young stars associated with the Trapezium and the BN/KL complex (McCaughrean, 1989). Using similar arguments as in the case of S 106, it can again be concluded that most stars in this cluster are not of extremely young age. Further to the south, the moderately populous cluster associated with the L 1641 N outflow source (Strom et al. 1989) (Fig.2 a) contains a larger number of objects associated with reflection nebulosity. Around some of the other outflow sources in L 1641 (Fig. 2 b, c, and d), still further south, the number of stars seen per unit area (to the same limiting magnitude) decreases, but the fraction of stars showing localized reflection nebulosity increases, as can be seen from the images in Fig. 2. We take this as evidence that in this southern part of L 1641, a large fraction of all stars visible are in fact very young and that star formation has been going on for only a relatively short time (a few 10^5 years, to give a rough estimate). Clearly, the presence of reflection nebulosity is a poor indicator for the age of a star, but it is the only one available to date. Emission line spectroscopy of individual stars in a cluster will hopefully be a more powerful indicator of accreation disk activity and age.

292

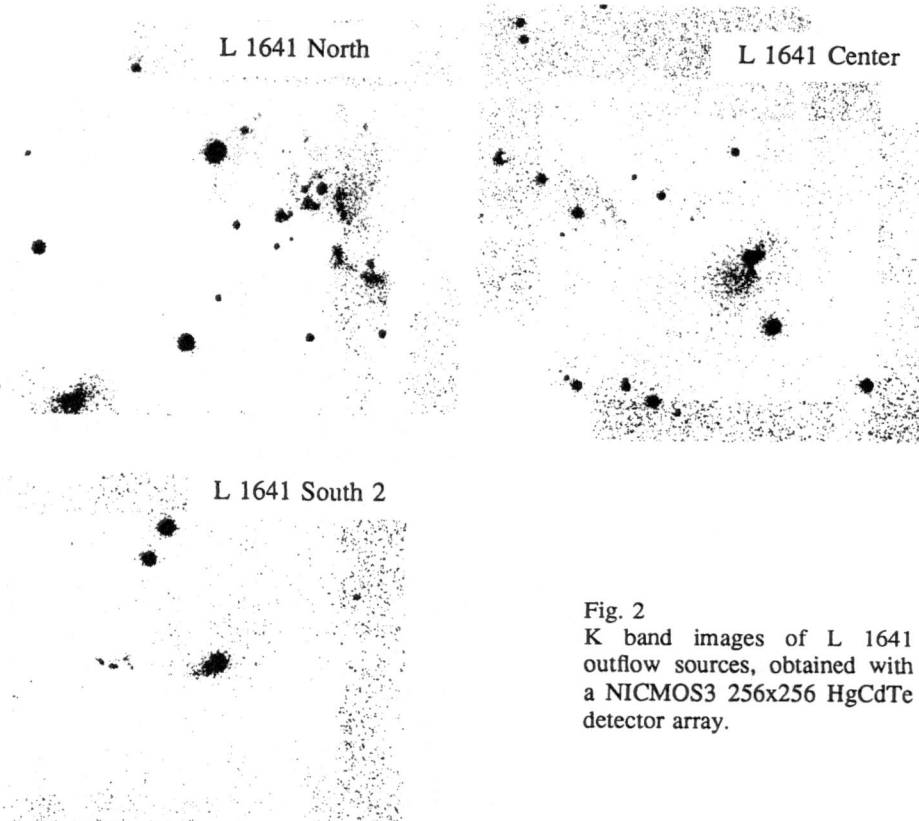

L 1641 North

L 1641 Center

L 1641 South 2

Fig. 2
K band images of L 1641
outflow sources, obtained with
a NICMOS3 256x256 HgCdTe
detector array.

References

Felli, M., Staude, H.J., Reddman, T., Massi, M., Eiroa, C., Hefele, H., Neckel, T., and Panagia, N., 1984, *Astr. Ap*, **135**, 261.

Fukui, Y. 1989, ESO Workshop on Low Mass Star Formation and Pre-Main Sequence Objects, ed. B. Reipurth, ESO Conference and Workshop Proceedings No. 33, p. 447.

McCaughrean, M. 1989, Proceedings of the Third Infrared Detector Technology Workshop, ed. Craig R. McCreight, NASA Technical Memorandum 102209, p. 201.

Rayner, J., Hodapp, K.-W., Zinnecker, H. 1990, Astrophysics with Infrared Arrays, Tucson, proceedings in press.

Strom, K.M., Margulis, M., Strom, S.E. 1989, *Ap. J. Letters*, **346**, L33.

STAR FORMATION IN THREE NEARBY CLOUD COMPLEXES

NEAL J. EVANS II

Astronomy Department,
The University of Texas, Austin Texas 78712, USA
E-mail NJE@ASTRO.AS.UTEXAS.EDU

ELIZABETH A. LADA

Harvard-Smithsonian Center for Astrophysics,
60 Garden Street, Cambridge, MA 02138, USA
E-mail LADA@CFA.BITNET

Abstract. We consider the distribution and nature of the gas and star formation in three nearby molecular cloud complexes : Taurus-Auriga, Ophiuchus, and Orion. Large-scale studies show that quite different distributions of dense gas and star formation exist in these regions, but that the global efficiency of star formation, so far, is about 1% in all of these complexes, very similar to the final efficiencies deduced for open clusters and associations. Furthermore, evidence for differences in the initial mass function among these regions is weak; all three regions may be compatible with the initial mass function for open clusters and field stars. Finally, we consider specific examples of star formation in each complex; the results indicate that the ideas of Adams, Lada, and Shu (1987) are reasonably consistent with the data on L1551 IRS 5, in the Taurus cloud, but that modifications are likely to be needed for IRAS 16293-2422 in Ophiuchus and for NGC2071 in Orion.

Keywords : star formation, molecular clouds, embedded clusters

1. Introduction

A major goal of star formation studies is to understand the physical processes by which stars originate. Consequently, current research has focused on determining the relationship between young stars and the physical and environmental conditions in the regions where they form. Two different approaches have been used to study this problem in our Galaxy. One approach involves investigations of the processes of star formation on global scales, while another approach concentrates on local processes, by studying individual sites of star formation. Both approaches seek to compare the raw materials with the recent products of star formation.

Global studies compare the large-scale distribution and properties of the gas and dust with those of young stellar objects (YSOs). Such knowledge helps us to determine what conditions are necessary for star formation and whether the nature of the star forming process varies with physical conditions or the environment of the cloud. For example, clouds with differing internal or environmental conditions may form stars with a different efficiency, or with a different initial mass function. Such differences would provide insights into the connection between the raw materials and the finished products.

On a local scale, we again compare the physical conditions with the nature of the objects. We are particularly interested in understanding the formation and very early evolution of individual objects; this interest launched the crusade to detect protostars (*e.g.*, Wynn-Williams 1982), generally characterized by infalling gas. The meager results of this crusade have led to suggestions that infall and outflow may occur simultaneously, and that infall may actually be intimately involved in the production of the ubiquitous outflows (Shu 1991), thus cleverly masking its own presence. Models of this type have made detailed predictions for the density distributions around individual forming stars (Adams, Lada, and Shu 1987). Observational studies of the nature of such objects allow us to test these predictions.

Nearby molecular cloud complexes provide excellent laboratories for investigations of the star forming process. Such regions have the advantage that they can be studied with high angular resolution and high sensitivity. There are several such complexes within about 150 pc, such as the Taurus-Auriga complex and the Ophiuchus complex, which are forming stars of moderate mass, but the nearest complex which is forming a significant number of truly massive stars is the Orion complex, at a distance of about 400 pc.

Both the Taurus-Auriga region and the Orion complex have been the subject of extensive optical study, including surveys for stars with Hα emission (see Herbig and Rao 1972 for a compilation) and follow-up spectroscopy (Cohen and Kuhi 1979). Comparing the results for these two clouds, Larson (1982) noticed some apparent differences between star formation in these clouds. In Taurus, where the gas is "relatively dispersed", the clouds are forming "scattered small groups" of stars, "predominantly of low mass". In contrast, in the Orion region, "massive condensed clouds", are forming stars "mostly in a single large cluster" which is "relatively depleted in low-mass stars". A comparison of the median mass of the T Tauri stars in the two clouds indicated a value about twice as high in Orion as in Taurus (1.1 $M_\odot$ versus 0.6 $M_\odot$), according to Larson (1986). No such conclusions could be drawn for Ophiuchus based on optical data because few stars were visible optically. Later infrared studies of Ophiuchus revealed many more stars, raising the question of whether similar studies of the other regions would change the conclusions based on optical studies.

In this paper, we will review recent observational results from both global and local studies of the star-forming processes in three nearby molecular cloud complexes : Taurus-Auriga, Ophiuchus, and Orion. In the next section, we compare the distributions of dense gas and stars; in section III, we compare rough estimates of the star formation efficiency and initial mass function in the three regions. Section IV contains a summary of detailed studies of a particular region of star formation in each complex.

2. Distribution of Gas and Star Formation

2. 1. TAURUS-AURIGA MOLECULAR CLOUD COMPLEX

The Taurus-Auriga complex is one of the best studied examples of a nearby star forming region. It is located at ∼140 pc (Elias 1978a) and contains ∼10^4 $M_\odot$ of material (Wouterloot and Habing 1985; Cernicharo, Bachiller and Duvert 1985; Ungerechts and Thaddeus 1987). Studies of the large scale structure of Taurus-Auriga, using ^{13}CO (Kleiner and Dickman 1984, 1985), star counts (Cernicharo, Bachiller and Duvert 1985), OH (Wouterlout and Habing 1985), CO (Ungerechts and Thaddeus 1987) and IRAS 100 μm optical depth (Scalo 1990), have revealed complex, irregular, and filamentary structures.

In addition to the diffuse, filamentary structure, small dense cores have been found in Taurus-Auriga by molecular line surveys of regions of optical obscuration (Myers, Linke, and Benson 1983; Myers and Benson 1983). The properties of these cores have been determined from observations of the NH$_3$ molecule, which is sensitive to gas densities $\geq 10^4$ cm^{-3} (Myers and Benson 1983; Benson and Myers 1989). These observations reveal that the cores are small, having sizes on the order of 0.1 pc and masses on the order of a few M$_\odot$. In addition the cores are cold (T$_{kinetic} \sim 10$ K) and the NH$_3$ lines have very narrow, nearly thermal, linewidths (FWHM ~ 0.3 km s^{-1}). However, the linewidths of other molecules are clearly not thermal (Zhou et $al.$ 1989). One should note that these properties are not restricted to Taurus cores but are typical of most dense cores found in nearby dark clouds (Myers and Benson 1983; Benson and Myers 1989).

One of the most notable characteristics of nearby dense cores is their association with star formation. Many dense cores are associated with known T Tauri stars (Myers and Benson 1983), bipolar outflows (Fuller and Myers 1987; Myers et $al.$ 1988) and low luminosity IRAS sources (Beichman et $al.$ 1986; Myers et $al.$ 1987). In fact, IRAS sources have been detected in approximately half of the ~ 100 cores which are known in Taurus and other nearby complexes (Benson and Myers 1989). Typically, one to a few stars are found in a core.

An understanding of the global processes of star formation in a given molecular cloud complex requires knowledge of the distribution of young stellar objects within that complex. The first large-scale (120 pc^2) infrared survey of the Taurus region was obtained at 2 μm (Elias 1978a) with a single detector. Consequently the spatial resolution was poor ($1' - 2'$) and only a modest number (1.8×10^4) of "pixels" could be obtained. In addition, sensitivity was poor, with a limiting magnitude of only m$_K = 7.5$. Nonetheless, Elias was able to find 215 sources, most of which he concluded were field stars. The stars which were associated with the complex appeared to be in small groups, scattered through the cloud, in agreement with conclusions based on optical studies (Cohen and Kuhi 1979; Larson 1982).

Recently, Kenyon et $al.$ (1990) have completed a survey of the embedded stellar population in the Taurus molecular cloud, using the IRAS data base, followed by near-infrared and optical studies. Their survey is complete for luminosities L $>$ 0.5 L$_\odot$, making it one of the most sensitive surveys of an embedded population in a large complex. They have identified at least seven new pre-main-sequence stars and six new deeply embedded objects. The distribution of the young stellar objects within the Taurus region is shown in Figure 1, superimposed on a map of CO integrated intensity (from Kenyon et $al.$ 1990). Roughly 100 objects are located within a 40 pc × 30 pc area. Although some concentrations are seen in this figure, the sources for the most part appear scattered throughout the cloud, in agreement with the early findings.

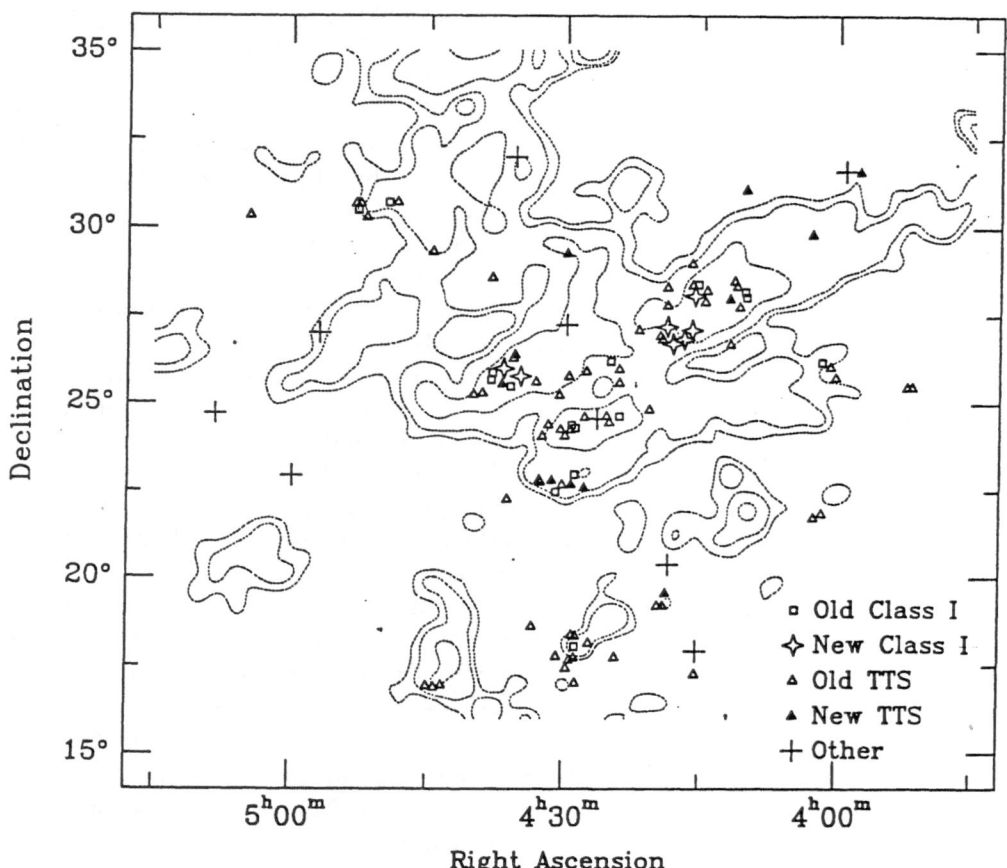

Fig. 1.   *Distribution of young stellar objects in the Taurus-Auriga Molecular Cloud Complex (from Kenyon et al. 1990). The distribution of YSOs is shown superimposed on a map of CO integrated intensity (Ungerechts and Thaddeus 1987). The CO contours are presented for integrated intensities equal to 3, 5, 10, 25 and 40 K km s^{-1}.*

2. 2. OPHIUCHUS MOLECULAR CLOUD COMPLEX

The Ophiuchus molecular cloud complex (see Wilking 1990 for a review) is another well studied nearby region of star formation, located near the Scorpius-Centaurus OB association. Distance estimates to the molecular complex range from the traditional value of 160 pc (Bertiau 1958, Whittet 1974, and Chini 1981) to a more recent value of 125 ±25 pc by de Geus *et al.* (1989), who argue that the near side of the cloud is as close as 80 pc, while the back side is at 170 pc. Many similarities exist between the Ophiuchus and Taurus complexes. The total mass of the Ophiuchus complex, measured in CO, is also about 10^4 $M_\odot$ (de Geus, Bronfman and Thaddeus 1990). Large scale molecular line surveys in CO (de Geus, Bronfman and Thaddeus 1990) and ^{13}CO (Loren 1989a, b) have revealed filamentary or clumpy structures. In addition to the 89 distinct ^{13}CO structures identified by Loren (1989a, b), several dense NH_3 cores, with properties similar to those found in Taurus, have been identified in this complex (Myers and Benson 1983; Benson and Myers 1989).

The Ophiuchus star forming region is distinguished from Taurus by the presence of a large centrally condensed core (Wilking and Lada 1983), located in the ρ Ophiuchi cloud, the westernmost cloud of the complex (see Klose 1986 for a review). The core is heavily obscured, with $A_v \sim 100$ mag (Vrba *et al.* 1975; Wilking and Lada 1983). $C^{18}O$ observations indicate a core 1 pc × 2 pc in size, containing ~ 600 $M_\odot$ (Wilking and Lada 1983). There are also concentrations of dense (n $\geq$ 10^4 cm^{-3}) gas in the core, as indicated by observations of H_2CO (Loren, Sandqvist and Wootten 1983) and DCO$^+$ (Loren, Wootten and Wilking 1990).

The Ophiuchus Complex is further distinguished from the Taurus Complex by the localized concentration of young stars in the ρ Ophiuchi cloud. Hα objective prism surveys of the Ophiuchus complex (Struve and Rudkjobing 1949; Haro 1949; Dolidze and Arakelyan 1959; Wilking, Schwartz and Blackwell 1987) have shown that the majority of young stars are associated with the western half of the ρ Ophiuchi cloud. In addition to optical studies, a large area (140 pc^2) in this complex was surveyed at 2.2 μm by Elias (1978b), with similar sensitivity, resolution, and number of pixels to those obtained in his study of Taurus-Auriga (Elias 1978a). Elias found nearly twice as many sources per unit area in Ophiuchus, though again most were field stars. In contrast to the distribution in Taurus, the associated sources in Ophiuchus were concentrated into a single group, later associated with the region of high molecular column density in the ρ Ophiuchi cloud (Wilking and Lada 1983). More recently, IRAS based studies of the embedded stellar population in Ophiuchus (Ichikawa and Nishida 1989) have also revealed a concentration of IRAS sources in the ρ Ophiuchi cloud. These results confirmed the earlier suggestions of the presence of a dense stellar cluster in the ρ Ophiuchi cloud (Grasdalen, Strom, and Strom 1973; Vrba *et al.* 1975).

The embedded stellar cluster in the ρ Ophiuchi core has been studied in detail using both near-infrared and IRAS observations (Wilking and Lada 1983; Lada and Wilking 1984; Young, Lada and Wilking 1986; Wilking, Lada and Young 1989).

The distribution of embedded sources is shown in Figure 2. A total of 78 objects have been identified as members of the cluster, and 39 additional infrared sources lie within the core boundary but remain unclassified (Wilking, Lada and Young 1989). Therefore, roughly 100 embedded young stellar objects are located within the 1 pc × 2 pc area of the core.

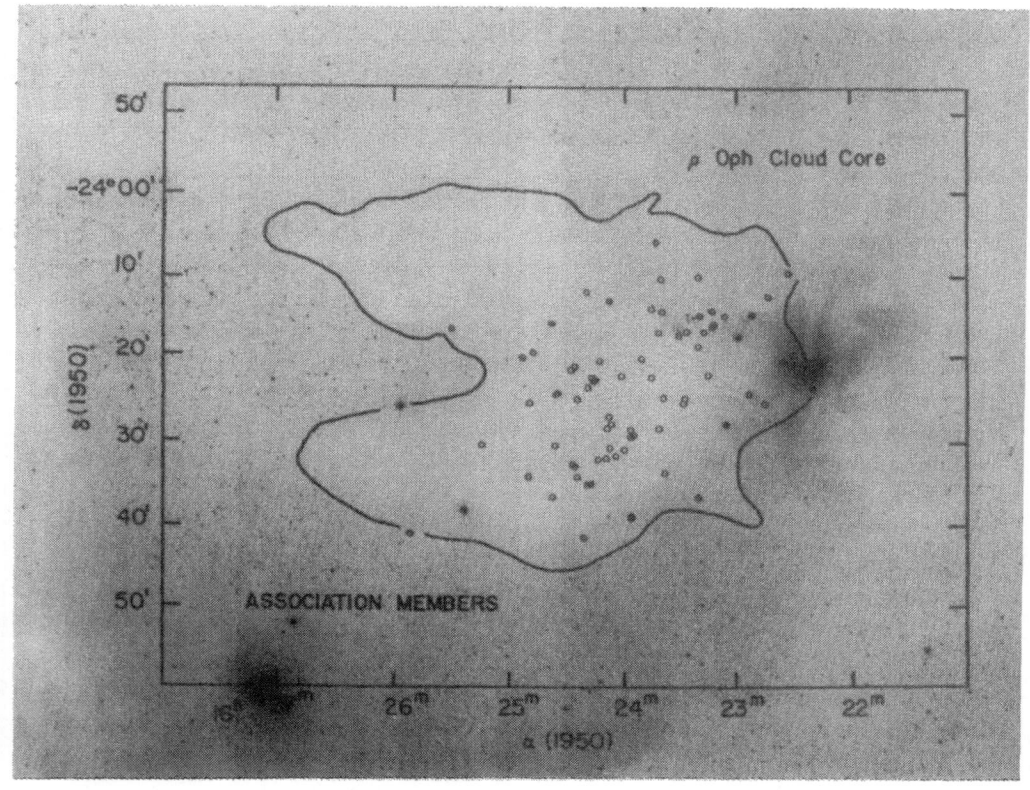

Fig. 2.   *The distribution of young stellar objects associated with the ρ Ophiuchi Molecular cloud (from Wilking, Lada and Young 1989). YSOs associated with the molecular cloud core are displayed as open circles, superposed on the red photograph of the Palomar Sky Survey. The boundaries of the molecular gas (^{13}CO, Loren 1989a) are presented by a solid contour.*

2. 3. THE ORION MOLECULAR CLOUDS - L 1641 AND L 1630

The nearest region of copious production of massive stars is in the direction of Orion. This complex is usually assumed to lie at a distance of 500 pc, but an extensive study of the distances to B stars indicates a distance of about 400 pc (Anthony-Twarog 1982); this estimate is consistent with the low end of the range (480±80 pc), based on proper motions of the H_2O masers in the OMC1 region (Genzel *et al.* 1981), but the uncertainties are considerably less. A large-scale map of the region in CO emission (Maddalena *et al.* 1986) reveals two major clouds : Orion A (L1641), with a mass of 1.0×10^5 $M_\odot$, including the Orion Nebula region and extending southward; and Orion B (L1630), extending northward from the Orion Nebula with a mass of 8×10^4 $M_\odot$. Extensive maps of ^{13}CO emission (Bally *et al.* 1987; Bally this vol.; Fukui this vol.) reveal complex filamentary structures throughout the L1641 and L1630 clouds. In addition, the L1641 cloud contains the very dense structures and prolific star formation regions near the Orion Nebula, marked by an intense concentration of OB stars and embedded infrared sources (Genzel and Stutzki 1989).

2.3.1 L1641

Recently, studies of both the dense cores and the young stellar population have been carried out in the L1641 molecular cloud. A survey of NH_3 emission towards IRAS sources in L1641 has identified a number of dense cores within the cloud (Wouterloot, Walmsley, and Henkel 1988). Maps of NH_3 reveal that cores in L1641 have significantly larger sizes and linewidths than those in Taurus (Harju, Walmsley and Wouterloot 1990). In addition the L1641 cores have higher kinetic temperatures and larger masses. IRAS-based studies of the young stellar population (Strom *et al.* 1989) have identified 93 IRAS point sources to be associated with the L1641 cloud. These sources are distributed throughout the cloud, very reminiscent of the distribution of sources seen in Taurus. One should note however that the IRAS selected sample is only complete to ~ 6 $L_\odot$, much higher than the completeness limit for comparable studies in Taurus.

Based on the above studies, it appears that star formation in L1641 resembles star formation in Taurus, at least in its distribution. However, we know from optical studies that several clusters are also present in this cloud. One well known cluster is the Trapezium cluster. Recently, McCaughrean *et al.* (1990) have found a total of 480 near-infrared sources in this area, indicating that many of the stars are buried in the dense molecular cloud immediately behind the Trapezium cluster. In addition to this cluster, Strom, Margulis, and Strom (1989b) have found a cluster of $\sim$20 sources associated with IRAS 05338-0624. It seems that star formation in L1641 also occurs in clusters, similar to star formation in the ρ Ophiuchi cloud. In fact, if we compare the number of embedded sources found in clusters to the total number of embedded sources known to be associated with the L1641 cloud, roughly 85% would be located in clusters. This number may be an overestimate, since the IRAS survey is not sensitive to the entire stellar mass spectrum; more low-mass stars may be distributed throughout the cloud.

2.3.2 L1630

Recently, unbiased, systematic and sensitive surveys for dense gas and embedded infrared sources have been obtained in the L1630 molecular cloud. These surveys provide the first complete census of dense cores and embedded young stellar objects (YSOs) within a single molecular cloud and allow us to investigate systematically the relationship between dense cores and their associated young stellar objects.

To identify the dense cores, the L1630 molecular cloud has been surveyed for dense gas in the J=2→1 transition of CS (Lada, Bally, and Stark 1991). Emission was detected, at a 3 σ level over 10% of the area surveyed, revealing a very clumpy structure in the dense gas on 1' – 2' scales. Forty-two individual cores were identified at a 5σ level above the noise, including all previously known regions of star formation (*i.e.*, NGC 2071, 2068, 2024, 2023, and core number 23, which includes the M78 (HH 19-27) star formation region). Approximately 50% of the total mass of dense gas is contained within the 5 most massive cores, each with a mass greater than 200 $M_\odot$. These 5 massive cores cover a total area of ~2 pc^2, only 1% of the total area surveyed, indicating that the dense gas is highly localized.

A near-infrared (2.2 μm) survey of a significant portion of the cloud (Lada *et al.* 1991), including both areas containing CS emission and areas without CS emission, revealed 912 sources having $m_K < 13$. At the distance to L1630, this limit corresponds to a 0.6 $M_\odot$ main sequence dwarf. Using statistical arguments, Lada *et al.* (1991) estimated that ~ 50% of the sources with $m_K < 13$ are background stars. Even with this large contribution from a uniform background distribution, the observed sources appear grouped or clustered. Four embedded clusters were identified, where an embedded cluster is defined as a region in the sky where the source density significantly increases over the background star density. These clusters are associated with the well known star formation regions, NGC 2071, 2068, 2024, and 2023. The embedded clusters exhibit a range in size (r = 0.3 pc to 0.9 pc) and number of sources (~ 20 - 300). The most spectacular cluster is associated with the HII region NGC 2024. Whereas optical images are dominated by a dark, obscuring dust lane, many sources are present in the 2 μm image, along with substantial nebulosity. Lada *et al.* (1991) count 309 sources having $m_K < 14$ located within the cluster boundary.

Surprisingly, Lada *et al.* (1991) find that the majority of the embedded sources associated with L1630 are located in the four embedded clusters. These clusters contain at least 58% (if no correction for background stars is made) and possibly as much as 96% (if correction for background stars is made) of the total number of sources in the molecular cloud! Furthermore, the total area covered by the four embedded clusters is only 18% of the total region surveyed, indicating that star formation in L1630 is a highly localized process even for stars whose masses are as low as the mass of the sun.

Comparing the distribution of dense gas and the distribution of embedded infrared sources (see Figure 3), Lada (1990) finds that the embedded clusters are coincident or nearly coincident with 4 of the most massive (M > 200 $M_\odot$) CS cores. Star formation, in this region, is occurring almost exclusively in the most massive cores of dense gas. There is no evidence for any significant star forming activity outside these localized centers. Moreover, 97% of the sources found in the embedded clusters are associated with only 3 massive CS cores.

The CS and 2.2 μm surveys have revealed that the formation of both high and low mass stars in L1630 is occurring mainly in clusters. These clusters in turn are produced by the largest and most massive dense cores in the cloud. These results suggest that in order to understand how most stars in this cloud form, one must understand how massive cores form and produce clusters of stars with relatively high efficiency.

3. Star Formation Efficiency and the Initial Mass Function

As described above, the studies of the Taurus, Ophiuchus and Orion complexes have revealed that the distribution of star formation is different in the three clouds. Taurus forms stars individually or in small groups dispersed through the cloud. Ophiuchus and L1641 (Orion A) are forming most of their stars in a single dense cluster, with other regions of more isolated star formation scattered through the cloud. Finally, L1630 (Orion B) is forming most of its stars in three rich clusters, which are spread through the cloud. It is now of interest to see if the differences between the distributions of star formation in these clouds are connected with other differences, such as the star formation efficiency and the initial mass function. Observational determination of both of these is very difficult, but we can make some preliminary comparisons.

In principle, it is straightforward to determine the global star formation efficiency of a molecular cloud complex. This can be done by comparing the total mass of stars with the total mass of stars plus gas [$SFE = M_*^{tot}/(M_*^{tot} + M_{gas})$]. Of course, the *final* star formation efficiency of a complex can only be determined when the cloud has finished forming stars. However we can compare the star formation efficiency *so far* by estimating the mass of already formed stars and the mass of the cloud. For example, in Taurus-Auriga, Kenyon *et al.* (1990) conclude that there are 100 – 140 pre-main-sequence stars with L$\geq$0.5 $L_\odot$; if the median mass is 0.6 $M_\odot$ (Larson 1986), then the total mass of stars (M_*^{tot}) is about 70 $M_\odot$. The total mass of the whole complex is 1×10^4 $M_\odot$, yielding a *SFE* of about 0.7%. In Ophiuchus, Wilking, Lada, and Young (1989) have done a very careful accounting of the stars in the ρ Ophiuchi cluster, concluding that $M_*^{tot} = 82$ $M_\odot$. Since the total mass of the Ophiuchus cloud is also 1×10^4 $M_\odot$, the overall efficiency is about 0.8% as long as there is not a large contribution from the rest of the complex. Similarly, we can estimate the star forming efficiency in L1641. Based on the infrared surveys of this cloud (Strom *et al.* 1989; Strom, Margulis and Strom 1989b; McCaughrean *et al.*

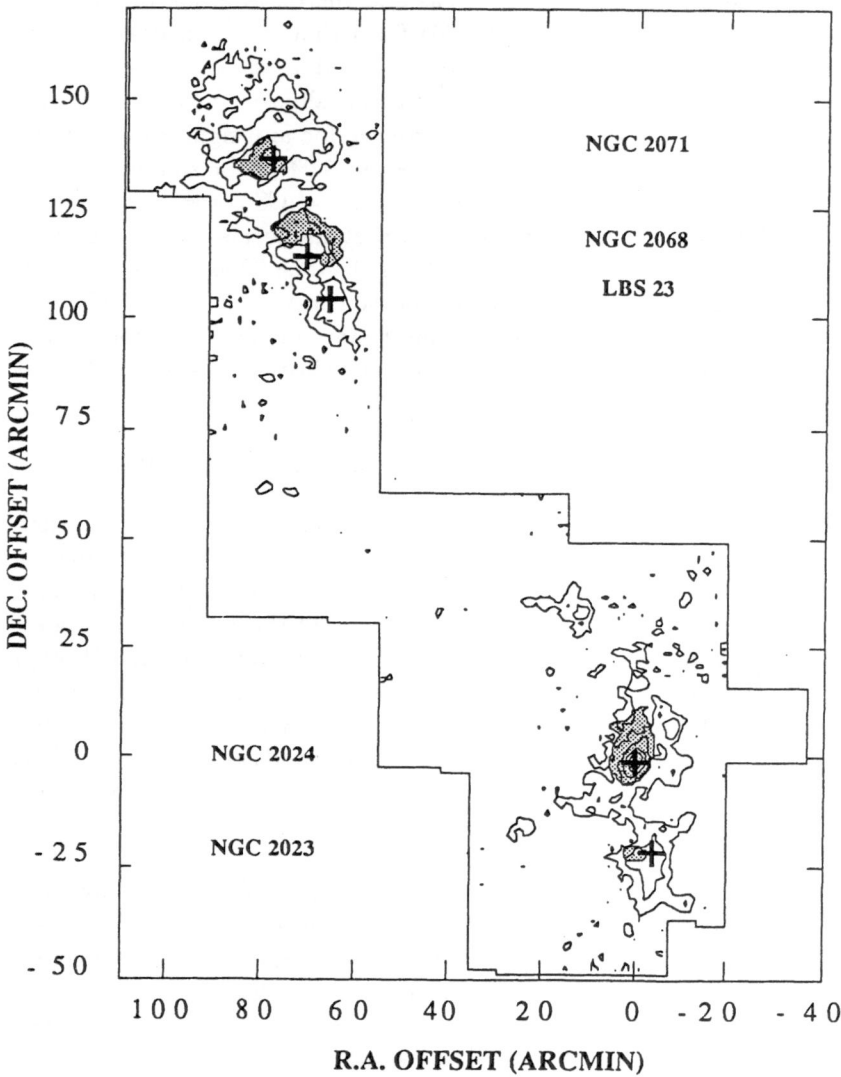

Fig. 3. *Locations of the embedded infrared clusters and dense CS cores in the L1630 Molecular Cloud (from Lada 1990). The locations and extents of the embedded stellar clusters are shown as shaded regions and the distribution of dense gas is presented as intensity contours of CS(2→1) emission. The lowest CS contour is equal to a 3σ detection level above the noise (0.8 K km s⁻¹. In addition, the peak intensity positions of the 5 most massive CS cores (M > 200 M⊙) are represented by crosses.*

1990), we estimate that ~ 600 young stellar objects are associated with the cloud. For a mean stellar mass of 1 $M_\odot$, and M_{gas} of 1×10^5 $M_\odot$, we again get an overall SFE of about 0.6%. Finally, estimates of the global star formation efficiency in the L1630 cloud (Lada 1990) are also low, with $SFE \sim 3\% - 4\%$.

Comparing the estimates for the global star formation efficiency and allowing for the very large uncertainties in all of them, we see that the star formation efficiencies of the three clouds are very similar, with $\sim 1\%$ of the gas in each complex forming stars so far. While these estimates are lower limits to the final SFE of these clouds, they are similar to the estimates for the *final SFE* of clouds which form open clusters (Leisawitz, Bash, and Thaddeus 1989) and of the cloud which formed the λ Orionis association (Duerr, Imhoff, and Lada 1982). These results indicate that only a small fraction of the gas of a typical molecular cloud ever has the conditions required for star formation.

The similarity of the global efficiencies of these three cloud complexes is somewhat surprising in view of the different distributions of star formation discussed earlier. If the only requirement for star formation is that the gas density rise above about 10^4 cm^{-3}, then the similarity in SFE would suggest that the fraction of the gas with n $\geq 10^4$ cm^{-3} is roughly the same for these three complexes. In this case, the difference between them would be that, in Ophiuchus and Orion, the dense gas is collected into much more massive cores than it is in Taurus, resulting in the formation of large clusters rather than small groups. However, the L1630 cloud suggests that the situation is more complex. Five massive (M > 200 $M_\odot$) dense cores exist in the cloud, yet only 3 cores seem to be producing most of the young stellar objects. This is reflected in estimates of the star formation efficiency of the cores. For example, the NGC 2024, NGC 2071 and NGC 2068 cores have SFE ranging from 20% - 40%, while the NGC 2023 core and core number 23 have $SFE \sim 7\%$ (Lada 1990). Since we do not yet know the final efficiencies, these differences may only be temporary. If they reflect the final efficiencies, then factors other than having a density $\geq 10^4$ cm^{-3} may be important. Detailed comparisons of the physical conditions in these different cores are needed to explore this question. It will also be important to assess the local efficiency for the NH$_3$ cores in Taurus, using the same criteria as in L1630 for determining the mass of gas with n $\geq 10^4$ cm^{-3}.

We can also ask whether the initial mass functions differ among these three regions. The initial mass function of field stars and fully-formed open clusters appears to be characterized by a power law slope in the $dN/dlog(M_*)$ versus M_* relation for large masses and a turnover below about 0.5 $M_\odot$ (*e.g.*, Scalo 1986, Larson, this vol.). The slope of the power law appears to be about the same (1.7 ± 0.5) in most clusters and in the field. Is this result consistent with the anecdotal statements from the star formation field that "Orion forms high mass stars" while "Taurus forms low mass stars"? The optical studies by Cohen and Kuhi (1979) found values for the slopes of 1.5 (1.2 – 2.5) in Taurus and 1.35 (1.3 – 1.5) in Orion, both consistent with the above range, as well as with each other. Is the difference then *only* that Orion is forming more stars overall, so the high end of the IMF is better sampled than in Taurus? This question is hard to answer at this point, but the lack of massive stars in Taurus is consistent with the total number of stars and a standard IMF.

The recent surveys at 2 μm have the potential for clarifying these issues, since they should be much more complete and less biased than the optical surveys. Such surveys can be used to construct cumulative luminosity functions for the clusters, in which the logarithm of N, the number of sources brighter than a given 2 μm magnitude, m_K, is plotted versus m_K. The four clusters in the L1630 cloud have luminosity functions which are well described by power laws for m_K above the completeness limit. The slopes in the $log(N) - m_K$ relation are consistent for the three rich clusters, with a value of 0.38±0.03. We can compare this result with what would be predicted from a given IMF with the following assumptions : the 2 μm emission arises only from stellar photospheres, the extinction at K is either negligible or at least does not depend on m_K, and the stars are on the main sequence. The result which would be predicted from a slope in the initial mass function of 1.7±0.5 would be 0.34±0.08, consistent with the data.

The near-infrared surveys can also be used to look for a turnover in the IMF at low masses, which would produce a flattening in the cumulative luminosity function, but considerable caution is needed. If the lowest mass stars are not yet on the main sequence, their luminosity will exceed the luminosity they will have when they reach the main sequence. Thus, a flattening in the cumulative luminosity function, *interpreted* in terms of a *main − sequence* mass–luminosity relation, would overestimate the mass at which the IMF turns over. There is some indication of a flattening in the luminosity function in the L1630 clusters, but it occurs around m_K of 12, near the completeness limit of 13, making it rather uncertain. DePoy *et al.* (1990) have obtained more sensitive ($m_K \sim 15$) data on one of the clusters (NGC2023) and confirm the flattening. Similarly, McCaughrean *et al.* (1990) found a possible deficit of stars fainter than $m_K = 11.5$ in the Trapezium cluster. They interpreted their data in terms of pre-main-sequence stars with ages of 10^6 years; in this interpretation, the mass function peaks strongly at 0.3 M$_\odot$, with little or no evidence for stars less massive than 0.1 M$_\odot$. While these conclusions may be sensitive to the age (and to the assumption of a single age for all the sources), they do suggest that the lower mass limit in the Trapezium cluster is not very different from that in Taurus, and that the initial mass function of this forming cluster is similar to that of more evolved clusters.

How does Ophiuchus fit into this question? Elias (1978a) commented that, while the most massive star in both Ophiuchus and Taurus is a B2 star, Ophiuchus boasts about 20 stars of spectral type A0 or earlier, compared to only 2 or 3 in Taurus. Wilking, Lada, and Young (1989) have constructed a luminosity function for the ρ Ophiuchi Cloud core and conclude that the luminosity function deviates from that expected from the IMF of field stars by 2.5 to 3.5 σ in the sense of a deficit of intermediate luminosity stars. They also find a deficit at low luminosities (L < 1 L$_\odot$), but attribute this to incompleteness. Recently, Kenyon *et al.* (1990)

compared the luminosity functions of Taurus and the ρ Ophiuchi core. They found that Ophiuchus contains many more deeply embedded (Class I) sources with L > 10 $L_\odot$ than does Taurus, while the T-Tauri stars (class II sources) in each cloud have similar luminosity functions.

The low end of the luminosity function in Ophiuchus has been somewhat more controversial recently. Based on a survey of a small region in the dense core of the cloud, Rieke, Ashok, and Boyle (1989) found a turnover at m_K of 9; considering the smaller distance to this cloud, this limit would correspond to a turnover in absolute magnitude at a value similar to that in Orion. However, Barsony et al. (1989), having surveyed a larger area, found no evidence of a turnover to m_K of 14. The discrepancy between the two surveys seems to be caused by an absence of faint sources in the region surveyed by Rieke, Ashok, and Boyle; whether this is simply due to chance or to spatial variations in the luminosity function is not yet clear. One possibility is magnitude segregation, in which the fainter sources are more widely distributed than the brighter sources; there is some evidence for this in the L1630 clusters (Lada et al. 1991); and it may correspond to mass segregation seen in some more evolved clusters (e.g., Scalo 1986).

If the extinction at K is negligible, the limit of Barsony et al. (1989) would correspond to an absolute K magnitude of about 8, implying very low mass stars, even in the unlikely event that they are on the main sequence. These sources may fill in the low luminosity part of the luminosity function in Ophiuchus, making it look more like the function in Taurus (see fig. 8 in Kenyon et al. 1990).

At this point, the evidence for variations in the IMF among these three regions is weak; many of the apparent differences can be traced to the larger cloud masses in the Orion cloud, which allow more stars to form, more fully sampling the IMF (see also Elmegreen 1983). Before suggested differences can be taken seriously, much more work needs to be done to correct for background sources, to rule out possible dust emission at 2 μm, and to pin down extinction corrections. In addition, more complete and sensitive surveys would help to remove some selection effects. In some dense regions, the extinction is clearly large, even at 2 μm, and observations at longer wavelengths are needed to determine the number of sources missed in the 2 μm surveys. The IRAS survey can help with this problem, but its low resolution implies that extensive follow-up studies are necessary. While evolutionary effects make it difficult to compare postulated IMF's to observed luminosity functions, the data so far seem to be consistent with all three regions having IMF's similar to those of evolved clusters and field stars.

4. Studies of Individual Sources

Having compared the star formation on a global scale in these three nearby clouds, we now turn our attention to studies of individual regions of star formation within each cloud. While many such regions have been studied, we will concentrate on one region in each cloud. Comparison between these regions will give us a preliminary look at the mode of star formation in the different clouds, but comparison of more regions in each cloud should eventually provide a more complete picture.

To provide a common ground for comparison of these individual regions in each of the three clouds, we will compare the observational results to the predictions of theoretical models of star formation, as developed recently by Shu and his collaborators (Shu, Adams, and Lizano 1987). In this picture, low mass star formation occurs in relatively isolated cores in molecular clouds. These cores contract slowly to form a density structure similar to that of an isothermal sphere; that is, $n(r) \propto r^{-2}$ (Shu 1977). At some point, collapse may begin in the center of the core; as a wave of infall propagates outward, an inside-out collapse occurs. If the infall wave propagates at speed a (in the simplest case, this is the isothermal sound speed), its radius is simply related to the time since collapse began by $r_2 = at$. Inside r_2, the density distribution will be modified to approximate $n(r) \propto r^{-1.5}$. These predictions are similar to those of earlier dynamical collapse calculations (Larson 1969; Penston 1969), but the predicted velocity field is significantly different. Observations of H_2CO in B335, a small star-forming globule, are in excellent agreement with the predictions of the Shu picture, while clearly in disagreement with those of Larson-Penston type collapse (Zhou et al. 1990). If rotation is initially present, it further modifies the collapse in the innermost regions; a disk is likely to form inside the centrifugal radius, R_c (Terebey, Shu, and Cassen 1984).

With this background in mind, we will compare three star formation regions : L1551 IRS 5 (in the Taurus complex); IRAS 16293-2422 (in the Ophiuchus cloud); and NGC2071 (in the Orion cloud, L1630). The first two regions are forming low mass stars, based on their luminosities of 30 $L_\odot$ and 27 $L_\odot$, respectively, while NGC2071 has a luminosity of 520 $L_\odot$, suggesting a star of considerably higher mass.

4. 1. L1551 IRS5 (TAURUS COMPLEX)

L1551 IRS5 is the prototypical bipolar outflow source (Snell, Loren, and Plambeck 1980) and the L1551 cloud (a component of the Taurus complex) appears to have a density distribution of $n(r) \propto r^{-2}$ on scales of a few tenths of pc (Snell 1981), consistent with the the initial configuration having been an isothermal sphere. The source has been modeled in detail by Adams, Lada, and Shu (1987), who predicted that infalling matter extends from 40 to 3500 AU, corresponding to angles of 0.3″ to 21″ at the distance of 160 pc to this particular cloud (Snell 1981). Inside 40 AU, they predicted that a disk should exist, extending inward to the stellar surface and argued that the emission from this disk compensated for a deficit of emission at mid-infrared wavelengths (5-20 μm) that appeared in spherically symmetric models (Adams and Shu 1985, 1986). The star itself has been studied by observing photospheric features at 2 μm (Carr, Harvey, and Lester 1987) and at shorter wavelengths via scattered light (Mundt et al. 1985). The spectra show the characteristic features of an FU Orionis object. These objects can be understood as pre-main-sequence stars surrounded by massive accretion disks, with the outbursts explained by disk instabilities (Hartmann and Kenyon 1985). Thus, there is some indirect evidence for a disk, as envisioned by Adams, Lada, and Shu (1987). The direct evidence is somewhat more equivocal : both the pattern of scattered light at 2 μm (*e.g.*, Campbell *et al.* 1988) and the high-resolution $C^{18}O$ map (Sargent *et al.* 1988) indicate elongated structures, but on scales considerably larger than the 40 AU size predicted by Adams, Lada, and Shu (1987).

The region predicted to contain infalling material has been studied recently with high-resolution (10" to 20") far-infrared observations (Butner *et al.* 1991a); the source is clearly resolved at 100 μm (with a beam-deconvolved size of about 11"), but not at 50 μm. Detailed comparison of the far-infrared scans with predictions of the the intensity distribution from the Adams, Lada, and Shu (1987) model (Adams, private communication) show good agreement. The data also agree well with the predictions of radiative transport models with spherically symmetric density distributions characteristic of infall ($n(r) \propto r^{-\alpha}$ with $\alpha = 1.5$); thus the departures from spherical symmetry, including the disk, in the detailed models of Adams, Lada, and Shu (1987) do not affect the far-infrared emission at the scales probed by the data. Similar models with other density laws ($\alpha = 1.0$ or 2.0) clearly do not fit the data well. Butner *et al.* (1991a) concluded that their data were in reasonable agreement with the model of Adams, Lada, and Shu (1987), specifically with the prediction of infalling material on scales of a few thousand AU. Confirmation of this result with spectroscopic data is clearly needed to establish firmly the presence of infall, but confusion with the outflowing gas makes this a tricky project.

The Adams, Lada, and Shu (1987) interpretation of the mid-infrared spectrum as evidence for a disk in L1551 is more problematic. In order to produce the observed emission at far-infrared wavelengths, the dust opacity must be very high at mid-infrared wavelengths, assuming the dust opacities used by Adams, Lada, and Shu (1987), which are similar to those of Draine and Lee (1984). The mid-infrared radiation from a disk will be completely attenuated and re-emitted at longer wavelengths. Spherically symmetric models of this source produce very deep silicate features and a pronounced mid-infrared deficit, *whether or not disks are included in the spectrum of the internal energy source* (Butner, Natta, and Evans 1991), as long as the dust opacities are assumed to be like those of Draine and Lee (1984). There are two possible answers to this problem : one is that the mid-infrared radiation is scattering out the poles of the outflow, in which case it should be polarized, like the near-infrared emission (*e.g.*, Hodapp *et al.* 1988); the second is that the actual dust opacities have a lower ratio of mid-infrared opacity to far-infrared opacity than do those of Draine and Lee (1984). Dust opacities with this property were advocated by Mathis, Mezger, and Panagia (1983), and Butner *et al.* (1991a) found that they provide better agreement with the spectral energy distribution of L1551 IRS 5 than do those of Draine and Lee (1984) or of Adams, Lada, and Shu (1987).

Leaving aside scattered light, with its own interpretational ambiguities (for example, compare the conclusions of Beckwith *et al.* (1984) with those of Beckwith *et al.* (1989) or those of Leinert and Haas (1987) with those of Koresko, Beckwith, and Sargent (1989)), the only way disks in deeply embedded sources, like L1551 IRS5, are likely to be seen directly is at very long wavelengths, where both the attenuation and the emission from dust in the infalling envelope are minimized. In fact, Keene and Masson (1990) have found a compact continuum source in millimeter-wavelength interferometer observations. This compact source cannot be reproduced in the infalling envelope models of Butner *et al.* (1991a), clearly indicating the need for a separate component. While Keene and Masson (1990) did not resolve the source, they interpreted it as a disk with an outer radius of 45 AU (0.3″), very similar to the predicted R_c in the Adams, Lada, and Shu (1987) model. The surrounding envelope does produce emission at these wavelengths, but this envelope emission is very extended. The most convincing observational signature of compact structures, plausibly interpreted on theoretical grounds as disks, is then a ratio between the emission in large and small beams at long (*e.g.*, millimeter) wavelengths which requires the existence of a substantial compact component.

4. 2. IRAS16293-2422 (L1689 – OPHIUCHUS)

This remarkable IRAS source lies in the L1689 cloud, east of the main body of the Ophiuchus cloud, where most of the near-infrared surveys discussed above were performed, but still part of the Ophiuchus complex (Loren 1989a). Since the source has a luminosity of 27 $L_\odot$ (Mundy, Wilking, and Myers 1986), very similar to that of L1551 IRS 5, it provides an interesting comparison. The source was found to have wings and self-absorption in the CS J=5→4 line, with a pattern suggestive of infall (Walker *et al.* 1986). Later studies suggested that the self-absorption was caused by an extended, foreground absorbing layer, calling into question the kinematic evidence for infall (Menten *et al.* 1987); instead, rotation appears to be present (Menten *et al.* 1987; Mundy, Wootten, and Wilking 1990). The source contains at least one outflow (Wootten and Loren 1987; Walker *et al.* 1988), with an axis roughly perpendicular to the axis of elongation of the continuum emission at 3 mm (Mundy, Wilking, and Myers 1986). The strong emission at millimeter wavelengths indicates that there is a very opaque core surrounding the source, making it invisible at wavelengths less than about 20 μm (Mundy, Wilking, and Myers 1986). Radio continuum measurements at 6 cm and 2 cm revealed two sources, located near peaks in the 3 mm continuum distribution and separated by about 5″ (750 AU), leading Wootten (1989) to suggest that the system was a protobinary star.

Using the same far-infrared scanning system as was used for L1551, Butner *et al.* (1991b) have resolved the source at both 50 and 100 μm. Unlike L1551 IRS 5, the far-infrared data are not consistent with a density distribution characteristic of infall, but instead suggest relatively flat density laws ($\alpha = 0$ to 0.5). This very different behavior, compared to that found in L1551 IRS5, may be related to the possible presence of a protobinary system (Wootten 1989; Mundy, Wootten, and Wilking 1990). Clearly, this source has much more circumstellar material, and large-scale rotation appears to play a more important role, compared to the case of L1551 IRS5. Certainly the Adams, Lada, and Shu (1987) picture does not apply well to this source, but this discrepancy may be more related to the rather unusual properties of this source than to differences between the Taurus and Ophiuchus regions.

4. 3. NGC2071 (L1630 – ORION)

Even though the models of Adams, Lada, and Shu (1987) were developed for low mass stars, it is interesting to test the basic ideas for stars of somewhat higher mass and in different environments. In particular, we may expect star formation to proceed somewhat differently in a clustered, rather than non-clustered, environment. The infrared source in NGC2071 offers such an opportunity. In a part of the L1630 cloud which is at a distance of 390 pc (Anthony-Twarog 1982), it lies in one of the clusters of 2 μm sources discussed above (Lada et $al.$ 1991); the region we are discussing lies north of the NGC2071 reflection nebula and contains a small group of 10 μm sources (Persson et $al.$ 1981), of which IRS 1 is clearly dominant at 10 μm. If all the far-infrared emission is attributed to IRS 1, the luminosity of this source is 520 $L_\odot$, suggesting a star of about 5 $M_\odot$ (Butner et $al.$ 1990).

Far-infrared scans of this source resolved it at both 50 and 100 μm (Butner et $al.$ 1990). Like L1551 IRS5, the far-infrared scans are consistent with a power law density distribution, decreasing with distance from the energy source. This source was not specifically modeled by Adams, Lada, and Shu (1987), but the basic ideas can still be tested. To make a more specific comparison, values for the centrifugal radius (R_c) and the radius of the infall region (r_2) can be estimated (see Evans 1991). Assuming a stellar mass of 5 $M_\odot$ (Zhou, Evans, and Mundy 1990), the same value for the isothermal sound speed as in the Adams, Lada, and Shu (1987) models of L1551 IRS5, and a rotation rate of $\Omega = 5.8 \times 10^{-14}$ s^{-1} (Takano et $al.$ 1986), one finds $R_c = 1900$ AU and $r_2 = 1.5 \times 10^4$ AU, which correspond to 5″ and 38″, respectively, at the distance of NGC2071. Thus, the far-infrared scans should once again be probing primarily the infall region. In fact, comparison of the data to radiative transport calculations with various choices for r_2 and α indicates that the best fit for the inner radius of the spherical distribution is 5″. The value of α depends somewhat on the choice of dust opacities : for the Draine and Lee (1984) opacities, the best-fitting α is 2.0, while the Mathis, Mezger, and Panagia (1983) opacities favor α of 1.5. The Mathis, Mezger, and Panagia (1983) dust opacities provide a somewhat better fit to the spectral energy distribution, suggesting that $\alpha = 1.5$ is the more likely case, thus making this source appear to be consistent with the ideas of Adams, Lada, and Shu (1987). However, the normalizing density can also be predicted from the sound speed and the mass of the star; the result is $n_i = 8 \times 10^4$ cm^{-3} ($n_i \equiv n(R_c)$), whereas the far-infrared data suggest considerably higher densities ($n_i = 7 \times 10^5$ cm^{-3} for the Mathis, Mezger, and Panagia (1983) opacities, and 8×10^6 cm^{-3} for the Draine and Lee (1984) opacities). At least some modifications to the Adams, Lada, and Shu (1987) picture appear to be necessary for this source. These modifications may be generally necessary for more massive stars or stars which are forming in clusters.

Since the far-infrared emission is proportional to the column density of dust, the derivation of n_i is rather indirect; it depends on the assumption of spherical symmetry, the assumed dust opacities, and the absence of clumpy structure. It is possible to probe the volume density directly, using molecular line observations. A study of 4 transitions of CS with resolution similar to that of the far-infrared data has been obtained (Zhou *et al.* 1991). All transitions peak strongly on the infrared position, and the maps are all quite round, indicating that spherical symmetry is a reasonable approximation. The data were compared with the predictions of non-LTE excitation and radiative transport models, using the microturbulent approximation, which assumed various density gradients. The temperature distribution was taken from the radiative transport models of the dust emission (Butner *et al.* 1990), since the densities are high enough to equilibrate the gas kinetic temperature to the dust temperature. The inner radius of the models was set to $5''$, based on the results of the dust modeling, leaving the normalizing density (n_i), the power law in the density gradient (α), and the CS abundance as free parameters. The best-fitting models had $n_i = 3 \times 10^6$ cm^{-3}, $\alpha = 1.3$, and a CS abundance of 2×10^{-9}. This model also predicted the correct strength of the C^{34}S J=7$\rightarrow$6 emission. These values are reasonably consistent with those found from the far-infrared analysis, considering the various uncertainties, although models with the density distributions found by the far-infrared analyses clearly gave worse fits to the CS data.

The CS data favors a density four times larger than that found by the far-infrared modeling with the dust opacities of Mathis, Mezger, and Panagia (1983); this difference may be caused by uncertainties in either the collision rates for CS excitation or the dust opacities. In particular, a smaller dust opacity in the far-infrared would require a larger column density to match the observed far-infrared emission. The other possible explanation is that the infalling cloud is not completely smooth, but contains some clumpy structure; since the molecular emission is sensitive to the volume density, as well as the column density, it will be enhanced relative to the dust emission in the clumps, leading to a higher density derived from the molecules than from the dust.

Lest it be thought that all molecular species trace the distribution of material equally well, it should be noted that a high resolution study of NH$_3$ shows a distribution very different from that of the dust or the CS (Kawabe *et al.* 1989; Zhou, Evans, and Mundy 1990). The NH$_3$ has a very elongated distribution. Based on the temperature and velocity structure in the NH$_3$, Zhou, Evans, and Mundy (1990) conclude that most of the NH$_3$ is distributed in a ring of radius 0.1 pc, outside the region of CS and dust emission. Only a small amount of NH$_3$ emission arises from gas close (about 0.01pc) to the star; unlike the bulk of the NH$_3$, this gas has a velocity structure consistent with Keplerian rotation about a 5 M$_\odot$ star. The abundance of NH$_3$ in the region of dust and CS emission must be about 100 times lower than in the outer parts or in typical low density regions, suggesting that it is frozen onto dust grains. A similar result was found in IRAS16293-2422 (Mundy, Wootten, and Wilking 1990). Thus, NH_3 is a poor tracer of the densest gas in these two sources and should be regarded with caution in other regions as well (see also Zhou et al. 1989).

5. Conclusions

The large-scale infrared surveys of the three clouds described above have confirmed many, but not all, of the ideas derived from earlier studies. The distribution of star formation does appear to be different in the three clouds : Taurus forms stars individually or in small groups dispersed through the cloud; Ophiuchus and L1641 (Orion A) are forming most of their stars in a single dense cluster, with other regions of more isolated star formation scattered through the cloud. However, L1630 (Orion B) is following yet another pattern, with most star formation occurring in 3 – 4 clusters, which are spread through the cloud.

Lada (1990) and Lada, Strom and Myers (1991) have suggested that Taurus and L1630 represent two distinct modes of star formation : isolated and clustered star formation. It is also possible that they represent extremes in a continuum of star formation styles, ranging from individualistic (Taurus) to gregarious (L1630). To understand the origin of these different modes or styles, we will need more complete studies of the physical conditions in these three clouds. For example, complete CS maps of Taurus-Auriga and Ophiuchus, comparable to those obtained for L1630 (Lada, Bally, and Stark 1991) would reveal whether the differences in star formation patterns can be traced to differences in the distribution and nature of the molecular gas. Even without these maps, there is evidence in all three regions that star formation requires fairly dense gas ($n > 10^4$ cm^{-3}); a substantial mass of such dense gas ($M > 200$ $M_\odot$) seems necessary for forming a cluster, but perhaps not sufficient, since two massive cores in L1630 are not yet forming substantial clusters.

Despite the differences in the distribution of star formation in these three complexes, the net result seems surprisingly similar. Each complex seems to have turned about 1% of its total mass into stars so far, resulting in a global efficiency which is very similar to those deduced from studies of fully formed open clusters and associations. Furthermore, the initial mass functions of open clusters and field stars, when used to predict the luminosity functions of embedded infrared sources, produce reasonable consistent results, within the substantial uncertainties. It may be that the vastly different stellar content of Orion (many O and B stars) and Taurus (mostly low mass stars) can be explained simply by the larger mass (and thus greater chance of forming massive stars) of the Orion complex.

Detailed studies of individual star formation regions in the three clouds reveal some interesting differences. While the density distribution around L1551 IRS 5 (in Taurus) agrees very well with theoretical models, the opacity is clearly higher in the nominally similar source, IRAS16293-2422 (in Ophiuchus) than would be expected from theoretical models or from comparison to Taurus. In the Orion complex, the NGC2071 region clearly has a higher density than does the gas around L1551 IRS 5; it is also higher than predicted from theory. This may be related to the fact that the forming star is more massive and that it is forming in the environment of one of the L1630 clusters. Similar studies of other regions in these three clouds are needed to distinguish which of these differences arise from differences between the clouds and which are peculiar to these three sources.

Acknowledgements

We thank the members of the Organizing Committee for their invitation to present this paper. We also thank Charles Lada, Lee Mundy and Harold Butner for useful comments and discussions. Portions of the research described here were supported in part by NSF Grant AST 88-15801, NASA grant NAG 2-420, NASA Training Grant NGT 50320 and by grants from the W. M. Keck Foundation and Texas Advanced Research Program to the University of Texas at Austin.

References

Adams, F. C., Lada, C. J., and Shu, F. H. 1987, *Ap.J.*, **312**, 788.

Adams, F. C., and Shu, F. H. 1985, *Ap. J.*, **296**, 655.

Adams, F. C., and Shu, F. H. 1986, *Ap. J.*, **308**, 836.

Anthony-Twarog, B. J. 1982, *Astron. J.*, **87**, 1213.

Bally, J., Langer, W. D., Wilson and R. W., Stark, A. A. 1987, *Ap. J. (Letters)*, **312**, L45.

Barsony, M., Burton, M. G., Russell, A. P. G., Carlstrom, J. E., and Garden, R. 1989, *Ap. J. (Letters)*, **346**, L93.

Beckwith, S. V. W., Sargent, A. I., Koresko, C. D., and Weintraub, D. A. 1989, *Ap. J.*, **343**, 393.

Beckwith, S., Zuckerman, B., Skrutskie, M. F., and Dyck, H. M. 1984, *Ap. J.*, **287**, 793.

Beichman, C. A., Myers. P. C., Emerson, J. P., Harris, S., Mathieu, R. D., Benson, P. J. and Jennings, R. E. 1986, *Ap. J.*, **307**, 337.

Benson, P. J. and Myers, P. C. 1989, *Ap. J. Suppl.*, **71**, 89.

Bertiau, F. C. 1958, *Ap. J.*, **128**, 533.

Butner, H. M., Evans, N. J., II, Harvey, P. M., Mundy, L. G. and Natta, A. 1990, *Ap. J.*, **364**, 164.

Butner, H. M., Evans, N. J., II, Lester, D. F., Levreault, R. M. and Strom, S. E. 1991a, *Ap.J.*, in press.

Butner, H. M., Evans, N. J., II, Lester, D. F., Harvey, P. M., Mundy, L. G., and Campbell, M. F. 1991b, *Ap.J.*, in preparation.

Butner, H. M., Natta, A., and Evans, N. J., II 1991, *Ap. J.*, in preparation.

Carr, J. S., Harvey, P. M., and Lester, D. F. 1987, *Ap. J. (Letters)*, **321**, L71.

Campbell, B., Persson, S. E., Strom, S. E., and Grasdalen, G. 1988, *Astron. J.*, **95**, 1173.

Cernicharo, J. Bachiller, R., and Duvert, G. 1985, *Astr. Ap. Suppl.*, **149**, 273.

Chini, R. 1981, *Astr. Ap.*, **99**, 346.

Cohen, M., and Kuhi, L. V. 1979, *Ap. J. Suppl.*, **41**, 743.

de Geus, E. J., de Zeeuw, P. T., and Lub, J. 1989, *Astr. Ap.*, **216**, 44.

de Geus, E., Bronfman, L., and Thaddeus, P. 1990, *Astr. Ap.*, in press.

DePoy, D., Lada, E. A., Gatley, I., and Probst, R. 1990, *Ap. J.*, in press.

Draine, B. T., and Lee, H. M. 1984, *Ap. J.*, **285**, 89.

Dolidze, M. V., and Arakelyan, M. A. 1959, *Soviet Astr. - AJ*, **3**, 434.

Duerr, R., Imhoff, C. L. and Lada, C. J. 1982, *Ap. J.*, **261**, 135.

Elias, J. H. 1978a, *Ap. J.*, **224**, 857.

Elias, J. H. 1978b, *Ap. J.*, **224**, 453.

Elmegreen, B. G. 1983, *Mon. Not. Roy. Astr. Soc.*, **203**, 1011.

Evans, N. J., II 1991, in *Frontiers of Stellar Evolution*, ed. D. L. Lambert (San Francisco : Astronomical Society of the Pacific), in press.

Fuller, G. A. and Myers, P. C. 1987, in *Physical Processes in Interstellar Clouds*, ed. G. E. Morfill and M. Scholer (Dordrecht : D. Reidel), p.137.

Genzel, R. Reid, M. J., Moran, J. M., and Downes, D. 1981, *Ap. J.*, **244**, 884.

Genzel, R., and Stutzki, J. 1989, *Ann. Rev. Astr. Ap.*, **27**, 11.

Grasdalen, G. L., Strom, K. M., and Strom, S. E. 1973, *Ap. J. (Letters)*, **184**, L53.

Harju, J., Walmsley, C. M. and Wouterloot, J. G. A. 1990, *Astron. & Astro.*, submitted.

Hartmann, L., and Kenyon, S. J. 1985, *Ap. J.*, **299**, 462.

Herbig, G. H., and Rao, N. K. 1972, *Ap. J.*, **174**, 401.

Hodapp, K., Capps, R. W., Strom, S. E., Salas, L., and Grasdalen, G. 1988, *Ap. J.*, **355**, 814.

Haro, G. 1949, *A. J.*, **54**, 188.

Ichikawa, T., and Nishida, M. 1989, *A. J.*, **97**, 1074.

Kawabe, R., Kitamura, Y., Ishiguro, M., Hasegawa, T., Chikada, Y., and Okamura, S. K. 1989, in *Structure and Dynamics of the Interstellar Medium, IAU Colloquium 120*, ed. G. Tenorio-Tagle, M. Moles, and J. Melnick (Heidelberg : Springer-Verlag), p.254.

Keene, J., and Masson, C. R. 1990, *Ap. J.*, **355**, 635.

Kenyon, S. J., Hartmann, L. W., Strom, K., M. and Strom, S. E. 1990, *A. J.*, **99**, 869.

Kleiner, S. C. and Dickman, R. L. 1984, *Ap. J.*, **286**, 255.

Kleiner, S. C. and Dickman, R. L. 1985, *Ap. J.*, **295**, 466.

Klose, S. 1986, *Ap. Space Sci.*, **128**, 135.

Koresko, C. D., Beckwith, S. V. W., and Sargent, A. I. 1989, *Astron. J.*, **98**, 1394.

Lada, C. J. and Wilking, B. A. 1984, *Ap. J.*, **287**, 610.

Lada, E. A. 1990, Ph.D. thesis, University of Texas at Austin.

Lada, E. A., DePoy, D., Evans, N. J., II, and Gatley, I. 1991, *Ap. J.*, in press.

Lada, E. A., Bally, J. and Stark, A. A. 1991, *Ap. J.*, in press, Feb. 20.

Lada, E. A., Strom, K. M. and Myers, P. C. 1991, in *Protostars and Planets III*, ed. E. H. Levy and J. Lunine (Tucson : University of Arizona Press), in press.

Larson, R. B. 1969, *Mon. Not. Roy. Astr. Soc.*, **145**, 271.

Larson, R. B. 1982, *Mon. Not. Roy. Astr. Soc.*, **200**, 159.

Larson, R. B. 1986, in *Stellar Populations*, ed. C. A. Norman, A. Renzini, and M. Tosi (Cambridge : Cambridge University Press), p.101.

Leinert, Ch., and Haas, M. 1987, *Astr. Ap. (Letters)*, **182**, L47.

Leisawitz, D., Bash, F. N., and Thaddeus, P. 1989, *Ap. J. Suppl.*, **70**, 731.

Loren, R. B. 1989a, *Ap. J.*, **338**, 902.

Loren, R. B. 1989b, *Ap. J.*, **338**, 925.

Loren, R. B., Sandqvist, A., and Wootten, H. A. 1983, *Ap.J.*, **270**, 620.

Loren, R. B., Wooten, H. A., and Wilking, B. A. 1990, *Ap. J*, submitted.

Maddalena, R. J., Morris, M., Moscowitz, J., and Thaddeus, P. 1986, *Ap. J.*, **303**, 375.

Mathis, J. S., Mezger, P. G., and Panagia, N. 1983, *Astr. Ap.*, **128**, 212.

McCaughrean, M., Zinnecker, H., Aspin, C., McLean, I. 1990, in *Astrophysics with Infrared Arrays* , ed. R. Elston (Tucson : University of Arizona Press), in press.

Menten, K. M., Serabyn, E., Güsten, R., and Wilson, T. L. 1987, *Astr. Ap. (Letters)*, **177**, L57.

Mundt, R., Stocke, J., Strom, S. E., Strom, K. M., and Anderson, E. R. 1985, *Ap. J. (Letters)*, **297**, L41.

Mundy, L. G., Wilking, B. A., and Myers, S. T. 1986, *Ap. J. (Letters)*, **311**, L75.

Mundy, L. G., Wootten, H. A., and Wilking, B. A. 1990, *Ap. J.*, **352**, 159.

Myers, P. C. and Benson, P. J. 1983, *Ap. J.*, **266**, 309.

Myers, P. C., Heyer, M., Snell, R. L., and Goldsmith, P. F. 1988, *Ap. J.*, **324**, 907.

Myers, P. C., Linke, R. A., and Benson, P. J. 1983, *Ap. J.*, **266**, 309.

Penston, M. V. 1969, *Mon. Not. Roy. Astr. Soc.*, **144**, 425.

Persson, S. E., Geballe, T. R., Simon, T., Lonsdale, C. J. and Baas, F. 1981, *Ap. J. (Letters)*, **251**, L85.

Rieke, G. H., Ashok, N. H., and Boyle, R. P. 1989, *Ap. J. (Letters)*, **339**, L71.

Sargent, A. I., Beckwith, S., Keene, J., and Masson, C. 1988, *Ap. J.*, **333**, 936.

Scalo, J. 1986, *Fundam. Cosmic Phys.*, **11**, 1.

Scalo, J. 1990, in *Physical Processes in Fragmentation and Star Formation*, ed. R. Capuzzo-Dolcetta, C. Chiosi, and A. Di Fazio (Netherlands : Kluwer Academic publishers), p.151-177.

Shu, F. H. 1977, *Ap.J.*, **214**, 488.

Shu, F. H. 1991, in *Frontiers of Stellar Evolution*, ed. D. L. Lambert (San Francisco : Astronomical Society of the Pacific), in press.

Shu, F. H., Adams, F. C., and Lizano, S. 1987, *Ann. Rev. Astr. Ap.*, **25**, 23.

Snell, R. L. 1981, *Ap.J. Suppl.*, **45**, 121.

Snell, R. L., Loren, R. B., and Plambeck, R. L. 1980, *Ap. J. (Letters)*, **239**, L17.

Strom, K. M., Margulis, M., and Strom, S. E. 1989a, *Ap. J. (Letters)*, **345**, L79.

Strom, K. M., Margulis, M., and Strom, S. E. 1989b, *Ap. J. (Letters)*, **346**, L33.

Strom, K. M., Newton, G., Strom, S. E., Seaman, R. L., Carrasco, L., Cruz-Gonzalez, I., Serrano, A., and Grasdalen, G. L. 1989, *Ap. J. Suppl.*, **71**, 183.

Struve, O. and Rudkjobing, M. 1949, *Ap. J.*, **109**, 92.

Takano, T., Stutzki, J., Fukui, Y., and Winnewisser, G. 1986, *Astr. Ap.*, **167**, 333.

Terebey, S., Shu, F. H., and Cassen, P. 1984, *Ap. J.*, **284**, 529.

Ungerechts, H. and Thaddeus, P. 1987, *Ap. J. Suppl.*, **63**, 645.

Vrba, F. J., Strom, K. M., Strom, S. E., and Grasdalen, G. L. 1975, *Ap. J.*, **197**, 77.

Walker, C. K., Lada, C. J., Young, E. T., Maloney, P. R., and Wilking, B. A. 1986, *Ap. J. (Letters)*, **309**, L47.

Walker, C. K., Lada, C. J., Young, E. T., and Margulis, M. 1988, *Ap. J.*, **332**, 335.

Whittet, D. C. B. 1974, *Mon. Not. Roy. Astr. Soc.*, **168**, 371.

Wilking, B. A. 1990, *Low Mass Star Formation in Southern Molecular Clouds*, in press.

Wilking, B. A. and Lada, C. J. 1983, *Ap.J.*, **274**, 698.

Wilking, B. A., Lada, C. J. and Young, E. T. 1989, *Ap. J.*, **340**, 823.

Wilking, B. A., Schwartz, R. D., and Blackwell, J. H. 1987, *Astron. J.*, **94**, 106.

Wootten A. 1989, *Ap. J.*, **337**, 858.

Wootten, A., and Loren, R. B. 1987, *Ap. J.*, **317**, 220.

Wouterloot, J. G. A., and Habing, H. J. 1985, *Astr. Ap. Suppl*, **60**, 43.

Wouterloot, J. G. A., Walmsley, C. M., and Henkel, C. 1988, *Astr. Ap*, **191**, 323.

Wynn-Williams, C. G. 1982, *Ann. Rev. Astr. Ap.*, **20**, 597.

Young, E. T., Lada, C. J., and Wilking, B. A. 1986, *Ap. J.*, **340**, 823.

Zhou, S., Evans, N. J. II, Butner, H. M., Kutner, M. L., Leung, C. M., and Mundy, L. G. 1990a, *Ap.J.*, **363**, 168.

Zhou, S., Evans, N. J., II, and Mundy, L. G. 1990, *Ap. J.*, **355**, 159.

Zhou, S., Evans, N. J., II, Mundy, L. G., Güsten, R., and Kutner, M. L. 1991, *Ap. J.*, submitted.

Zhou, S., Wu, Y., Evans, N. J., II, Fuller, G. A., and Myers, P. C. 1989, *Ap. J.*, **346**, 168.

WAVE DYNAMICS AND STAR FORMATION IN TAURUS

Ralph E. Pudritz and Ana I. Gomez de Castro
Dept. of Physics, McMaster University, Hamilton, ON L8S 4M1

1. Introduction

The mechanism underlying the formation of cores and larger scale structures in molecular clouds must play a fundamental role in the physics of star formation since young stellar objects are usually found within or very near cores (Myers et al 1987, Beichman et al 1986). The Taurus cloud is an ideal object to study in this regard because of its proximity (160 pc), and because only low mass star formation is presently occurring there. Barnard's (1927) beautiful optical photograph of the region reveals that the obscuring gas and dust has filamentary structure that is comparable to the size of the cloud complex (several 10's of pc). This structure is clearly seen in CO maps of the region as well (eg. Duvert et al 1986) where it is apparent that structure on much larger size scales than cores is common. In addition to the filamentary structure one also observes that there are small dark clouds present such as L1489, L1495, etc.

The CO clumps in the Taurus cloud and others appear to be like raisins embedded within the stringy filaments. The dense NH_3 cores mapped by Benson and Myers (1989) however, are concentrated in the dark clouds (such as L1495). It is standard procedure to characterize the mass distribution of clumps and cores in molecular clouds by fitting the number of clumps per unit mass range with a power law spectrum; $dN(m)/dm \propto m^{-\alpha}$. We have done this for the collection of the 14 NH_3 cores in Taurus that were studied by Benson and Myers (1989) and find that the core mass spectrum has the value $\alpha \simeq 1.5$. This is somewhat different value than has been determined for ρ Oph where $\alpha \simeq 1.1$ (Loren 1989) but similar to that for the Rosette nebula $\alpha \simeq 1.5$ (Blitz 1989). It is presently unclear whether or not these differences in core mass spectra from cloud to cloud reflect experimental uncertainty or real differences in the physics or evolutionary states of the different well studied clouds. We present simulations which indicate that the structure and mass spectra of clouds is expected to evolve rather strongly during the life of a molecular cloud.

In accounting for the large scale features of the Taurus clouds we can rule out the damaging effects of low mass star formation within the cloud (such as the production of cavities and shells blown by the bipolar outflows) since such regions would not exceed a few pc. is size. It is therefore likely that the large scale structure in Taurus was formed by global processes involving the whole cloud. An important point in this regard is that the Taurus region appears to be associated with a bubble in the HI survey of Heiles and collaborators (eg. Kulkarni and Heiles,

1988). Expanding shells are likely to break up into filaments and this may be a factor in understanding the Taurus cloud. There are several other important clues about the origin of global structure in Taurus. The HI study of Shuter et al (1987) shows that the velocity field of the region as determined by 21 cm observations (in absorption) has a wavelike character with a wavelength of order 10 pc and a mean velocity of 2.7 km s^{-1}. The magnetic field in Taurus as revealed by optical polarimetry (Moneti et al 1984, Heyer et al 1987) is also ordered on the scales of 10's of pc. In fact there is a distinct "waviness" to the field. Since the magnetic field is comparable to gravitational energy densities (see eg Heiles, these proceedings, Myers and Goodman 1988) the observational evidence supports the idea that hydromagnetic waves whose wavelength is comparable to cloud size, may in fact be playing an important role in global cloud physics.

What would be explained if such waves (either nonlinear Alfvén waves or even linear fast magnetosonic waves) were present? There are at least four important aspects of cloud dynamics and structure of interest:

(i) long wavelength transverse waves can in principle stir up the largest scales in the cloud and thereby provide support against rapid global collapse (Arons and Max 1975),

(ii) scales intermediate between the Jeans length in the cloud and the cloud scale can still collapse for plausible wave spectra (Bonazzola et al 1987, Pudritz 1990),

(iii) non-linear Alfvén waves and even linear fast magneto-sonic modes can produce density fluctuations, which in principle could be identified with structure observed on supra Jeans scales (Carlberg and Pudritz 1990),

(iv) large scale waves with periods of several million years would constantly change the apparent alignment of the cloud magnetic field direction with respect to the major axis of the filaments thereby accounting for the fact that these alignments can be found to vary from filament to filament in Taurus, and indeed other molecular clouds.

The last point on this list is rather important in that the *lack of alignment* between field and filament directions has been interpreted by some authors to imply that the magnetic field on these scales is not dynamically important for cloud physics (eg Heyer 1988), the implication being that ambipolar diffusion has weakened the field substantially by time densities of 10^4 cm^{-3} have been achieved. However, since waves on the scales of cores must damp while on larger scales they may propagate for nearly the life of the cloud, one *expects misalignments in clouds stirred by hydromagnetic waves!*

In this contribution, we demonstrate using linear instability analysis that wave spectra strongly affect the stability of molecular clouds. We then present recent numerical simulations on the effect that a single long wavelength mode has upon the formation of large scale structure in a cloud that is supported against collapse by a subcritical, isotropic magnetic field.

2. Stability of Clouds Traversed by Hydromagnetic Waves

The physics of wave propagation in molecular clouds has been recently extensively discussed (eg. Pudritz 1990, Carlberg and Pudritz 1990, henceforth CP) and only the main points will be reviewed here. An essential point is that not all physical scales in a molecular cloud can be supported by the pressure of hydromagnetic waves. Waves of short enough wavelength damp by ion-neutral friction. Magnetic force is only communicated to the neutrals by the intermediary of ion-neutral collisions. The neutrals will suffer the effect of magnetic forces only if collisions with ions are sufficiently frequent. If we imagine ions to be "riding" an Alfvén wave, then if a given neutral does not collide with any ion in a wave period, large velocity differences between the ions and the neutrals are established so that friction between the ion and neutral "fluids" are large (the latter depends upon the relative velocity between the two components). Large friction always means wave damping and so one finds the critical condition (Braginskii 1965, Kulsrud and Pearce 1969) that Alfvén waves with angular velocities larger than $2\omega_{n,i}$ damp completely where $\omega_{n,i} = 0.50 \times 10^{-14}(n_3\zeta_{-17})^{1/2}$ s^{-1} is the frequency with which a given neutral collides with any ion. The associated damping length λ_{min} is such that all waves with $\lambda < \lambda_{min}$ completely damp, while larger wavelength waves can propagate suffering some damping as they go. The ratio of this damping length to the non-magnetic Jeans length in the gas is

$$\frac{\lambda_{min}}{\lambda_J} = 0.62\frac{V_o}{5c_s}\frac{B}{B_c}\zeta_{-17}^{-1/2} \tag{1.}$$

where the virial speed V_o in a molecular cloud is typically five times the sound speed, B_c is the critical magnetic field strength in a molecular cloud (about 30 μG) that could support the cloud against gravitational collapse, and where ζ_{-17} is the ionization rate per hydrogen atom measured in units of 10^{-17} s^{-1} per hydrogen atom. Thus for clouds that are magnetically subcritical, the damping length for waves is somewhat larger that the Jeans length and of order the observed size of individual cores seen in molecular clouds. For supercritical clouds, the damping scale is less than the Jeans length so effectively all supra Jeans scales could be stirred (but not supported) by hydromagnetic waves. The damping scale is roughly the size of cores observed in molecular clouds, and CP proposed that a likely reason why cores form is that wave pressure is unavailable to support gas on scales $\lambda_J < \lambda < \lambda_{min}$. Thus magnetically subcritical clouds start to produce cores on a scale determined by the magnetic damping length.

The time scale for the damping of linear waves is

$$\tau_D = 3.2 \times 10^6 \frac{1}{(n_3\zeta_{-17})^{1/2}} \left(\frac{\lambda}{\lambda_{min}}\right)^2 yr \tag{2.}$$

which is very sensitive to the wavelength. Applying equation (2.) to the scales characterizing cores ($n_3 = 10$, $\lambda \simeq \lambda_{min}$)), we find that wavelengths of order the core size damp in 1.0×10^6 yr whereas long wavelength waves of order the size of the cloud $\lambda \simeq 10\lambda_{min}$ damp in 3.2×10^8 yr. Thus, while the longest modes in a cloud "ring" for the life of the cloud, modes on scales of cores damp in about a cloud free-fall time (a million years). The energy losses on small scales do not pose a great problem for maitaining waves in the cloud since the power dissipated on the scales of cores represents only 1% of the power in the wave field at the longest wavelengths (most of the power must be found at the largest scales if global gravitational collapse is to be staved off). The wave power dissipated on the scale of cores in fact heats such regions to temperatures of 10-20 K (CP). Thus, both the formation of cores, and their heating are natural consequences of wave damping in molecular clouds.

We now turn to consider the question of the stability of molecular clouds which are traversed by a spectrum of MHD waves (see Pudritz 1990). Bonazzola et al (1987) were the first to point out that a spectrum of waves with enough power on the largest scales could stir up the gas sufficiently to prevent global gravitational collapse of the cloud. While these authors based their argument on the virial theorem, one can derive the results in a straightforward fashion from the equations of motion of a cloud that is traversed by slowly varying (spatially and temporally) waves. In such a situation, it is well known that the background wave field contributes an additional stress on the fluid, $-\nabla.\mathbf{P_w}$ where $\mathbf{P_w}$ is the wave pressure tensor. Introducing this into the equations of motion describing a molecular cloud, and linearizing, one derives the dispersion relation for waves in warm, partially ionized, self gravitating, magnetized gas stirred by a background wave field,

$$\omega_\pm = \frac{i}{2}\left(-\frac{\omega_{A,n}^2}{\omega_{n,i}} \pm [(\omega_{A,n}^4/\omega_{n,i}^2) + 4\Omega_n^2]^{1/2}\right) \qquad (3.)$$

where $\omega_{A,n}$ is the angular velocity of Alfvén waves in the neutral gas. In this expression, the Jeans function $\Omega_n^2 \equiv \omega_g^2 - k^2[V_A^2 + c_s^2 + \delta v_A(k)^2]$ shows that gravitational instability (the gravitational response frequency is $\omega_g \equiv (4\pi G\rho_n)^{1/2}$) is offset by the mean magnetic field (whose associated Alfvén speed is V_A), gas pressure (c_s is the sound speed in the cloud), and wave pressure (the amplitude of the Alfvén velocity fluctuations due to a spectrum of waves is assumed to obey a power law, $\delta v_A(k) = V_A(k_o/k)^\beta$). The dispersion relation (3.) reduces, in the absence of gravity, gas pressure, and background MHD waves, to that first discovered by Kulsrud and Pearce (1969) for the propagation of Alfvén waves in partially ionized gas. Gravitational instability sets in for any scale for which $\Omega_n^2 > 0$ exactly as the Jeans analysis would suggest. One difference with the traditional Jeans result is that the growth time for the unstable modes is strongly modified by the rate of ambipolar diffusion of the magnetic field out of the gas.

The major consequence of having a spectrum of waves stabilize the cloud is that the bandwidth of stable scales depends upon the index β of the wave spectrum. One readily finds from (3.) that for wave spectra with $\beta > 1$, the largest scales in the cloud are stabilitzed against collapse. Figure 1 is a plot of the growth time scale of an unstable mode as a function of the mode wavelength in the case that the energy density in the waves equals that of gravity ($a \equiv \omega_g^2/k_o^2 V_A^2 = 1$). The scale λ_o is taken to be the size of the molecular cloud and the instability time scale is in units of $\omega_g^{-1} = 0.72 \times 10^6 n_3^{-1/2}$ yr. As one increases the spectral index β, there is a steeper fall off in the wave energy as one goes to smaller scales. This in turn implies that there is an increasingly larger band width of scale lengths that are without support by wave pressure, and hence gravitationally unstable.

3. Nonlinear Effects: Simulations of Single Alfvén Modes

The effect of a spectrum of large amplitude Alfvén modes on a molecular gas is strikingly reminiscent of structure observed in Taurus. Simulation of CP showed that small scale fragmentation of the cloud was later followed by the appearance of larger scale, filamentary structure. The mass spectrum of clumps in the CP simulations constantly evolved to smaller values of α, the reason being that significant agglomeration of smaller lumps into larger entities occurs. The clumps that appeared reasonably obeyed Larson's relations.

In order to more fully understand the effect that finite amplitude Alfvén waves can have on molecular clouds, we have completed a number of numerical

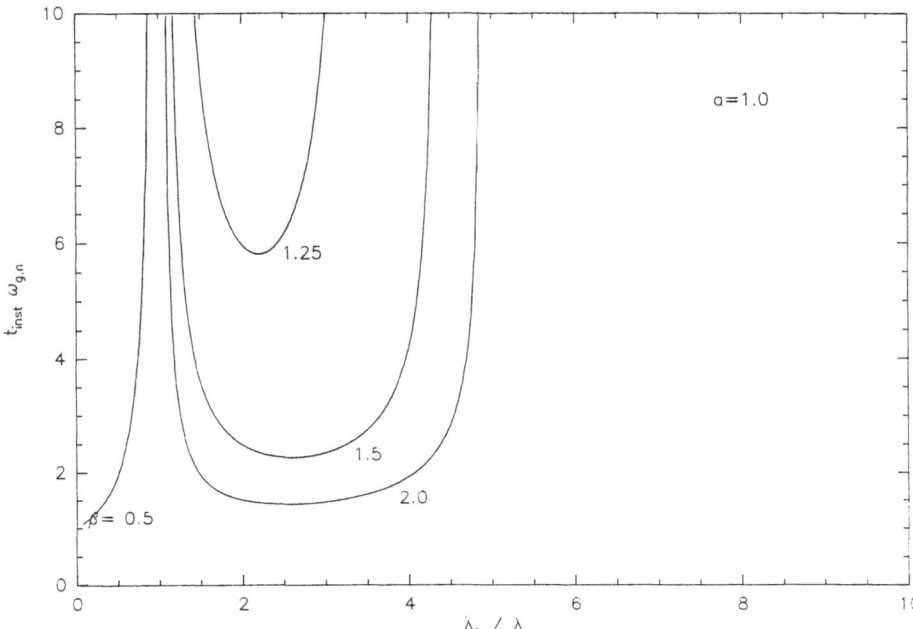

Figure 1. Growth times of modes in clouds stirred by power law wave spectra; the case of $a = 1$ (see text); adapted from Pudritz (1990).

experiments in which the evolution of a cold molecular cloud supported by strong isotropic field is compared with an identical model which has a single, long wavelength, Alfvén wave propagating along the z-direction in the model as an additional ingredient. We employed the 3-D, N-body, gas dynamics code ("sticky particles") described in CP and Pudritz and Carlberg (1989) to which we refer the reader for the details.

In Figure 2, we show a comparison of two simulations, one without any waves and one with a single Alfvén mode whose energy was only 25% of the gravitational energy density. The Jeans length in this cold cloud calculation is $\lambda_J/R = 0.081$ while the wavelength of the single mode we studied is $\lambda/R = 0.33$, where R is the initial radius of the cloud. In all calculations, the initial state of the cloud is a uniform sphere. Slow contraction of the cloud occurs because of ambipolar diffusion. Our models evolve quasi-statically because our initial state is strongly subcritical. The Figure shows quite plainly that large scale structure is much more apparent in the single long mode calculation than in the mode-free model. Evidently, the large scale wave has been resposible for gathering together many small fragments into larger, filamentary entities.

This impression is made precise in Figure 3 which compares the clump mass spectra in the no-mode versus long-mode simulations. The figure shows histograms

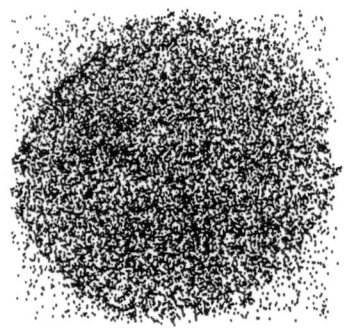 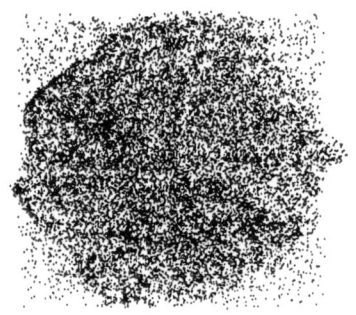

Figure 2. Comparison of simulations of magnetically supported, cold molecular clouds with (a.) no Alfvén wave and (b.) with a single long wavelength mode. These figures show the two respective simulations after $\simeq 4$ free-fall times. The nonmagnetic Jeans length in the cloud is tiny (0.081 of the cloud's original radius).

of the clump masses at different times (a cloud free-fall time is 0.4 time units). In both cases, the cloud breaks up into fragments whose mass is nearly the Jeans mass, corresponding to the strong peak at mass $m = 3.3 \times 10^{-4} M_c$. As time goes by, we see that power appears at the larger mass scales at the expense of power on the smallest scale; ie, agglomeration is occurring. However, agglomeration is not the whole picture. By the end of the calculation, the long mode calculation has significantly more power on the largest scales than does the no-mode simulation. Clearly the long wavelength wave has swept together gas and created significantly larger scale structure than a model in which long wavelengths are absent. This illustrates the basic point namely, that waves generate structure.

In order to give better insight into how larger scale structure develops, we show in Figure 4 the spatial distribution of clumps with a minimum overdensity of 30 over the mean cloud density at two different times, namely at 3 and 3.5 cloud free-fall times. We project these clump positions onto the three different planes x-z, x-y, and y-z. One sees the very striking amount of agglomeration that is occurring between these two time steps. It is apparent that while a filamentary pattern is difficult to discern at $t = 1.2 = 3t_{ff}$ it is clearly emerging a half free-fall time later. The reason for the delay in the formation of the larger scale structure is also clear; the wave period of an Alfvén wave is directly proportional to its wavelength, so larger scale filamentation should start in at about 4 free-fall times for our particular choice of a mode.

Finally, we present a log-log plot of the number of clumps $N(m)$ vs the mass of the clump as shown in Figure 5. This again illustrates the trend of spectral evolution of the mass spectrum of clumps, with a new feature. One notes that

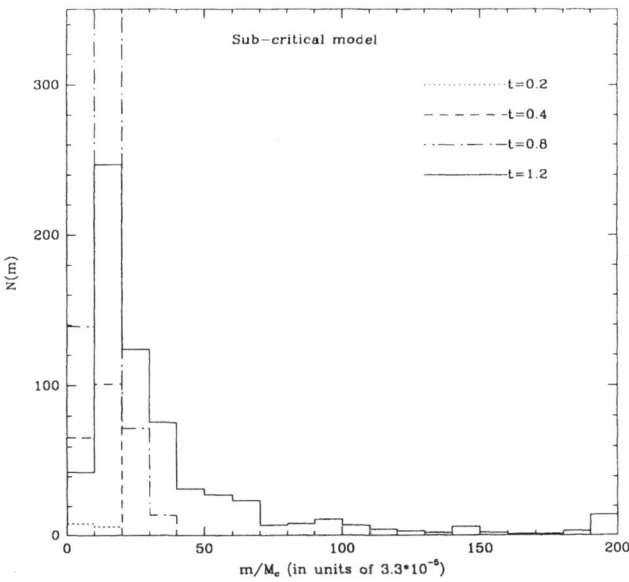

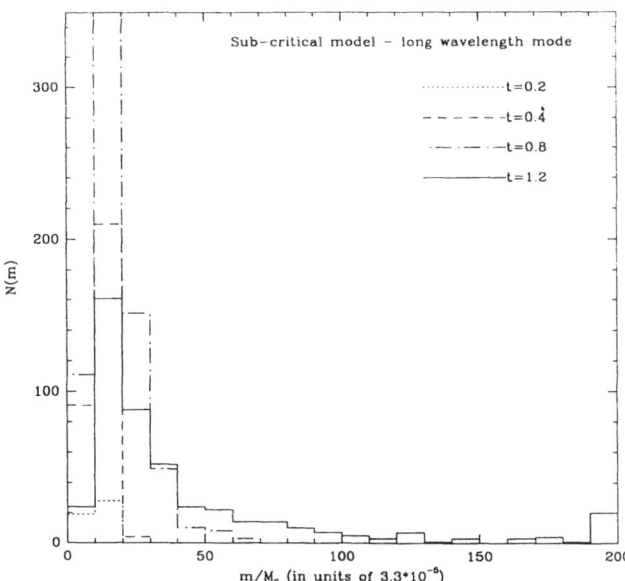

Figure 3. Mass spectra of no mode versus single long wavelength modes in magnetically subcritical cloud. One free fall time is 0.4 time units.

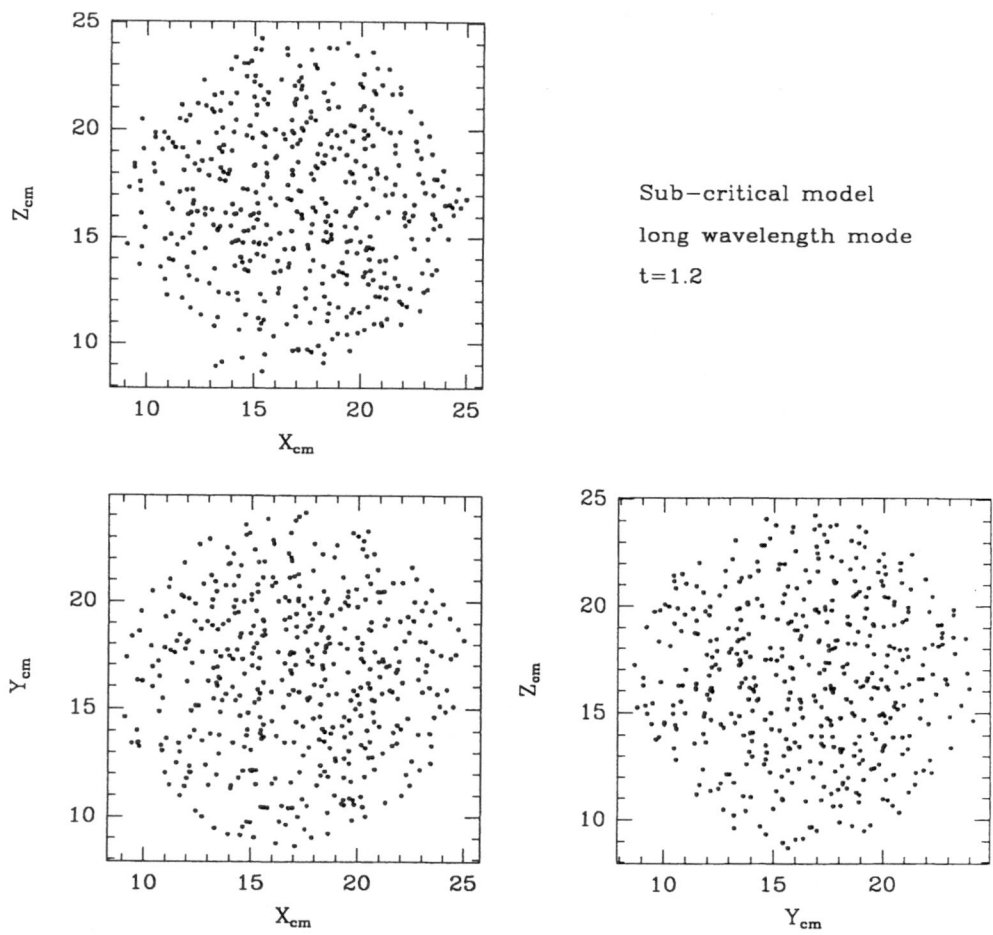

Figure 4a. Evolution of structure in long wavelength mode simulation; projected spatial distribution of clumps with minimum overdensity of 30 at 3 free-fall times.

the spectrum appears to turn over at the small mass end. While something like this has been seen observationally, we note that it is an expected feature of clouds which first fragment into small pieces. We also see that the slope of the log-log plot becomes ever shallower. We have not yet found a model which reaches steady state. This suggests that the spectral index constantly evolves in real clouds, so that cloud to cloud differences may be a real effect, possibly reflecting cloud *age*.

4. Origin and Evolution of Cloud Structure

Given the evidence that there may be long wave-length waves sloshing around inside the Taurus cloud, what may have been the likely origin for such phenomena? Of course there are many possibilities; any agent that shakes a molecular cloud will

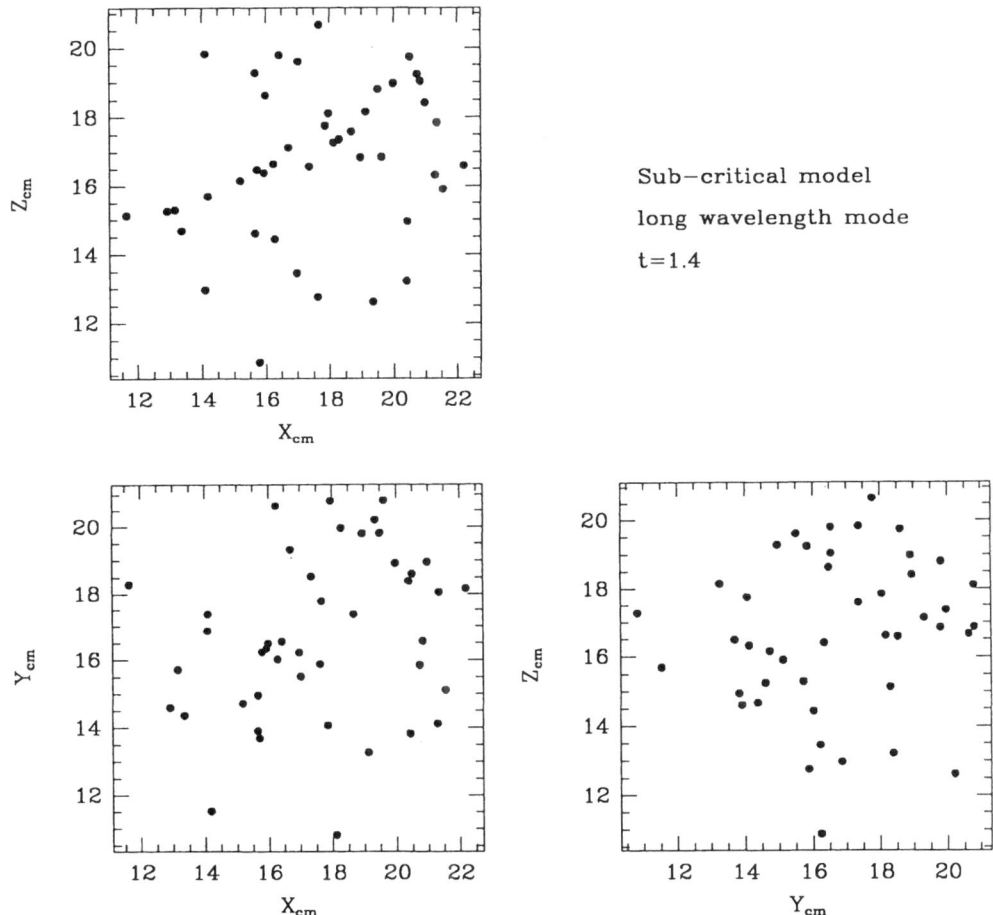

Sub−critical model

long wavelength mode

t=1.4

Figure 4b. Same as Figure 4a except at 3.5 free-fall times.

set off a train of Alfvén waves. However, given the long-livedness of the largest wavelengths, it is entirely possible that large scale wave power has something to do with the formation mechanism of a molecular cloud. We propose that in fact such modes are produced during the process of cloud formation by the Parker-Jeans instability (see eg Blitz and Shu 1979, and Elmegreen 1982). The nonlinear development of the Parker instability consists of gas flowing down the field lines back to the galactic plane, and there colliding with the gas in the magnetic valley. This must lead to oscillations and shocks, as recent numerical experiments have confirmed (eg. Matsumoto et al 1989). Thus the very act of forming a molecular cloud is likely to produce sufficient large scale wave power to keep it supported against subsequent gravitational collapse (Gomez de Castro and Pudritz, in preparation).

Thus, we arrive at the view that it is highly likely that molecular cloud structure evolves rather strongly over the life of a cloud, and that hydromagnetic waves

326

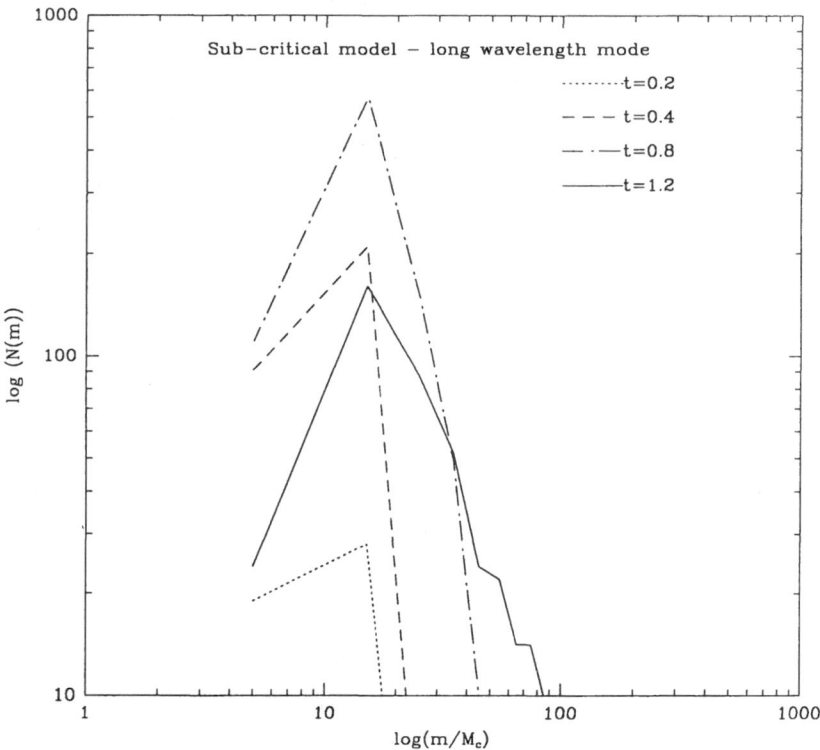

Figure 5. Log-log plot of clump mass spectrum of long-mode simulation.

play an essential role in cloud physics. A sequence of events that summarizes this process is presented below:

1. A Jeans-Parker instability triggers the formation of a molecular cloud or cloud complex; long wavelength modes generated in this process stir the newly forming clouds for their subsequent life.

2. Since molecular clouds are very cold, the smallest fragments are much below the mass of the cloud; the preliminary cloud fragmentation produces pieces of order the Jeans length λ_J or alternatively λ_{min}.

3. The larger scale modes sweep together larger scale structures several free-fall times into the life of the cloud. Agglomeration of the small pieces occurs all the while.

4. The filaments are themselves temporary structures; they end up by draining into the most massive clumps, which ultimately dominate the gravitational field of the cloud.

5. Since the agglomeration time of the clumps is as short as the free-fall time, star formation in the clumps is always disrupted by subsequent merger with other clumps. This implies that star formation does not predominantly occur in small cores scattered through the volume of the cloud, but rather within the most massive clumps that are formed the agglomeration process. Thus a cluster mode of star formation is favoured (see E. Lada's and P. Myers contributions, these

proceedings).

6. The Taurus cloud may be quite young; it may yet form a massive clump(s) in which a cluster will form. A good candidate for the site of a future cluster is within L1495.

7. The spectral index α characterizing the mass spectrum of clumps in a cloud likely evolves over the life of a cloud; its value may correlate with the age of a cloud.

We are indebted to Ray Carlberg and Patricia Monger for many stimulating conversations and help in implementing some changes of our numerical code. It is a pleasure to thank Edith Falgarone and the scientific organizing committee for their invitation to speak at this interesting meeting, and for financial support given towards this end. This research was supported by NSERC of Canada.

REFERENCES

Arons, J., and Max, C.E. 1975, *Ap. J.*, **196**, L77.

Barnard, E.E. 1927, in Carnegie Inst. of Washington Publications No. 247,I

Beichman, C.A., Myers, P.C., Emerson, J.P., Harris, S., Mathieu, R., Benson, P.J., and Jennings, R.E. 1986, *Ap. J.*, **307**, 337.

Benson, P.J., and Myers, P.C. 1989, *Ap. J. Suppl.*, **71**, 89.

Blitz, L. 1987, in *Physical Processes in Interstellar Clouds*, G.E. Morfill and M. Scholer eds. (Kluwer: Dordrecht)35

Blitz, L. and Shu, F. 1980, *Ap. J.*, **238**, 148.

Bonazzola, S., Falgarone, E., Heyvaerts, J., Perault, M., and Puget, J.L. 1987, *Astr. Ap.*, **172**, 293.

Braginskii, S.I. 1965, in *Reviews of Plasma Physics, Vol 1* , M.A. Leontovich ed. (New York: Consultants Bureau)205

Carlberg, R.G., and Pudritz, R.E. 1990, *M. N. R. A. S.*, , in press.

Duvert, G., Cernicharo, J., and Baudry, A. 1986, *Astr. Ap.*, **164**, 349.

Elmegreen, B.G. 1982, *Astr. Ap.*, **253**, 655.

Heyer, M.H. 1988, *Ap. J.*, **324**, 311.

Heyer, M.H., Vrba, F.J., Snell, R.L., Schloerb, F.P., Strom, S.E., Goldsmith, P.F., and Strom, K.M. 1987, *Ap. J.*, **321**, 855.

Kulkarni, S.R., and Heiles, C. 1988, in *Galactic and Extragalactic Radio Astronomy*, G.L. Verschuur and K.I. Kellermann eds. (Springer-Verlag: New York)95

Kulsrud, R., and Pearce, W.P. 1969, *Ap. J.*, **156**, 445.

Loren, R.B. 1989, *Ap. J* , **338**, 902, 925.

Matsumoto, R, Horiuchi, T., Hanawa, T., and Shibata, K. 1989, *Pub. Astr. Soc. Japan*,

Moneti, A., Pipher, J.L., Helfer, H.L., McMillan, R.S., and Perry, M.L. 1985, *Ap. J.*, **282**, 508.

Myers, P.C., and Goodman, A. 1988, *Ap. J.*, **326**, L27.

Myers, P.C., Fuller, G.A., Mathieu, R.D., Beichman, C.A., Benson, P.J., Schild, R.E., and Emerson, J.P. 1987, *Ap. J.*, **319**, 340.

Pudritz, R.E. 1990, *Ap. J.*, **350**, 195.

Shuter, W.L., Dickman, R.L., and Klatt, C. 1987, *Ap. J.*, **322**, L103.

VII - EARLY STAGES OF STELLAR EVOLUTION

THEORETICAL AND OBSERVATIONAL ASPECTS OF
YOUNG STARS OF INTERMEDIATE MASS

FRANCESCO PALLA

Osservatorio Astrofisico di Arcetri,
L.go E. Fermi, 5, I-50125 Firenze, Italy

Abstract. An account is given of the observational and theoretical properties that characterize young stars of intermediate–mass ($2 \leq M_*/M_\odot \leq 10$), known as Herbig Ae/Be stars. The mass range of these objects is an interesting one, since it involves the occurrence of complex phenomena associated with the transition from fully convective configurations, typical of T Tauri stars, to radiatively stable objects, as in the case of massive stars. An overview of relevant observations testifying the variety of surface phenomena associated with the Herbig Ae/Be stars is presented. Recent developments in the theory of the formation on intermediate–mass protostars will also be discussed.

Keywords : Protostellar and pre–main-sequence evolution

1. Introduction

There has been a relative lack of interest in recent years in the subject of protostars and pre–main-sequence (PMS) stars of intermediate mass ($2 \leq M_*/M_\odot \leq 10$). Modern approach to the problem of star formation and PMS evolution has privileged the discussion of objects of solar and sub–solar mass, and a general, yet synthetic, account on the observational and theoretical foundations of the formation of sunlike stars has been given by Lada and Shu (1990). The reasons for the scarce attention paid to the stars under exam here are several.

Observationally, low mass stars outnumber more massive objects, a well known property epitomized by the Salpeter's law (1955) of the stellar distribution function. From the form of the initial mass function, we know that the probability of finding a star with $M_* = 5M_\odot$ is about 400 times smaller than that of the typical star, whose mass is $\approx 0.4M_\odot$ according to current estimates (e.g. Scalo 1990). In addition to the statistical argument, the discovery that the dense and cold cores of molecular clouds are predominantly the sites of current low–mass star formation, has provided to the observers a powerfool tool for the detailed study of the physical conditions in individual regions that are relatively close to the Sun, and for comparative studies on a reasonably large sample of objects. At the same time, the observations have now reached enough resolution to test some specific predictions of the theoretical models so far developed, as convincingly illustrated in the contribution by N. Evans (this volume).

Theoretically, the reason for concentrating the largest effort on modelling low–mass star formation is easily explained. Low–mass protostars follow an evolution that is *qualitatively* different from that of their more massive counterparts. From a comparison of the relevant timescales, the accretion time onto the protostellar core $t_{acc} = M_*/\dot{M}$, and the time needed to the core to reach internal equilibrium $t_{KH} = GM_*^2/R_*L_*$, we see that, unlike the former, the Kelvin–Helmoltz timescale is strongly dependent on mass. As long as $t_{acc} < t_{KH}$, the protostar keeps accreting matter without changing dramatically its internal structure; but when $t_{acc} > t_{KH}$ a transition occurs where the infall lags behind the gravitational contraction of the core : thus, the star will end its PMS phase of quasi–static contraction and join the main–sequence while still accreting. This transition is quite rapid and, due to the strong dependence of t_{KH} on mass, interests only a rather narrow range of masses between $2M_\odot$ and $3M_\odot$. As a consequence, there should be no optically visible PMS stars above these masses, since the protostellar cores have already reached the conditions for hydrogen burning in the center and should quietly evolve as main–sequence stars. These qualitative analysis was soon confirmed by the first hydrodynamical calculations of the gravitational collapse of interstellar clouds (Larson 1972), and has remained unquestioned thereafter (cf. the discussion in the review article by Shu, Adams and Lizano 1987).

As a summary of the theoretical studies at the base of the classical protostellar and PMS theories for intermediate–mass objects, two results stand out :

1 - accreting protostars join the main–sequence at a mass $M_* = 2 - 3M_\odot$ (Larson 1972; Appenzeller 1980);

2 - PMS evolutionary tracks predict that intermediate–mass stars have fully radiative configurations (Iben 1965; Ezer and Cameron 1965).

These predictions must be confronted with two well established observational results, namely :

1 - PMS stars of intermediate–mass are known to exist : following Herbig's (1960) original suggestion about their nature, they are classified as Herbig Ae/Be stars (HAEBE);

2 - these stars show the same evidence of complex surface activities observed in T Tauri stars that, on the contrary, possess convective interiors.

Clearly, the classical theoretical models are insufficient to explain these two facts. It is the aim of this contribution to present an overview of the evolution of the field, in the light of recent results that help removing the serious inconsistencies between theory and observations. It is also hoped that the reader will be convinced that the subject of young intermediate–mass stars is a highly interesting one, sharing many, if not all, of the exciting theoretical and observational properties that have made the study of low–mass star formation so fashionable.

2. Observational background

There are about 60 known stars that satisfy the criteria suggested by Herbig to belong to the class of intermediate-mass PMS objects. This number is much smaller than that predicted by Herbig from a consideration of the relative PMS contraction time to the main–sequence lifetimes : accordingly, there should be several hundred contracting stars of large mass within 1 kpc of the Sun. The large discrepancy is partly due to the fact that a systematic survey has not been carried out yet, and also that not all of them should be actually observable or fulfill all of Herbig's spectroscopic criteria. The latter were established entirely by analogy with the T Tausi stars, namely presence of emission lines in the spectra and association with regions of nebulosity and heavy obscuration.

The recognition that the stars proposed by Herbig (a sample of only 26 stars, originally) were indeed young stars still in the PMS phase came from the work of Strom et al. (1972), who determined the effective temperatures and surface gravities for some of them, and of Cohen and Kuhi (1979) who showed that their location in the HR diagram is well above the ZAMS. The most recent compilation of HAEBE is that of Finkenzeller and Mundt (1984) and includes 57 entries. Their catalogue has served as the basis for the detailed study of individual sources, and is slowly being expanded by new additions that mainly come from the recognition of new members selected according to their far–infrared (FIR) colours (e.g. Hu et al. 1989). Figure 1 shows the HR diagram of the HAEBE stars, together with the region occupied by T Tauri stars for comparison. The figure gives schematically the magnitude of the projected rotational velocities for the two samples, a property that will be commented upon below.

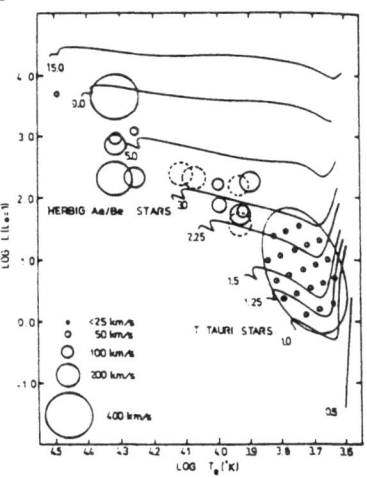

Fig. 1. Location of the HAEBE and T Tauri stars in the HR diagram, with the evolutionary tracks of Iben (1965). The magnitude of the projected rotational velocities $v \sin i$ is given by the size of the circles. (From Finkenzeller 1985).

The concentrated effort of several groups on few prototypes of the HAEBE class has resulted in the derivation of a coherent, although still incomplete, picture of their main properties. In the context of this review, it suffices to mention briefly the most important of them, but for a full account the interested reader is suggested to refer to the thorough summary given by Catala (1989). Schematically, HAEBE stars are characterized by :

- extended chromospheres and strong stellar winds, as indicated by the high resolution profiles of several spectral lines (Hα, Mg II h and k, CaII K, HeI and NaID, CIV etc.);
- significant variability and periodicity in the lines, interpreted in terms of rotational modulation (Praderie et al. 1986) or, alternatively, nonradial pulsations (Baade and Stahl 1989);
- strong UV depletion and IR excesses up to 100 μm, the latter indicative of the presence of warm circumstellar dust shells;
- molecular flows, highly collimated optical jets and Herbig–Haro objects, and maser emission;
- radio continuum emission.

The similarities of these phenomena with those commonly observed in T Tauri stars seems to indicate that star formation and early stellar evolution are all marked by the occurrence of the same phenomena over an extended mass interval. The major difference between HAEBE and T Tauri stars is in that the former rotate faster, typically with $v \sin i$ between 100 and 200 km s^{-1}, and are depleted of slow rotators (cf. Fig. 1) : the problem of the removal of the excess of angular momentum, althoug still present, is therefore less compelling than in the low–mass case.

There have been some interesting developments in the observations since Catala's review, and they will be discussed below.

2. 1. FIR EMISSION

Berrilli et al. (1990) have presented the results of a search in the IRAS–PSC for FIR counterparts of the 57 HAEBE stars listed by Finkenzeller and Mundt. 35 sources are recognized to have most probable identification and their location in the [60-25]/[25-12] colour–colour diagram is shown in panel a) of Figure 2. For comparison, panel b) shows the corresponding position of a sample of T Tauri stars. Surprisingly, the two distributions are quite similar, despite the large difference in the luminosity output of the associated stars (a factor 10^2 to 10^4). Emission from circumstellar dust shells at temperatures varying between 90 and 170 K is responsible for the colours shown in the figure. Both classes of stars have FIR colours that are quite distinct from those of sources associated with highly luminous objects (ultracompact and more evolved HII regions), indicated by the small dots in fig. 2.

Berrilli et al. present the spectral energy distribution (SED) extending from the visual to 100 μm for the 35 sources with FIR counterparts. Interestingly, the large majority of the sources (25 out of 37) have flat or very broad spectra, reminiscent of the T Tauris SEDs : the remaining 8 have spectra that decrease all the way to the longest wavelengths and that can be explained in terms of free-free emission. It would be interesting to see how the constraints set by the models developed to explain the infrared emission from T Tauri stars apply also to the HAEBE stars.

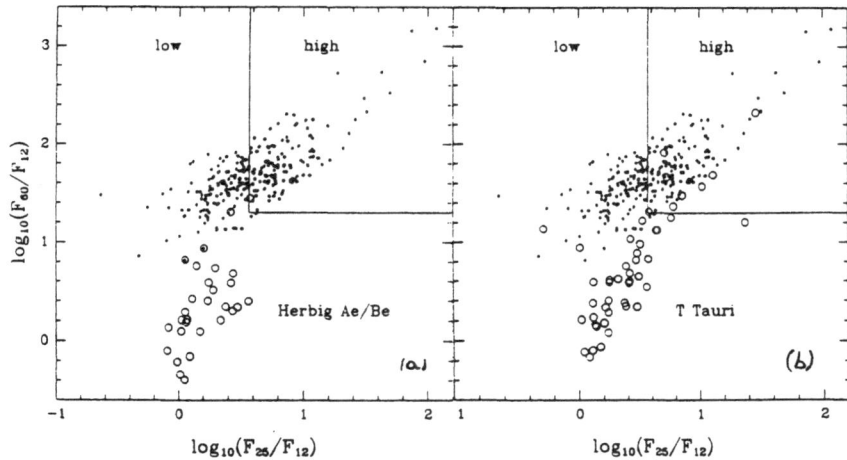

Fig. 2. Colour-colour diagrams of HAEBE (a) and T Tauri (b) stars. Small dots give the location of sources associated with early type stars. Sources within the box satisfy the colour criteria of Wood and Churchwell (1989) to discriminate ultracompact HII regions. (From Palla *et al.* 1990)

2. 2. RADIO CONTINUUM EMISSION

Searches at radio wavelengths have been performed in the hope of gaining indirect evidence of the presence of surface magnetic fields, obviously required to help explaining the chromospheric activity unexpected in fully radiative stars. The most ambitious project is being carried out by Skinner, Brown, and Linsky who have recently published (Skinner *et al.* 1990) the initial results of a complete survey of the 57 known HAEBE stars done with the VLA. Continuum emission at 3.6 and 6 cm was detected at a rather low level (0.1–1.9 Jy) in 4 out 14 sources, namely : TY CrA, HD 200775, HD 259431, MWC 1080. In addition to these results, two other HAEBE stars have been detected : AB Aur by Güdel *et al.* (1989), and V645 Cyg by Curiel *et al.* (1989). As to the nature of the emission, unfortunately the results are not clear cut. In one case (HD 200775), the flux is consistent with free-free emission from a fully ionized, perhaps not spherically symmetric wind; while in another case (TY CrA) the spectrum has a negative slope, implying nonthermal emission. For the prototype of the class of HAEBE stars, AB Aur, the situation is more confusing, since the data at the shortest wavelength are consistent with thermal emission from a wind with characteristics derived from visible and UV data, but not those at 6 cm. Obviously, any definite answer should await the completion of the survey.

2. 3. JETS AND H_2O MASERS

Ray *et al.* (1990) and Poetzl *et al.* (1989) have reported the discovery of HH-like objects in a large fraction (about 30%) of young stars of intermediate to high luminosity. These are the first results of this kind, since previous searches have only dealt with sources of low luminosity (cf. Reipurth 1989). From the limited amount of information available at the present, it seems that the jets emanating from intermediate-mass PMS stars share the same characteristics (radial velocities, inferred mass loss rates, energy and momentum fluxes) of those associated with T Tauri stars.

Previous searches for H_2O maser emission at 22.2 GHz have all been characterized by a low detection rate. For example, the survey by Thum *et al.* (1981) toward 81 members of the Orion population yielded only 1 detection, T Tauri. This prompted the authors to conclude that no H_2O emission above the sensitivity limit of the survey ($\sim$ 1Jy) is associated with stars in the mass range $1 \leq M_*/M_\odot \leq 10$. More recently, I have completed a survey of comparable sensitivity (Palla 1990) of 43 HAEBE stars and detected emission in 5 sources : 3 of them represent first time detections (T Ori, BD +40°4124, Lk Hα 198), while the other two were already known : V645 Cyg (Lada *et al.* 1981), and Lk Hα 234 (Rodriguez and Cantó 1983). Given the highly variable character of the maser emission, it can be concluded that, as in the case of more luminous stars, the occurrence of masers associated with intermediate–mass PMS is not an uncommon phenomenon (cf. also Rodriguez *et al.* 1987).

Overall, these new set of observations, together with those described in Catala's review, show with little doubt the complexity of the environment associated with the HAEBE stars.

3. Theoretical overview

3. 1. PROTOSTELLAR EVOLUTION

From a theoretical point of view, intermediate–mass stars pose the challenging problem of studying in detail the transition from a state purely dominated by the accretion of interstellar gas onto the core to the gravitational contraction of the core itself, that can no longer remain fully convective and reverts to a radiatively stable configuration. However, the literature on the subject is scant. Since Larson (1972) classical paper on the simulation of the spherical accretion of protostellar cores, only the work of Yorke (1979) deals with the structure, evolution and appearance of stars in the mass interval of interest here. [1] Yorke studied the hydrodynamical collapse of gas clouds of 3 $M_\odot$ and 10 $M_\odot$, respectively, from the onset of the gravitational instability to the end of the accretion phase, but did not follow the actual growth of the protostellar core, that was assumed to evolve in a parametric manner following Larson's results. Therefore, the latter remain the only self–consistent calculations available until now. It is important to remind that Larson's models, in

---

[1] For the sake of completeness, Shustov and Tutukov (1987) quote a paper by Tutukov and Chiefi (1985) on the evolution of accreting stars with mass $0.1 \leq M_*/M_\odot \leq 6.6$, but unfortunately the detailed results are not available.

addition to the usual limitation of assuming spherical symmetry, did not include the contribution in the energy equation of deuterium and hydrogen burning, thus missing, in the latter case, a precise estimate of the arrival of the contracting cores on the ZAMS. As for deuterium burning, Stahler (1988) has shown its fundamental role in establishing the main properties of low–mass protostars.

According to Larson's analysis, the formation and early evolution of intermediate-mass stars is characterized by the occurrence of the following phenomena :

1 - for $M_* \geq 2M_\odot$, the occurrence of a *"luminosity wave"* from the deep interiors of the core to the outer layers, accompained by a temporary and minor increase of the radius and of the total luminosity (about a factor of 2) ;

2 - for $M_* \sim 3M_\odot$, the radiatively stable core goes through the entire PMS phase while still gaining mass from the infalling envelope, as shown in the theoretical HR diagram of Figure 3;

3 - for $M_* \geq 3M_\odot$, the action of radiation pressure on the infalling grains should also be able to reverse the flow and halt the accretion.

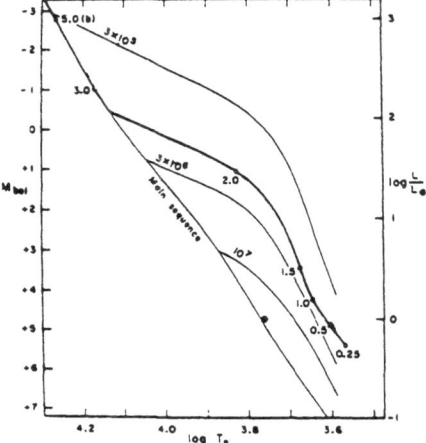

Fig. 3. The predicted locus (*heavy curve*) of newly formed stars in the HR diagrams according to Larson's calculations. Open circles mark the mass of the accreting protostellar core. Light curves are the isochrones from Iben and Talbot (1966).(Adapted from Larson 1972)

The occurrence of a luminosity wave inside the core has been reanalyzed by Stahler (1989b), in the light of the results of modern protostellar theory. Stahler has shown that the luminosity increase due to the eruption of the wave at the surface of the core can be substantial, up to 30 times the initial luminosity for a star of 5 $M_\odot$, and suggests that this mechanism could be partly responsible for the occurence of the wild variations observed in PMS stars of comparable mass, such as the FU Ori objects.

As for the second aspect, the fact that the accreting cores reach the ZAMS at a rather low mass (2-3 $M_\odot$) is a consequence of having neglected deuterium burning, whose main effect is to establish a mass–radius relation for low–mass cores such that the radius increases proportionally with mass. We shall return at length on the importance of deuterium burning in more advanced stages of the protostellar evolution.

Finally, the estimate of the critical mass where radiation pressure effects become relevant has been revised upward, using better values of the dust opacity at visual wavelengths. It is found that radiation pressure takes over gravity when the luminosity to mass ratio satisfies the condition : $L_*/M_* \approx 700 L_\odot/M_\odot$ (Shu et al. 1987), a value reached by main–sequence stars more massive than about 7 $M_\odot$ (see also the discussion in Nakano 1989). As we shall see below, protostars of comparable masses never attain such high values, so that the argument that the accretion phase might terminate due to radiation pressure is very weak. Given the amount of observational evidence of winds and energetic outflows associated with HAEBE stars, the latter seem to offer a more efficient mechanism.

3. 2. PRE–MAIN-SEQUENCE EVOLUTION

The basic results found by Iben (1965) and Ezer and Cameron (1965) have remained unquestioned, despite the approximations in what concerns the initial conditions of the model calculations. A review of the various sources of uncertainties in the computation of the evolutionary tracks has been given by Mazzitelli (1989), but it is limited to PMS stars less massive than 1.2 $M_\odot$ and does not discuss the modifications in the initial conditions introduced by the results of the protostellar models. A discussion of this point can be found in Stahler (1989a). The main point is that using the so-called "Hayashi initial conditions", that require fully convective configurations, the stars are forced to have extremely high values of the radius at the beginning of the PMS evolution. Based on simple energetic arguments, Cameron (1962) obtained an initial value of the radius given by : $R_{PMS}^{init} \approx 50 R_\odot (M_*/M_\odot)$. According to the results of the accretion phase, that should provide the correct initial conditions, the protostellar cores never attain such large radii, but remain a factor of $\sim$10 below Cameron's estimate (Winkler and Newman 1980; Stahler, Shu and Taam 1980). However, this large discrepancy does not affect in a qualitative manner the evolution of low–mass PMS stars, since they satisfy the assumption about internal convection, driven by the action of deuterium burning in the center during the accretion phase. Indeed, the location of the theoretical birthline (Stahler 1983) and the position of the most luminous T Tauri stars in the HR diagram accords well with the beginning of the evolutionary tracks in the Hayashi phase computed by Iben. On the contrary, more massive protostellar cores develop radiative interiors : thus, they will start the PMS phase strongly departing from the classical assumptions and the use of the evolutionary tracks to interpret the observed properties of intermediate–mass PMS stars will become less and less appropriate. We will now discuss the results of some recent calculations that can help to answer some of the fundamental questions raised in the Introduction.

3. 3. RECENT DEVELOPMENTS

Palla and Stahler (1990a,b) have extended previous calculations of the spherical accretion of protostellar cores limited to $M_{core} = 1M_\odot$, to cover the entire mass interval up to the point where the cores reach the ZAMS. The novel feature of the evolution is the discovery of the onset of deuterium burning in a shell at the time of the transition to radiative stability. This shell burning alters the succession of events prior to the arrival onto the ZAMS in a dramatic way. Schematically, the entire burning process can be divided into 3 main stages :

Central D–burning : it represents the main property of the evolution of protostellar cores up to $M_{core} \sim 1M_\odot$. The steady–state burning occurs near the center and keeps the star fully convective. The thermostatic nature of the burning ensures that the core radius grows steadily with mass.

The radiative barrier and the fading of central D–burning : as more matter is added to the core, the interior temperaure increases, the opacity drops, and the radiative luminosity becomes adequate in carrying away that produced by the steady–state burning of deuterium. Therefore, an internal zone will first become radiatively stable, so that the newly accreted deuterium can no longer be transported to the center of the star. This way, the whole region interior to the radiative barrier quickly burns the remaining deuterium and becomes radiatively stable.

The onset of D–shell burning : as soon as the temperature of the elements outside the barrier reaches the critical value of $\sim 10^6 K$, deuterium ignites in a shell maintaining convection in the outer layers. The luminosity of the burning shell rises quickly to attain its steady–state value, as illustrated in Figure 4. This energy source persists as the star accretes matter, but slowly retreats to the surface, due to the increased rate of energy released by the gravitational contraction of the inner parts. The situation is reminiscent of the "luminosity wave" discussed earlier, but in this case it has a completely different nature and its effects on the structure of the core become much more dramatic. It is found that, as the shell moves toward the surface, the outer layers expand considerably and the core doubles its size. This behavior can be appreciated by an inspection of Figure 5 that shows the run of the core radius vs. mass, obtained in the case of spherical accretion at a representative accretion rate of $10^{-5} M_\odot yr^{-1}$. This result is quite general and has a weak dependence on the details of the accretion process, whether from direct infall, as in the example presented here, or mediated by the presence of a circumstellar disk. The rapid increase in the core radius is followed by a slow decline, due to the increased strength of the gravitational pull as more matter is added onto the core. The major consequence of this behavior is that the interior temperature remains rather low, thus postponing the time of the onset of nuclear burning in the center. According to Fig. 5, hydrogen burning starts at a core mass $M_{core} \sim 6M_\odot$, but the ZAMS is only reached at $M_{core} \sim 8M_\odot$. In analogy with the low–mass protostars, the knowledge of the mass–radius relation allows one to derive the birthline of intermadiate–mass stars in the HR diagram, as shown by the heavy curve in Figure 6.

The agreement between the location of the birthline and the distribution of the observed HAEBE stars (the T Tauri stars have been omitted for clarity) is remarkable : in accordance with the prediction, there is no optically visible PMS phase for $M_* \geq 9M_\odot$. The comparison with the similar curve obtained by Larson, and shown in Fig. 3, underlines the crucial role that D–shell burning plays in setting the proper conditions in the accreting cores, and in reconciling the theoretical predictions with, at least a set of, the observations.

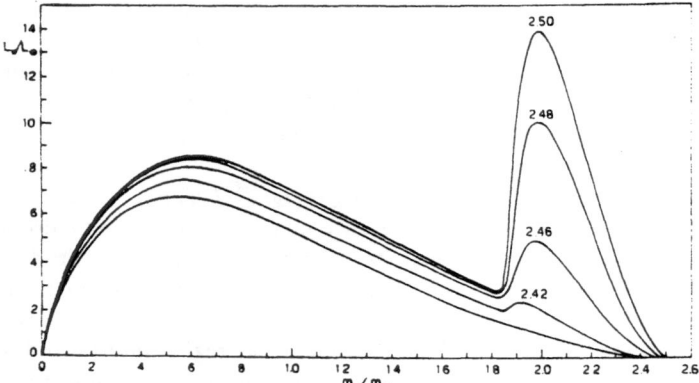

Fig. 4.   The rise of the luminosity in the burning shell of an accreting protostar a. different epochs. Curves are labelled by the core mass. (From Palla and Stahler 1990b)

4. Discussion

The implications of the results presented in the previous section are numerous. First of all, given the position of the birthline vis à vis the evolutionary tracks of Iben, it is evident that the classical assumptions about intermediate–mass PMS stars must be revised. Not only the initial radii, set at the end of the accretion phase, are vastly different from those commonly assumed, but also the *internal configuration of the stars* presents a much more complicated structure. For example, a PMS star of 3 $M_\odot$ will begin its evolution with a radiative core and a convective region extending all the way to the surface, due to the action of D–burning. The internal luminosity distribution will show a double peak, the off-center maximum due to the radiative luminosity and that at the position of the burning shell, and the ensuing nonhomologous contraction will not occur in a simply predictable manner.

Another implication of the birthline concerns the *age estimate* of the HAEBE stars, that in reality should be much younger than expected on the basis of the evolutionary tracks :   the difference can be appreciable, up to even 50% of the

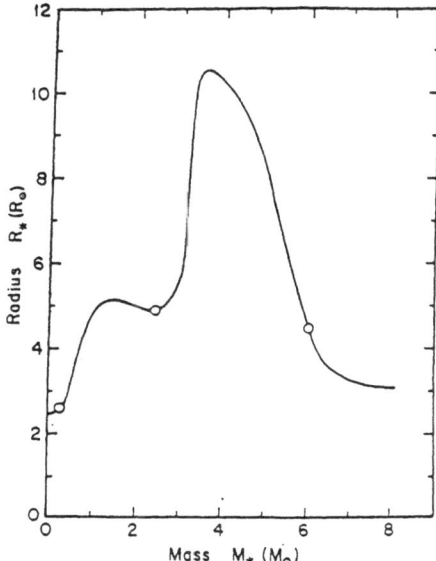

Fig. 5. The core mass–radius relation for accreting protostars. The various phases of the accretion phase and the main burning events are schematically indicated. (Adapted from Palla and Stahler 1990a)

traditional value. An indirect sign of their relative youth is given by the association with HH–like objects, molecular outflows, and masers previoulsy discussed. This in turn implies that the stars have been active for a shorter period of time, thus releasing some constraints on the mechanism(-s) responsible for the activity.

Finally, the role of D-burning in promoting *surface convection* in stars that were thought to have none, can indeed resolve the paradox brought about by the observations. Not only that, but the onset of nuclear reactions near the surface can also excite instabilities that will possibly develop (non)radial oscillations that could account for the observed rapid variability in the spectral lines. This possibility was alluded to by Herbig (1983), who noted that the PMS tracks of stars more massive than $2M_\odot$ must cross the post–main-sequence instability strip. In this regard, it is important to remind the work of Toma (1972) who studied the radial pulsational instability of low–mass PMS stars due to *central D-burning* in fully convective configurations (see also the recent discussion of Gahm and Liseau 1988). The fact that intermediate–mass protostars possess radiative cores and *D-burning shells* makes a simple estimate of the predicted oscillation periods much more difficult : a re–analysis of the problem with more realistic conditions is warranted.

Obviously, the questions and problems left to be answered are numerous. The few issues that have been stressed here reveal how the field is still in its infancy, compared to the more mature subject of low–mass star formation, where the role

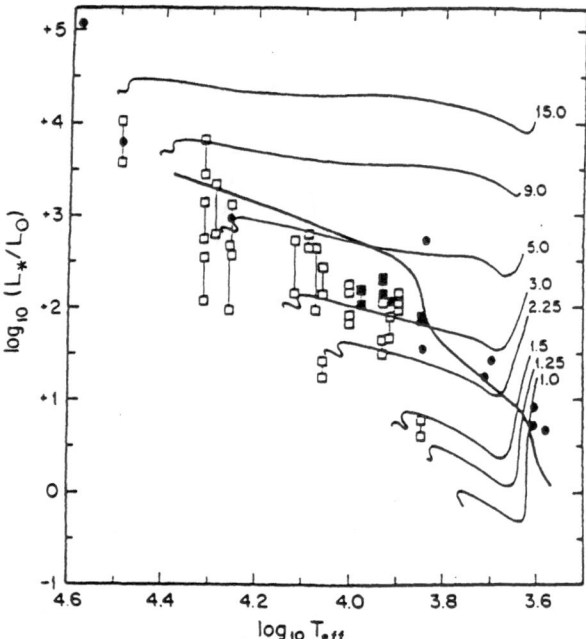

Fig. 6.   The theoretical birthline (*heavy curve*) of intermediate–mass PMS stars. The squares are observed HAEBE stars with the uncertainty in the luminosity, due to different assumptions about the extinction (Finkenzeller and Mundt 1984). The filled circles are optically visible stars with CO outflows (Levreault 1988). The evolutionary tracks are from Iben (1965). (From Palla and Stahler 1990a)

of rotation and magnetic fields , the presence of discs etc., not discussed in this context, have been incorporated already in a more coherent way. On the observational side, the question of the occurrence of binaries among HAEBE stars should be attacked soon, as well as the study of these stars in the youngest galactic clusters to learn about possible evolutionary trends. On the theoretical side, much more refinement in the current models is demanded, and the calculation of updated PMS evolutionary tracks should have the highest priority, especially in the light of the wealth of observational material brought about by the study of the embedded star clusters in star forming regions.

Acknowledgements

It is a pleasure to thank the Organizing Committee, and in particular Edith Falgarone, for allocating ample time to present and discuss at the Meeting the main ideas collected in this contribution. All of the numerical calculations presented here come from a collaborative effort with Steve Stahler, an endeavor that started a long time ago. Finally, special thanks go to Claude Catala for his kind hospitality at the DESPA–Observatoire de Meudon and for sharing the excitment about Herbig stars.

References

Appenzeller, I. : 1980, in *Star Formation*, A. Maeder and L. Martinet eds., p.3.

Baade, D. and Stahl, O. : 1989, *Astron. Astrophys.*, **209**, 268.

Berrilli, F., Ceccarelli, C., Lorenzetti, D., Nisini, B., Saraceno, P. and Strafella, F. : 1990, *Il Nuovo Cimento*, **13C**, 293.

Cameron, A.G.W. : 1962, *Icarus*. **1**, 13.

Catala, C. : 1989, in *Low Mass Star Formation and Pre–Main-Sequence Evolution*, ESO Conf. and Work. Proc., Bo Reipurth ed., p.471.

Cohen, M., Kuhi, L.V. : 1979, *Astrophys. J. Suppl.*, **41**, 743.

Curiel, S., Rodriguez, L.F., Cantó, J., Bohigas, J., Roth, M. and Torrelles, J.M. : 1989, *Astro. Lett. Comm.*, **27**, 299.

Ezer, D. and Cameron, A.G.W. : 1965, *Canadian J. Phys.*, **43**, 1497.

Finkenzeller, U. : 1985, *Astron. Astrophys.*, **151**, 340.

Finkenzeller, U. and Mundt, R. : 1984, *Astron. Astrophys. Suppl.*, **55**, 109.

Gahm, G.F. and Liseau, R. : 1988, in *Activity in Cool Star Envelopes*, O. Havnes *et al.* eds., Kluwer Acad. Publ., p.99.

Güdel, M., Benz, A., Catala, C. and Praderie, F. : 1986, *Astron. Astrophys.*, **217**, L9.

Herbig, G.H. : 1960, *Astrophys. J. Suppl.*, **4**, 337.

Herbig, G.H. : 1983, in *Birth and Infancy of Stars*, Les Houches–Session XLI, R. Lucas, A. Omont and R. Stora eds., p.535.

Hu, J.Y., Thé, P.S. and de Winter, D. : 1989, *Astron. Astrophys.*, **208**, 213.

Iben, I. : 1965, *Astrophys. J.*, **141**, 993.

Iben, I and Talbot, R.J. : 1966, *Astrophys. J.*, **144**, 968.

Lada, C.J., Blitz, L., Reid, M.J., and MOran, J.M. : 1981, *Astrophys. J.*, **243**, 769.

Lada, C.J., and Shu, F.H. : 1990,*Science*, **248**, 564.

Larson, R.B. : 1972, *Monthly. Not. Roy. Astron. Soc.*, **157**, 121.

Levreault, R.M. : 1988, *Astrophys. J.*, **330**, 910.

Mazzitelli, I. : 1989, in *Low Mass Star Formation and Pre–Main-Sequence Evolution*, ESO Conf. and Work. Proc., Bo Reipurth ed., p.433.

Nakano, T. : 1989, *Astrophys. J.*, **345**, 464.

Palla, F. : 1990, *Astrophys. J. Lett.*, in press.

Palla, F. and Stahler, S.W. : 1990a, *Astrophysical. J. Lett.*, **360**, L47.

Palla, F. and Stahler, S.W. : 1990b, *Astrophysical. J.*, in press.

Palla, F., Brand, J., Cesaroni, R., Comoretto, G. and Felli, M. : 1990, *Astron. Astrophys*, in press.

Poetzl, R., Mundt, R. and Ray, T.P. : 1989, *Astronom. Astrophys.*, **224**, L13.

Praderie, F., Simon, T., Catala, C. and Boesgard, R.M. : 1986, *Astrophys. J.*, **303**, 311.

Ray, T.P., Poetzl, R., Solf, J. and Mundt, R. : 1990, *Astron. Astrophys.*, **357**, L45.

Reipurth, B. : 1989, in *Low Mass Star Formation and Pre–Main-Sequence Evolution*, ESO Conf. and Work. Proc., Bo Reipurth ed., p.247.

Rodriguez, L.F. and Cantó, J. : 1983, *Rev. Mex. Astron. Astrof.*, **8**, 163.

Rodriguez, L.F., Haschick, A.D., Torrelles, J.M. and Myers, P.C. : 1987, *Astron. Astrophys.*, **186**, 319.

Salpeter, E.E. : 1955, *Astrophys. J.*, **121**, 161.

Scalo, J.M. : 1990, in *Windows on Galaxies*, A. Renzini, G. Fabbiano and J. Gallagher eds., Kluwer–Dordrecht, in press.

Shu, F.H., Adams, F. and Lizano, S. : 1987, *Ann. Rev. Astron. Astrophys.*, **25**, 23.

Shustov, B.M. and Tutukov, A.V. : 1987, in *Star Forming Regions*, J. Jugaku and M. Peimbert eds., IAU Symposium 115, Reidel–Dordrecth, p.440.

Skinner, S.L., Brown, A. and Linsky, J.L. : 1990, *Astrophys. J. Lett.*, **357**, L39.

Stahler, S.W. : 1983, *Astrophys. J.*, **274**, 822.

Stahler, S.W. : 1988, *Astrophys. J.*, **332**, 804.

Stahler, S.W. : 1989a, *Pubb. Astron. Soc. Pac.*, **100**, 1474.

Stahler, S.W. : 1989b, *Astrophys. J.*, **347**, 950.

Stahler, S.W., Shu, F.H. and Taam, R.E. : 1980, *Astrophys. J.*, **241**, 637.

Strom, S.E., Strom, K.M., Yost, J., Carrasco, L., Grasdalen, G.L. : 1972, *Astrophys. J.*, **173**, 353.

Thum, C., Bertout, C., Downes, D. : 1981, *Astron. Astrophys.*, **94**, 80.

Toma, E. : 1972, *Astron. Astrophys.*, **19**, 76.

Tutukov, A.V. and Chiefi, A. : 1985, *Nauchn. Inf.*, **58**, 40.

Winkler, K.H. and Newman, M.J. : 1980, *Astrophys. J.*, **236**, 201.

Wood, D.O.S. and Churchwell, E. : 1989, *Astrophys. J.*, **340**, 265.

Yorke, H.W. : 1979, *Astron. Astrophys.*, **80**, 308.

A SURVEY OF CIRCUMSTELLAR STRUCTURE
AROUND YOUNG LOW MASS STARS

S. TEREBEY, C. A. BEICHMAN, T. N. GAUTIER, J. J. HESTER

Infrared Processing and Analysis Center
and Palomar Observatory;
Jet Propulsion Laboratory and California Institute of Technology,
MS 100-22, Caltech, Pasadena CA 91125
E-mail ST@IPAC.CALTECH.EDU

P. C. MYERS

Harvard-Smithsonian Center for Astrophysics,
60 Garden St., MS 42, Cambridge, MA 02138

S. N. VOGEL

Astronomy Program, University of Maryland,
College Park, MD 20742

Abstract. We present results from a near-infrared array, CO interferometer, and H_2O maser interferometer survey of the circumstellar environments of 26 young low-luminosity embedded stars located in nearby molecular clouds. About 75% of the sample show evidence for stellar winds/outflows in the near-infrared or CO data indicating that most of these sources are in the early wind clearing phase of their evolution. Close to 15% are multiple on the scale of $20''$, suggesting that fragmentation of their surrounding dense cloud cores is important before or during gravitational collapse. Roughly 10% have H_2O maser emission and the kinematics imply the masers arise in gravitationally unbound gas (i.e., a stellar wind or outflow) rather than in a circumstellar disk.

Keywords : infrared sources, stars-accretion, circumstellar gas, star formation, pre-main-sequence stars, stellar winds, binaries

1. Introduction

We present high spatial resolution observations showing the circumstellar environments of young low mass stars on scales of roughly a 1000 AU. The sources are members of the class known as IRAS-Dense cores; highly obscured objects with steep IRAS spectra that are found near the peaks of dense gas emission (Myers and Benson 1983; Beichman et al. 1986; Myers et al. 1987; Benson and Myers 1989). These low-luminosity infrared sources embedded in dense cloud cores are suspected protostars, i.e. stars deriving part or all of their luminosity from accretion (Terebey, Shu, and Cassen 1984; Shu, Adams, and Lizano 1987; Adams, Lada, and Shu 1988).

Our current theoretical understanding of low mass star formation indicates that young stars form from the collapse a dense cloud core. During the wind clearing phase a vigorous young stellar wind erodes the infalling cloud core, eventually halts the infall and in the process reveals the young star. During the infall phase a protostellar disk–of uncertain size and mass– probably forms. Models of the collapse region (Terebey, Shu, and Cassen 1984) find it typically extends out to $10^{17} cm$, or 10,000 AU, roughly the region encompassed by our infrared and interferometer data. Matter inside the collapse zone will not be at rest; thus the gas observed inside this region should trace either the infall, outflow, or disk components.

Despite the successes of the theoretical picture, many important details of the process must be observationally confirmed. We use the morphology of the gas-rich circumstellar environment around young stars to investigate the importance of the different protostellar phenomena : winds, infall, disks, and stellar multiplicity.

2. Data

2. 1. SAMPLE

The sample consists of 26 sources which have an IRAS source close to the center of a dense molecular cloud core. These sources are all highly obscured–inferred visual extinctions are on the order of 30 magnitudes. Most sources have no optical counterpart and all have the steep IRAS spectra indicative of deeply embedded young stars. The dense cloud cores are typically 0.1 pc in size with kinetic temperatures of $10 - 20$ K, and contain about $1 - 10$ M$_\odot$ of gas. The sample is not a complete list of dense cores for every cloud but can be considered representative in that the sources satisfy well-defined criteria, span a range in luminosity from 0.5 to 100 L$_\odot$ with a typical value of 2 L$_\odot$, and come from a variety of nearby molecular clouds. The properties of the sample are described in more detail in Terebey, Vogel, and Myers (1989; hereafter TVM 1989). The source TMR-1 (IRAS 04361+2547), an embedded star in the Taurus molecular ring (TMC-1/Heiles Cloud 2 region) has been added to the original sample.

2. 2. OBSERVATIONS

We have surveyed the sources for compact CO outflows using visibility data from the Owens Valley Millimeter Interferometer. Follow-up synthesis maps for one-third of the sources confirm the initial survey results reported in TVM (1989). The sample was also searched for H_2O maser emission in a 2′ field of view using the VLA in the C/D array configuration, with follow-up high spatial resolution observations in the A or B arrays for the detected maser sources (Terebey, Vogel, and Myers 1990).

We report initial results for a near-infrared camera survey using the PFIRCAM on the Palomar Observatory Hale 5m telescope. The camera utilized a 128×128 array of HgCdTe detectors operating between 1 and 2.5 μm with 0.765″ pixels and covering a 98″ field of view. Each source was observed at K (2.2 μm); sources showing structure were also observed at J (1.25 μm) and H (1.65 μm). Three-quarters of the sample have been observed; the sources with IRAS upper limits at 12 microns were excluded.

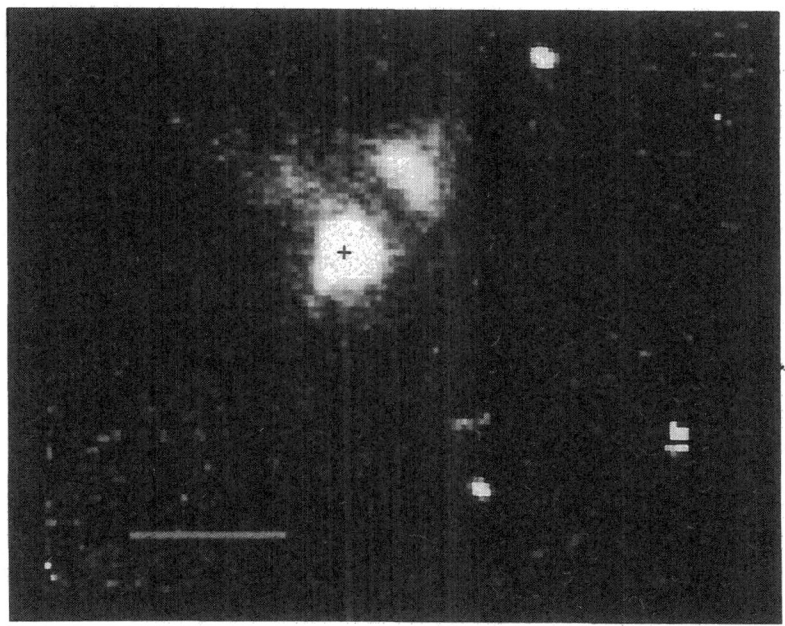

Fig. 1. A 2.2 µm image of the young embedded source TMR-1. The cross marks the infrared point source position. The reference line corresponds to 20″ (2800 AU). We interpret the spindle of emission extending to the NW to be a stellar wind cavity. Perpendicular to the direction of the spindle is a deep absorption feature. Two unrelated point sources lie to the NW and SW. *(See Color-plates section)*

2. 3. NEAR-INFRARED DATA

Figures 1 and 2 illustrate the phenomena present in the near-infrared images.

The 2.2 μm image (Fig. 1) of the Taurus cloud source TMR-1 reveals a bright point source and an elongated spindle approximately 20" (2800 AU) in length that extends to the NW. We interpret the extended emission as scattered light that traces a stellar wind cavity. The CO interferometer data shows an outflow coincident with the near infrared spindle. Cutting across the spindle at a perpendicular angle lies a dust band in absorption. The dust band appears to be part of an incomplete ring or disk of circumstellar material, 1100 AU in radius, that is inclined to the line of sight. There is faint emission visible from the edge of the dust band to the NE, as if illuminated by the embedded source. Apparently we are seeing high density material that is detected because of its favorable placement with respect to the background continuum emission. This striking dust absorption feature is unique to TMR-1 of all the objects observed. Two other point sources are also located in the field, to the NW and to the SW. Further discussion of TMR-1 can be found in Terebey et al. (1990).

Figure 2 shows a near-infrared color composite image of the ρ Oph cloud source L1681B. L1681B is revealed to be a triple system; the three embedded sources are separated by about 25" and are nearly equal in brightness. Our VLA data detect water maser emission towards both the central star and the star to the NE. This indicates both stars (projected separation of 4000 AU) are very young. Probably all three stars have formed from the same dense molecular cloud core which has fragmented from its original size of roughly 0.1 pc.

The infrared nebulosity associated with the central star we interpret as scattered light that is tracing a stellar wind cavity. The molecular outflow detected in the CO interferometer data (Fig. 3) extends south of the infrared source, and apparently traces only one outflow lobe (TVM 1989).

3. Discussion

3. 1. OUTFLOWS

From the CO data there is evidence for outflows in 69% (18/26) of the sources (TVM 1989 and references therein). About two-thirds of the outflow sources that have been observed in the near-infrared also show extended emission at 2.2 μm. In most cases the near-infrared colors of the extended emission are bluer than the colors of the embedded stars, consistent with the interpretation that the extended emission is due to scattered stellar light. Comparing the near-infrared images and CO interferometer maps shows that the near-infrared emission is less extended than the CO emission but is otherwise closely related to the CO outflow structure. In a few cases the near-infrared images show complementary structure as in L1681B (Fig. 2 and 3); the CO data show only one outflow lobe, while the near-infrared images detect another outflow lobe located on the opposite side of the embedded infrared source.

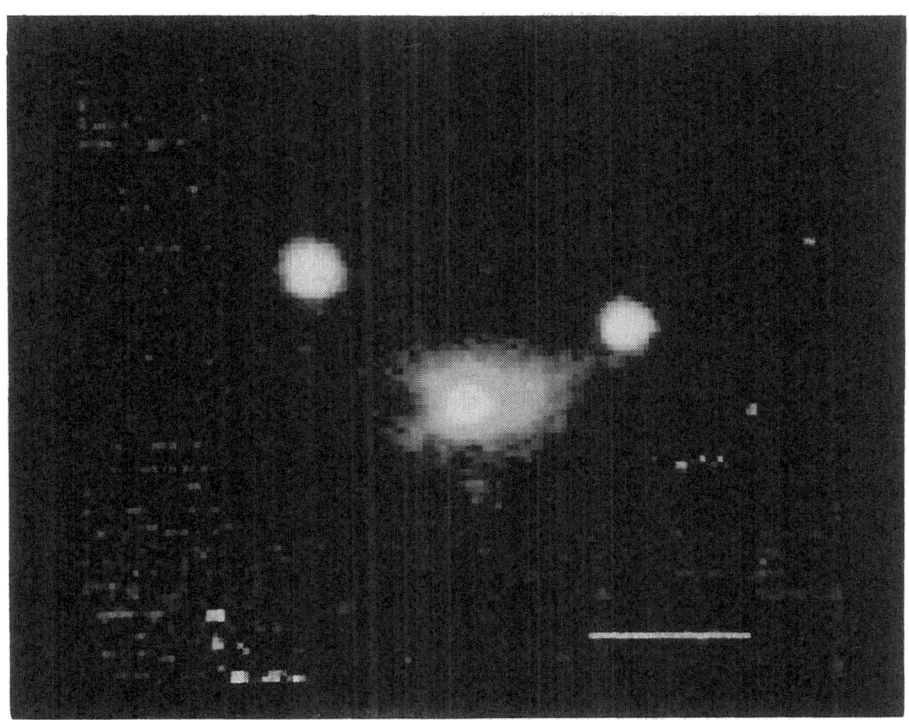

Fig. 2.   A near infrared color composite image of the embedded source L1681B. Blue, green, and red correspond to 1.25, 1.65, and 2.2 μm, respectively. The reference line shows 20″ (3200 AU). There are three embedded sources separated by 3000 - 5000 AU. H_2O maser emission is detected coincident with the central and NE sources, thus demonstrating both sources are young. We interpret the nebulosity as scattered light that is tracing a poorly collimated stellar wind cavity.   *(See Color-plates section)*

The higher resolution of the near-infrared images ($\sim 1.5''$) compared with the CO data allows us to trace the outflow closer to the embedded star. Generally there is only moderate collimation of the outflow to within several hundred AU of the star. Given the close correspondence between CO outflows and extended near-infrared emission we identify two new outflow sources on the basis of their extended near-infrared emission alone. This increases the outflow frequency to about 75% of the sample.

The outflow frequency has implications for the the relative durations of the the infall and stellar wind clearing phases. Assume that the ratio of time spent in each phase equals the ratio of the number of sources in each phase. Then the purely infall phase lasts no more than 30% the length of the outflow phase. Kinematic ages for the oldest outflows give ages on the order of $1 - 2 \times 10^5$ yr (Myers et al. 1988), which imply an infall phase lasting less than 60,000 years. Given typical accretion rates ($\sim 5 \times 10^{-6}$ $M_\odot yr^{-1}$) this time scale appears uncomfortably short to accrete a solar mass size star. The time scale problem has a simple resolution if the infall and outflow phases overlap for some moderate length of time.

3. 2. MULTIPLICITY AND FRAGMENTATION

Roughly 60% of the near-infrared images show more than one star on the object frame. To distinguish embedded stars from background stars requires additional photometry or spectroscopy. However our object frames and background sky frames are quantitatively similar (excluding the observed source) in terms of source brightness and star density, suggesting that many of these additional objects are background stars. One class of sources appear associated– they are separated by roughly 30'' and are nearly equal in brightness. For this class the near-infrared images indicate about 15% of the sources are multiple on the scale of $20 - 30''$, or about $3000 - 5000$ AU. Since this is smaller than the typical 0.1 pc dense core size it implies that fragmentation of the dense cores is important and that more than one star can form per dense core.

Theoretical models of gravitational cloud collapse find that the velocity shear inhibits the growth of small density perturbations. This suggests that fragmentation occurs at an early stage before the dense cores become gravitationally unstable.

3. 3. H$_2$O MASERS

About 10% of the sources show water maser emission (Terebey, Vogel, and Myers 1990). The detected sources are preferentially those with the highest luminosity and steepest spectra (rising toward longer wavelengths). The maser emission is confined to within 100 AU of the star. Maser emission from circumstellar disks was searched for since the physical parameters expected in disks such as high density ($> 10^6$ cm^{-3}) and temperature (> 300 K at 1 AU) correspond to those in known masers and the protostar provides a nearby energy source. However the kinematics imply the masers originate in gravitationally unbound gas (i.e. a stellar wind or

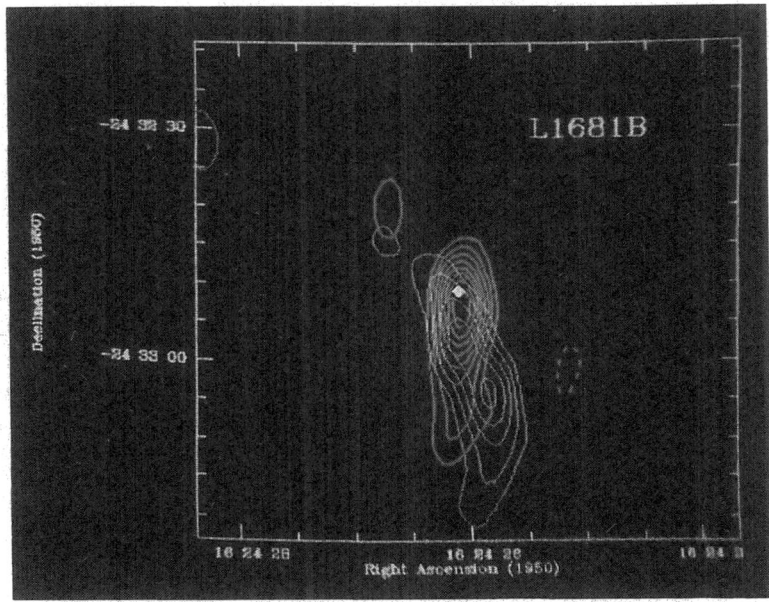

Fig. 3.   An interferometer map of the CO outflow seen towards the embedded source L1681B. The diamond marks the position of the infrared point source. Red-shifted and blue-shifted gas defines a stellar wind cavity extending south of the infrared source. *(See Color-plates section)*

outflow). The collimation of the outflow on small scales, as traced by the masers, matches the collimation seen in the CO data on a much larger scale.

Acknowledgements

This paper is dedicated to the memory of Gary Bailey who died on January 26. 1990. Gary pioneered the use of infrared arrays for astronomical use and was a good friend who will be sorely missed. Thanks to the Palomar crew for their usual good cheer as the new camera was installed and debugged. CAB thanks J. Bahcall for the hospitality of the Institute of Advanced Study during his sabbatical leave. Palomar Observatory is supported by a grant from the National Science Foundation. This work was carried out in part at the Jet Propulsion Laboratory, California Institute of Technology, under a contract with the National Aeronautics and Space Administration. This work was also supported by NASA grant 188-44-24-10, NSF grant AST 84-12473, and by a grant from the Caltech President's fund. The junctions used in the SIS receivers at OVRO were provided by R. E. Miller of AT&T Bell Labs.

References

Adams, F. C., Lada, C. J., and Shu, F. H. : 1988, *Ap. J.*, **312**, 788.
Beichman, C. A., Myers, P. C., Emerson, J. P., Harris, S., Mathieu, R., Benson, P. J., and Jennings, R. E. : 1986, *Ap. J.*, **307**, 337.
Benson, P. J. and Myers, P. C. : 1989, *Ap. J. Suppl.*, **71**, 89.
Myers, P. C., and Benson, P. J. : 1983, *Ap. J.*, **266**, 309.
Myers, P. C., Fuller, G. A., Mathieu, R. D., Beichman, C. A., Benson, P. J., Schild, R. E., and Emerson, J. P. : 1987, *Ap. J.*, **319**, 340.
Myers, P. C., Heyer, M., Snell, R., and Goldsmith, P. : 1988, *Ap. J.*, **324**, 907.
Shu, F. H., Adams, F. C., and Lizano, S. : 1987, *Annual Rev. Astron. Astrophys.*, **25**, 23.
Terebey, S., Beichman, C. A., Gautier, T. N., and Hester, J. J. : 1990, *Ap. J. Letters*, **362**, L63.
Terebey, S., Shu, F. H., and Cassen, P. C. : 1984, *Ap. J.*, **286**, 529.
Terebey, S., Vogel, S. N., and Myers, P. C. : 1989, *Ap. J.*, **340**, 472.
Terebey, S., Vogel, S. N., and Myers, P. C. : 1990, in preparation.

Aperture Synthesis CS and 98 GHz Continuum Observations of Protostellar *IRAS* Sources in Taurus

N. OHASHI[1], R. KAWABE[1], M. HAYASHI[2], AND M. ISHIGURO[1]

1. *Nobeyama Radio Observatory, National Astronomical Observatory, Japan.*
2. *Department of Astronomy, University of Tokyo, Japan.*

ABSTRACT   The CS ($J = 2 - 1$) line and 98 GHz continuum emission have been observed for 11 protostellar *IRAS* sources in the Taurus molecular cloud with resolutions of 2.6″–8.8″ (360 AU–1200 AU) using the Nobeyama Millimeter Array (NMA). The CS emission is detected only toward embedded sources, while the continuum emission from dust grains is detected only toward visible T Tauri stars except for one embedded source, L1551-IRS5. This suggests that the dust grains around the embedded sources do not centrally concentrate enough to be detected with our sensitivity ($\sim$4 mJy r.m.s), while dust grains in disks around the T Tauri stars have enough total mass to be detected with the NMA. The molecular cloud cores around the embedded sources are moderately extended and dense enough to be detected in CS, while gas disks around the T Tauri are not detected because the radius of such gas disks may be smaller than 70 (50 K/T_{ex}) AU. These results imply that the total amount of matter within the NMA beam size must increase when the central objects evolve into T Tauri stars from embedded sources, suggesting that the compact and highly dense disks around T Tauri stars are formed by the dynamical mass accretion during the embedded protostar phase.

1. Introduction

Recent observations of dust emission and near-infrared excess revealed that many T Tauri stars have compact ($\lesssim$100 AU) and highly dense disks around them (Sargent and Beckwith 1987; Strom *et al.* 1989; Beckwith *et al.* 1990). Theoretical fitting to infrared spectra also required non-spherical structures like disks within a radius of $\sim$100 AU even around embedded sources (Adams, Lada, and Shu 1987; Myers *et al.* 1987). It is, however, not yet clear when and how such compact disks are formed.

We have observed 11 protostellar *IRAS* sources in Taurus in the CS ($J = 2 - 1$) and 98 GHz continuum emission with the NMA in order to investigate the evolution of compact structures around protostellar sources. Interferometric observations are sensitive only to small scales, so that the confusion from extended foreground and background envelopes is negligible.

2. Results

The results of our observations are summarized in Table 1. We achieved the spatial resolutions of 2.6″–8.8″, corresponding to 360 AU–1200 AU at a distance of 140 pc to

Table 1. Observations with the NMA

Observed Source	Optical Appearance	$\log[F_{12}/F_{25}]$	L_{IR} ($L_\odot$)	Continuum (mJy)	CS(J=2–1) (Jy km s^{-1})
L1551-IRS5	invisible	−1.02	25	130	15
L1489	invisible	−0.64	3.5	< 9.6	5.6
04361+2547	invisible	−1.02	3.0	< 14	1.4
04368+2557	invisible	< −0.47	1.8	< 11	11
04169+2702	invisible	−0.84	1.1	< 14	2.6
04108+2803	invisible	−0.65	0.68	< 21	1.4
HL Tau	visible	−0.48	6.0	74	< 0.56
DG Tau	visible	−0.32	3.8	57	< 0.39
FS Tau	visible	−0.42	0.66	< 11	< 0.34
GG Tau	visible	−0.11	0.39	41	< 0.44
DL Tau	visible	−0.14	0.23	23	< 0.37

Taurus. The sample consists of 6 embedded *IRAS* sources and 5 visible T Tauri stars. The embedded sources tend to have colder color between 12 and 25 μm than the visible T Tauri stars, as was pointed out by Beichman *et al.* (1986). Observed 98 GHz continuum fluxes are consistent with the broad band spectra of dust emission, indicating that the 98 GHz continuum is the thermal emission from dust grains (Adams, Emerson, and Fuller 1990; Keene and Masson 1990).

Figures 1 shows the 98 GHz continuum maps. The continuum emission was detected toward 4 T Tauri stars and one embedded source, L1551-IRS5. The maps show quite compact sources not resolved by our spatial resolution, being consistent with the continuum emission arising from disks whose radii are less than 100 AU (Beckwith *et al.* 1990; Adams, Emerson, and Fuller 1990). The mass of these dust disks is estimated to be 1.6×10^{-1}–2.9×10^{-2} $M_\odot$ under the assumption of the $\kappa_\nu \propto \nu^1$ emissivity law. The density in such disks is extremely high. For example the lower limit to the H_2 number density in the dust disks is estimated to be several times 10^9 cm^{-3} assuming the emitting region to be a sphere of 100 AU in radius.

Figures 2 shows the CS ($J = 2 - 1$) maps. In contrast with the continuum emission, CS emission was detected toward all the embedded sources, while no visible T Tauri star shows significant CS emission. The CS maps are extended with their FWHM sizes being ∼1500–2000 AU.

The lower limit to the gas mass in the CS envelopes is estimated to be 1.8×10^{-2}–1.7×10^{-3} $M_\odot$, assuming the optically thin CS emission with T_{ex}=10 K. The upper limit to the gas mass within the NMA beam size is 1.5×10^{-2} $M_\odot$ (see Discussion). The H_2 number densities are thus larger than 10^5–10^6 cm^{-3}, if the emitting region is a sphere of 1500 AU in diameter. The linear size and high density suggest that the observed CS envelopes are very inner part of the extended molecular cloud cores.

3. Discussion

The remarkable point of the present results is that there is a clear difference between the embedded sources and visible T Tauri stars: the visible T Tauri stars are predominantly detected in the continuum emission which arises from the dense disks in the vicinity of stars, while the embedded objects are only detected in CS whose emission comes from the inner portion of molecular cloud cores. This indicates that the embedded sources do not

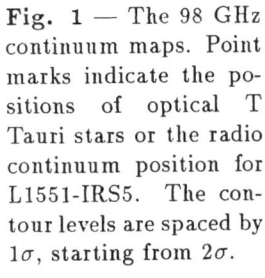

Fig. 1 — The 98 GHz continuum maps. Point marks indicate the positions of optical T Tauri stars or the radio continuum position for L1551-IRS5. The contour levels are spaced by 1σ, starting from 2σ.

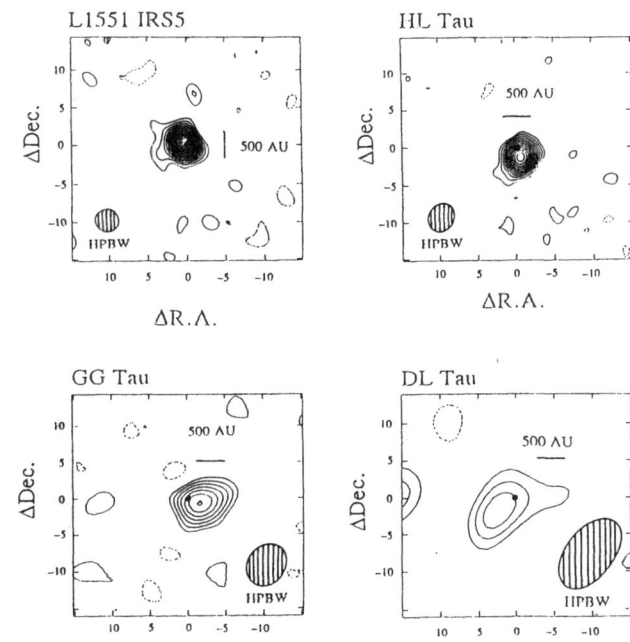

have the central mass concentration large enough to be detected in the 98 GHz continuum, while the visible T Tauri stars and L1551-IRS5 do.

For example, let us consider the following 2 cases. Most of the mass around the embedded sources may be widely distributed in their extended envelopes with sizes $\sim$0.01 pc–0.1 pc. Thus we assume a spherical envelope with the size, mass, and density profile of 0.1 pc, 1 $M_\odot$, and r^{-2}, respectively, for embedded sources, then the total gas mass within a radius of 500 AU is only 2.5×10^{-2} $M_\odot$. The 98 GHz continuum emission from such a region is close to our sensitivity of $\sim$12 mJy (3σ), corresponding to 1.5×10^{-2} $M_\odot$, and is rather difficult to be detected with high significance. On the other hand, a disk whose radius and mass are 100 AU and 0.1 $M_\odot$, respectively, produce the 98 GHz continuum emission enough to be detected by the 1000 AU beam. Total mass within the NMA beam size for disks around T Tauri stars is larger than that for embedded sources, so that the 98 GHz continuum emission is detected only toward T Tauri stars. Hence the detection of the 98 GHz continuum emission depends on how much mass is contained within the NMA beam size.

The CS emission is detected only toward embedded sources, because for those objects the detectable CS emission easily arises from the moderately extended envelopes with sizes $\sim$1500–2000 AU. The non-detection of the CS emission toward visible T Tauri stars sets an important upper limit to the size of gas disks if we consider that the CS emission was not detected because the beam averaged gas column density is not large enough as a result of beam dilution. The upper limit to the radius of gas disks is then 70 AU for the assumed excitation temperature of 50 K.

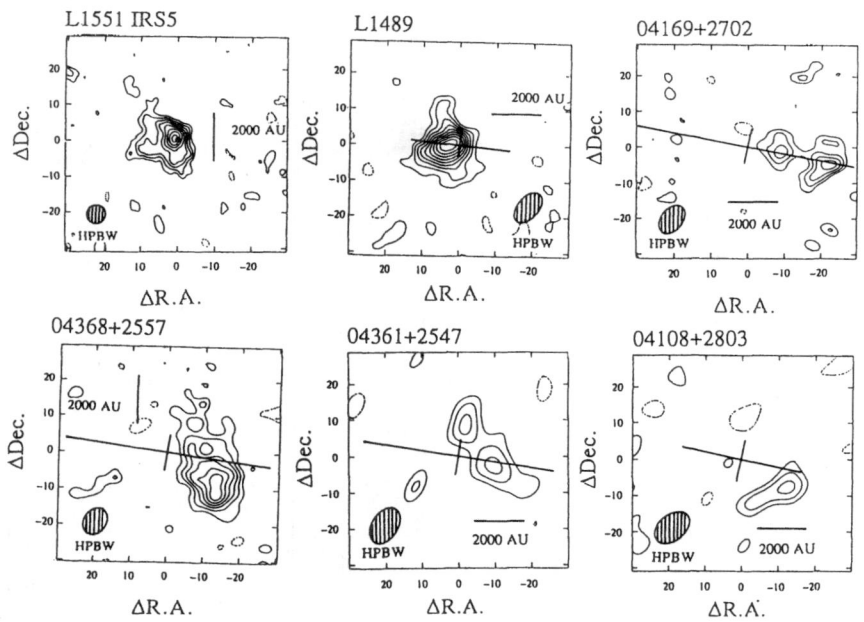

Fig. 2 — The CS $(J = 2 - 1)$ maps. The contour levels are spaced by 1σ, starting from 2σ. Plus marks indicate the positions of *IRAS* sources.

We conclude from the above results that the embedded objects in the younger evolutionary phase do not have enough mass concentration to be detected with the NMA beam size of ~500 AU around the central objects, although they have extended massive envelopes with steep density gradient toward the center, and that the more evolved T Tauri stars have enough amount of material to be detected within the ~500 AU radius beam. This suggests that such mass concentration in disks around T Tauri stars occurs through the dynamical mass accretion in the embedded phase of evolution. This result would provide important evidence that embedded sources are accreting protostars.

References

Adams, F. C., Lada, C. J., and Shu, F. H. 1987, *Ap. J*, **312**, 788.

Adams, F. C., Emerson, J. P., and Fuller, G. A. 1990, *Ap. J.*, **357**, 606.

Beichman, C. A., Myers, P. C., Emerson, J. P., Harris, S., Mathieu, R., Benson, P. J., and Jennings, R. E. 1986, *Ap. J.*, **307**, 337.

Beckwith, S. V. W., Sargent, A. I., Chini, R. S., and Güsten, R. 1990, *A. J.*, **99**, 924.

Myers, P. C., Fuller, G. A., Mathieu, R. D., Beichman, C. A., Benson, P. J., Schild, R. E., and Emerson, J. P. 1987, *Ap. J.*, **319**, 340.

Sargent, A. I., and Beckwith, S. 1987, *Ap. J.*, **323**, 294.

Strom, K. M., Strom, S. E., Edwards, S., Cabrit, S., and Skrutskie, M. F. 1989, *A. J.*, **97**, 1451.

Keene, J and Masson, C. R., 1990, presented at this conference.

Aperture Synthesis Observations of Orion-KL in CS lines and 3mm Continuum.

Y. MURATA[1], R. KAWABE[2], M. ISHIGURO[2], T. HASEGAWA[3], AND M. HAYASHI[1]

[1] Department of Astronomy, University of Tokyo
[2] Nobeyama Radio Observatory, National Astronomical Observatory
[3] Institute of Astronomy, University of Tokyo

Summary: We have made aperture synthesis maps toward the Orion-KL nebula in CS(1-0), CS(2-1) and 3mm continuum with the Nobeyama Millimeter Array (NMA). The CS(1-0) maps were obtained across the central 2′ region of the Orion-KL nebula. Our maps show the rotating disk around the KL nebula with the diameter of 0.3 pc and the mass of 25 $M_\odot$. The disk clearly shows the Keplerian rotation, which indicates that the mass of the KL nebula is 100 – 150 $M_\odot$. The shell structure around the molecular outflow is also found in CS(1-0) map. This structure also appears in CS(2-1) map.

From 3mm continuum map, we have found at least six continuum sources. One of them is f-f emission from the ultra-compact HII region around BN. Other sources are the dust clumps whose masses are estimated to be 1 – 4 $M_\odot$. The distribution of the dust clumps shows clear anti-correlation with the 20 μm emissions. This indicates that the 20 μm emissions are obscured by these dust clumps.

1. Introduction

The Orion-KL nebula is a massive star forming region which has been studied extensively (Genzel and Stutzki 1989). It is certain that IRc2 is the central star of this region from observations concerning the outflow (Genzel *et al.* 1981; Wright *et al.* 1983). Around the Orion-KL nebula, we can see the typical structures of young stellar objects. These are the bipolar molecular outflow (Erickson *et al.* 1981) and the rotating disk perpendicular to the axis of the outflow (Plambeck *et al.* 1982; Hasegawa *et al.* 1984; Vogel *et al.* 1985). We have made CS observations for the Orion-KL nebula with NMA, to understand the detailed structure and dynamics of this region. A 3 mm continuum map have been also obtained at the same time.

2. Observations

The CS(J=1-0),(2-1) lines and 3mm continuum were observed with the Nobeyama Millimeter Array (NMA; Ishiguro *et al.* 1984). We used IRc2 as the phase center for the observation. The front-end receivers were the dual-channel SIS receivers, whose typical system noise temperature are 300 K at 49 GHz and 400 K at 98 GHz. The back-end was a digital FFT spectro-correlator (FX), which has 1024 frequency channels. The CS(1-0) observations were carried out from 1988 March to 1989 April with 7 array configurations. The synthesized beam had an angular size of 6″ × 8″ (FWHM), and the primary beam size was 140″. The bandwidth of FX was 80 MHz, giving a velocity resolution of 0.48 km s^{-1}. The CS(2-1) and 3mm continuum observations were carried out from 1989 December to 1990 March with 4 configurations. The synthesized and primary beam size were 2″ × 3″ and 70″, respectively. The bandwidth of FX was 320 MHz, giving a velocity resolution of 0.96 km s^{-1}. The total bandwidth for the continuum observation was 250 MHz.

The bandpass calibration was made from the observation of 3C84 or 3C273. The visibility gain calibration was made with 0605-085, assuming its flux density of 3.2 Jy at 48 GHz, and 1.8 Jy at 98 GHz.

3. Disk with Keplerian rotation

Figure 1 shows the spatial distributions of the CS(1-0) emission in the LSR velocity ranges of 5.5 – 12.2 km s^{-1}. Three features are prominent in Fig. 1: (*1*) a strong compact source at the center, corresponding to the hot core, (*2*) the large rotating disk around Orion-KL, which is seen as the elongated structure from northeast to the southwest, and (*3*) two ridges extending toward the northwest and perpendicular to the disk.

The disk has the diameter of 0.3 pc and the thickness of 0.03 pc. The density of this disk is $\sim 10^5$ cm^{-3} derived by LVG model (Linke and Goldsmith 1980) using the CS(2-1) data (Mundy *et al.* 1988). The gas kinetic temperature was assumed to be $\sim$ 40 K derived from the CH$_3$CN results (Andersson 1985). The total mass of the disk component is $\sim$ 25 M$_\odot$ assuming the disk structure. The disk is seen as the elongated structure with the position angle of 30°. There is a weaker CS emitting region in a few arcsecond northeast of IRc2, which divide the disk to NE and SW parts. The center line of the disk are shown in Fig. 1.

We made a position-velocity map along the center line of the disk (Figure 2). The center position is the nearest point to IRc2. The hot core shows the broad line emission ($\gtrsim$ 20 km s^{-1}) in $0''$ – $-20''$ from the center. The other part in this map corresponds to the disk component. The Keplerian rotation curves with the central mass of 50, 100, 150, and 200 M$_\odot$ are shown in Fig. 2, for the systemic velocity of V$_{LSR}$ = 8.5 km s^{-1}. When considering about the effects of the random gas motion ($\sim$ 0.5 – 1.0 km s^{-1}), the rotation of the outer disk is well fitted by the Keplerian motion with the central mass of between 100 – 150 M$_\odot$.

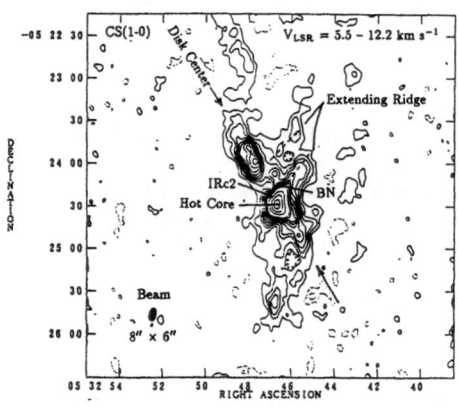

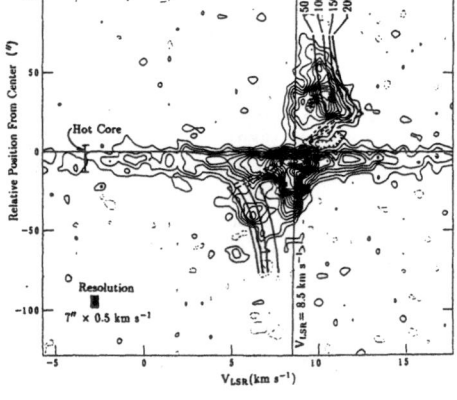

Figure 1 Map of the average emission in the CS (1-0) line in the velocity interval 5.5 < V$_{LSR}$ < 12.2 km s^{-1}. Thick lines indicate the center of the disk.

Figure 2 Position Velocity map of the CS disk. Thick lines show the Keplerian rotation curve for each central masses.

The motion in the inner part of the disk ($r < 25''$) departs from the Keplerian rotation, and has no high velocity gas motion except for the hot core. This indicate that there is no CS emission in the center of the disk. The SO expanding ring (Plambeck *et al.* 1982) and the proper motion of H_2O masers correspond to inner edge of the disk. These observational results indicates that the dynamics of the molecular gas change dramatically at the inner edge of the disk. Therefore, the shock surface is expected to exist between rotating disk and expanding ring.

4. Shell structures around the molecular outflow

Two ridges in Fig. 1 correspond to the CS shell in previous result (Murata *et al.* 1989). The extension is about $30'' - 40''$ and the southern ridge (extending from the hot core emission) is stronger than the northern one. The molecular outflow is confined between these two ridges.

We can also see these structures in CS(2-1) map in $V_{LSR} = 9.5 - 11.4$ km s^{-1} (Figure 3) with the resolution of $2'' \times 3''$. Because the resolution is twice as high as that of CS(1-0) map, we can resolve some clumps on the shell. The shell is surrounding the molecular outflow like as CS(1-0). The H_2 emissions in this region is the evidence of the shock between the outflow and the ambient gas (Beckwith *et al.* 1978). These results support the hydrodynamical models of molecular outflows that high velocity winds accelerate neutral gas which in turn sweeps up the ambient material to form a shock-compressed shell (Königl 1982).

Two NH_3 filaments are located on the extension of the ridges in Figure 1. This means that the high velocity winds expand out through the CS shell (or extending ridge) toward the directions along which the NH_3 filaments and the HH objects are formed (Murata *et al.* 1990).

5. Continuum sources at λ 3mm

Figure 4 shows the map of the λ3mm continuum with $2'' \times 3''$ resolution. Six clumps (A–F) are identified. The clump A corresponds to BN object. The flux density of the clump A is consistent with that of f-f emission from the ultra compact HII region around BN, whose spectrum was proposed from the lower frequency observations by Moran *et al.* (1983).

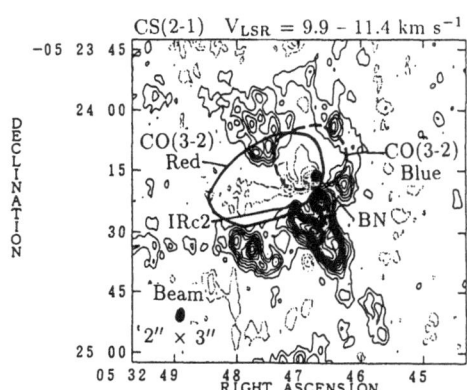
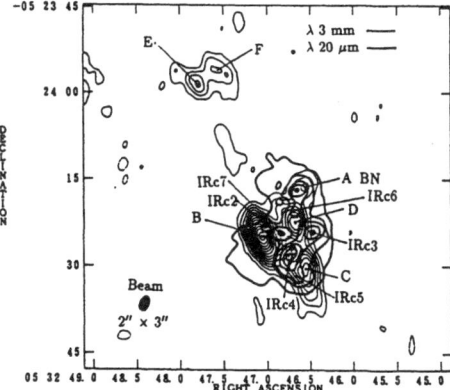

Figure 3 Map of the average emission in the CS (2-1) line in the velocity interval $9.5 < V_{LSR} < 11.4$ km s^{-1}. Thick solid and broken lines show the molecular outflow observed in CO (3-2) (Erickson *et al.* 1982)

Figure 4 Map of the 3 mm and 20 μm (Dowens *et al.* 1981). A–F are the identified 6 clumps.

Though the other clumps are more intense than the clump A, they have not detected at lower frequency (Garay, Moran and Reid 1987). Therefore, the emissions of these five clumps come from the dust. The parameters of the clumps are in table 1. The mass and the optical depth is derived following the appendix in Mezger et al. (1990). Typical mass and size of the clumps are 1–3 $M_\odot$ and $4'' - 7''$, respectively.

We compared the distribution of these dust clumps with the 20 μm continuum map (Downes et al. 1981; Fig. 4). The maps show clear anti-correlation between 20 μm and 3 mm continuum emissions. Though the optical depth of the dust is small ($\sim 10^{-2}$) at λ 3 mm, these clumps will become optically thick at 20 μm. This anti-correlation indicates that the dust clumps obscure the 20 μm emission reflecting the radiation from IRc2 and BN.

Table 1: Parameters of clumps.

Clumps	Size ($''$)	Flux (mJy)	Mass* ($M_\odot$)	τ_{3mm}* ($\times\ 10^{-3}$)
A(BN)	4.2 × 4.0	52	—	—
B	7.6 × 4.0	370	3.7	7.4
C	7.8 × 4.7	260	2.6	4.3
D	6.3 × 4.5	207	2.1	7.3
E	4.0 × 3.0	70	0.7	3.5
F	5.2 × 5.5	124	1.2	2.7
B+C+D	———	836	8.4	6.2

* Assuming T_d = 200 K, $\tau \propto \lambda^{-2}$.

References

Andersson,M. 1985, Proc. on "(Sub)Millimeter Astronomy", pp353.
Beckwith,S., Persson,S.E., Neugebauer,G.,and Becklin,E.E. 1978, Ap. J., **223**, 464.
Downes,D., Genzel,R., Becklin,E.E., and Wynn-Williams,C.G. 1981, Ap. J., **244**, 869.
Erickson,N.R. et al. 1982, Ap. J.(Letters), **261**, L103.
Garay,G., Moran,J.M., and Reid,M.J., 1987, Ap. J., **314**, 535.
Genzel,R., Reid,M.J., Moran,J.M., and Dowens,D. 1981, Ap. J., **244**, 884.
Genzel,R., and Stutzki,J., 1989, Ann. Rev. Astr. Ap., **27**, 41.
Hasegawa,T. et al. 1984, Ap. J., **283**, 117.
Ishiguro,M. et al. 1984, in Proc. Int. Symp. of Milli-meter and Submillimeter Wave Radio Astronomy, ed. J.Gomez-Gonzales (Granada:URSI), p.75
Königl,A. 1982, Ap. J., **261**, 155.
Linke,R.A., and Goldsmith,P.F., 1980, Ap. J., **235**, 437.
Mezger,P.G., Wink,J.E., and Zylka,R. 1990, Astr. Ap., **228**, 95.
Moran,J.M.et al. 1983, Ap. J.(Letters), **271**, L31.
Mundy,L.G. et al. 1988, Ap. J., **325**, 382.
Murata,Y., et al. 1989, in Structure and Dynamics of the Interstellar Medium, IAU Coll. No 120, ed. G.Tenorio-Tagle, M.Moles, and J.Melnick, (Berlin:Springer), pp.327
Murata,Y. et al. 1990, Ap. J., in press.
Plambeck,R.L. et al. 1982, Ap. J.(Letters), **259**, 617.
Vogel,S.N., Bieging,J.H., Plambeck,R.L., Welch,W.J., Wright,M.C.H. 1985, Ap. J., **296**, 600.
Wright,M.C.H. et al. 1983, Ap. J.(Letters), **267**, L41.
Wynn-Williams,C.G., Genzel,R., Becklin, E.E., and Downes,D. 1984, Ap. J., **281**, 172.

A 45 AU RADIUS SOURCE AROUND L1551–IRS 5:
A POSSIBLE ACCRETION DISK

J. Keene[1] and C. R. Masson[1,2]
[1]*California Institute of Technology, 320-47, Pasadena, CA, 91125, USA*
[2]*Center for Astrophysics, 60 Garden Street, Cambridge, MA, 02138, USA*

Abstract. Our interferometric and single-dish observations of the continuum emission from L1551–IRS 5 show that, at millimeter wavelengths, there are two distinct components to the source, an envelope with a radius $\geq$ 2000 AU, and a compact core with a radius $\leq$ 64 AU. The compact core has a large optical depth, indicating a high column density (~ 1000 g cm^{-2}). By modeling the temperature in the region of the compact core, we show that its size must lie in the range 45 ± 20 AU. The compact core is most plausibly identified with an accretion, or preplanetary, disk around the star, although the present observations do not have sufficient angular resolution to rule out other structures.

1. Introduction

The immediate circumstellar environment of protostars is the most important and least understood influence on the process of star formation. During gravitational collapse of material with nonzero angular momentum, it is likely that a disk will form around the protostar (e.g., Cameron 1985; Safronov and Ruzmaikina 1985; Lin and Papaloizou 1985; Hayashi, Nakazawa, and Nakagawa 1985). The rate of accretion on to the forming star is then controlled by the viscosity in the disk which permits angular momentum to be transferred outward and allows some of the disk material to fall on to the star. At some stage during the process, a strong wind develops, giving rise to a bipolar outflow of material, probably terminating accretion on to the star. Since, for energetic reasons, these winds must originate very close to the stellar surface, the accretion disk almost certainly plays an important role in producing the strong collimation often observed. In the final stages of evolution, planetary systems are believed to be formed from condensations in the remnant disk material.

Despite the vital role expected for such disks in all aspects of the formation and growth of low-mass stars, there is very little direct evidence about their nature. The most direct observation of a circumstellar disk is provided by the observations of Smith and Terrile (1985) of a thin, ~ 400 AU radius disk around the star β Pic, presumably the remnant of a more massive protostellar accretion disk. Estimates of outer radii of accretion disks give values of $\sim 10 - 100$ AU, (e.g., Adams, Lada, and Shu 1987), corresponding to angular radii of $0''07 - 0''7$ in the nearest star-forming regions, a scale which is difficult to resolve with most astronomical instruments. Indirect evidence for their nature comes from the existence of the solar system. From the elemental abundances found in the planets, it can be estimated that the disk of the presolar nebula must have extended at least to a radius of 40 AU and must have contained at least 0.02 $M_\odot$, just to account for the material in the planets (e.g., Safronov and Ruzmaikina 1985 and references therein).

One of the best known examples of a young star which is still deeply embedded in its parent cloud is the object IRS 5 in the L1551 dark cloud. This object has one of the most spectacular bipolar molecular outflows observed (Snell, Loren, and Plambeck 1980). The outflow lobes extend to a distance of $\gtrsim 10'$ ($\gtrsim 0.5$ pc at the assumed distance of 160 pc; Snell 1981) from the central star and are well collimated on this scale. Near-infrared observations (Campbell *et al.* 1988) show that the outflow is well collimated even at a distance of $1''$ (160 AU) from the central star. VLA cm continuum observations (Bieging, Cohen, and Schwartz 1984; Bieging and Cohen 1985; Rodríguez *et al.* 1986) show that there is a well-collimated jet of ionized material, with a diameter of $0''.5$ extending right to the center of the source.

Most of the material in the immediate environment of IRS 5 is molecular and is not detected by continuum observations at the VLA, while high obscuration impedes optical and infrared studies. The most sensitive tool for investigation of such material is continuum emission from dust grains, which is usually optically thin in the millimeter wavelength range and therefore provides the best tracer of the column density of material. Spectral line observations can provide velocity information, but the available resolution is limited by the sensitivity available in the small line widths.

2. Observational Results

We have mapped L1551 in the continuum at 2.73 mm at high spatial resolution with the Owens Valley Radio Observatory (OVRO) millimeter-wave interferometer. We have measured the 1 mm continuum flux density with the 5 m Hale telescope at Palomar Observatory and mapped the 1.25 mm continuum with the 10.4 m Caltech Submillimeter Observatory (CSO).

The observed flux densities are presented in Table 1. We have estimated the calibration uncertainties to be 20% and have added that in quadrature to the statistical noise.

TABLE 1

L1551 OBSERVATIONS

Wavelength (mm)	Peak Flux Density (Jy beam^{-1})	Beam Size ($''$)
1.0	5.7 ± 1.3	55
1.25	2.37 ± 0.48	27
2.73	0.13 ± 0.03	2.7×2.6

2.1. CSO Map at 1.25 millimeters

Figure 1 shows the 1.25 mm map of L1551 made at the CSO in 1988 September with a $27''$ beam (FWHM). Although no structural details can be discerned in Figure 1, it shows that L1551 is spatially resolved; Gaussian deconvolution gives a diameter of $24''$ (FWHM). Our measurements are in good agreement with similar recent data by Walker, Adams, and Lada (1990) and indicate that a substantial amount of the dust emission arises in an envelope with a radius of ~ 2000 AU.

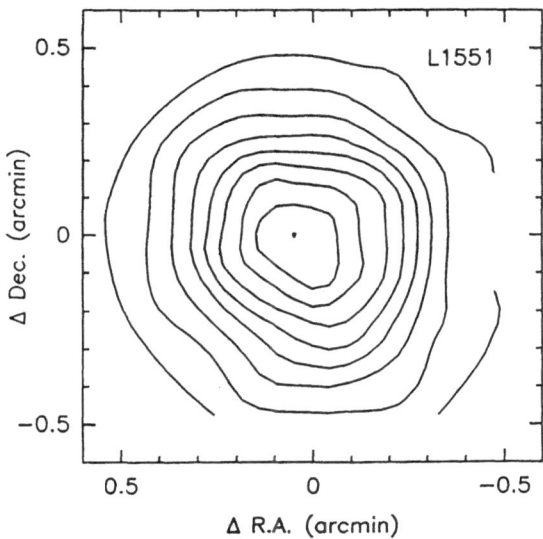

Fig. 1. 1.25 mm map of L1551–IRS 5 taken with 27″ resolution at the CSO. The peak flux in this map is 2.37 Jy per beam and the rms noise is ∼ 0.06 Jy. The Gaussian deconvolved source size is 24″ (FWHM). The contour levels are spaced by 10%, starting at 20% of the peak flux; the 10% contour lies outside the mapped area. The lowest contour is 8.5 σ.

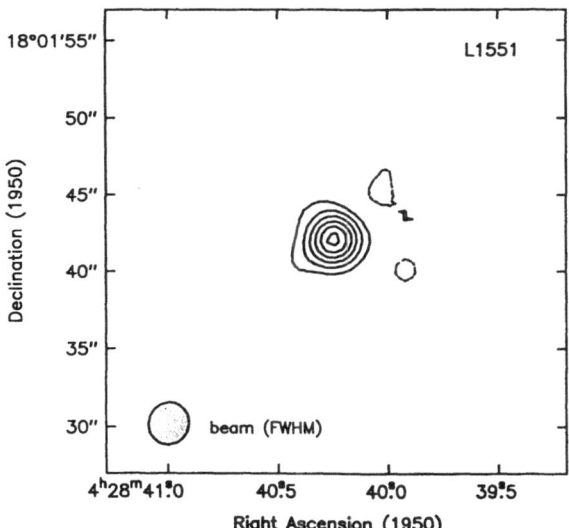

Fig. 2. 2.73 mm map of L1551–IRS 5 taken with 2.″7 × 2.″6 resolution at OVRO. The peak flux density in this map is 130 mJy per beam and the integrated flux is 150 mJy; the contours are 20 mJy. The rms noise in the map is 5 mJy per beam.

2.2. OVRO Interferometer Map at 2.73 millimeters

Figure 2 shows the map of the 2.73 mm continuum emission of the L1551–IRS 5 area made from only the best 3 days of data. The beam size is 2″.7 × 2″.6. Because the minimum projected baseline in this dataset was 35 m, structures with scale sizes $\gtrsim 10''$ are resolved out. Thus the interferometer map shows only a strong pointlike source despite the evidence presented above from the single-dish measurements that there also exists a large envelope. The center of the 2.73 mm source is at $04^h 28^m 40^s.25$, $+18° 01' 42''.2$ (1950) and is coincident with the position of the radio continuum sources mapped by Rodríguez *et al.* (1986).

The presence of the extended envelope, seen at 1.25 mm with the CSO, complicates attempts at deconvolution to determine the size of the compact core seen by the interferometer. A limit on the source size can be found by considering the measured visibility amplitudes plotted in Figure 3. In this figure, the observed visibility amplitudes from the entire 1986–87 season were vector averaged over 1 hr periods and then binned at regular u-v intervals (and thus azimuthally averaged). The error bars represent only the scatter among the vector-averaged points. At the shortest baselines the visibility amplitudes increase due to the extended envelope which is heavily resolved by the interferometer; at larger baselines the visibility amplitudes are approximately constant, although there is a slight falloff with increasing baselines.

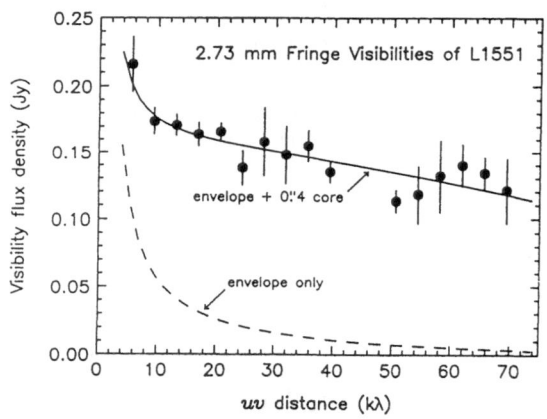

Fig. 3. The observed 2.73 mm visibilities. The dashed line shows the visibilities resulting from an envelope with 0.29 Jy total flux and with a volume emissivity $\propto r^{-2}$. The solid line shows the visibilities resulting from a similar envelope with a flux of 0.14 Jy plus a compact Gaussian source with a flux of 0.15 Jy (for a total flux of 0.29 Jy) and a half-power radius of 0″.4. It is obvious that the two-component model (solid line) fits the data much better than a simple power-law source.

For comparison we have computed the visibility spectrum of a $\rho \propto r^{-3/2}$ power-law envelope as predicted by Terebey, Shu, and Cassen (1984) for the collapse of a slowly rotating, spherical, isothermal cloud. The temperature profile was assumed to have a power-law form, $T \propto r^{-1/2}$. In the Rayleigh-Jeans limit, the source volume emissivity, therefore, is proportional to r^{-2}. The total flux density (0.29 Jy within a 60″ beam) was interpolated between our 1.0 mm measurement and that of Walker, Adams, and Lada (1990) at 2.9 mm (Table 2). The lower curve in Figure 3 (*dashed line*) shows the visibility for this power-law envelope; it falls far short of the data at large baselines. Varying the power-law index changes the shape of the curve slightly, but for no power-law index can the predicted flux densities be made to match both the visibility amplitudes and the large beam flux. The large visibility amplitudes seen at large baselines therefore demonstrate that the compact core detected by the interferometer is a discrete structure and cannot be simply the dense, central part of a power-law envelope.

A good fit to the data in Figure 3 can be obtained, however, by using a two-component model, with a power-law envelope to account for the rise in flux density at short baselines and a compact core to account for the large flux in the long-baseline data. The solid curve plotted through the data points consists of a power-law envelope plus a compact circular Gaussian source. This two component model fits the observations very well. The small decrease in the visibility amplitudes at large baselines gives an estimate of the size of the compact core; the plotted curve is derived from a compact source with a half-power radius of $0\rlap{.}''4$ and a flux density of 0.15 Jy. Again, the total flux in a $60''$ beam was taken to be 0.29 Jy (Table 2). To fit the data only the size and total flux of the compact core were allowed to vary. This size for the compact core derived from Figure 3 is much more realistic than a simple Gaussian fit to the interferometer map because the effect of the envelope is taken into account, but it is still subject to observational errors which are likely to reduce the measured amplitudes on the longest spacings. Therefore we regard $0\rlap{.}''4$ (64 AU) as an *upper limit* to the radius of the compact core.

Using the upper limit for the size of the compact core, we can calculate the lowest brightness temperature required to produce the observed flux densities. For a circular Gaussian source with a radius of $0\rlap{.}''4$ and an integrated flux density at 2.73 mm of 150 mJy, the peak Rayleigh-Jeans brightness temperature is 24 K. If the source size is the same at 1.36 mm, the required brightness temperature is 28 K to produce the observed flux density of 0.7 Jy (Woody *et al.* 1989). If the compact core is actually smaller than $0\rlap{.}''4$ in radius, then the brightness temperature must be correspondingly higher. In §4.2 below we use an upper limit to the source temperature to derive a lower limit of $0\rlap{.}''16$ to the source radius.

3. Continuum Energy Distribution of L1551

The continuum energy distribution of L1551–IRS 5 (Fig. 4) shows a large far-infrared luminosity characteristic of thermal emission from dust grains and a relatively flat centimeter wavelength spectrum characteristic of optically thin free-free emission from ionized gas. Integration under a smooth curve fitted through only the points measured with large beam sizes ($> 25''$) gives a total luminosity for L1551–IRS 5 of $33 \pm 3\ L_{\odot}$ for the assumed distance of 160 pc.

In Table 2 we separate out the contributions of the compact core and the envelope to the total flux in the mm wavelength region. To do this we have used the observed flux densities to calculate the spectral indices of the total flux and of the compact core, then interpolated or extrapolated to the unobserved frequencies. For the compact core we have used only the 1.36 and 2.73 mm measurements for the interpolation since they were made with the smallest beam sizes. Subtraction of the compact core from the total flux gives us the envelope flux.

The compact core has an observed spectral index of 2.4 ± 0.5 over the range from 1.36 to 2.73 mm, and the Planck correction should be negligible at the temperatures expected 64 AU from the central star. This spectral index is close to the value of 2 expected for optically thick thermal emission. The inferred spectral index of the envelope between 1.0 and 2.9 mm is 3.3 ± 0.7. The dust in the envelope has a relatively low temperature and a Planck correction must be made to calculate the emissivity law of the dust. If the mean temperature of this material is taken to be 25 K, the emissivity law has a slope of 1.5 ± 0.7, in adequate agreement with the theoretically expected value of 2.0 (see e.g., Draine and Lee 1984).

TABLE 2

ADOPTED FLUX DENSITIES FOR L1551

Wavelength (mm)	Total[a] (Jy)	Core[b] (Jy)	Envelope (Jy)
1.0	5.7[c]	1.5	4.2
1.36	2.3	0.7[d]	1.6
2.73	0.29	0.13[e]	0.16
2.90	0.24[f]	0.11	0.13

[a] Assumes, unless measurement is indicated, that $F_\nu \propto \nu^{3.0}$.
[b] Assumes, unless measurement is indicated, that $F_\nu \propto \nu^{2.4}$.
[c] This work, 55″ beam.
[d] Woody et al. 1989; 3″ beam.
[e] This work, 2″.7 × 2″.6 beam.
[f] Walker, Adams, and Lada 1990; 60″ beam.

4. The Structure of L1551–IRS 5

The most detailed modeling of the emission from sources such as L1551–IRS 5 has been performed by Adams and Shu (1986), who have successfully reproduced the observed energy distribution with a model of an infalling envelope resulting from the symmetrical collapse of a rotating, spherical, isothermal cloud (Terebey, Shu, and Cassen 1984). In these models, the infalling envelope has a power-law density distribution outside the radius where rotation becomes significant. Inside this radius, the infalling material forms an accretion disk as it spirals in toward the central star. The luminosity arises, in general, from three different sources, nuclear burning in the central star, dissipation of gravitational potential energy in the accretion disk, and the accretion shock, where the orbiting material at the inner edge of the disk meets the surface of the star. In their model, the emission from the central star and accretion disk was used only to heat the envelope, and the appearance of the disk itself was not considered in detail. Since the interferometer data presented here show evidence for a compact core in addition to the envelope, we have identified the compact core with the disk and have attempted to model the appearance of this disk as well as the extended envelope. In this section, we analyze the available spectral and mapping data to identify what constraints they place on models of the source, and we present a simplified model which has the essential characteristics required to account for the observations.

We assume that dust emission is represented by an opacity law of the form described by Keene, Hildebrand, and Whitcomb (1982) and Hildebrand (1983). The low-frequency opacity ($\lambda > 100\ \mu$m) is taken to be $\tau = 1.3 \times 10^{-26} N_H \lambda_{mm}^{-2}$ (cm^2 atom^{-1} mm^2), which gives rise to an energy distribution of the form $F_\nu \propto \nu^4$ in the Rayleigh-Jeans limit and which is in good agreement with the characteristics of the extended envelope discussed in the previous section. Although we find a shallower slope for the emission from the compact core in L1551, we show below that this slope can plausibly be accounted for by the effects of high optical depth. In general, the sources which are detectable by interferometers at low frequencies are those in which brightness temperatures and optical depths are high; opacity effects are therefore important and the dust emissivity law cannot simply be inferred from the slopes of the observed spectra.

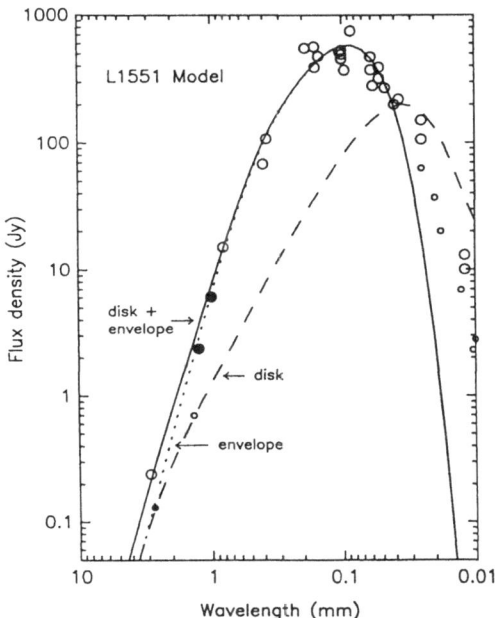

Fig. 4. Curves representing simple models for the compact core and envelope and the combined energy distribution plotted with the observations. The dashed curve shows the flux from the model disk; the dotted curve shows the flux from the model envelope; the solid curve shows the flux from the disk/envelope combination. Filled symbols indicate flux densities obtained in this work; large symbols indicate flux densities obtained with beam sizes $> 25''$; small symbols indicate flux densities obtained with beam sizes $< 25''$.

4.1. Envelope

In Figure 4 we plot the observed fluxes of L1551, where we take the large beam measurements to be observations of the core/envelope combination and the interferometric measurements at mm wavelengths to be measurements only of the core, along with curves representing simple models for the core and envelope separately. For simplicity, the plotted model for the envelope consists of a homogeneous sphere of material with a total mass of 1.12 $M_\odot$, a radius of $1''5$ (240 AU), and a temperature of 57 K. The total mass is constrained by the long wavelength data points where the envelope emission is optically thin and the radius and temperature were taken to be those where the Planck-averaged optical depth of the envelope becomes small (i.e., to fit the observed emission peak). This model does not account for the extended nature of the envelope, which makes single-dish millimeter wavelength maps appear extended (e.g., Fig. 1), but it does give a rough estimate of the temperature and mass of envelope material. However, the total mass of the envelope is probably larger by a factor of 2–3 than assumed in the model because the bulk of the envelope material is far from the central star and has a lower temperature than that derived from the emission peak. The size and mass of the envelope are poorly defined, since it blends into the surrounding cloud, but the total mass is probably 2–3 $M_\odot$ within a radius of ~ 2000 AU.

4.2. Compact Core

The compact core has been measured only between 1.36 and 3.4 mm, and in this wavelength region has a spectral index of ~ 2.4. The simplest explanation for this index is that the compact component is so condensed that the material is optically thick at a wavelength of 1.36 mm and has $\tau \sim 1$ at 2.73 mm. An argument in favor of this explanation is the large brightness temperature calculated for the compact core. At a radius of $0\rlap{.}''4$ (64 AU) from the central star, the characteristic temperature, $(L_*/4\pi\sigma r^2)^{1/4}$, should be ~ 120 K. Based on our upper limit of $0\rlap{.}''4$ for the radius of the compact core, we calculate a Rayleigh-Jeans brightness temperature of > 24 K at 2.73 mm, indicating that the optical depth through the compact core must be *at least* 0.2; it could be significantly greater if the compact core radius is much smaller than $0\rlap{.}''4$.

If our adopted opacity law (Keene, Hildebrand, and Whitcomb 1982; Hildebrand 1983) is valid for the extreme density conditions found in the compact core then the column density required for an optical depth of $\tau \sim 1$ at 2.73 mm is $\sim 6 \times 10^{26}$ atoms cm^{-2}, or 1000 g cm^{-2}. This is an extraordinarily high value by the standards of molecular clouds, but it is within the range, albeit at the high end, of surface densities required for models of accretion disks around young stars (see Lin and Papaloizou 1985; Cameron 1985). The volume density of molecules is also extremely high. Even if the gas in the core had a spherical distribution, with a radius of $0\rlap{.}''4$, the number density would be $n_{H_2} \sim 10^{11}$ cm^{-3}. Since the material is probably flattened into a disk, the density should be significantly higher.

The size of the compact core is easy to determine from the observations. As discussed before, the upper limit from the 2.73 mm flux visibilities (Fig. 3) is 64 AU. If we assume that the 1.36 mm flux density of 0.7 Jy is due to an optically thick source, a rough lower limit to the size of the emitting region is also easily obtained. If the source temperature falls from the stellar surface as $r^{-1/2}$, more slowly than is expected for accretion disks thus giving the minimum size source, then the source radius is $\sim 0\rlap{.}''16$, or 25 AU. Combining this with the upper limit from observations, the radius of the compact core is therefore fairly well constrained to the range 45 ± 20 AU.

Since we have demonstrated that the compact core is a discrete, dense structure, resembling the accretion disks expected to exist around young stars, we have chosen to model it as such a thin accretion disk in order to define its characteristics in more detail. The deduced size and column density would change only slightly, however, if some other structure were assumed. This disk is embedded in the cooler envelope (of somewhat higher mass; §4.1) which serves to obscure it at short wavelengths. The structure of the source is sketched in Figure 5.

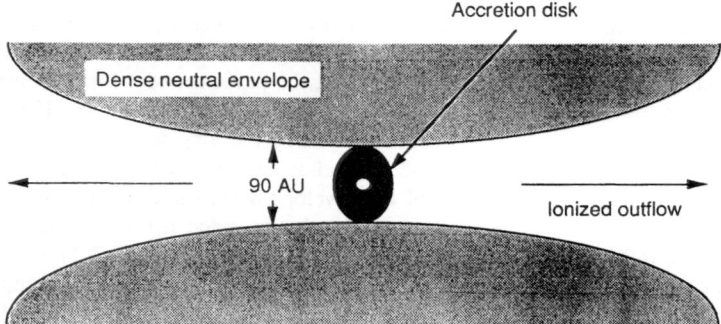

Fig. 5. Sketch of a cross-section of the neutral envelope around L1551–IRS 5, showing the embedded dense neutral disk surrounding the central star and a well-collimated ionized jet.

The theory of accretion disks around young stars is poorly understood, due to the lack of observational constraints; many different models have been presented, with quite different physical characteristics (e.g., Cameron 1985; Safronov and Ruzmaikina 1985; Lin and Papaloizou 1985; and Hayashi, Nakazawa, and Nakagawa 1985). The temperature structure in an ideal, self-luminous, optically thick accretion disk should be a power law, with a dependence given by $T \propto r^{-3/4}$ (Lynden-Bell and Pringle 1974). Adams and Shu (1986) have shown that this distribution will be modified by radiative interaction between the star and the disk. In a deeply embedded source like L1551–IRS 5, there is a further important effect, which is due to the optically thick envelope. The ideal disk models assume that the disk radiates freely into empty space, while the presence of the envelope means that the disk is effectively inside an oven at the temperature of the inner edge of the envelope. Because radiated power is proportional to T^4, the envelope does not have much effect on those parts of the disk which are significantly hotter than the envelope, but it sets a minimum temperature, below which the disk temperature cannot fall. We assume that the inner boundary of the envelope falls at the outer edge of the disk and, as an approximation, we assume that the envelope temperature is given by $T_{env} = T_* \times (r_*/r_{env})^{1/2}$, where T_* is the effective temperature of the star, r_* is its radius, and r_{env} is the inner radius of the envelope. Since the disk should match the stellar temperature at the surface of the star and the envelope temperature at the boundary of the envelope, we assume that the surface temperature of the disk falls as $r^{-3/4}$ out to the point where it reaches the envelope temperature and is constant thereafter. The majority of the disk area is at the temperature of the envelope. For simplicity, we ignore the details of the inner region, where grain destruction occurs, since the millimeter wave emission is insensitive to the inner boundary for temperature laws flatter than $T \propto r^{-1}$.

The calculated emission from our disk model is plotted in Figure 4, where it can be seen to reproduce the small-beam interferometric measurements quite well. The inclination of the disk was taken to be 45° (Stocke et al. 1988). In this model, the disk has a radius of 40 AU, a constant surface density of 840 g cm^{-2}, and a total mass of 0.6 $M_\odot$. The temperature at the outer edge of the disk was 150 K. The short wavelength emission of the disk exceeds the observed values, but at these wavelengths the disk is obscured by the envelope. To demonstrate that the combination of a compact disk and a surrounding envelope can fit the observations, we have computed the continuum energy distribution of a model which consists of the disk as described above, embedded in a homogeneous spherical envelope (§4.1). The combined energy distribution is also plotted in Figure 4, where it can be seen to match the large beam measurements very well at wavelengths longer than about 40 μm. The single-temperature envelope in our simple model does not reproduce the measurements at shorter wavelengths but more realistic envelope models have been discussed by Adams, Lada, and Shu (1987) and Butner et al. (1990).

5. Discussion

We have shown that there is a compact core to the L1551–IRS 5 cloud, centered on the radio continuum source, with a radius of ~ 45 AU, an optical depth at 2.73 mm of ~ 1, corresponding to a column density of ~ 1000 g cm^{-2}, and we estimate the temperature at its outer edge to be ~ 150 K. The density in this core is $n_{H_2} \gtrsim 10^{11}$ cm^{-3}, if its shape is spherical, and is higher if the structure is flattened. This source has not been seen in any other observations because of the extremely high dust opacity. We have shown that there also exists a dense envelope of material around this source with a radius > 2000 AU. The two components are distinct, as can be seen in the 2.73 mm fringe visibilities.

5.1. Accretion Disks

The properties expected for circumstellar accretion disks have been discussed by many authors. Reviews of the chief competing theories are given by Cameron (1985), Safronov and Ruzmaikina

(1985), Lin and Papaloizou (1985), and Hayashi, Nakazawa, and Nakagawa (1985). Based on the properties of the planets in the solar system, it is believed that the preplanetary nebula must have had a surface density of 1000 g cm^{-2} at a radius of ~ 1 AU at the epoch of planet formation, but the total disk mass inferred depends on the radial structure. In the most parsimonious model, discussed by Hayashi, Nakazawa, and Nakagawa (1985), the surface density falls steeply with radius, and the total mass required is 0.013 $M_\odot$. At the other extreme, the models discussed by Cameron (1985) have high surface densities which fall more slowly with radius, and correspondingly larger masses. The surface density which we estimate, ~ 1000 g cm^{-2}, is comparable with the values used in the models described by Cameron (1985) and Lin and Papaloizou (1985), and the mass which we deduce within the radius of Saturn (9.5 AU) is ~ 0.03 $M_\odot$, in agreement with the models, but the disk radius is 45 ± 20 AU, and the total mass is therefore much larger than that of a minimum mass solar nebula.

Thus, the compact core centered on L1551–IRS 5 has properties very close to those expected for accretion disks around young stars, and we therefore believe that the most plausible hypothesis is that it is such a disk. Definite confirmation of this hypothesis will require further observations to map the structure of the source and to detect the velocity signature of Keplerian rotation to demonstrate that the material is in orbit around IRS 5.

5.2. Envelope Structure

It is obvious from optical, near-infrared, molecular line, and radio observations that the envelope surrounding IRS 5 does not have the spherical symmetry we have assumed for simplicity in our modeling. IRS 5 lies at the center of a large molecular outflow (Snell, Loren, and Plambeck 1980) and at the center of an ionized jet, visible at radio (Bieging and Cohen 1985) and optical (Mundt and Fried 1983; Campbell et al. 1988) wavelengths. Near-infrared observations reveal a conical structure that has been interpreted to be reflection from the inner edge of a thick disk (Strom et al. 1985; Hodapp et al. 1988). This infrared structure is larger than the disk revealed by our interferometer observations but could easily be the inner edge of the dense neutral envelope. The implied lack of spherical symmetry of the envelope does not affect our analysis of the compact core, nor does it change our estimate of the envelope mass. Our 1.25 mm CSO observations do not reveal any departure from spherical symmetry of the envelope on the $\sim 20''$ scale nor do the high-resolution 50 and 100 μm observations of Butner et al. (1990).

6. Conclusions

We have made high-resolution observations of the millimeter-wave continuum emission from L1551–IRS 5. At a wavelength of 2.73 mm we have detected a compact core with a radius of $\leq 0\rlap{.}''4$ (≤ 64 AU) which accounts for about half of the total continuum flux density at this wavelength. The most likely explanation for this source is that it is an accretion disk surrounding the central star.

From a single-dish map at a wavelength of 1.25 mm, we demonstrate that the compact core is surrounded by an envelope with a radius $\geq 12''$. We estimate that the mass of the envelope is $2 - 3$ $M_\odot$.

The spectrum of the continuum emission from the compact core is very flat and is consistent with optically thick thermal emission for wavelengths shorter than 2.73 mm. If our adopted opacity law is valid for the compact core and if the 2.73 mm optical depth is ~ 1, then the column density in the compact core is $N_H \approx 6 \times 10^{26}$ atoms cm^{-2}, or ~ 1000 g cm^{-2}. This is much larger than any column densities previously measured in neutral material, but it is comparable with column densities predicted in models of accretion/preplanetary disks around young stars. The radius of the compact core closely matches the expected size of an accretion disk around the young star and the size of our own solar system.

With the aid of a simple disk model we have shown that the radius of the compact core is 45 ± 20 AU, and that the total mass is $\sim 0.6\ M_\odot$. This mass is larger than expected in most models of presolar systems, largely because the surface density is somewhat high, but in the course of the evolution of such a system, much of the material may yet be accreted on to the star or blown off in the wind.

The OVRO millimeter interferometer is supported by NSF grant AST 87-14405 to Caltech; the CSO is supported by NSF grant AST 88-15132 to Caltech.

References

Adams, F. C., Lada, C. J., and Shu, F. H. 1987, *Ap. J.*, **312**, 788.

Adams, F. C., and Shu, F. H. 1986, *Ap. J.*, **308**, 836.

Bieging, J. H., and Cohen, M. 1985, *Ap. J. (Letters)*, **289**, L5.

Bieging, J. H., Cohen, M., and Schwartz, P. R. 1984, *Ap. J.*, **282**, 699.

Butner, H. M., Evans, N. J., II, Lester, D. F., Levreault, R. M., and Strom, S. E. 1990, *Ap. J.* in press.

Cameron, A. G. W. 1985, in *Protostars and Planets II*, ed. D. C. Black and M. S. Matthews (Tucson: University of Arizona Press), p. 1073.

Campbell, B., Persson, S. E., Strom, S. E., and Grasdalen, G. L. 1988, *A. J.*, **95**, 1173.

Draine, B. T., and Lee, H. M. 1984, *Ap. J.*, **285**, 89.

Hayashi, C., Nakazawa, K., and Nakagawa, Y. 1985 in *Protostars and Planets II*, ed. D. C. Black and M. S. Matthews (Tucson: University of Arizona Press), p. 1100.

Hildebrand, R. H. 1983, *Quart. J. R. A. S.*, **24**, 267.

Hodapp, K.-W., Capps, R. W., Strom, S. E., Salas, L., and Grasdalen, G. L. 1988, *Ap. J.*, **335**, 814.

Keene, J., Hildebrand, R. H., and Whitcomb, S. E. 1982, *Ap. J. (Letters)*, **252**, L11.

Lin, D. N. C., and Papaloizou, J. 1985, in *Protostars and Planets II*, ed. D. C. Black and M. S. Matthews (Tucson: University of Arizona Press), p. 981.

Lynden-Bell, D., and Pringle, J. E. 1974, *M. N. R. A. S.*, **168**, 603.

Mundt, R., and Fried, J. W. 1983, *Ap. J. (Letters)*, **274**, L83.

Mundt, R., Stocke, J., Strom, S. E., Strom, K. M., and Anderson, E. R. 1985, *Ap. J.*, **297**, L41.

Rodríguez, L. F., Cantó, J., Torrelles, J. M., and Ho, P. T. P. 1986, *Ap. J. (Letters)*, **301**, L25.

Safronov, V. S., and Ruzmaikina, T. V. 1985, in *Protostars and Planets II*, ed. D. C. Black and M. S. Matthews (Tucson: University of Arizona Press), p. 959.

Sargent, A. I., and Beckwith, S. 1987, *Ap. J.*, **323**, 284.

Smith, B. A., and Terrile, R. J. 1985, *Science*, **226**, 1421.

Snell, R. L. 1981, *Ap. J. Suppl.*, **45**, 121.

Snell, R. L., Loren, R. B., and Plambeck, R. L. 1980, *Ap. J. (Letters)*, **239**, L17.

Stocke, J., Hartigan, P. M., Strom, S. E., Strom, K. M., Anderson, E. R., Hartmann, L. W., and Kenyon, S. J. 1988, *Ap. J. Suppl.*, **68**, 229.

Strom, S. E., Strom, K. M., Grasdalen, G. L., Capps, R. W., and Thompson, D. 1985, *A. J.*, **90**, 2575.

Terebey, S., Shu, F. H., and Cassen, P. 1984, *Ap. J.*, **286**, 529.
Walker, C. K., Adams, F. C., and Lada, C. J. 1990, *Ap. J.*, **349**, 515.
Woody, D. P., Scott, S. L., Scoville, N. Z., Mundy, L. G., Sargent, A. I., Padin, S., Tinney, C. G., and Wilson, C. D. 1989, *Ap. J.* (*Letters*), **337**, L41.

STRONG MAGNETIC FIELDS IN BIPOLAR OUTFLOWS

M. D. SMITH*, P. W. J. L. BRAND† & A. MOORHOUSE‡,
*Dept. of Physics, University of Durham, †Dept. of Astronomy, University of Edinburgh
& ‡Dept. of Mathematics, UMIST

ABSTRACT. A supersonic wind from a young star will produce regions of strong magnetic field in the stellar environment. The associated shocks compress the molecular gas, increasing the density n, pressure p, and field B. Crucially, the Alfvén speed, $v_A \propto B/n^{1/2}$, is also increased since the total shock compression is approximately of the form $B \propto n$. But is there any evidence for such high v_A- or 'active cloud' - regions within bipolar outflows? We indicate below one implication which has important observable consequences: fast shocks of low Alfvén number (v/v_A) now arise. With a low ionization level, the C-shock structure is qualitatively different from the high Alfvén number flows which are common to 'quiescent cloud' conditions. The magnetic-field cushioning now allows molecular hydrogen to survive very fast shocks and broad H_2 lines are feasible. We display results which show that the resolved broad lines and line ratio properties in the OMC-1 outflow can be explained with fast bow shocks moving through such active regions.

1. Line profiles from C-shocks

Shocks in dense molecular clouds are expected to be C-type, in which the low ionized fractions allow ion-magnetosonic waves to propagate upstream, spreading the transition and avoiding the discontinuous jump of a J-shock (see Draine, this volume). This does assume that the ionization level remains low all across the shock - typically limiting the velocity difference between the neutrals and ions/field to $v_{in} < 40$ km s^{-1} (Draine et al 1983, Smith & Brand 1990a).

We shall mainly look at shocks with $v > 30$ km s^{-1} (or curved shocks in which this is satisfied over part of the shock surface) for which line profiles could be resolved by present observations. 'Quiescent' and 'active' cloud regions are here defined by the Alfvén speed. In the former, $v_A < 4$ km s^{-1}. Hence the shock Alfvén number is large in quiescent regions. This implies that the ion/field component possesses only a small fraction of the flow momentum, most of which is in the neutral component. Thus the neutrals are very slowly accelerated and lag way behind the ions. Hence C-shock velocities cannot far exceed

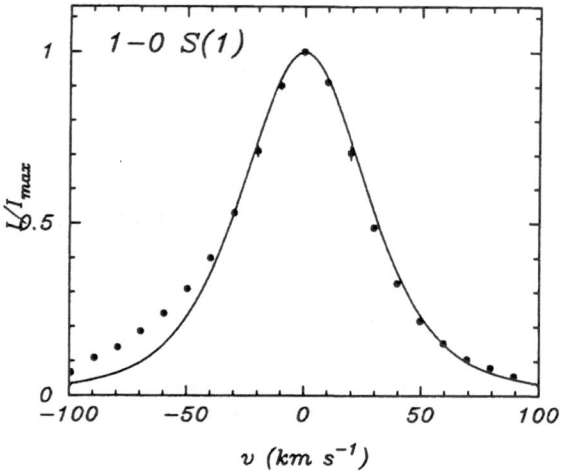

Figure 1: Calculated line profile with the OMC-1 Peak 1 data superimposed.

v_{in} and the molecular lines produced are narrow.

In contrast, in active regions ($v_A > 10$ km s^{-1}) the low Alfvén number implies that the ions/field contains considerable momentum. The neutrals do not lag far behind the ions - the velocity difference (i.e. the streaming velocity) can be a small fraction of the shock velocity. Hence the shock velocity can be far in excess of the streaming velocity in active regions (Smith et al 1990a,b). Moreover, since the ions heat the neutrals via streaming collisions, the temperature of the molecular hydrogen is tied to the streaming velocity and not the shock velocity. Thus molecular dissociation is similarly inhibited in strong field regions.

We have employed a shock model to determine if the above scenario can provide an explanation for the extremely wide lines observed at peak 1 of OMC-1 ($v \sim 140$ km s^{-1}). No previous modelling of these lines has proved possible (see Smith & Brand 1990b). Indeed, we find a reasonable fit without the need of manipulating parameters. Figure 1 displays the data and model for a paraboloidal bow shock of velocity $v_w = 250$ km s^{-1} moving in the plane of the sky. (The bow calculations take into account the dissociated cap region (bespeckled) and the field direction at each point on the bow surface (figure 2). Each part of the bow is treated as a planar shock element. UV radiation from the leading edge is discussed by Smith & Brand (1990b).) The magnetic field required is 25 times larger than the 'standard' quiescent value; with $v_A = 46.4$ km s^{-1} the field is 30μG for an H_2 density of 10^6 cm^{-3}. With an ionized fraction $2.5\ 10^{-7}$ the breakdown speed of a planar shock is 130 km s^{-1} . The H_2O abundance was set to zero throughout the shock - this value, however, is not crucial to the profiles.

2. Molecular hydrogen line ratios

There is now extensive accurate H_2 line ratio data for OMC-1. Two factors are apparent:

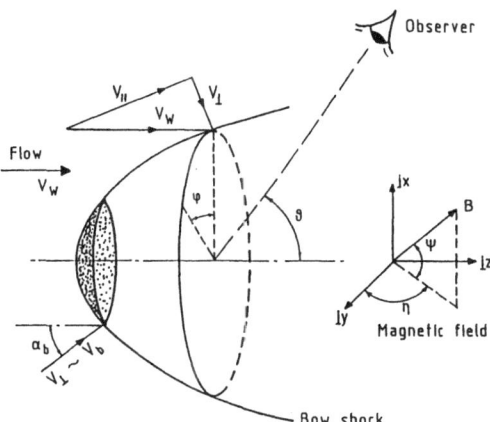

Figure 2: The bow shock.

(1) the ratios cannot be modelled by a constant temperature slab since the excitation temperature increases with the upper energy level of the transitions (Brand et al 1988) and (2) the ratio of 0-0 S(13) (17445K) to 1-0 O(7) (8365K) is independent of the position in the outflow as observed by Brand et al (1989a).

We employ the column density ratio (CDR) method to interpret the data. A CDR expresses the column of H_2 in a particular energy level to that in the strong 1-0 S(1) line, all the numbers being normalised to that of a 2000K slab of size which would reproduce the 1-0 S(1) line intensity. This method is necessary to make accurate quantitative fits in which error bars are of a recognisable size. A column analysis assumes that the H_2 is in LTE ($n \geq 10^6$) but it remains a useful tool at lower densities.

Remarkably, the bow shock model not only fits the data but possesses a narrow upper envelope (figure 3) provided the bow velocity exceeds the breakdown velocity ($v > 130$ km s^{-1} in the example). The CDR envelope is fixed by the shock shape (changing v_w, B or χ shifts the position of the emitting region without changing the ratios), and the cooling function (a high H_2O production rate leads to quicker downstream cooling and a steeper CDR envelope). A perfect fit is not apparent, with a little too much emission predicted for T_j near 12,000K, but the scope for detailed modelling is enormous e.g. non-uniform field or distorted shock fronts.

3. Discussion

Any 'direct' observation of the magnetic field strength will prove difficult. Evidence through this model may prove irrepressible once CO and H_2O line predictions and H_2 line mapping is completed. Masers as dense clumps in which the Alfvén speed is large may well be related. Faraday rotation measurements of background extragalactic radio sources limit the ionized fraction rather than the field. Ion-neutral line profile differences are measurable only when

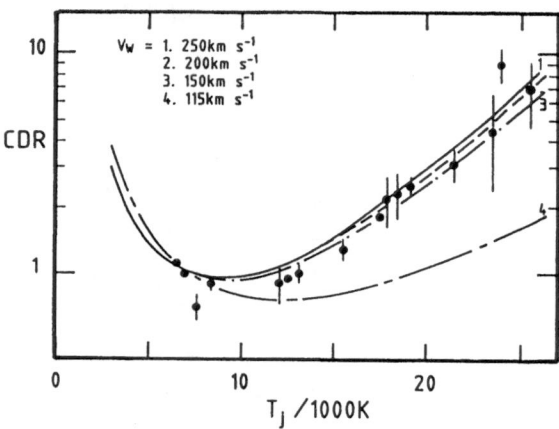

Figure 3: Line ratios expressed as column density ratios for the strong field example compared to OMC-1 Peak 1 data.

high ionization levels occur. High fields cannot be inferred to hold upo the active cloud - such reasoning applies only to quiescent cloud regions.

Two-component outflows are predicted by this 'shock absorber' model. One outflow produces a region with $v_A \sim$ 10 - 60 km s^{-1} - hence tentatively related to the CO bipolar outflows. The second high-velocity system drives through this medium. The speeds of 80 - 300 km s^{-1} correspond to HH objects and jets.

A J-shock model can reproduce the line ratio data in OMC-1 with the pressure as the one modelling parameter. A low H_2O abundance is required. Also a large number of pre-accelerated clumps or high-velocity turbulence is postulated in an attempt to explain the line widths (Brand et al 1989b). The pre-acceleration (after which H_2 must reform) is especially difficult to achieve within Peak 1 - a strong peak which subtends a small angle with the distant outflow centre.

References

Brand, P.W.J.L., Moorhouse, A., Burton, M.G., Geballe, T.R., Bird, M., & Wade, R., 1988. *Astrophys. J. Letters 334,* L103.

Brand, P., Toner. M.P. et al. 1989a. *M.N.R.A.S.* **236,** 929.

Brand, P., Toner. M.P. et al. 1989b. *M.N.R.A.S.* **237,** 1009.

Draine, B.T., Roberge, W.G. & Dalgarno, A., 1983. *Ap. J.,* **264,** 485.

Smith, M.D. & Brand, P.W.J.L., 1990a. *M.N.R.A.S.* **242,** 495.

Smith, M.D. & Brand, P.W.J.L., 1990b. *M.N.R.A.S.* **245,** 108.

Smith, M.D., Brand, P.W.J.L. & Moorhouse, A., 1990a & 1990b. *M.N.R.A.S.* submitted

VIII - SYMPOSIUM SUMMARY

Symposium Summary: Fragmentation and Star Formation
in Molecular Clouds:

A.E. Glassgold
New York University
New York, NY 10463, U.S.A.

This Symposium on fragmentation and star formation has
dealt with the heart of the study of molecular clouds, which
is how they form stars. This problem is one of the most
profound and challenging problems in all of astrophysics.
The complexity of the interstellar medium adds to its
difficulty and we cannot expect a quick and easy solution.
Nonetheless, the reports presented at this Symposium
indicate that substantial progress is being made in this
field.

The basis for current research in star formation has
been developed over several decades. During the 1980s, large
scale surveys with the mm lines of CO (see, e.g., the
reviews by Thaddeus 1990, Scoville and Sandars 1988) have
determined the distribution of the molecular gas near the
plane of the Milky Way and how that gas is made up of giant
molecular cloud complexes. The unusual properties of the
immediate neighborhoods of recently formed stars were also
elucidated. Of particular interest in connection with low-
mass star formation has been the systematic investigation of
the properties of the cores of dark clouds (e.g., Myers
1988). Near infrared studies of molecular clouds have
provided a powerful method for detecting the young stellar
objects ("YSOs") embedded in molecular clouds. The IRAS
satellite has been particularly useful here because of its
all-sky coverage. Another striking discovery of the last
decade was the detection of powerful molecular outflows
around YSOs (reviewed by Lada 1985). The outflows were a
surprise in view of the widely held belief that star formation involves gravitational collapse.

The measurement of the magnetic field strength and
direction is one of the most important goals of research in
the interstellar medium. Magnetic fields almost certainly
control many aspects of the structure and dynamics of

molecular clouds and the formation of stars. Perhaps the
main point to be drawn from Heiles' Symposium review (Heiles
1990) is that we seem to be finally emerging from a long
developmental period, characterized by fairly uncertain
results, to one where reliable but limited measurements of
the parallel component of the magnetic field are now
becoming available. For example, there is now good evidence
for a square root dependence of the field on the density.
Another obvious conclusion is that there are tremendous
opportunities here for additional observational programs.

Many diverse observations of low-mass star formation
have been integrated into a coherent conceptual framework by
the theory of star formation developed by Shu and his
collaborators (reviewed by Shu, Adams, and Lizano 1987, Lada
and Shu 1990). The theory starts from the idea that the
cores are in a quasi-equilibrium state with magnetic fields
balancing gravity. Gravitational collapse does not occur
until the field (slowly) slips away by ambipolar diffusion
and becomes decoupled from the bulk of the gas. The collapse
then proceeds from inside out at the sound speed, with the
infalling gas spiraling inward through a disk onto the
protostar. The corotation of the young protostar with the
disk at essentially breakup speed leads to the generation by
magnetic stresses of a fast wind that emanates from the
equatorial regions of the protostar. The wind is fast (of
order 100 km s^{-1}) and neutral and powers dramatic events far
from the protostar (100-1000 AU) in directions perpendicular
of the accretion disk. In addition to its unified conceptual
basis, the theory makes predictions that can be tested
observationally. Although not discussed extensively at the
Symposium, the record so far is very promising. For example,
Evans (Evans and Lada 1990) showed that the theory can
account for the broad infrared spectrum of protostellar
sources (Adams, Lada, and Shu 1987) and for the large scale
density distribution of the infalling matter. A review of
the current situation is given by Lada and Shu (1990).

A basic objective of current observational studies of
star formation that was discussed at length at the Symposium
is to relate the properties of the molecular gas in star-
forming regions to the distribution of protostars and other
YSOs. The basic tools are molecular line maps of the gas and
near infrared imaging of the embedded sources, e.g., with
the new array cameras. On the largest scales, the line
surveys are now emphasizing the study of individual cloud
complexes. The idea is to obtain the internal physical
properties of complexes on all scales, starting from the
largest (the size of the complex, 10-100 pc) down to the
smallest generally accessible with current techniques
(~ 0.01 pc). The probing of the internal structure requires

telescopes with suitable spatial resolution and a multi-transition analysis. One of the molecules now in vogue for probing the higher density regions is CS, which is assumed to have a constant abundance (independent of region) ~ 10^{-9}.

Striking results of this type of research were reported in the review talks by Myers (1990) and Evans and Lada (1990), who reported in-depth comparison of the core and YSO populations in different cloud complexes. By studying six nearby complexes, Myers and his collaborators are able to demonstrate systematic progressions in core and YSO properties with complex mass, e.g., an increase in the number of stars formed and in the probability for forming a massive star. Lada (Evans and Lada 1990) discussed the differences between the Taurus and Ophiuchus complexes. Not only is Ophiuchus more efficient at star formation, but the process seems to be more localized. The observations in Orion (Lada 1990) illustrate star formation in clusters. Lada also discussed the observational problems in determining the mass distributions of the cores and of newly formed stars. The hope is to definitively determine these distributions in nearby complexes and thus understand the basis for the initial mass function (Larson 1990). Perhaps the main, interim conclusion from these studies is that the molecular line and near infrared mapping of clouds have matured to the point where quantitative analyses of the star-forming regions are possible.

A closely related topic is the small scale structure of the interstellar medium and its connection with star formation. Small scale observations of molecular gas are now being made with angular resolutions of a few arc seconds, using existing mm interferometers. At the distance of the nearby cloud complexes, this corresponds to a spatial scale of the order of 6×10^{15} cm (0.002 pc or 500 AU). It is common to observe blobs of this size in star-forming regions with aspect ratios of a few. Although symptomatic of the dynamics of star formation, the blobs by themselves do not provide direct evidence of "thin accretion disks", as sometimes claimed, although we all suspect if not believe that such disks exist. Detailed modeling of multi-wavelength observations are considerably more convincing, as in the study of L1151-IRS5 (Keene 1990, Keene and Masson 1990).

The many new results on gas structure and YSOs represent a new stage in the observational study of star formation that will be significantly enhanced by the expansion of existing interferometer capabilities and infrared imaging in the next few years. In this context, it was entirely appropriate that some aspects of the methodologies of this research were discussed critically throughout the Symposium and in a panel discussion on fragmentation. For example, the selection and substantiation of high gas density (molecular line) tracers were hotly debated. Many observational studies tacitly assume that particular species measure the local volume density, but examples of contradictory and misleading information were cited as well as cases of close agreement between different tracers. It is obvious that proposed density tracers should not be overly sensitive to temperature and chemical effects. Until a badly needed calibration of the most useful molecular lines is effected, mapping of gas structure will have to be made in several lines, chosen so that the same angular resolution and sensitivity limits are achieved. It would also be desirable to model the local physical conditions, but this requires determining the kinetic temperature as well as the density. Considering the intrinsic inhomogeneous nature of the clouds, this may require a larger measure of faith in current radiation transfer techniques than they deserve. Theoretical work on this problem is badly needed.

Related questions on the interpretation of the measurements were raised but not pursued at length, For example, results of intercomparisons between gas and YSO distributions are expressed in terms of the "star formation efficiency", usually defined as the fraction of a region's gas that has been converted to stars. We can see that this definition is rather tricky when we consider that both distributions (gas and stars) are inhomogeneous in space and time. In particular, observed correlations between neighboring cloud cores and YSOs refer to different times so that, within the relevant time difference, the gas will have been perturbed by the newly formed stars. A related issue is the connection between fragmentation, conventionally viewed as a dynamical process, and the observations of clumpy molecular gas and star formation reported at the Symposium. Theories of cloud fragmentation (and growth) have to make specific predictions about cloud condensations, including their spatial correlations, that can be tested observationally.

It was completely appropriate that interstellar turbulence should occupy a prominent place on the program of the Symposium. I was particularly impressed by the stimulating report of Falgarone and Phillips (1990) on the possible connection between the extended wings, observed in molecular line shapes measured with high signal to noise, and the structure functions in the statistical theory of turbulence. In particular, they ascribe the wings to the phenomenon of "intermittency", familiar in laboratory and atmospheric turbulence. The existence of the these wings in regions without molecular outflows has been known for some time but Falgarone and Phillips are the first to examine them in a systematic fashion. Additional observational tests and theoretical underpinning have to be made before their interpretation can be accepted as definitive. Also, other theoretical interpretations will have to judged inferior. Elmegreen (1990) presented an interesting explanation of the roughly constant ratio of the wing to core widths (about 3) in terms of the nonlinear interaction of MHD waves.

The poster paper by O'Dell (1990) on measurements of spatial fluctuations of line emission from HII regions highlighted the long history of interstellar turbulence (see, also, O'Dell and Castaneda 1987). At the 1955 "Symposium on The Dynamics of Cosmic Clouds", Courtes (1955) presented evidence for the Kolmogorov scaling relation, $v_l \sim l^{1/3}$, near Lambda Orionis and a steeper law for the Orion Nebula. O'Dell's modern results for a variety of Galactic HII regions lead to line width - size correlation with an exponent of 1/4, whereas the molecular cloud exponent is close to 1/2. Understanding this difference would appear to be an important goal for the theory of interstellar turbulence. Moreover, O'Dell's measurements refer to the neutral transition regions of HII regions as well as to the ionized gas; both are intimately related to the subjects of star formation and cloud fragmentation.

An intriguing result discussed by Falgarone (1990) is that the fractal dimensions of interstellar clouds, determined on different scales with different lines, is the same (1.4). The significance of this result is unclear. It suggests that the physics of the boundaries is independent of scale. One must ask, however, how common or easy is it to achieve the dimension 1.4. It remains to be seen whether the fractal dimension provides the key to understanding the small scale structure of interstellar clouds.

Although fully developed, incompressible turbulence, the source of many of the most useful concepts in turbulence, shouldn't apply to the interstellar medium (see e.g., Larson 1980, Scalo 1988), some of the concepts (such as scaling) are very useful. The powerful observational techniques now available for studying the interstellar medium should make it possible to test the applicability of turbulence to the problem of interstellar clouds and star formation.

Except for a very elegant review of the chemistry of "translucent" clouds by Black (1990), there were few reports on (or relevant to) realistic chemical modeling of the inhomogeneous clouds found in nature. Indeed the most stimulating reports on interstellar chemistry came from the observers. For example, Joncas (1990) and Feldt and Wendker (1990) described HI measurements of molecular clouds that are informative on the H_2-H transition regions of molecular clouds. Stutzki (1990) described the status of modeling the more familiar CO-C-C$^+$ transition regions, in the context of further observations of the CO/HII interface observed edge on in M17. In order to account for the CI emission, a strongly clumped model of the cloud has been introduced which permits diffusion of UV radiation into an inter-clump medium, thus enabling each clump to have its own photodissociation region. However, certain aspects of the model are troubling, in particular the very large clump-interclump density contrast (~ 10^2) and the failure to explain the observed emission in the J = 7-6 line of CO. High spatial resolution maps of the M17SW cloud core reveal about 200 high density clumps and provide important structural information for future modeling of this situation (Stutzki and Gusten 1990).

A novel chemical point was made by Boulanger (1990) on the basis of observations of rapid, (small-scale) fluctuations in the 12/100 micron IRAS colors. His interpretation is that there are large changes in the amount of carbon in small particles (or large molecules), presumably produced by rapid cycling of material between dense and diffuse phases. Important modifications in the thermal and chemical properties of the interstellar medium are implied and, presumably, subject to observational tests.

One of the problems in developing more realistic chemical models is the sheer technical difficulty in capturing the essence of real (inhomogeneous and evolving) clouds. Of course, many informative and useful programs have been developed that deal effectively with either the chemical, thermal, or radiative-transfer aspects of interstellar clouds, but rarely with more than one of these essential ingredients. Almost always, the geometry is grossly oversimplified and steady-state is sometimes assumed without justification; real clouds seldom resemble a sphere or a slab and are likely to be variable temporally. In this context, it was refreshing to see the promising approach adopted in the poster paper by Keto, Lattanzio, and Monaghan (1990). They modeled the observations of the ring of star formation observed in W49A (Welch et al. 1987) with a 3-d hydrodynamics code that does include some chemistry, thermal effects, and radiative transfer. Similarly, some of the problems with shock models discussed by Draine (1990) may be overcome by abandoning planar shocks and considering instead bow shocks (Smith et al. 1990).

Certainly chemical modeling of fragmenting and star-forming regions of interstellar clouds is important. At the empirical level, it is crucial to know whether a molecule used for mapping is chemically active and whether its abundance is sensitive to local physical conditions. It is also generally accepted that the chemistry determines the abundances of the coolants and of the ions that couple the gas to the magnetic field. Of course chemistry includes the solid as well as the gas phase. The criticality of the chemistry of dust for understanding the infrared radiation of star-forming clouds is also widely recognized. For these and other reasons, it is important that interstellar chemistry becomes better integrated into general research on star formation.

It is usual in a conference summary to identify or suggest important problems for future research, a custom that I find difficult to follow. However, the Symposium did give me the sense that this field is in the process of transition in which the remarkable discoveries of the last decade are being extended by greatly improved technical capabilities, such as interferometers, infrared array cameras and, hopefully, new space observations at far infrared and submm wavelengths. In other words, areas of research on star formation that will be important in the next several years have been identified and are being pursued vigorously. These include the improved measurements of magnetic fields, systematic studies of the properties of star-forming cores and young stellar objects, and the probing of the actual star-forming entities e.g., the infalling cloud, the accretion disk, and the various outflow components, on ever decreasing size scales.

This does not mean that we are close to understanding even a small fraction of the complex of problems that make up the subject of star formation. The Symposium focused on the particularly difficult, basic problem of the small scale structure of interstellar clouds and its relation to star formation. This subject is of course related to more global ones such as the formation and evolution of the giant molecular cloud complexes. Although many good qualitative ideas were discussed, our understanding of this subject needs to become more quantitative and, eventually, organized according to some general principles. Perhaps the concepts and methods developed in the study of turbulent fluids will be useful. Certainly, the Symposium did an excellent job in giving a current account of the complex of challenging issues and of some methods that should help us understand the subject better.

386

On behalf of all the participants, I would like to thank Edith Falgarone and the members of the Scientific Organizing Committee for arranging a truly stimulating meeting. The beauty of the city and the surrounding areas made Grenoble an ideal meeting place. Many hanks are also due the Local Organizing Committee and, particularly, the excellent supporting staff, for all their generous help and hospitality. On a personal note, I would also like to thank the organizers, for making it possible for me to return to Grenoble and to participate in the Symposium, and NASA (Grant NAGE-630) for additional support.

References

Adams, F.C., Lada, C.J., and Shu, F.H. 1987, Ap.J. 312, 836.
Black, J.H. 1990, this meeting.
Boulanger, F. 1990, this meeting.
Courtes, G. 1955, in "Gas Dynamics of Cosmic Clouds", (Interscience, 1955), p.131.
Draine, B.T. 1990, this meeting.
Elmegreen, B.G., 1990, this meeting.
Evans, N.J. II and Lada, E.A., 1990, this meeting.
Falgarone, E., 1990, this meeting.
Falgarone, E. and Phillips, T.G., 1990, this meeting.
Feldt, C. and Wendker, H.J. 1990, this meeting.
Heiles, C. 1990, this meeting.
Joncas, G. 1990, this meeting.
Keene, J. 1990, this meeting.
Keene and Masson, C.R. 1990, Ap.J. 355, 635.
Keto, E.R., Lattanzio, J.C., and Monaghan, J.J. (1990), this meeting.
Lada, E.A. 1990, thesis, Univ. of Texas.
Lada, C.J. 1985, Ann. Rev. Astr. and Ap., 23, 267.
Lada, C.L. and Shu, F.H. 1990, Science, 248, 564.
Larson, R.B. 1990, this meeting.
Larson, R.B. 1980, M.N.R.A.S. 186, 479.
Myers, P.C., 1988, in "Interstellar Processes", Eds. D.J. Hollenbach and H.A. Thronson, (Reidel), p.71.
Myers, P.C., 1990, this meeting.
O'Dell, C.R. and O. Castaneda, O. 1987, Ap.J. 317, 686.
O'Dell, C.R. 1990, this meeting.
Scalo, J.M. 1988, in "Interstellar Processes", Eds. D.J. Hollenbach and H.A. Thronson, (Reidel), p.349.
Shu, F.H., Adams, F., and Lizano, S. 1987, Ann. Rev. Astr. and Ap., 255, 23.
Smith, M.D., Brand, P.W.J.L., and Moorhouse, A. 1990, this meeting.
Stutzki , J. 1990, this meeting
Stutzki , J. and Gusten, R. 1990, Ap.J. 356, 513.
Thaddeus, P. 1990, this meeting.
Welch, W.J. et al., 1987, Science 238, 1550.

IX - POSTER CONTRIBUTIONS
(in alphabetical order)

HIGH-VELOCITY MOLECULAR BULLETS IN
BIPOLAR OUTFLOWS : L1448 AND HH7-11

R. BACHILLER, J. CERNICHARO, J. MARTIN-PINTADO, M. TAFALLA

Centro Astronomico de Yebes (I.G.N.),
Apartado 148, E-19080 Guadalajara, Spain.

B. LAZAREFF

IRAM,
300, Rue de la Piscine, F-38406 St. Martin d'Heres CEDEX, France.

On the main axis of two bipolar outflows (L1448 and HH7-11), we have discovered molecular clumps ("bullets") moving at extremely high-velocities. The observations were made in CO lines (mainly $J = 2 - 1$) with the IRAM 30-m telescope at Pico Veleta, near Granada (Spain). We present in Figure 1 some profiles observed toward the axis of the two outflows. In addition to the ambient line and to the extremely high-velocity wings, we observe several very well defined peaks at high velocities. Our maps show that these features are well delimited in the space, indicating that they arise in small high-velocity clumps or "bullets".

In L1448, the bullets have typical sizes of about 0.03 pc and masses of a few $10^{-4} M_O$. They appear in pairs (one red-shifted for each blue-shifted) at symmetrical positions with respect to the position predicted for the driving source (the "U-star"). Such a high symmetry indicates that the bullets have been ejected from the immediate vicinity of the U-star. In the HH7-11 outflow, we have detected high-velocity molecular bullets traveling at velocities > 100 km/s, however the CO emission from the HH7-11 bullets is significantly weaker than that from the L1448 bullets. This difference in intensity could be related to the different degree of evolution of both sources, since the kinematical timescale of the L1448 outflow is shorter than that of HH7-11 (both sources are at about 350 pc of distance).

Our observations suggest that the CO bipolar outflows are driven by jet-like neutral winds at extremely high-velocities. The existence of high-velocity CO bullets in those winds could be a common phenomenon, at least in the youngest outflows. It remains unclear whether those neutral winds are of stellar origin or if they arise in the surface of circumstellar disks. However, the models in which the wind originates from the two sides of an accretion disk seem to be favored by the L1448 observations, because the coupling between the two sides of the disk leads naturally to simultaneous fluctuations of the mass ejection rate in the two opposite directions. In the frame of the MHD disk-driven wind models, the instabilities in the mass ejection rate of the accretion disk (and associated MHD wind) can generate periodical ejections of matter which are bipolar in origin, similar to the ejections observed toward L1448. More detailed reports on our observations can be found in Bachiller et al. (1990, Astron. Astrophys., 231, 174) and in Bachiller and Cernicharo (1990, Astron. Astrophys., in press).

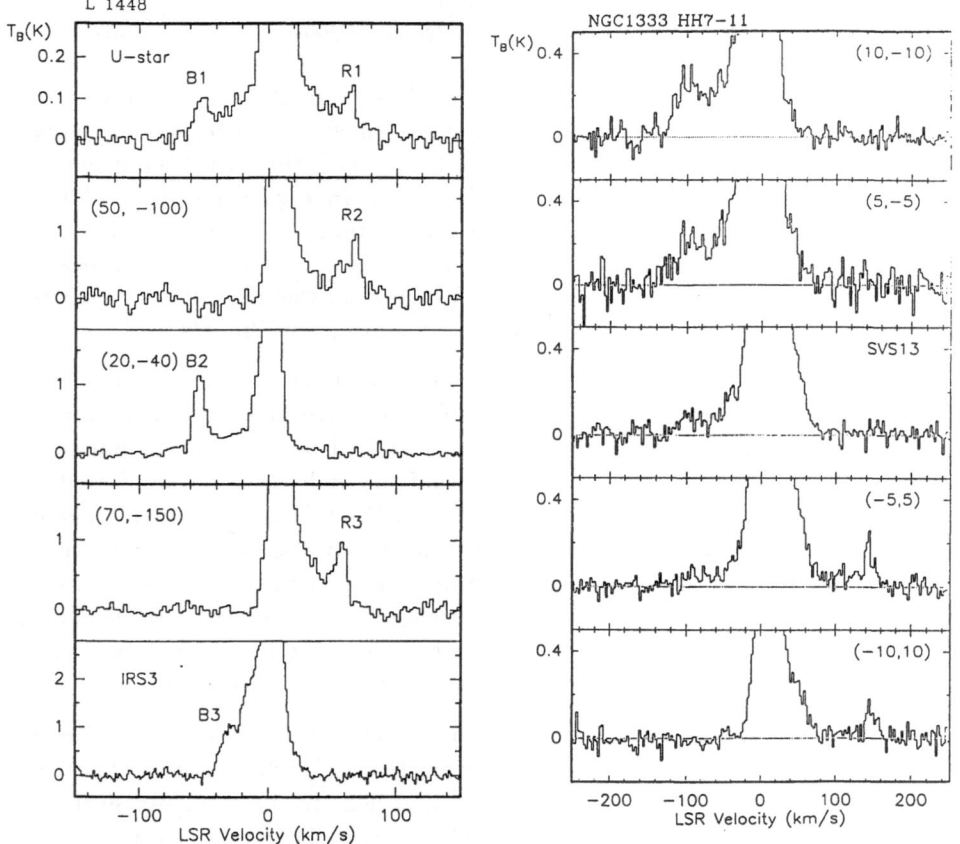

Fig. 1. Left : Some CO ($J = 2-1$) profiles observed toward the main axis of the L1448 outflow. Position offsets are measured with respect to the IRS3 position (in arcsec). Right : CO profiles observed toward the HH7-11 outflow. Position offsets are with respect to SVS13.

THE EFFECTS OF IONIZING RADIATION ON STAR FORMATION IN MOLECULAR CLOUDS

F. BERTOLDI[1], C.F. McKEE[2], and R.I. KLEIN[2,3]
[1] *Princeton University Observatory*
[2] *Astronomy Department, University of California at Berkeley*
[3] *Institute of Geophysics and Planetary Physics, L.L.N.L.*

ABSTRACT. The gravitational stability of molecular cloud clumps before and after the onset of massive star formation is discussed. We suggest that the most massive clumps are magnetically supercritical but gravitationally stabilized by the hydromagnetic turbulence caused by FUV photoionization-regulated low-mass star formation in their interiors. The ionizing radiation of an O star can trigger star formation in initially sub- and supercritical clumps.

1. Initial Clump Stability

Giant Molecular Clouds (GMCs) are observed to be clumpy, with most of the molecular mass in clumps of much higher than mean density. Individual clumps within GMCs have been mapped and cataloged in the Rosette (Blitz 1987), Ophiuchus (Loren 1989), and M17 SW (Stutzki and Güsten 1990) molecular clouds. The observed highly supersonic (but presumably sub-Alfvénic) internal turbulence and the strong magnetic field are believed to be of comparable importance in supporting a clump against gravitational collapse (Myers and Goodman 1988). It follows that the magnetic critical mass of a clump, M_Φ (Mouschovias and Spitzer 1976), is approximately equal to its Jeans mass, M_J (including thermal and turbulent contributions): $M_\Phi \simeq M_J$. The most massive clumps (that contain a large fraction of the total mass) within a molecular cloud are observed to be close to virial equilibrium, whereas the smaller clumps have masses far below their virial masses. Since a clump that is in virial equilibrium is also close to its gravitational critical mass, $M \simeq M_{cr} \simeq M_\Phi + M_J$ (McKee 1989), it follows that virialized, magnetic, turbulent clumps are magnetically supercritical, i.e., $M > M_\Phi$. Thus the magnetic field alone cannot support such clumps against gravitational collapse; the turbulent support is crucial.

This poses an interesting problem: The turbulent energy in a magnetically supercritical clump is dissipated in 2 to 10 free-fall times (typically $10^6 - 10^7$ yr)(McKee 1989). Unless this energy is constantly replenished by some internal source, the clump will slowly contract and eventually collapse. If a significant fraction of their mass is turned into stars when the supercritical clumps collapse, the resulting star formation rate for molecular clouds would exceed the values implied from observations (e.g., McKee 1989) by an order of magnitude.

We propose that the magnetically supercritical clumps are kept gravitationally stable over the lifetime of a GMC through a photoionization-regulated equilibrium: the continuous dissipation of the turbulent energy is offset by the steady injection of energy through the winds of newly formed stars in the clumps' interior. The balance between dissipation and star formation is regulated by the fact that star formation (via ambipolar diffusion) can only proceed in the interior fraction of the clumps that is shielded from the diffuse interstellar FUV radiation, where the gas ionization and thereby the ambipolar diffusion rate is dominated by cosmic rays. Excessive star formation would cause the clump to expand, thereby allowing the FUV radiation to penetrate deeper into the clump, which reduces the star formation rate again. This self-regulating equilibrium yields star formation rates and clump column densities that are consistent with observations. The onset of massive star formation will eventually disperse the clumps and the GMC (for a detailed model, see Bertoldi and McKee [in preparation]). McKee (1989) has shown that a similar mechanism can explain the moderate star formation rates in GMCs as well as their observed mean extinctions.

2. Induced Star Formation

The idea that star formation is somehow triggered by external events has received serious consideration for more than a decade. Recent observations (Reipurth 1983; Nakano et al. 1989; Sugitani et al. 1989, [these proceedings]; Duvert et al. 1990; Bally [these proceedings]) established that cometary-shaped, photoevaporating clumps in the vicinity of massive stars frequently show signs of active star formation, suggesting that the dynamical effects of the massive stars' ionizing radiation may be quite efficient in triggering star formation in the exposed neutral clumps. We have investigated this question and here describe our preliminary results (for a detailed discussion, see Bertoldi, McKee and Klein [in preparation]).

When a massive star is born in a clumpy molecular cloud, the dynamical effects of its ionizing radiation will implode the clumps that are exposed to the star's ionizing radiation (Klein et al. 1983; Bertoldi 1989a,b). During this *radiation-driven implosion* a fraction of a given clump's initial mass is evaporated off the ionization front that embraces the clump; the remaining mass is compressed to high density and accelerated away from the star; the imploded clump will settle into an cometary-shaped quasi-equilibrium configuration (*equilibrium cometary cloud:* ECC) that is characterized by the pressure balance between the (magnetic) neutral gas and the ionization front (Bertoldi and McKee 1990).

Before the exposure to a massive star's ionizing radiation, the molecular cloud clumps may be presumed stable against gravitational collapse, so that their masses are smaller than each clump's gravitational critical mass, $M_{cr} \simeq M_J + M_\Phi \simeq 2M_\Phi$. Magnetically supercritical $(M > M_\Phi)$ clumps require significant kinetic (turbulent) support, whereas magnetically subcritical clumps $(M < M_\Phi)$ are sufficiently stabilized by their magnetic fields.

As discussed above, a magnetically supercritical clump can be stabilized by the self-regulating input of turbulent energy from the clump's intrinsic low-mass star formation. The dynamical effects of its sudden exposure to ionizing radiation can induce the gravitational collapse of a magnetically supercritical clump as the rocket effect accelerates it and thereby strips off the embedded young low-mass stars that provided for the clump's internal turbulent energy support; deprived of the young stars and star-forming cores, the clump's equilibrium is severely disturbed. Due to the steady acceleration of the clump, any newly formed low-mass stars that could provide for the turbulent support are rapidly ejected from the clump. Furthermore, the mean density of the clump is increased, which enhances the energy dissipation. These effects, taken together, are likely to gravitationally destabilize the clump in a dissipation time.

The smaller, magnetically subcritical clumps are initially supported by their magnetic fields. They can form stars only via the slow, self-gravity driven ambipolar diffusion of the neutral gas through the magnetic field to the point where the clump, or a density enhancement within it, becomes magnetically supercritical and collapses. When a subcritical clump is imploded and slowly evaporates as an ECC, the ambipolar diffusion rate in its interior can be significantly enhanced and a central core could eventually become magnetically supercritical and collapse. The timescale for ambipolar diffusion is directly proportional to the fractional ionization of the molecular gas, which in turn is governed by both the photoionization by the FUV radiation of the ionizing star and cosmic ray ionization. Close to the surface of an ECC, the photoionization of the chemically nonreactive metals by the FUV radiation dominates the gas ionization, whereas in the deep interior, the cosmic ray ionization of H_2 dominates; here the ionization will be a function of the gas density. The compressive effect of the photoevaporation enhances the clump density and can thereby reduce the gas ionization to such an extent that the ambipolar diffusion timescale is significantly reduced. In certain cases star formation is thereby enhanced in photoevaporating clumps.

We conclude that the efficiency of induced star formation in the clumpy environment of a newly formed ionizing star mostly depends on the clumps' magnetic field and the upper mass cutoff of the clump mass distribution; star formation can be induced easily if the most massive clumps are initially magnetically supercritical, which seems to be the case in the observed molecular clouds in which the clump structure has been resolved. The massive, magnetically supercritical clumps will be the most vulnerable to a rapid triggering of (possibly massive, self-propagating) star formation.

REFERENCES

Bertoldi, F. 1989a, Ph.D. thesis, University of California at Berkeley.
Bertoldi, F. 1989b, *Ap. J.*, **346**, 735.
Bertoldi, F., and McKee, C. F. 1990, *Ap. J.*, **354**, 529.
Blitz, L. 1987, in *Physical Processes in Interstellar Clouds*, ed. G. Morfill and M. Scholer (Dorndrecht: Reidel), p. 35.
Duvert, G., Cernicharo, J., Bachiller, R., and Gómez-González 1990, *Astr. Ap.*, **233**, 190.
Klein, R. I., Sandford, M. T., II, and Whitaker, R. W. 1983, *Ap. J. (Letters)*, **271**, L69.
Loren, R. B. 1989, *Ap. J.*, **338**, 902.
McKee, C. F. 1989, *Ap. J.*, **345**, 782.
Mouschovias, T. Ch., and Spitzer, L. 1976, *Ap. J.*, **210**, 326.
Myers, P.C., and Goodman, A.A. 1988, *Ap. J. (Letters)*, **326**, L27.
Nakano, M., Tomita, Y., Ohtani, H., Ogura, K., and Sofue, Y. 1989, *Publ. Astr. Soc. Japan*, **41**, 1073.
Reipurth, B. 1983, *Astr. Ap.*, **117**, 183.
Sugitani, K., Fukui, Y., Mizuno, A., and Ohashi, N. 1989, *Ap. J. (Letters)*, **342**, L87.
Stutzki, J., and Güsten, R. 1990, MPI Preprint.

TRANSFER OF CONTINUUM UV RADIATION INSIDE FRAGMENTED CLOUDS

P. BOISSE
Radioastronomie Millimétrique
E.N.S.
24 Rue Lhomond
F-5231 Paris cedex 05
France

1. INTRODUCTION

The penetration of visible and UV continuum radiation is a governing factor for many processes inside interstellar clouds. It determines for instance: 1) the overall chemical equilibrium (formation/destruction of molecules and neutral or ionized species, fractional ionization of the gas which is directly related to the coupling with the magnetic field); 2) the overall energy balance (heating of the dust, heating of the gas through collisions with electrons extracted from grains by the photoelectric effect or with grains; 3) cooling of the gas due to fine structure line emission (OI,CI,CII).

Ample evidence is given in these proceedings that real interstellar clouds (the giant molecular ones as well as the much more diffuse high latitude Cirrus) are very far from homogeneity. This structure must be taken into account when treating the radiation transfer problem because **several orders of magnitudes** differences may result.

2. METHODS

Analytical approach: It is a generalization of that presented in Boissé (1990)(hereafter paper I). The main assumptions are: 1) absorption and isotropic scattering by dust grains; 2) clouds are described as a statistical medium made of 2 randomly mixed discrete phases (cf. paper I for details). This model introduces as few parameters as possible: the overall cloud average opacity, the clump opacity, the clump volume filling factor, the density contrast and the dust albedo. Quantities to be determined are the average intensity at any point M, $<I(M, \theta, \varphi)>$, the mean intensity $<J(M)>$ (average of $I(M, \theta, \varphi)$ over θ, φ) and the fluxes $<\Phi^+(M)>$, $<\Phi^-(M)>$ (along an axis normal to the assumed plane-parallel layer).

For unscattered photons an exact solution can be obtained (cf. paper I and Natta and Panagia, 1984). Including multiple scattering, a "non-local approximation" (valid at least when the clump opacity is not $>> 1$), the following system can be obtained (i and j refering to the 2 phases; z is the depth in the layer of thickness L):

$$<J_i(z)> = \int_0^L F_{ii}(z-z') <J_i(z')> dz' + \int_0^L F_{ij}(z-z') <J_j(z')> dz'$$

<u>Various types of solutions:</u> 1) *full solution* of the system (obtained by iteration); 2)*"effective" solution:* in most cases, the full solution is very close to that for a uniform layer characterized by a modified effective extinction coefficient K_e (with $K_e < <K>$) and albedo ω_e; explicit algebraic expressions for K_e and ω_e are available;- 3) *approximate solution:* for media with moderate average opacities one obtains an explicit form for $<J(z)>$ by assuming only 2 or 6 discrete flight directions for photons.

Monte-Carlo approach: the medium is a rectangular array of cells filled by material with the 2 above discrete densities. Simulations are used to validate the analytical method and estimate other parameters which cannot be computed (fluctuations ...).

3. RESULTS

I) Media with SAME clump and interclump dust albedo

Summary of results from paper 1: the layer behaves like a uniform one with a *reduced* opacity and albedo; as the medium gets more and more inhomogeneous, the observable radiation is more and more determined by the structure of the medium and less by the physical properties of the dust (cf. also Natta and Panagia, 1984 for extinction of an extended source by a non-uniform cloud).

II) Media with DIFFERENT clump and interclump dust properties.

Grain properties are likely to be density dependent (cf.Désert et al.,1990). To investigate specific effects due to different albedo values in the dense and diffuse phases, the layer to be used for comparison is the one characterized by the (uniform) mass-weighted albedo and the same density structure.

Good agreement is obtained between both methods. As expected, when material of high albedo is "hidden" in opaque clumps, the transmission factor is lower (the medium behaves like one of lower albedo).

For instance with a clump filling factor of 0.1, a clump opacity of 2.5, a mass fraction in the interclump phase of 37%, a total average layer opacity of 10 and dust albedos of 0.7 (clumps) and 0.17 (interclump: the mass-weighted albedo is then 0.5) we get an effective opacity and albedo for the "equivalent uniform layer" of 6.5 and 0.29 respectively.

III) Media with highly opaque clumps

For average layer opacities larger than 10, the validity of the assumptions required by the analytical method is questionable (cf. paper I). To discuss this point, we consider the asymptotic regime reached when the clump opacity goes to infinity (all other parameters remaining constant). We have shown that the limit given by the analytical method does exist. Both methods indicate that the asymptotic values are reached for relatively moderate clump opacities (about 6 in the case considered). For this reason, the analytical limit (although it cannot be correct) stills leads to good estimates, given the uncertainties in real cloud parameters. In such cases, an alternative approach would be to consider the transfer through a random distribution of completely opaque and partially reflecting obstacles (cf. Kamiuto, 1989).

4. CONCLUSIONS

We have shown that the analytical method described in paper I still leads to satisfactory results
    - when both the density and the dust albedo too are subject to fluctuations,
    - for large clump opacities.
From results obtained in the context of radiative transfer one may speculate that beyond this field, similar effects due to inhomogeneities will occur. Indeed, many quantities other than intensities involve a statistical average over regions characterized by a broad range of densities, temperatures, radiation fields ... These carry information not only on the physical conditions, but *also* (and possibly *mainly* !) on the relative proportion of gas in given conditions (i.e. on the *scatter* of physical conditions). Then, as far as the ultimate structure has not been attained, determining physical parameters from large scale data (e.g. temperatures or densities from emission line ratios as is currently done) can be very misleading.

References
Boissé, P.: 1990, *Astron . Astrophys.* **228**, 502
Désert, F.X., Boulanger, F., Puget, J.L.: 1990, preprint
Kamiuto, K.: 1989, *J. Quant. Spectrosc. Transfer* **41**, 23
Natta, A. and Panagia, N.: 1984, *Astrophys. J.* **287**, 228

DO RANDOM VELOCITY FIELDS INHIBIT OR INDUCE GRAVITATIONAL COLLAPSE AND FRAGMENTATION ?

PHILIPPE CHANTRY, ROLAND GRAPPIN, JACQUES LÉORAT
Observatoire de Meudon, DAEC, 92190 Meudon - France
E-mail CHANTRY@FRMEU51.BITNET

Recent numerical works on self-gravitating turbulence (Bonazzola et al., 1987; Léorat, Passot, Pouquet 1990) tend to show that turbulent velocity fields could act as a "turbulent" pressure, i.e. inhibit the collapse of self-gravitating clouds. This effect is efficient when the caracteristic scale of the turbulent flow is small compared to the Jeans's length. When these two scales are comparable, we show (Chantry, Grappin, Léorat, 1990) that the turbulence do no more inhibit the collapse, but seems to trigger it.

We performed two- and three-dimensional simulations of self-gravitating isothermal fluids, using a pseudo-spectral method. In the three-dimensional simulations, turbulence is sustained by a large-scale external shear force, so that a statistically stationary state is reached before gravitational term is put on. With 32^3 resolution, we create such stationary states with Mach numbers around 1, Reynolds numbers around 40 and with less than 10% of compressible kinetic energy. When self-gravity is added, we observe that

a) when the system is initially linearly *unstable* (following Jeans's criterion), the turbulence accelerates the gravitational collapse;

b) when the system is initialy linearly *stable* (or marginally stable), the turbulence trigger the collapse of scales near but smaller than the Jeans's scale.

In both cases, we are able to follow numerically the gravitational collapse up to density contrasts ($dens_{max}$ / $dens_{min}$) between a few hundreds and 10^6.

We also simulated two-dimensional decaying flows with very simple initial conditions :

$$U_x = -U \cos x \sin y;\ U_y = U \sin x \cos y;\ density = 1. \qquad (1)$$

These simulations show that a purely incompressible flow is able to generate very quickly density fluctuations at the hyperbolic stagnation points of the flow (saddle points). When self-gravity is present and the system is Jeans unstable, this non-linear growth of the density is a starting point for a further phase of gravitational growth which leads to collapse (Sasao, 1973). When the system is gravitationally stable and isothermal, the behaviour of the fluid depends on the initial Mach number. If the Mach number is low, the condensations decay after a few turnover times (Fig. 1a); when it is high enough, the condensations become frozen and the system reaches a quasi-stationary state (Fig 1b) with high density contrast (greater than 300). When the system is gravitationally stable and the polytropic index is smaller than unity, the quasi-stationary state may evolve into a further collapsing phase (Fig 1c) leading to density contrasts greater than 5000.

A simple analysis predicts that if one compresses a given Jeans stable mass of polytropic gas in a space of dimension D, the mass may become Jeans unstable after the compression if the polytropic index γ of the gas is below a limit depending on D. If D=2, this limit is $\gamma = 1$ and if D=3, it is 4/3. This could explain the previous results. In all the cases, the initial growth of the resulting condensed objects is due to non-linear compression at the caracteristic scale of the flow, so that turbulence seems to be an effective mechanism for the fragmentation of the clouds. This effect, which favors collapse at the integral scale of the turbulent flow could act together with the "turbulent pressure" effect, which inhibits collapse at scales much larger than that of the flow.

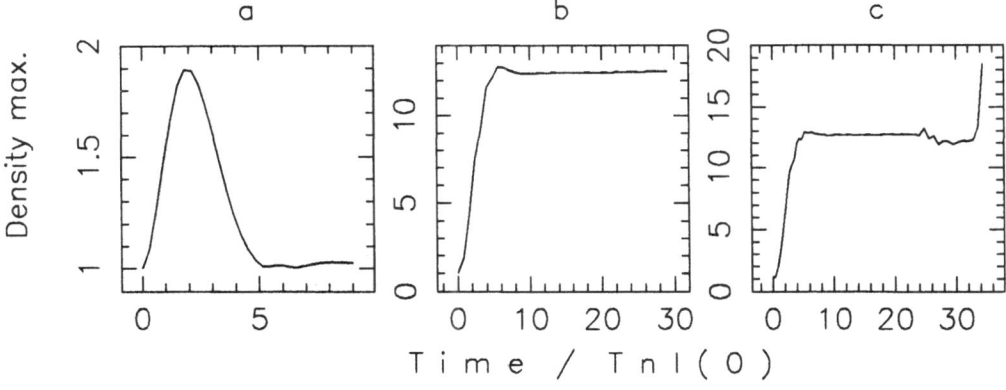

Fig. 1. Time evolution of the maximum density reached in two-dimensional simulations with initial conditions (1). a) initial rms Mach number $\bar{M}_0 = 0.35$; polytropic index $\gamma = 1$; b) $\bar{M}_0 = 0.94$; $\gamma = 1$; c) $\bar{M}_0 = 0.94$; $\gamma = 0.9$

References

Bonazzola, S., Falgarone, E., Heyvaerts, J., Pérault, M. and Puget, J.-L. : 1987, "Jeans collapse in a turbulent medium." *Astron. Astrophys.*, **172**, 293–298
Chantry, P., Grappin, R. and Léorat, J. : 1990, in preparation.
Léorat, J., Passot, T. and Pouquet, A. : 1990, "Influence of supersonic turbulence on self–gravitating flows." *Mon. Not. R. astr. Soc.*, **243**, 293–311
Sasao, T. : 1973, "On the generation of density fluctuation due to turbulence in self–gravitating media." *Publ. Astron. Soc. Japan* , **25**, 1–33

Fragmentation in Molecular Clouds

Frank O. Clark, AFGL; Timo Prusti, SRON;
R.J. Laureijs, IPAC

<u>B209</u>. Fragments are detected in the outer parts of the B209 molecular cloud in a region of density 1000 cm^{-3} using ^{12}CO, ^{13}CO, H$_2$CO, IRAS 60 and 100 μm images, discernable as velocity components, and have characteristics: $<R_{fragment}>\sim$ 0.4 pc; $<M_{fragment}>\sim$ 2 M$_\odot$; $<MVR_{fragment}>\sim$ 1.6 M$_\odot$ km/s pc; T$_{gas}\sim$ 11 K; T$_{dust}\sim$ 16 K??. The system of fragments has $<separation>\sim$ 0.5 pc and $<MVR>\sim$ 1.6 M$_\odot$ km/s pc. These are illustrated in Figure 1, which contains an IRAS 100 μm map made with Geisha and attendant U. Köln Gornergrat CO spectra for each respective fragment. The differing characteristic velocities of each fragment are readily apparent.

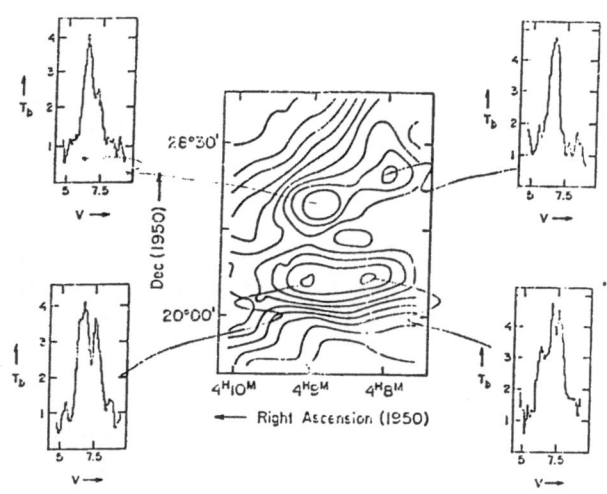

These lower density fragments in the outer part of the cloud are not redistributing their MVR. Although no apparent star formation has occurred within these fragments, there is prodigious star formation within the central environs of B209 itself, where the gas density is only marginally higher, $\sim$3 x 10^3 cm^{-3}. Here one finds: $<R>\sim$ 0.3 pc; $<M>\sim$ 15 M$_\odot$; $<MVR> < $ .4 M$_\odot$ km/s pc; T$_{gas}\sim$ 11 K; T$_{dust}\sim$14 K. The central star forming region has redistributed a significant amount of angular momentum.

L1563 is a cloud of lower opacity than B209, with no evidence of star formation, detected using ^{12}CO, ^{13}CO, H_2CO, and all four IRAS bands. The cloud center has two clear fragments resolved in molecular and 100 μm data, and discernable in velocity. These have characteristics: within each fragment: $<R_{fragment}>\sim 0.2$ pc; $<M_{fragment}>\sim 10$ $M_\odot$; $<MVR_{fragment}>\sim <0.2$ $M_\odot$ km/s pc; $T_{gas} = 11.6$ K; $T_{dust} = 14.2$ K; and for the cloud as a whole: $<fragment separation>$: 0.5 pc; $<system\ MVR>$: 5 $M_\odot$ km/s pc; $<T_{gas}> = >11.6?$; $<T_{dust}> = 16$ K. There has been a large reduction in MVR. Figure 2 shows the two detected fragments in the central core of L1563, in the form of a map of $I(100) - I(60)/0.22$, which effectively removes the cloud envelope and all cirrus, leaving only the dense region.

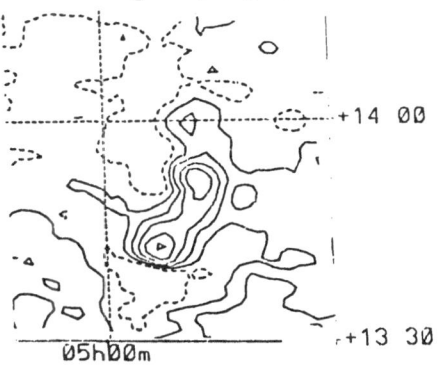

Conclusions. In these clouds, fragmentation is revealed in both infrared emission and molecular spectral lines. Redistribution of angular momentum has occurred in those fragments closest to cloud "cores", with or without accompanying star formation. The dust has a higher temperature than the gas in the regime probed. A combined dust temperature is shown in Figure 3. These data reveal T_d falling quickly in the outer cloud envelope, as predicted by Falgarone and Puget (1985), and remaining near 14 K up to extinctions of $\sim 8^m$.

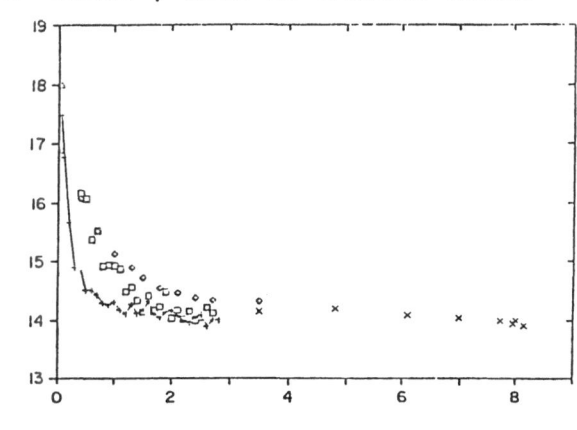

Blue Extinction (m)

VELOCITY CORRELATIONS IN TURBULENT MOLECULAR CLOUDS

B.M. DEISS, A. JUST, W.H. KEGEL
Institut für Theoretische Physik, J.W. Goethe- Universität
Robert-Mayer-Strasse 10, D-6000 Frankfurt a.M.
Federal Republic of Germany

The observed widths of molecular lines indicate that molecular clouds are in a highly turbulent state with supersonic internal velocities. At the same time molecular clouds are highly fragmented into substructures exhibiting large density contrasts. These substructures seem to be arranged in a hierarchical order scheme where one can think of newly formed stars representing the smallest spatial scale of ordering. Stars, protostars and even dense clumps can be regarded as being hydrodynamically decoupled from the ambient gas. Thus, in an idealized way, one can think of molecular clouds as being a gravitationally coupled system consisting of gas (continuous component) and 'point-masses' (particle component) randomly moving through the gas. The random motion of these 'point-masses' is accompanied by spatial and temporal fluctuations of their gravitational potential which, in turn, induce velocity and density fluctuations of the gas [1],[2],[4],[5].

Using fluctuation theory we derived in quasilinear approximation analytical expressions for the resulting power spectrum of the velocity fluctuations V_k. To this end, the dynamics of the system of the 'point-masses' was described by a Vlasov equation; the gas was treated in two ways: (i) as a viscous, polytropic gas (ξ being the dynamical viscosity coefficient), and (ii) as a gas with finite cooling time τ_{cool} [1]. V_k was calculated in the time asymptotic limit, which means that we had to restrict ourselves to stable modes only. The respective shapes of the power spectra are sketched in fig.1. For a given physical situation V_k shows a maximum at a characteristic wavenumber k_g. For wavenumbers smaller than k_g Landau damping of the particle component is the relevant damping mechanism of the coupled system. For larger wavenumbers the spectrum is determined by the damping mechanisms of the gaseous component. Hence, that wavenumber where the fluctuation spectrum reaches a maximum depends on how the various damping effects of the gas vary with k. In example (ii) the damping effect due to heating and cooling is much stronger than that of purely visous damping (example (i)). Hence, $k_g' > k_g$ and $V_k(k_g') > V_k(k_g)$. The latter relation ensues from the fact that Landau damping of the particle component becomes increasingly inefficient for increasing k in damping fluctuations of the gas.

The two-point autocorrelation function $\alpha(x)$ is the Fourier transform of the power spectrum V_k. We were interested especially in the correlation function of the stationary stable velocity fluctuation field. Since k_g, and thus the maximum of V_k, lies within the stable mode regime, $\alpha(x)$ depends only weakly on the lower limit of the wavenumber integration. We approximated $\alpha(x)$ by extrapolating V_k to $k \to 0$ and integrated over the whole k axis

[3]. From fig. 2 one finds for the correlation length, given as the usual e-fold length of $\alpha(x)$, $x_{correl} = 1.8\,k_g^{-1}$. Thus the largest turbulent eddies are of the order of the inverse of the characteristic wavenumber which, for cases (i) and (ii), is given by

$$k_g = \frac{1}{0.17pc}\exp\left\{-\frac{1}{5}\left(\frac{c_s}{\sigma}\right)^2\right\}$$

$$\times\left[\left(\frac{\rho_g}{10^{-21}g\,cm^{-3}}\right)\left(\frac{\rho_p}{10^{-22}g\,cm^{-3}}\right)\left(\frac{0.3km\,s^{-1}}{c_s}\right)^2\left(\frac{km\,s^{-1}}{\sigma}\right)^3\left(\frac{400yr}{\tau_{cool}}\right)\right]^{\frac{1}{5}}$$

and

$$k_g' = \frac{1}{0.06pc}\exp\left\{-\frac{1}{5}\left(\frac{c_s}{\sigma}\right)^2\right\}$$

$$\times\left[\left(\frac{\rho_g}{10^{-21}g\,cm^{-3}}\right)^2\left(\frac{\rho_p}{10^{-22}g\,cm^{-3}}\right)\left(\frac{0.3km\,s^{-1}}{c_s}\right)\left(\frac{km\,s^{-1}}{\sigma}\right)^3\left(\frac{c_s/\xi}{1.3\,10^{-9}g\,cm^{-2}}\right)\right]^{\frac{1}{5}}$$

respectively. ρ_g and c_s denote mean mass density and sonic velocity of the gas, and ρ_p and σ denote mean mass density and velocity dispersion of the particle component, respectively.

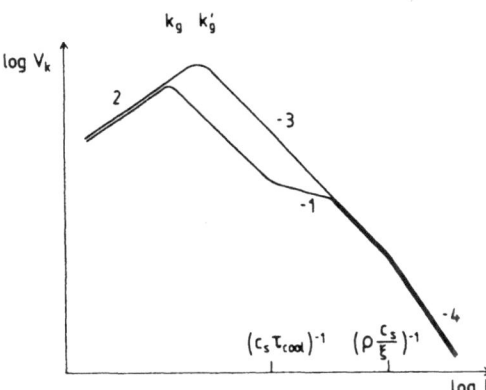

Fig.1: Power spectrum of velocity field (upper line: case i; lower line: case ii) Labels denote the respective power indices.

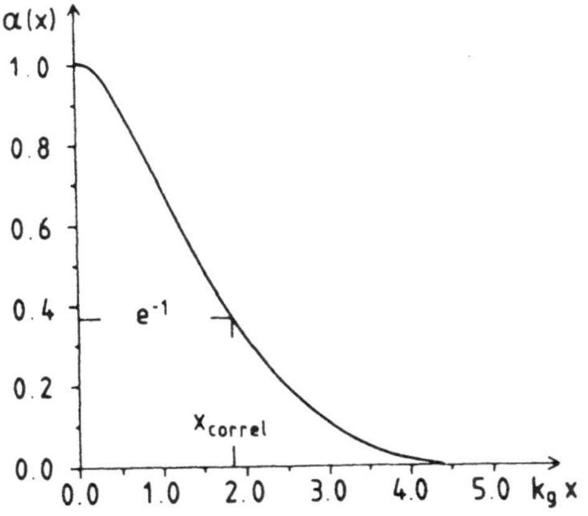

Fig.2: Two-point autocorrelation function of the velocity field

[1] Deiss,B.M.,Kegel,W.H.: 1986, Astron.Astrophys. **161** , 23
[2] Deiss,B.M.,Just,A.,Kegel,W.H.:1990, Astron.Astrophys, in press
[3] Deiss,B.M.,Just,A.,Kegel,W.H.:1990, in preparation
[4] Jacobi,S.,Just,A.,Deiss,B.M.,Kegel,W.H.: 1990, Astron.Astrophys,in press
[5] Kegel,W.H.:1987, In *Phys.Proc. in Interstellar Clouds*, eds. G.E.Morfill, M.Scholer;Reidel

MHD INSTABILITIES AND FRAGMENTATION OF MOLECULAR CLOUDS

A.E.Dudorov
Physical Department of Moscow State University,
Sternberg State Astronomical Institute,
13,Universitetskij prospect, 119899, Moscow, V-234, USSR

The properties of the interstellar magnetic clouds, their hierarchy, turbulence and their origin in a processes of the rising of MHD-instabilities are discussed.

1. Observations. The observational characteristics of basic molecular cloud structures in the interstellar medium of Galaxy are given in Table 1 (see Scalo 1985; Goldsmith 1987; Dudorov 1990). The exponents q_σ, k, q_ω of scaling relations between the velosity dispersion and cloud dimension $(s - R^{q_\sigma})$, between the magnetic field and the density $(B - n^k)$ and between the angular velocity and cloud dimension $(\omega-R^{-q_\omega})$ for every hierarchical level are presented also in the Table. Besides that the Table display the values of the ratio of magnetic energy to a module of gravitational one, ϵ_m and of plasmic beta $\beta=8\pi P/B^2$, where P - gas pressure.

2. MHD turbulence. Observational parameters of scaling relations agree with theoretical ideas of the development of turbulence in magnetized conductive medium (Vainshtein 1983). In a strong magnetic field (levels 1,2) the turbulence can be regarded as a two-dimensional one with the index $q_\sigma=0.75-1.0$ being dependent on the compressibility of the medium. In this case the strength of magnetic field does not depend on the density.

The turbulence of molecular and protostellar clouds (levels 3,4) is three-dimensional one. Magnetic field is not active in this case, its strength depend on the density. The weakening of the magnetic field influence on the turbulence correlate with the decreasing of spectral index q_σ along the hierarchical sequence. Note, that the index of Kolmogorov turbulence $q_\sigma=1/3$ and passive Kraichnan MHD turbulence is characterized by index $q_\sigma=0.25$.

The index $q_\sigma=0.5$ can be explained by the intermittent compressible turbulence of the conductive medium in variable magnetic field. This value is appeared also in a picture of magnetogravitational turbulence (Falgarone and Puget 1986).

3. MHD gravitational instabilities. The discussion of the last point shows, that hierarchical levels may correspond to real turbulent structures of interstellar magnetic clouds.

Superclouds (SC) are represented frequently by shells,

Table 1
Characteristics of interstellar magnetic clouds

Hie-rarchy	SC Super-cloud	MCC Molecul. cloud complex	MC Molecul. cloud	PC Proto-stellar cloud	References W-warm C-cold
T (K)	100	15-40 10	15-40 8-40	30-100 10	W C
n (cm-3)	1	100-300 1E2-1E3	1E2-1E3 1E3-1E4	1E3-1E5 1E4-1E6	W C
M (Mo)	1E7-3E7	8E4-2E6 1E3-2E4	1E3-1E5 20-500	10-1E3 0.3-20	W C
R (pc)	0.5E3-2E3	30-80 3-20	3-30 0.2-4	0.5-3 0.05-0.5	W C
σ (km/s)	10-20	6-15 1-3	4-12 0.5-1.5	1.5-3 0.2-0.4	W C
q_σ	1	0.7-1 0.6-0.7	0.4-0.6 0.3-0.4	0.2-0.4 0.2-0.3	W C
q_ω	0	0	0.3	0.4	W, C
k	0	1/3-1/2	1/2-2/3	1/2-2/3	W, C
ϵ_m	1	1	1/2	<1	W, C
β	<<1	<1	1	>1	W, C

Note to table: $1E4=1*10^4$ and so on.

envelopes of superbubbles, clumps of spiral arms and other two-dimensional structures, oriented along a force line of galactic magnetic field. The basic dimension of supercloud is comparable with length of unstable disturbances in MHD shock wave and with the radius of superbubbles (≈ 1 kpc). The masses of superclouds may be limited by gravitational instability in the magnetic sheets. In this case mass $M=\rho\lambda_{cr}\lambda_j H$, wher the magnetic critical length $\lambda_{cr}=2-4$ kpc, Jeans lengs $\lambda_j=0.2-0.4$kpc, the magnetic pressure scale $H=200-400$pc. Some superclouds may be formed by Parkers instability. When $\beta=2$,

$y=1.4$, the ratio of cosmic ray and magnetic pressure $P_{cr}/P_m=0.2$, then the unstable wavelength $\lambda_p=10H$.

The Supercloud may contain several complexes of molecular clouds (CMC), each of those may contain many molecular clouds (MC). The complexes of molecular clouds are observed often as a filamental or cilindrical structures. They may be a consequence of formation of magnetic flux tubes by channel gravitational instability in superclouds after their relaxation (Oganesyan 1960).

The formation of small scale turbulent magnetic field in turbulized gas of molecular clouds may lead to decrease of total pressure (Kleeorin et. al. 1990) and therefore to formation of long magnetic flux tubes with radius of crossection, $\lambda=H_\rho=100pc$. This MHD instability has the maximum increment, when $H_\rho \ll H_B$, where H_ρ and H_B are homogenity scales of the density and the magnetic field.

The molecular clouds may be formed by gravitational division of magnetic cilinders orin a process of gravitational relaxation due to the magnetic ambipolar diffusion. The ambipolar diffusion in quasistatic regime leads to the temporal collapse of molecular clouds and subsequent fragmentation in protocluster clouds with mass $M=500-1000$ Mo (Dudorov 1990).

The magnetic braking of rotation is very efficient in the case hierarchical levels 1,2 (c.f. Table 1). In course of the evolution of molecular and protocluster clouds the angular velocity rise upto 10-50 of its initial values.

Fagmentation of magnetostatic protocluster clouds may be induced by thermal, ionizational, resistive instabilities, leading to formation of protostellar clouds. The last stage of fragmentation of collapsing protostellar clouds may be induced by anomal ambipolar diffusion, when the magnetic protostars are formed with masses $M>0.1-0.3$ Mo, and with the ratio $e_m=0.1-0.3$. If the gas clouds are ionized by XR, UV the masses of protostars may be as large as several tenth of solar masses. Cosmic rays allow the formation of protostars with the masses more than 10 Mo.

References

Dudorov A. E. 1990, Astronomie (VINITI, in Russian), 39, 77
Falgarone E. , Puget J. L. 1986, Astron. Astrophys. 162, 235
Goldsmith P. F. 1987, in: Interstellar Processes, eds. D. J. Hollenbach and H. A. Thronson, Dordrecht, Reidel P. C. , 51
Kleeorin N. I, Rogachevskiy I. V. , Ruzmaikin A. A. 1990, JETF (in Russian), 97, 1555
Oganesyan R. S. 1960, Astronomical J. (in Russian), 37, 665
Scalo J. M. 1985, in: Protostars and Planets II, eds. D. C. Black and M. S. Matthews, Tucson, University of Arizona, 201
Vainshtein S. I. 1983, Magnetic Fields in Cosmos, Moscow, Nauka

A MULTI-TRANSITION AND MULTI-ISOTOPE STUDY OF CO
IN THE GIANT MOLECULAR CLOUD ORION-A

A. DUTREY, A. CASTETS, G. DUVERT

Groupe d'Astrophysique de Grenoble,
B.P.53X 38 041 Grenoble, France.

J. BALLY, W.D. LANGER, R.W. WILSON

AT&T Bell laboratories,
Box 400 Holmdel, New Jersey 07, USA.

In order to study the excitation conditions in the Orion-A region, we applied an LVG code to ^{12}CO , ^{13}CO and $C^{18}O$ data obtained with the AT&T Bell Laboratories 7-meter telescope in USA (CO isotopes : $J = 1 - 0$, CS : $J = 2 - 1$) and the radiotelescope of the "Groupe d'Astrophysique de Grenoble" in France (CO isotopes : $J = 2 - 1$).

Figure 1 presents the CS (2-1) integrated area map superimposed to the $C^{18}O$ (2-1) integrated area. Figure 2 corresponds to the N(^{13}CO) map of the main filament (referred as the $\int$ shape filament) derived from our ^{13}CO LVG analysis (Castets et al. 1990). They show that contrarily to $C^{18}O$, ^{13}CO do not probe the dense cores. However, some high H_2 density, low column density features are revealed, not seen in CS and located in the envelope. Conversely the $C^{18}O$ map reveals that the same dense cores are seen in CS and $C^{18}O$ map suggesting in the Orion-A region a high density in the cores.

In the south (below $Dec = +2.5'$, see the location in the maps) the ^{13}CO reveals several components with different excitation temperature and line opacities. The $C^{18}O$ analysis is consistent with the existence of four components ($V = 7, 8.3, 9.1, 10.1 km/s$), especially below $Dec = -10'$. Due to the opacity of ^{13}CO lines, they are not always clearly separated in the ^{13}CO data. Below $Dec = -10'$, the ^{13}CO LVG analysis of the "main component" gives an average density of $1200 cm^{-3}$ and N(^{13}CO) of $2 \cdot 10^{16} cm^{-2}$ while preliminary $C^{18}O$ results show an average n(H_2) of $5000 cm^{-3}$ and N($C^{18}O$) of $8 \cdot 10^{14} cm^{-2}$ for the brightest components.

The n(H_2) density obtained from $C^{18}O$ represents an average value along the line-of-sight, while the density derived from ^{13}CO probes only external layers. To get the density and to determine the excitation conditions in the cores, a CS study is on progress.

Reference

Castets A., Duvert G., Dutrey A., Bally J., Langer W.D., Wilson R.W *Astron. Astrophys.* **1990, in Press.**

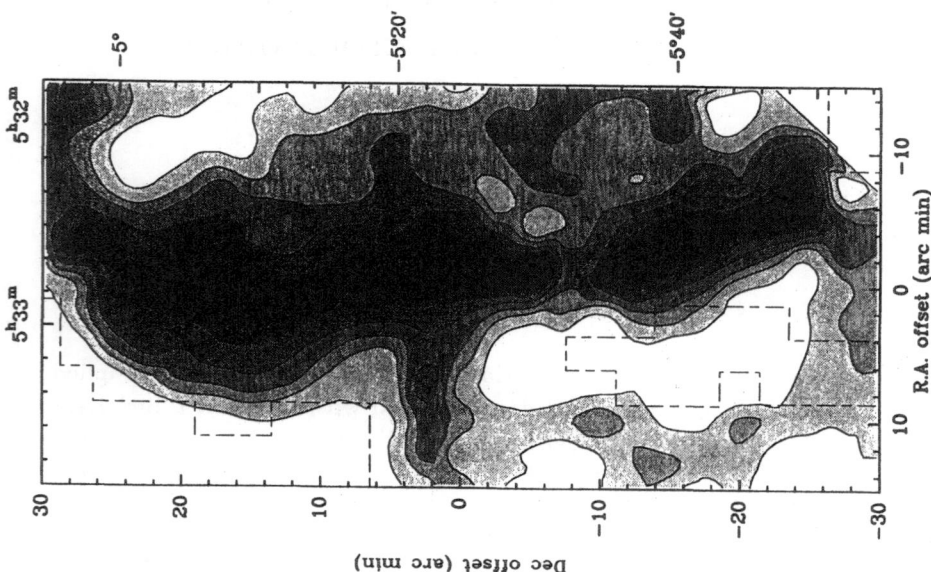

Fig 2 - ^{13}CO column density map derived from LVG analysis. Contours levels are $5 \cdot 10^{15}$, 10^{16}, $1.6 \cdot 10^{16}$, $2.5 \cdot 10^{16}$, $4 \cdot 10^{16}$, $6.5 \cdot 10^{16}$, cm^{-2}.

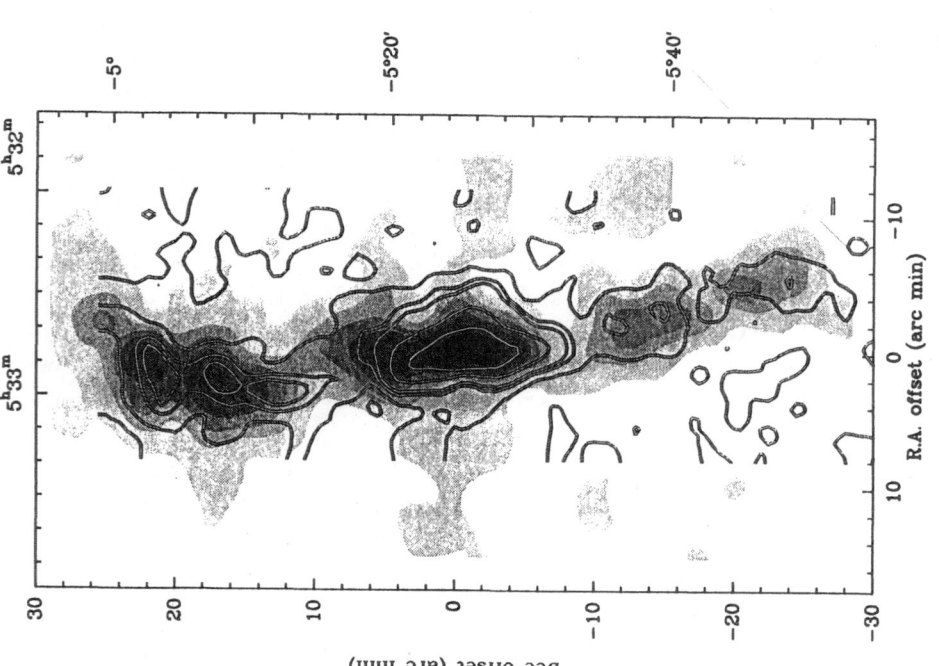

Fig 1: CS 2-1 integrated area map (lines) superimposed to C^{18}O 2-1 integrated area (grey levels). Contours levels are CS: 0, 2, 4, 5, 9, 15 K.km/s; C^{18}O : 1, 2.5, 3.5, 4.5, 5.5, 7 K.km/s.

PROCESSES CONTROLLING THE INITIAL MASS FUNCTION: INTERSTELLAR TURBULENCE AND MAGNETIC FIELDS

R. C. FLECK, JR.
Embry-Riddle Aeronautical University
Daytona Beach, FL USA

ABSTRACT. The observed flattening of the initial stellar mass function at low mass can be accounted for in terms of the different interstellar cloud size-mass scaling and different ambipolar diffusion time scaling for small, thermally-supported clouds and larger clouds supported primarily by turbulent pressure.

Following Reddish and Sloan (1971; see also Reddish 1978), we expect the mass spectrum for stars, $N_*(M)$, and for dense, interstellar cloud cores, $N_c(M)$, to be related by the rate at which these clouds form stars, t^{-1}:

$$N_*(M) \ \alpha \ N_c(M) \ t^{-1},$$

where the cloud mass spectrum (cf. Scalo 1985; Benson and Myers 1989)

$$N_c(M) \ \alpha \ M^{-1.5 \pm 0.5},$$

and the cloud-to-star conversion rate t^{-1}, assumed by Reddish and Sloan to be the gravitational free-fall time, is here taken to be equal to the ambipolar diffusion time, inasmuch as interstellar clouds are now known to be both turbulent and magnetic (cf. Fleck 1988; Myers and Goodman 1988).

A recent, comprehensive model of ambipolar diffusion in turbulent magnetic clouds predicts that $t \ \alpha \ R^{0.5} \ \alpha \ M^{0.25}$ for larger clouds (radius $R \geq 0.1$ pc; $M \geq$ few $M_\odot$), while $t \ \alpha \ R^{-1} \ \alpha \ M^{-1}$ for smaller clouds which are supported against gravity primarily by thermal gas pressure (Myers and Goodman 1988). Thus, the predicted stellar initial mass function (IMF), counting stars in logarithmic mass intervals, is

$$N_*(\log M) \ \alpha \ N_*(M) \ M \ \alpha \ M^{0.5 \pm 0.5} \quad \text{for } M \leq M_\odot$$

$$\alpha \ M^{-0.75 \pm 0.5} \quad \text{for } M \geq \text{few } M_\odot.$$

408

The agreement shown here in the accompanying diagram (subject to arbitrary normalization) with the field star IMF (Scalo 1986) is certainly suggestive and is, it appears, the first attempt to discern quantitatively the possible role of magnetic fields in determining the IMF. The flattening/ turnover at low mass can be accounted for in terms of the different size-mass scaling and ambipolar diffusion time scaling for small, thermally-supported clouds and large clouds supported primarily by "turbulent" pressure. This transition is characterized by a minimum in the diffusion time at a fragment mass of order 1 $M_\odot$, which translates to a maximum, cloud-to-star conversion rate at the transition mass and a concomitant tendency for the IMF to

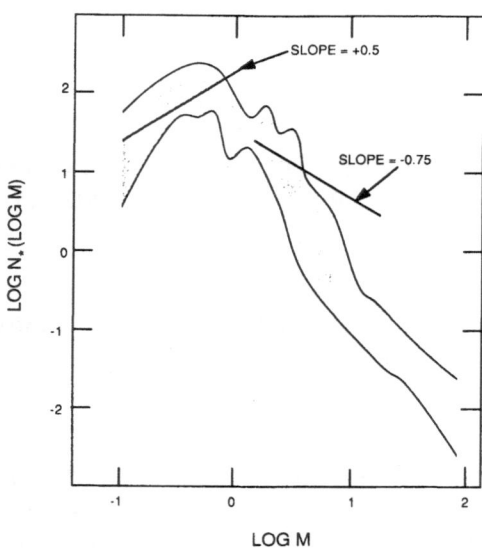

FIELD STAR IMF (FROM J. SCALO, 1986)

"peak" (i.e., turnover) in that vicinity. The influence of this "sonic" transition on the IMF turnover has been discussed previously by Fleck (1982).

While other processes such as mass accretion (loss) during the protostellar stage, which would flatten (steepen) the IMF, may also operate to "fine tune" the predictions here for the IMF, the evidence presented suggests that mechanisms associated with interstellar turbulence and magnetic fields may in fact be controlling factors for star formation in the Galaxy.

REFERENCES

Benson, P. J., and Myers, P. C. (1989) *Ap. J. Suppl.*, **71**, 89.
Fleck, R. C. (1982) *M.N.R.A.S.*, **201**, 551.
_____ . (1988) *Ap. J.*, **328**, 299.
Myers, P. C., and Goodman, A. A. (1988) *Ap. J.*, **329**, 392.
Reddish, V. C. (1978) *Stellar Formation* (Pergamon, Oxford).
Reddish, V. C., and Sloan, C. (1971) *Observatory*, **91**, 70.
Scalo, J. M. (1985) in *Protostars & Planets II*, eds. D. C. Black and M. S. Mathews
    (University of Arizona Press, Tucson), p. 201.
_____ . (1986) *Fund. Cosmic Phys.*, **11**, 1.

EFFECTS OF THE UV RADIATION ON
THE SURROUNDING GAS AND DUST

A. FUENTE, J. MARTIN-PINTADO

Centro Astronómico de Yebes.Apdo. 148. E-19080 Guadalajara(Spain)

J. CERNICHARO

Centro Astronómico de Yebes.Apdo. 148. E-19080 Guadalajara(Spain)
IRAM. Divina Pastora, 7. E-18012 Granada(Spain)

N. BROUILLET

Observ. de Bordeaux,B.P. 21,F-33270 Floirac(France)

G. DUVERT

Observ. de Grenoble,B.P. 68,F-38402 Saint-Martin-d'Heres Cedex(France)

Abstract. Two molecular clouds associated with well known reflection nebulosities have been studied in order to stablish the effects of the UV radiation on the surrounding gas and dust. Observations of the J=2→1 and J=1→0 transitions of CO and ^{13}CO have been carried out with the telescopes POM 1 and POM 2 towards the clouds associated with LKHα 198 and HD200775. The IRAS maps at 12, 25, 60 and 100 μm towards these globules have been also examined.

Fig. 1 shows the contours of the integrated intensity emission of the J=2→1 transition of CO, and of the 100 μm flux density superposed on the B POSS plate of LKHα 198. CO emission is not detected towards the reflection nebulosity and presents a sharp border this direction. Comparing the J=1→0 and J=2→1 transition of CO and ^{13}CO we found that the excitation temperature of the gas increases toward the reflection nebula. A higher kinetic temperature and a H$_2$ density of at least 5 10^3 cm^{-3} are required to explain ^{13}CO ratios. These densities are larger than the mean hydrogen density inside the cloud (500 - 10^3 cm^{-3}). The lack of CO emission toward the reflection nebulosity is not associated with a minimum in infrared emission. In fact, the infrared emission at 100 μm peaks between the star and the reflection nebulosity. Higher dust temperatures are also found towards this position where the dust seems to be heated by LKHα 198 and by the UV radiation illuminating the reflection nebulosity (see Fig. 1). Higher dust and gas kinetic temperatures towards the reflection nebulosity suggest that the lack of CO emission is due to be photodissociated the CO and ^{13}CO by the UV radiation.

Fig. 2 shows the B POSS plate of the nebulosity illuminated by HD200775 (NGC7023). The contours of the J=2→1 transition of ^{13}CO and of the 100 μm flux have been superposed on it. Note that the ^{13}CO contours delineate the regions with high visual extinction and that the star is actually located within a region of low extinction. However, the 100 μm flux density peaks at the position of the star because of the heating of the dust by the star. High angular resolution observations of the (1,1) and (2,2) lines of NH$_3$ (40") reveals that the gas is also being heated by the star and that ammonia is photodissociated in the regions closer to it (Fuente et al. 1990) . The ammonia clumps have also H$_2$ densities rather high ($\approx 10^4$ cm^{-3}.)

In summary, sharp boundaries containing small fragments with high density seems to appear in the interface between reflection nebulosities and molecular clouds, just adjacent to the photodissociation region produced by the UV radiation. Gas and dust with high excitation conditions are observed at these edges.

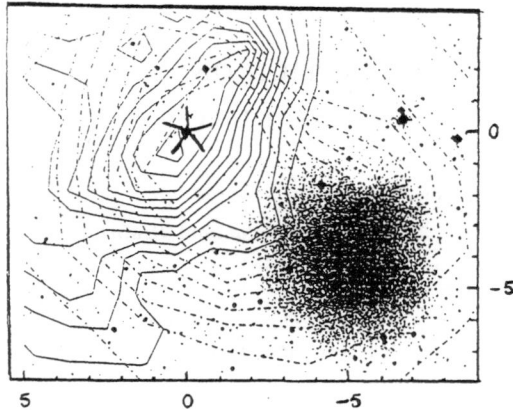

Fig. 1.   POSS B plate of the reflection nebulosity located to the southwest of LKHα
198. The contours of the integrated intensity map of the J=2→1 transition of CO (solid
line) and of the 100μm flux density map are superposed (dashed line). CO contours start
with 2 K kms^{-1} and increase by steps of 2 K kms^{-1}. 100μm contours are 30, 40, 50, 60,
70, 80, 90, 100 MJy sr^{-1}.

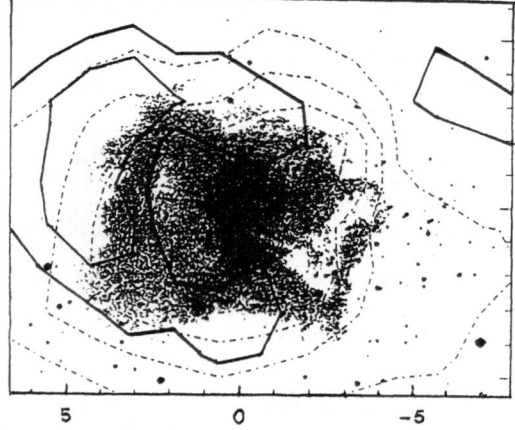

Fig. 2.   POSS B plate of the reflection nebulosity illuminated by HD200775 (NGC7023).
The map of the integrated intensity emission of the J=2→1 transition of ^{13}CO (solid line)
and of the flux density at 100μm are superposed (dashed line). ^{13}CO contours are 6, 8 K
kms^{-1} and those of 100μm are 100, 200, 300, 400, 500 MJy sr^{-1}.

References

Fuente, A., Martín-Pintado, J., Cernicharo, J., Bachiller, R.: 1990, in press (to be published in *Astron. Astrophys.*)

Acknowledgements

This work has been partially supported by the Spanish CICYT undergrant number PB88-0453.

HIGH RESOLUTION INFRARED MAPS OF IRAS GALAXIES

S.K. Ghosh, T.N. Rengarajan and R.P. Verma
Tata Institute of Fundamental Research
Homi Bhabha Road, Bombay 400 005, India

IRAS has provided infrared maps of a large number of galaxies. However the size of IRAS survey detectors is quite large (~5' in the cross-scan direction) and there is a need for improving the resolution of these maps. In order to understand the spatial structure of galaxies, we have generated high resolution maps of 18 large galaxies by deconvolving the IRAS pointed observations from survey detectors as well as from Chopped Photometric Channel (CPC) using Maximum Entropy Method (MEM).

The present sample of galaxies has been selected from "A catalog of IRAS observations of large optical galaxies" (LGC : Rice, W. et. al., Ap.J. Suppl., **68**, 91, 1988). The LGC presents IRAS observations of 85 large galaxies (with blue-light major diameters greater than 8'), out of which 49 were observed in pointed mode with multi-scans for each galaxy. We have selected 18 galaxies which were observed in the pointed mode with DPS02B Macro with the angle between different scans for the same galaxy less than 2 degrees. We have deconvolved the data for each of the 4 IRAS survey bands (hereafter referred as AO) 12, 25, 60 and 100 μm, as well as the CPC data at 50 and 100 μm. The point source functions were obtained from the respective AO and CPC observations of the planetary nebula NGC6543. The local average background as well as the noise for each map were estimated from the data in the immediate neighbourhood around the galaxy, for use during the deconvolutions.

We have obtained the deconvolved maps from AO for all the 18 sample galaxies and CPC maps for some of the galaxies. Deconvolved sizes at 10% level of a known point source are (in-scan x cross-scan): 0'.7 x 2'.2 (12 μm & 25 μm); 1'.2 x 2'.5 (60 μm); 3' x 3' (100 μm) for AOs and 1' dia (50 μm) and 1'.4 dia (100 μm) for CPC maps. The deconvolved maps of our sample of galaxies show consistent structures at the scale of 2-5'. The dynamic range for strong galaxies is ~ 300 for the AO and ~20 for the CPC. Many of these galaxies show multiple components. Almost all the galaxies show extended emission which is several times larger than the nuclear emission. On the average, the size of the galaxy along major axis (at 60 and 100 μm), at 2.5% contour level is about two thirds of the optical size. There is a very good correlation between AO and CPC structures. For illustration the deconvolved maps (AO and CPC) for NGC 891 are shown in Figure 1.

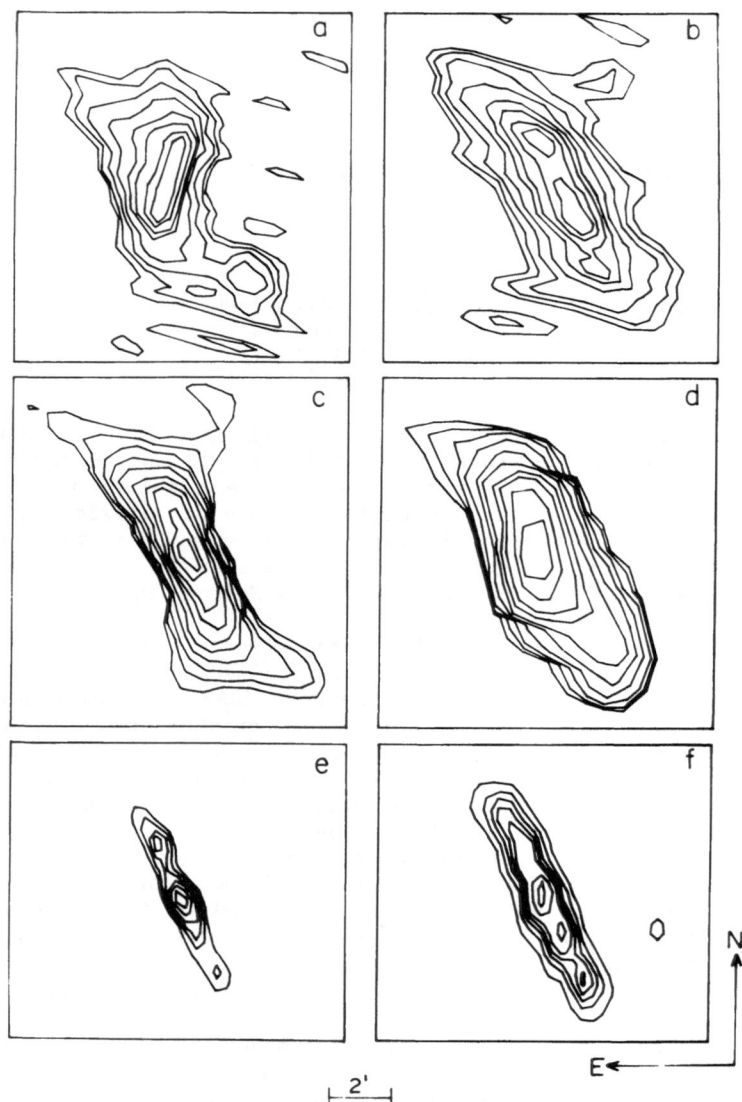

Fig. 1 Deconvolved IRAS maps at a) 12 µm, b) 25 µm, c) 60 µm, and d) 100 µm from AO and e) 50 µm and f) 100 µm from CPC. The contour levels (as percent of the peak) are a), b) 70, 50, 30, 20, 10, 5, 2.5 and 1.25; c) and d) 70, 50, 30, 20, 10, 5, 2.5, 1.25, 0.63 and 0.31; e) 90, 70, 50, 40, 30 and 20; f) 90, 70, 50, 40, 30, 20 and 10. The peak intensities in Jy/(sq. arc. min.) for a) to f) are 1.23, 1.2, 18.6, 27.5, 39.3 and 75.8 respectively.

INTERPRETATION OF LOW J ^{12}CO AND ^{13}CO OBSERVATIONS OF ORION A BY MEANS OF AN ONION SHELL RADIATIVE TRANSFER MODEL

KLAUS M. GIERENS

I. Physikalisches Institut, Universität zu Köln
Zülpicher Str. 77, D-5000 Köln, Fed. Rep. of Germany

^{12}CO and ^{13}CO $J = 1 \to 0$ and $J = 2 \to 1$ observations of a $1 \deg \times 2 \deg$ region centered on the BN/KL nebula in Orion A showed almost everywhere surprising intensity ratios. According to the standard interpretation of CO lines the ^{13}CO $T_A(2 \to 1)/T_A(1 \to 0)$ ratio meant optically thick and thermalized ^{13}CO emission whereas at the same positions the ^{12}CO/^{13}CO intensity ratios indicated optically thin ^{13}CO (Castets et al. 1989). Castets et al. (1990) suggested that temperature gradients in the observed clumps caused by external UV heating could be responsible for these results.

Using an Onion shell radiative transfer model (Gierens 1990) we show that the apparently contradictory intensity ratios can be reproduced over a large range in average density and column density. A temperature gradient is not sufficient to explain the intensity ratios, we must take into account also the density profile and the abundance gradients in the clumps. The radial dependences of kinetic temperature and abundance are taken from the models of photodissociation regions by Tielens and Hollenbach (1985). A $n \propto r^{-3/2}$ density law is applied according to the results of starcounts (Cernicharo, Bachiller, Duvert 1985) and hot edged polytropic models (Dickman and Clemens 1983).

We find the following structure of the Orion A clumps : A cold core ($T_{kin} \leq$ 15 K, $R \approx 0.3$ pc) is surrounded by a photodissociation region (PDR). The ^{12}CO $J = 1 \to 0$ line emerges mainly from a layer at the inner edge of the PDR where the kinetic temperature begins to raise. A typical temperature in this layer is 30 K. The density is high enough to keep the transition thermalized. The $J = 2 \to 1$ line emerges farther out in the clump (because of the higher opacity). This transition is not thermalized there but its excitation temperature is also around 30 K in the PDR and does not change very much with radius. So the ^{12}CO $T_A(2 \to 1)/T_A(1 \to 0)$ ratio is approximately one.

The main contribution to the ^{13}CO lines comes from the cold isothermal clump cores, where both transitions are thermalized. This explains the two other intensity ratios, namely ^{13}CO $T_A(2 \to 1)/T_A(1 \to 0) \approx 1$ and ^{12}CO/^{13}CO ≈ 3.

We derive as typical clump properties (including the PDR) :

(1) average density 500 - 3000 cm^{-3} (this is twice the density at the edge of a clump),
(2) H_2 column density $5 \cdot 10^{20}$ - 10^{22} cm^{-2},
(3) core kinetic temperatures less than 15 K,
(4) radii 0.25 - 1.5 pc,
(5) masses 10 - 1000 solar masses.

These values agree well with those given by Bally et al. (1987) for some typical clumps in Orion A.

References

Bally, J., Langer, W.D., Stark, A.A. and Wilson, R.W. : 1987, *Astrophys. J. Letters* **312**, L45

Castets, A., Duvert, G., Bally, J., Wilson, R.W. and Langer, W.D. : 1989, in *The Physics and Chemistry of Interstellar Molecular Clouds*, (G. Winnewisser and J.T. Armstrong, Eds.), Springer, Heidelberg

Castets, A., Duvert, G., Dutrey, A., Bally, J., Langer, W.D. and Wilson, R.W. : 1990, *Astron. Astrophys.* **234**, 469

Cernicharo, J., Bachiller, R. and Duvert, G. : 1985, *Astron. Astrophys.* **149**, 273

Dickman, R.L. and Clemens, D.P. : 1983, *Astrophys. J.* **271**, 143

Gierens, K.M. : 1990, thesis, Universität zu Köln

Tielens, A.G.G.M. and Hollenbach, D. : 1985, *Astrophys. J.* **291**, 722

GLOBAL PROPERTIES OF STAR FORMATION IN TAURUS

[1]Ana I. GOMEZ DE CASTRO[1,2] and [2]Ralph E. PUDRITZ[2]
[1]Observatorio Astronomico   [2]Department of Physics
de Madrid   McMaster University
Alfonso XII, 3   Hamilton, L8S 4M1
28014-Madrid, Spain   Ontario, Canada

SUMMARY: A careful analysis of the data available in the literature has been carried out to determine the global properties of star formation in Taurus. Three main trends arise in the data:
(1) the magnetic field in Taurus is distorted in the direction of the proper motions of the T Tauri Stars (TTS) (see below).
(2) the groups with different proper motions seem to have different stellar contents. An analysis of the stellar population based on the equivalent width of Hα, W(Hα), shows that groups I and III (Jones and Herbig, 1979) have approximately the same population (TTS show either large or small W(Hα)). However in group II (central part of the cloud) there are also a large number of stars with intermediate W(Hα) (Gomez de Castro and Pudritz, 1990).
(3) the mass function of the NH_3 cores (thought to be "protostellar cores") is $dN/dM \propto M^{\alpha}$ with $\alpha=-1.5$, in contrast with the IMF of the TTS given by Cohen and Kuhi (1979), $\alpha=-2.5$.
Only point (1) will be outlined in this contribution.

1. MAGNETIC FIELD GEOMETRY AND STELLAR MOTIONS

The magnetic field in Taurus, as delineated by the observed optical polarization of background stars, changes its orientation across the cloud (Moneti et al., 1984; Hsu, 1984). In Figure 1, we have superposed the proper motions for 75 TTS (Jones and Herbig, 1979) on the magnetic map. This clearly shows that the stars move approximately parallel to the filaments and nearly perpendicular to the field direction. The velocity dispersion in one coordinate is typically 2-3 km/s; less than the escape velocity from the clouds (Jones and Herbig, 1979).

There is no evidence of any systematic velocity differences between the stars and the molecular cloud. The radial velocity dispersion respect to the gas is <1.5 km/s (Herbig, 1977; Hartmann et al., 1986). This is consistent with the internal motions of the gas (Ungerechts and Thaddeus, 1987).

If clouds and stellar motions are coupled as these data suggest, then the filaments that constitute the Taurus clouds are moving in the direction of the field distortion.

416

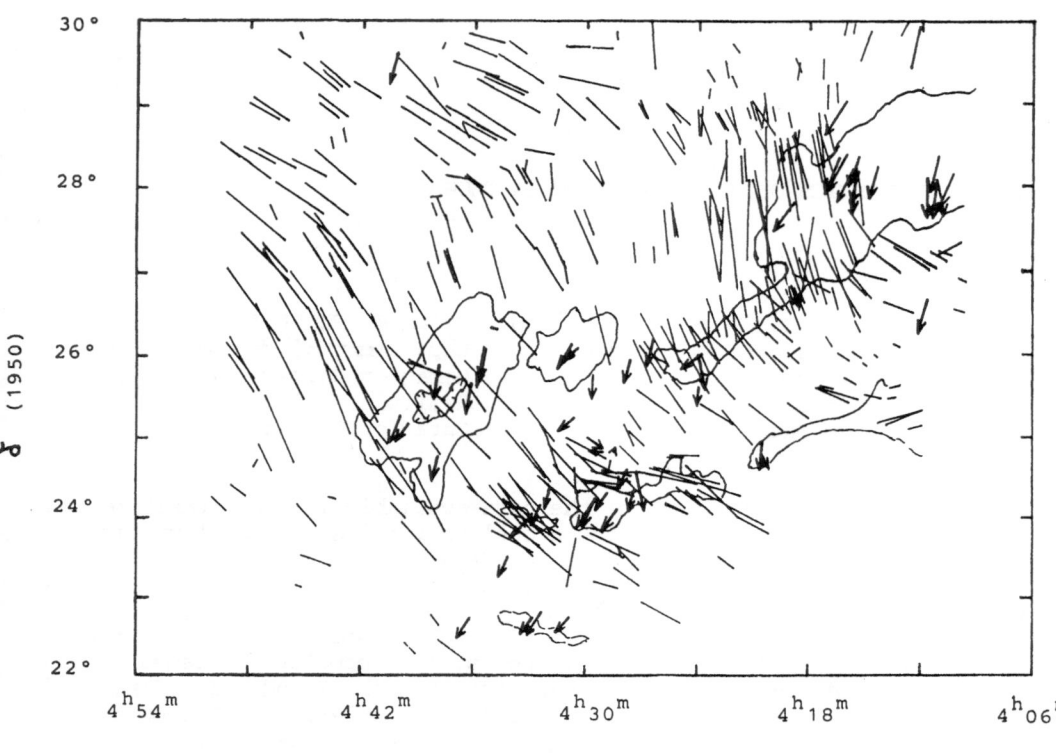

α (1950)

Figure 1. The magnetic field in Taurus delineated by the polarization vectors taken from Hsu (1984), Moneti et al. (1984) and Heyer et al. (1987). The proper motions of the TTS are indicated by arrows. The aspect of the Taurus clouds is sketched.

REFERENCES:
Cohen, M., Kuhi, L.V., 1979; Ap.J. Suppl., 41, 743
Gomez de Castro, A.I., Pudritz, R.E., 1990; preprint
Hartmann, L., Hewett, Stahler, S., Mathieu, R.D., 1986; Ap.J., 309,275
Herbig, G.H., 1977; Ap.J., 214, 747
Heyer, M.H., Vrbs, F.J., Snell, R.L., Schloerb, F.P., Strom, S.E., Goldsmith, R., Strom, K.M., 1987; Ap.J., 312, 855
Hsu, J.C., 1984; Ph.D. Thesis, University of Texas, Austin
Jones, B.F., Herbig, G.H., 1979; A.J., 84, 1872
Moneti, A., Pipher, J.L., Helfer, H.L., McMillan, R.S., Perry, M.L., 1984; Ap.J., 282, 508
Ungerechts, H., Thaddeus, P., 1987; Ap.J. Suppl., 63, 645

POLARIZATION OF BACKGROUND STARLIGHT AND THE STRUCTURE OF THE INTERSTELLAR MAGNETIC FIELD

ALYSSA A. GOODMAN
Astronomy Department, University of California,
Berkeley, CA 94720, USA
E-mail AGOODMAN@UCBBKYAS.BITNET

PHILIP C. MYERS
Harvard-Smithsonian Center for Astrophysics,
60 Garden Street, MS 42, Cambridge, MA 02138, USA
E-mail MYERS@CFARG1.BITNET

PIERRE BASTIEN
Department de Physique, Université de Montreal,
BP 6128, Succursale A, Montreal, PQ H3C 3J7, CANADA
E-mail BASTIEN@CC.UMontreal.CA.BITNET

1. Abstract/Introduction

We discuss the use of polarization maps of background starlight in studying the structure of the interstellar magnetic field. We make the assumption that the polarization observed is due to magnetically aligned dust grains associated with interstellar clouds along the line of sight, and that the position angle (Θ_E) of polarization observed gives the direction (modulo 180°) of $\vec{B}_\perp$, the plane-of-the-sky projection of the (line-of-sight-averaged) magnetic field.

There are two basic points in this paper. (1.) Out of context, the projected orientation of an elongated dark cloud may appear special (e.g. roughly "parallel" or "perpendicular") in relation to the local (plane-of-the-sky) field direction given by polarization observations, but, when the view is expanded to include an entire complex of dark clouds, the shape and orientation of clouds within a complex often appears unrelated to the field structure. (2.) The dispersion in the postion angle of polarization observed in a region of the sky contains information about the ratio of the strength of the uniform (straight) component of the local magnetic field as compared to the dispersion (nonuniform component) in the field. Furthermore, in a region where Zeeman measurements covering the same region as polarization observations have been made, the uniform-to-nonuniform ratio deduced for the field from polarization data, can be combined with information about the line-of-sight field and an estimate of the correlation length of the field, to describe the magnetic field in three dimensions. We discuss the results of such an analysis for the dark cloud Lynds 204 (L204).

Keywords : Magnetic fields, dark clouds, L204

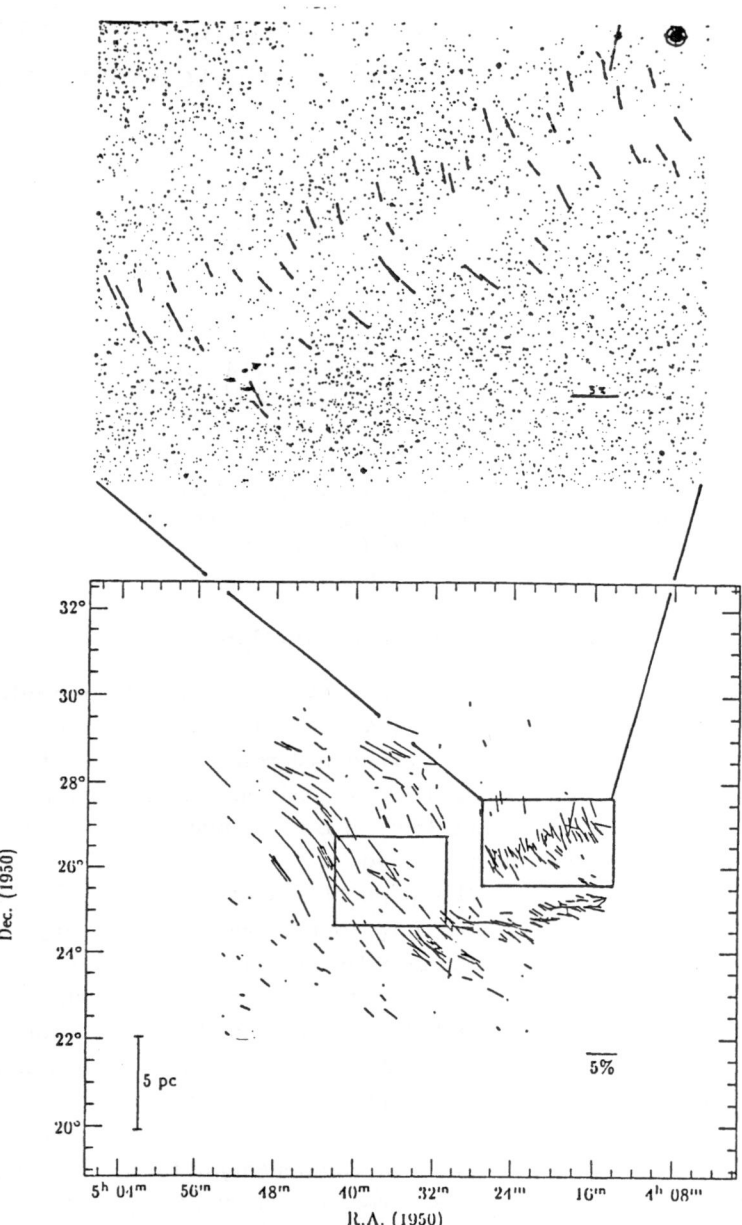

Figure 1 : *Upper Panel–* Optical polarization of background starlight in the B216-217 dark cloud region, superposed on a reproduction of the Palomar Sky Survey photograph (Heyer *et al.* 1987). *Lower Panel–* Optical polarization map of the Taurus molecular cloud complex (Moneti *et al.* 1984; Heyer *et al.* 1987; Goodman *et al.* 1990). The boxed region not enlarged above shows the Heiles Cloud 2 area studied (at 2 μm) by Tamura *et al.* 1987.

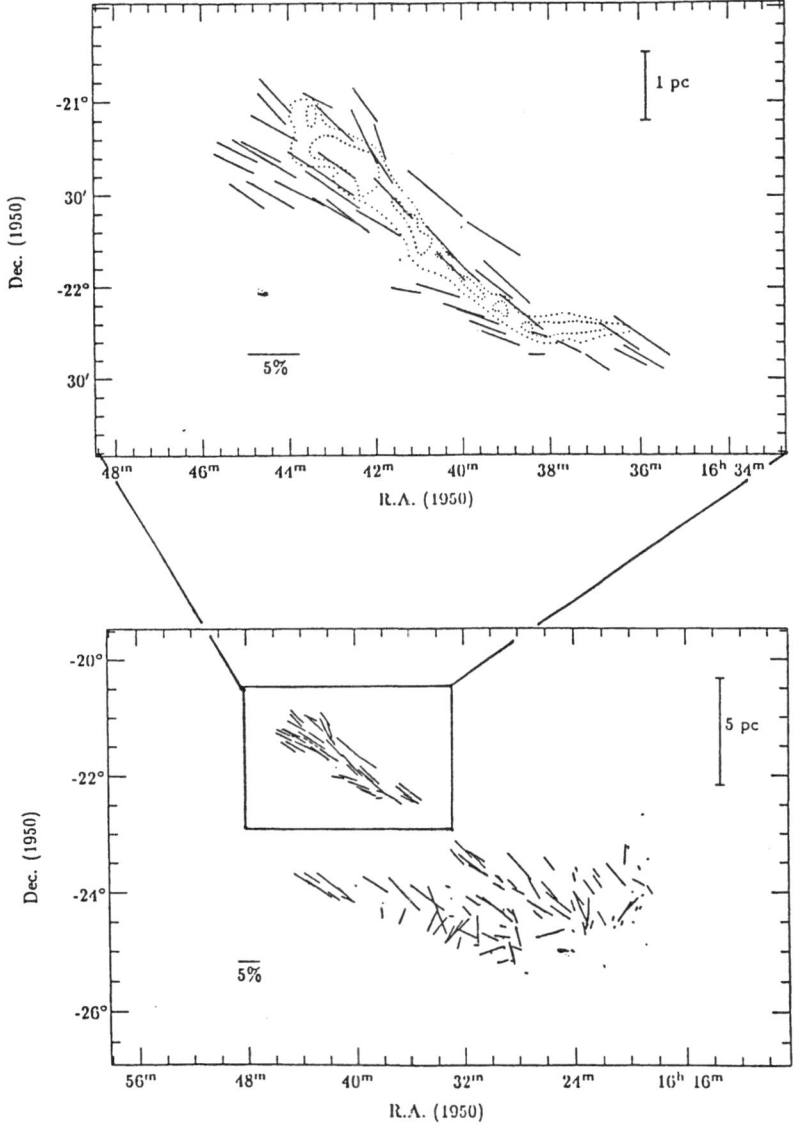

Figure 2 : *Upper Panel–* Optical polarization of background starlight in the L1755 dark cloud region (Goodman *et al.* 1990), superposed on a ^{13}CO contour map (Loren 1989). *Lower Panel–* Optical polarization map of the Ophiuchus molecular cloud complex (Vrba, Strom, and Strom 1976; Goodman *et al.* 1990).

2. Orientation of Cloud Features with Respect to $\vec{B}_\perp$

To date, one of the most prevalent uses of polarization maps of the regions around dark clouds has been to directly compare the projected magnetic field structure with the projected cloud structure (e.g. a molecular-line map). When one elongated cloud is examined alone, out of the context of the surrounding cloud complex, often one is led to believe that the cloud axis is oriented in a special way with respect to the field. Figures 1 and 2 illustrate this point. Based on the top panel of Figure 1 (B216-217) alone, we might deduce that filaments are elongated perpendicular to field lines : while, based on the top panel of Figure 2 (L1755) alone, we would come to an "orthogonal" conclusion–that elongation is parallel to field lines. In truth, in most cases, the orientation of a filament is, when viewed alone, arbitrary with respect to $\vec{B}_\perp$. However, when we expand our view to include an entire cloud complex, we find that the field has a smooth structure on scales of at least tens of pc, and that the orientation of several-pc-long filaments within the complex is not necessarily obviously governed by the structure of the magnetic field. The lower panels of Figures 1 and 2 show polarization maps of the entire Taurus and Ophiuchus molecular cloud complexes, respectively, and it is apparent from these maps that the "special" orientation of B216-217 and/or L1755 with respect to the magnetic field, could be fortuitous.

It is important, also, to keep in mind that the orientations we observe on the plane of the sky represent projections of true orientations in three-dimensional space. The probability distribution for the true angle between two directions in space, given an observed projected (2D) angle, is only weakly peaked at the true (3D) angle, with a long tail extending to larger values, and a small chance that the 2D angle observed is actually greater than the true 3D angle. Given the relatively large number of filamentary clouds for which there are polarization maps (~ 10), and given that they show relatively random orientation of cloud axis with respect to $\vec{B}_\perp$, it is safe to assume that the true 3D angle between a filament and the field is likely to be larger than the 2D angle observed.

3. Analysis of the Dispersion in Polarization Angle

Given that the direct (i.e. morphological) interpretation of polarization maps is apparently less than straightforward–especially concerning the relation between the field direction and the elongation of filamentary clouds, we may perhaps wish to extract information about the magnetic field from polarization data in less direct, but more quantitative, ways.

What polarization maps definitely show is a tremendous degree of coherence or "uniformity" in the structure of the interstellar magnetic field. In other words, if one plots the number distribution of polarization angle over a given region, the dispersion in the distribution is small. Just how large a region will still show a small dispersion is a question which must at least partially concern the correlation length of the field. We know, however, that even over an entire molecular cloud complex, such as Taurus (see Figure 1, lower panel), the distribution of Θ_E remains relatively narrowly peaked (Goodman *et al.* 1990; Myers and Goodman 1990).

Myers and Goodman (1990; hereafter MG90) describe a method by which the observed distribution of Θ_E is modelled as arising from the sum of a uniform (straight) field plus a nonuniform field characterized by a one-dimensional dispersion σ_B.[1] We refer the reader to MG90 for a detailed description of this model, and we will present only the results here.

The MG90 model fits observed distributions of polarization angle in terms of two parameters : the mean polarization angle $\bar{\Theta}_E$, and the dispersion $s = \frac{\sigma_B}{N^{\frac{1}{2}} B_{oz}}$, where N is the number of correlation lengths along the line of sight through the cloud, and B_{oz} represents the plane-of-the-sky component of the uniform magnetic field. An example of the model fit to the distribution of polarization angle for the dark cloud L204 is illustrated in Figure 3a. (MG90 presents similar fits for 14 other dark clouds, five cluster regions, and six dark cloud complexes.) For the fit shown in Figure 3a, $\bar{\Theta}_E = 70 \pm 2°$ and $s = 0.44 \pm 0.02$.

If several Zeeman measurements have also been made in a region where optical polarization data is available, the line-of-sight field information provided by the Zeeman data can be combined with the value of $\bar{\Theta}_E$ and s from a fit to the distribution of polarization angle, to estimate the three-dimensional uniform field component, and its inclination to the line-of-sight. In L204, Heiles (1988), has made Zeeman measurements on a grid of 27 positions covering approximately the same region of the sky as the McCutcheon *et al.* (1986) polarization measurements used to construct the distribution in Figure 3a (see Figure 3b). MG90 combine the polarization and Zeeman data to find a uniform field in L204 of 6 μG, inclined at 47° to the line of sight, projected along the line 70° E of N in the plane of the sky.

4. Discussion : For the Future

Since one must still be able to "see" a star through the gas along the line of sight in order to make an optical polarization observation, the information from optical polarization maps technically only pertains to gas with column density less than 2×10^{21} cm^{-2}, corresponding to about 2 mag A_V, or the periphery of a dark cloud. It is possible that the structure of the field traced by optical polarization observations is not representative of the denser gas associated with the visually opaque portions of a dark cloud.

Existing infrared polarization observations, which can observe stars through far more extinction than optical observations, would seem to indicate that the field structure in higher density regions is not markedly different than what is found optically. For example, in the Heiles Cloud 2 region (marked by a box in the lower panel of Figure 1), Tamura *et al.* (1987) find a similar direction and dispersion in

---

[1] Chandrasekhar and Fermi (1953), Jokipii and Parker (1969), and Zweibel (1990) also present relevant analyses of the disperion in the interstellar magnetic field.

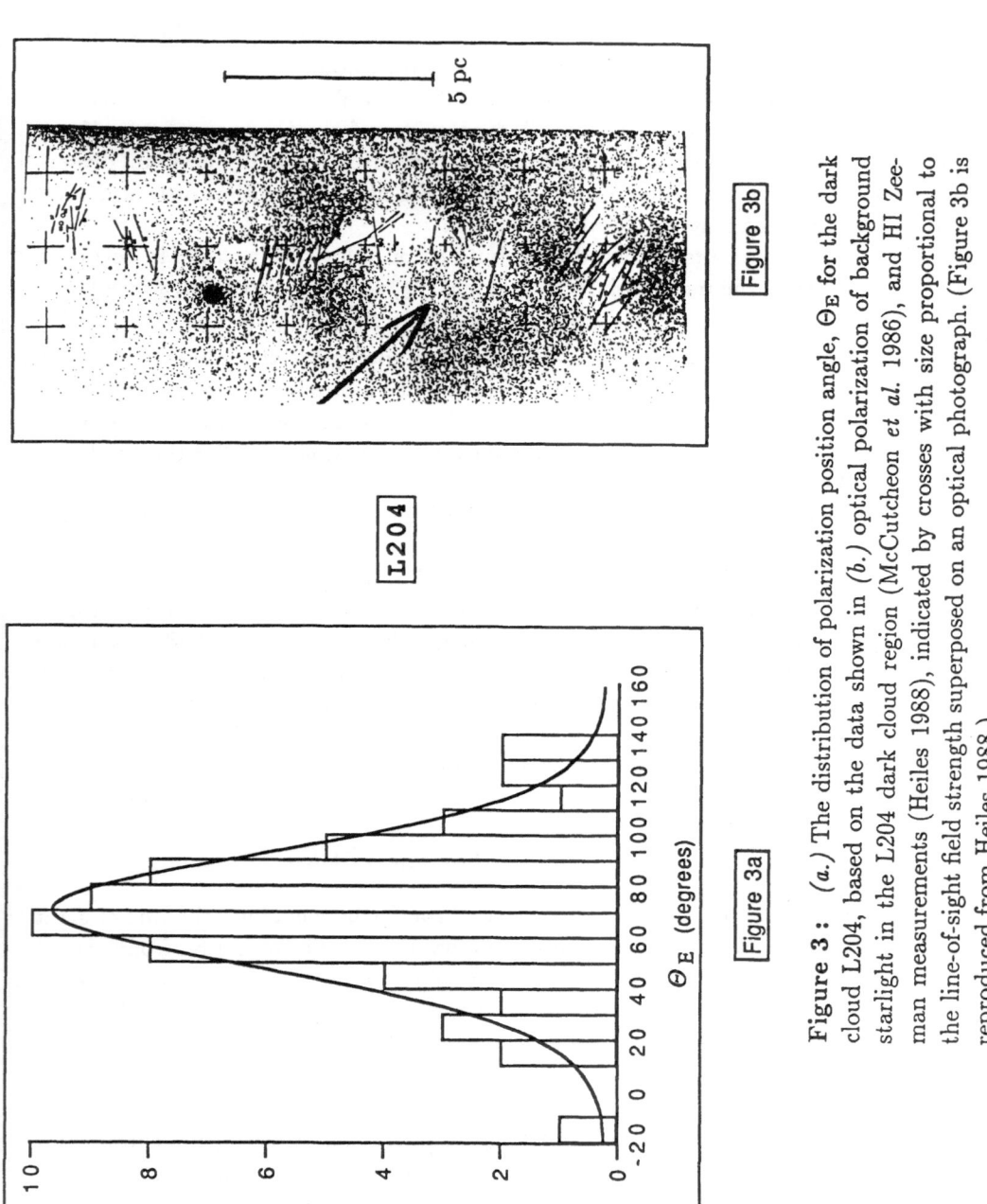

Figure 3b

L204

Figure 3a

Figure 3: (*a.*) The distribution of polarization position angle, Θ_E for the dark cloud L204, based on the data shown in (*b.*) optical polarization of background starlight in the L204 dark cloud region (McCutcheon *et al.* 1986), and HI Zeeman measurements (Heiles 1988), indicated by crosses with size proportional to the line-of-sight field strength superposed on an optical photograph. (Figure 3b is reproduced from Heiles 1988.)

the field as measured by 2 μm polarization observations to what is indicated by the Moneti *et al.* (1984) (optical) data shown in Figure 1.

More infrared polarization data is necessary before we can say, with statistical evidence, what changes take place in the nature of the field over a range of densities.

References

Chandrasekhar, S. and Fermi, E. : 1953, "Problems of Gravitational Instability in the Presence of a Magnetic Field." *Ap. J.*, **118**, 113.

Goodman, A.A., Bastien, P., Myers, P.C., and Ménard, F. : 1990, "Optical Polarization Maps of Star Forming Regions in Perseus, Taurus, and Ophiuchus." *Ap. J.*, **359**, 363.

Heiles, C. : 1988, "L204 : A Gravitationally Confined Dark Cloud in a Strong Magnetic Environment." *Ap. J.*, **324**, 321.

Heyer, M.H., Vrba, F.J., Snell, R.L., Schloerb, F.P., Strom, S.E., Goldsmith, P.F. and Strom, K.M. : 1987, "The Magnetic Evolution of the Taurus Molecular Clouds I. Large Scale Properties." *Ap. J.*, **321**, 855.

Jokipii, J.R., and Parker, E.N. : 1969, "Stochastic Aspects of Magnetic Lines of Force with Application to Cosmic-Ray Propagation." *Ap. J.*, **155**, 799.

Loren, R.B. : 1989, "The Cobwebs of Ophiuchus. I. Strands of ^{13}CO : The Mass Distribution." *Ap. J.*, **338**, 902.

McCutcheon, W.H., Vrba, F.J., Dickman, R.L., and Clemens, D.P. : 1986, "The Lynds 204 Complex : Magnetic Field Controlled Evolution?" *Ap. J.*, **309**, 619.

Moneti, A., Pipher, J.L., Helfer, H.L., McMillan, R.S., and Perry, M.L. : 1984, "Magnetic Field Structure in the Taurus Dark Cloud." *Ap. J.*, **282**, 508.

Myers, P.C. and Goodman, A.A. : 1990, "On the Dispersion in Direction of Interstellar Polarization." submitted to *Ap. J.*.

Tamura, M., Nagata, T., Sato, S., and Tanaka, M. : 1987, "Infrared polarimetry of dark clouds–I. Magnetic field structure in Heiles Cloud 2." *M. N. R. A. S.*, **224**, 413.

Vrba, F.J., Strom, S.E., and Strom, K.M. : 1976, "Magnetic Field Structure in the Vicinity of Five Dark Cloud Complexes." *A.J.*, **81**, 958.

Zweibel, E. : 1990, "Magnetic Fieldline Tangling and Polarization Measurements in Clumpy Molecular Gas." submitted to *Ap. J.*.

First Detection of 661 GHz ^{13}CO J=6→5:
Large Amounts of Warm Molecular Gas

U.U. Graf[1], R. Genzel[1], A.I. Harris[1], R.E. Hills[2], A.P.G. Russell[1,3],
and J. Stutzki[1,4]

[1] MPI für extraterrestrische Physik, Garching (FRG)
[2] MRAO, Cavendish Laboratory, Cambridge (UK)
[3] Joint Astronomy Centre, Hilo, Hawaii (USA)
[4] I. Physikalisches Institut, Universität zu Köln, Köln (FRG)

I. Introduction

Submillimeter and far-infrared observations of carbon monoxide (Jaffe, Harris, and Genzel 1987; Genzel, Poglitsch, and Stacey 1988; Schmid-Burgk et al. 1989; Boreiko, Betz, and Zmuidzinas 1989) have indicated the presence of warm, dense molecular gas near regions of recent star forming activity. Estimates based on the comparison of mid-J (submm) and high-J (far-IR) ^{12}CO lines in M17 and S106 (Harris et al. 1987a) gave a lower limit of $\approx 10^{18}$ cm^{-2} ($\tau(^{12}$CO 7→6) ≈ 1) to the CO column density of quiescent ($\Delta v \leq 10$ km/s) gas at temperatures of at least 100 K and H$_2$ densities of 10^4 to 10^6 cm^{-3}. The mid-J ^{12}CO lines are likely to be optically thick in most sources. In order to obtain a better estimate of the column densities, it is thus of great interest to observe isotopic mid-J CO lines, which are likely to be optically thin.

II. Observations and Results

The ^{13}CO J=6→5 (661.0673 GHz) detection was made on 1989 November 27 with the MPE cooled Schottky submillimeter heterodyne receiver (Harris et al. 1987b) mounted at the Nasmyth focus of the James Clerk Maxwell Telescope (JCMT), on Mauna Kea, Hawaii. On the same observing run we obtained complementary ^{12}CO J=6→5 and ^{13}CO J=3→2 data.

Strong ^{13}CO J=6→5 emission was detected from a number of positions toward Orion IRc2 (Fig. 1a)), Θ^1C, the Orion Bar, and NGC 2024 (Fig. 1b)). Both the ^{13}CO and the ^{12}CO J=6→5 spectra taken on Θ^1C and the Bar are affected by self-chopping due to the limited chop throw of 120$''$. These intensities are thus only lower limits.

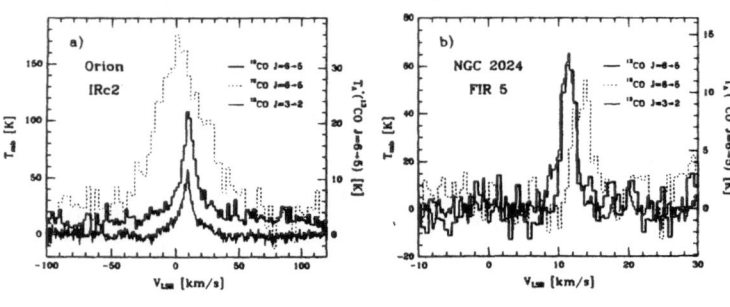

Fig. 1: CO lines detected toward Orion IRc2 (a)) and NGC 2024 (FIR 5, Mezger et al. 1988) (b)). The apparent velocity shift between the ^{13}CO lines and the main isotope in NGC 2024 is most likely caused by foreground absorption (Fig. 3).

III. Discussion

The ^{13}CO J=6 level lies 111 K above ground state. The J=6→5 transition has an A-coefficient of 1.9×10^{-5} s^{-1}, and in optically thin gas the critical density is $\approx 2 \times 10^5$ cm^{-3} (based on ^{12}CO collision rates given by Flower and Launay 1985). The high brightness temperature seen in this transition is an immediate proof of the presence of large amounts of warm, dense gas. In the optically thin limit a typical integrated line intensity of 120 K km/s (NGC 2024) implies a ^{13}CO column density in the J=6 level of 5×10^{15} cm^{-2}. Assuming LTE conditions, the maximum fractional population of the J=6 level is $\approx$11%. This constrains the total column density in warm ^{13}CO to $\geq 5\times10^{16}$ cm^{-2}.

We compared the results from an escape probability, radiative transfer program (Stutzki and Winnewisser 1985) with the observed source parameters (Fig. 2). We base our radiative transfer analysis on the ^{13}CO lines only, since these are likely to have only moderate optical depths. The similarity of the ^{13}CO 6→5 and 3→2 line profiles supports the assumption that they arise from the same material. At 70 K $\leq$ T$_{kin}$ $\leq$ 400 K and densities $\geq 10^5$ cm^{-3}, the ^{13}CO 6→5/3→2 ratio is a good measure of the kinetic gas temperature (Fig. 2). For the observed range of brightness temperatures the column density is constrained by the intensity of the ^{13}CO J=6→5 emission to N(^{13}CO) = 4×10^{16}-5×10^{17} cm^{-2}. The 6→5/3→2 intensity ratio constrains the temperature to $\geq$100 K.

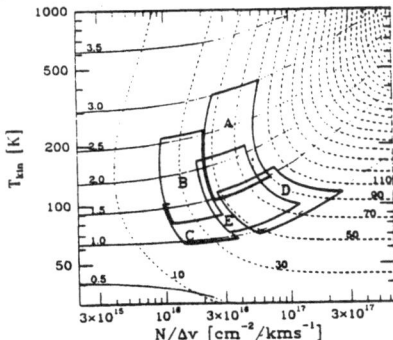

Fig. 2: Radiative transfer model of the ^{13}CO emission compared with the observed source parameters (Graf et al. 1990a). The model is calculated for an H$_2$-density of 10^6 cm^{-3}. We present the results in the two dimensional parameter space spanned by the kinetic gas temperature and the ^{13}CO column density per velocity interval. The dashed lines are curves of constant ^{13}CO J=6→5 peak brightness temperatures, the dotted lines represent constant ratios of 6→5/3→2 peak brightness temperatures. The heavy boxes outline the range of parameters consistent with the observed values for the indidual sources: A: Orion IRc2 (spike), B: Orion IRc2 (plateau), C: Orion Bar (lower limits), D: NGC 2024 (0″,0″), E: NGC 2024 (-15″,15″).

IV. NGC 2024: Two Gas Components

The observations show an apparent velocity shift between the ^{12}CO and ^{13}CO J=6→5 emission from NGC 2024. In addition, low-J, rare isotopic CO transitions (obtained with the IRAM 30m telescope with similar angular resolution) show rather complicated profiles (Fig. 3). the C^{17}O J=2→1 profile is additionally affected by the presence of nuclear quadrupole hyperfine structure. We simultaneously fitted three lines (C^{17}O J=2→1, C^{18}O J=2→1 and ^{13}CO J=2→1, Fig. 3) with two velocity components $T(v) = T_{ex} \times \{1 - \exp[-\tau \times \sum s_j \Phi(v - v_j)]\}$ along the line of sight. The foreground component can partially absorb the background component. The relative strength s_j of the different lines are given by the abundance ratios of the different isotopes (plus LTE excitation factors), or in the case of the C^{17}O hyperfine components by the spectroscopic factors.

A two velocity component model fits the data very well (Table 1). The fit identifies a cold foreground component with parameters very similar to the ones for the OH absorption component along the dust bar (Barnes et al. 1989). The warm background component agrees in velocity, velocity width and column density very well with the bulk emission from the molecular ridge (e. g. Moore et al. 1989 for CS observations). The column density is about twice that derived above for the warmer, ^{13}CO J=6→5 emitting gas component. A temperature gradient in the warm component, with about 2/3 of the column density at temperatures around 50 K and 1/3 at temperatures around 200 K gives a consistent picture for all observed CO lines. In particular the apparent velocity shift between the ^{12}CO and ^{13}CO J=6→5 lines can be explained by foreground absorption: with the excitation parameters of Table 1 the cold foreground component is optically thick ($\tau \approx 40$) in ^{12}CO J=6→5 and has a low enough temperature to absorb essentially all of the ^{12}CO emission with velocities lower than $\approx$10 km/s. Due to the low temperature the opacity of the cold component drops rapidly with increasing rotational quantum number and indeed the observed ^{12}CO J=7→6 line (Graf et al. 1990b) is much brighter, slightly wider, and less "shifted" than the J=6→5 line.

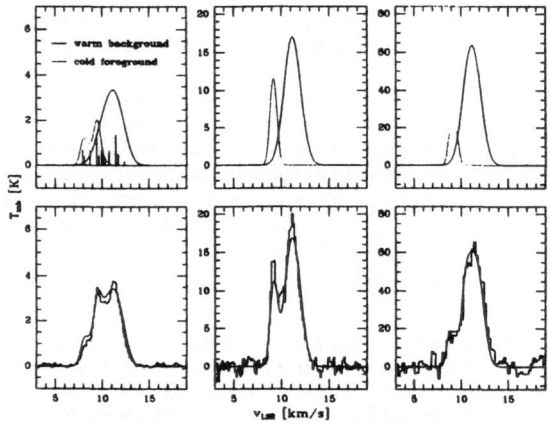

Fig. 3: Two component source model for NGC 2024 FIR 5, as derived from a simultanous fit to three low–J isotopic CO lines: $C^{17}O$ 2→1 (left; the upper panel shows the relative positions and intensities of the hyperfine components), $C^{18}O$ 2→1 (middle) and ^{13}CO 3→2 (right). The upper panels show the calculated emission from the two individual components: The two emission components add up to the line profiles shown in the lower panels, overlaid by the measured line profiles.

component	T_{ex} [K]	v_{LSR} [km/s]	Δv [km/s]	N_{H_2} [cm^{-2}]
foreground	18	9.2	0.8	4.0×10^{22}
background	70	11.2	2.0	2.2×10^{23}

Table 1: Fitted parameters of the two source components

V. What Heats the Gas?

The strong line emission detected in the ^{13}CO J=6→5 transition from both dynamically active and quiescent material confirms the presence of large amounts of warm, dense gas in star forming regions. Shocks, which are the dominant heating mechanism in the Orion plateau source (Draine and Roberge 1984, Chernoff, Hollenbach and McKee 1982), can be ruled out as a significant heating source in the other cases because of the narrow lines observed. Collisional heating of the gas by warm dust would require dust temperatures above 100 K which are highly unlikely because of the rapid increase of the dust cooling with temperature. Warm molecular gas is predicted to exist in photodissociation regions. However, all present models (e. g. Burton, Hollenbach, and Tielens 1990; Sternberg 1990), including recent extensions to higher densities, fail by about an order of magnitude to explain the observed ^{13}CO mid-J emission. See Graf et al. (1990a) for a more detailed discussion of the possible heating mechanisms.

References:

Barnes, P. J., et al. 1989, Ap. J., **342**, 883.

Boreiko, R. T., Betz, A. L., and Zmuidzinas, J. 1989, Ap. J., **337**, 332.

Burton, M., Hollenbach, D., and Tielens, A. G. G. M. 1989, 22nd ESLAB Symposium, "Infrared Spectroscopy in Astronomy"

Chernoff, D. F., Hollenbach, D. J., and McKee, C. F. 1982, Ap. J., **259**, L97.

Draine, B. T., and Roberge, W. G. 1984, Ap. J., **282**, 491.

Flower, D. R., and Launay, J. M. 1985, M.N.R.A.S., **214**, 271.

Genzel, R., Poglitsch, A., and Stacey, G. J. 1988, Ap. J., **333**, L59.

Graf, U. U. et al. 1990a, Ap. J. (Letters), in press.

Graf, U. U. et al. 1990b, in prep.

Harris, A. I., et al. 1987a, Ap. J., **322**, L49.

Harris, A. I., et al. 1987b, Internat. J. Infrared Millimeter Waves, Vol. 8, No. 8, p.857.

Jaffe, D. T., Harris, A. I., and Genzel, R. 1987, Ap. J., **316**, 231.

Mezger, P. G., et al. 1988, Astr. Ap., **191**, 44.

Moore, T. J. T., et al. 1989, M.N.R.A.S., **237**, 1p.

Schmid-Burgk, J., et al. 1989, Astr. Ap., **215**, 150.

Sternberg, A. 1990, in prep.

Stutzki, J., and Winnewisser, G. 1985, Astr. Ap., **148**, 254.

"THE MASS DISTRIBUTION OF THE YOUNG STELLAR POPULATION IN CHAMAELEON   I
AND IN Rho   OPHIUCHI"

Jane C. Gregório Hetem, J.R.D. Lépine, R. Ortiz

Instituto Astronômico e Geofísico
Universidade de São Paulo, Brazil.

Abstract
      We obtain the mass distribution and the age distribution of the
young stars associated with Chamaeleon I and Rho Ophiuchi, two   nearby
sites of star formation. Our method consists in determining the tempera
ture and the luminosity of each object in order to locate it on the   HR
diagram, and then comparing the position on the HR diagram   with   the
evolutionary tracks and isochrones presented by Cohen and Kuhi (1979) .
The star-formation process is found to have started more recently in ω
Oph than in Cham I.

I. Introduction
      The mass distribution of the pre-main-sequence   (PMS)   stars
embedded in a molecular cloud is determined by the process of fragmentation of the cloud which leads to star formation, in a way which   is   not
well understood. It would be of interest to know the mass distribution of
the young stars in different clouds in order to investigate the influence
of the physical parameters of the clouds in this process. In the present
work we obtain the mass distribution of the young stars embedded in   the
Chamaeleon I and Rho Ophiuchi molecular clouds, two prominent   nearby
sites of star formation.
      The only way to estimate the mass of PMS stars is to   compare
their position on the H-R diagram with a grid of evolutionary tracks for
protostars of different mass. The position of the stars relative to   the
isochrones also provides an estimate of their age. A determination of the
temperature and luminosity of the stars is thus required as a first step.
      The luminosity function of the young stars in ρ Oph has recen-
tly been investigated by Wilking, Lada and Young (1989), and in a previous
work we have compared the luminosities of the objects embedded in Rho Oph
and in Cham I (Gregório-Hetem et al., 1990). We use here the same method,
based on model fitting of the observational data, to estimate both   the
luminosity and the temperature of the stars. While the luminosity distri
butions of the stars of the two regions are similar, striking   differences appear between the mass distributions.

II. The Method

The spectral energy distribution of each star was constructed from the visible and near-infrared photometric data found in the literature, and from the IRAS far-infrared data of the Point Source Catalog.We found enough data for 39 pre-main-sequence stars in each cloud; these are probably the most luminous stars of each association. The list of stars and the references for photometric data are basically the same used by Gregório-Hetem et al. (1990).

For each object, the energy distribution was fitted by a model of a central star represented by a blackbody with temperature T_*, surrounded by a series of spherically symmetric dust shells forming an envelope with internal radius R_0 and temperature T_0, and with density decreasing outwards like $r^{-1.5}$ and temperature decreasing outwards like $r^{-0.4}$. Both the extinction and the emission of the shells are taken into account; the adopted dependence of opacity on wavelength is $\tau(\lambda)=\tau(1\mu m)\bar{\lambda}^{-1}$. The adjustable parameters are T_*, T_0, R_0, and the optical depth of the envelope. This simple model provides good fits to the energy distributions: the temperature T_* obtained from the fit is in good agreement with the temperature corresponding to the spectral type, when it is known.The luminosity of the objects is obtained by integrating the flux over the whole wavelength range, and assuming a distance of 140 pc to Cham I (Rydgren, 1980) and of 160 pc to ρ Oph (Bertiau , 1958).

We used the temperature and luminosity derived from our model to plot the positions of the stars on the HR diagram, to compare them with the convective-radiative evolutionary tracks of pre-main sequence stars collected by Cohen and Kuhi (1979), and with the corresponding isochrones. The number of stars in a given mass interval is simply obtained by counting the stars situated between the evolutionary tracks of the corresponding mass. We selected for convenience the logarithmic mass intervals with limits 0.1, 0.2, 0.4, 0.8, 1.6 M_0. Cohen and Kuhi did not present evolutionary tracks for 0.4 M_0 and 1.6 M_0, but instead for the neighbouring values 0.35 and 1.5 M_0; however an approximate interpolation is easily done. We somewhat arbitrarily, corrected the 0.2 M_0 track of Cohen and Kuhi originaly taken from Grossman and Grabosky (1971),turning it similar to the tracks for lower and for larger mass and thus removing a cooling phase during the contraction, which seems unlikely to exist.

III. Results and Discution

We present in figures 1 to 4 the mass distribution and the age distribution of the young stars associated with the two regions. One can see that Cham I contains stars which are considerably more massive than those of ρ Oph. The maximum of the mass distribution is around .3 M_0 in ρ Oph and 1.2 M_0 in Cham I. The age distributions of the objects belonging to the two associations are also considerably different. While in ρ Oph most of the stars are younger than $3*10^5$ years, in Cham I most of the stars are older than $3*10^6$ years. This does not mean that the star formation process has ceased in Cham I; the presence of Herbig-Haro objects indicates that some star formation may be still on-going about 10^7 years in clouds with characteristics similar to those of ρ Oph and Cham I.

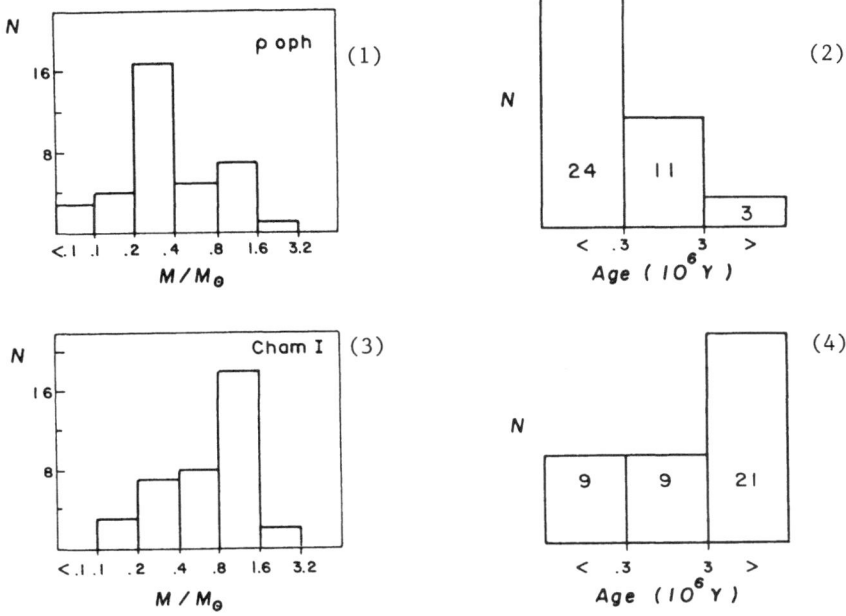

Figures 1 to 4

References

Bertiau , F.C. 1958, Astrophys. J. 128, 533.

Cohen, M., Kuhi, L.V., 1979, Astrophys. J. Suppl. 41, 743.

Gregório-Hetem, J.C., Lépine, J.R.D., Ortiz, R. 1990,
      Proceeding of the VI Latin-American Regional Meeting of the
      International Astronomical Union, Revista Mexicana de Astronomia,
      in press.

Grossman,A.S., and Grasboske,H.C.,1971,Astrophys. J. 164, 475.

Rydgren, A.E., 1980, Astron. J. 85, 444.

Wilking, B.A., Lada, C.J., Young, E.T., 1989, Astrophys. J. 340, 823.

DENSE CLOUD CORES

R.GÜSTEN AND A.SCHULZ

Max-Planck-Institut für Radioastronomie, Auf dem Hügel 69, D-5300 Bonn 1, FRG

E.SERABYN

Division of Physics, Mathematics and Astronomy
California Institute of Technology, Pasadena, CA 91125

In order to search for ultra-dense gas condensations as possible sites of future star formation, observations with highest possible angular resolution in high-gas density probing molecular species are required. Here we report first results from two-telescope (IRAM 30-m & CSO 10.4-m) multi-transition excitation studies of the linear molecules CS and HCN, both with critical densities $n_c \sim 10^7$ cm^{-3}. We obtained maps towards a number of galactic cloud cores in the CS(J=5-4;7-6) and the HCN(3-2;4-3) mm/submm transitions and those of their optically thin isotopic species (C^{34}S; H^{13}CN) accessible from ground. For a proper excitation analysis, the gas temperature has been determined independently from the symmetric top molecules NH$_3$ or CH$_3$CN. Here we present first results obtained towards the NGC 2024 molecular cloud core (Schulz et al. 1990).

The Dense Condensations in NGC 2024

Several recent mm/submm studies focused on the dense gas ridge buried within the optically prominent NGC 2024 dust lane. From their submm continuum observations Mezger et al. (1988) identified 6 small-scale dust clumps to which, in view of the very low dust temperatures ($\sim$16 K) and extremely high gas densities ($\geq 10^8$ cm^{-3}) deduced, they refer to as *genuine protostellar condensations*. As these clumps had not been recognized in earlier lower-resolution molecular line observations, the authors suggested strong molecular depletion onto dust grains.

Our detailed NH$_3$ and CS excitation studies revealed the density and temperature structure across the dense ridge which appears separated into two subregions by the disrupting effect of the southern HII-region. The global temperature structure derived closely corresponds with profiles of the dust color temperature given by Thronson et al. (1984), and is strongly suggestive of a heating scenario with the bulk sources of energy located outside the dense bar. The cloud material must be highly clumpy to allow the deep penetration depth of energetic photons.

Our 8.5-11" resolution CS maps show a general correspondence with the 1.3mm dust continuum emission. Embedded within the dense bar we find a number of small-scale clumps (size : 2-4 10^{-2} pc) which partly could be identified with submm dust knots. But basic dicrepancies are faced when physical characteristics are compared. Densities and column densities inferred from our LVG analysis of the CS excitation are high ($\sim$1-5 10^6 cm^{-3}; $\sim$6 10^{23} cm^{-2}) but fall short of estimates from the dust emission by an order of magnitude. The overall gas mass of the dense bar amounts to 100-150 $M_\odot$, much lower than previous determinations. In the molecular line data, there is no evidence for the cold massive clumps deduced by Mezger et al., and the high CS brightness temperatures observed towards the clumps ($\sim$35-40 K) closely match the larger-scale NH_3 temperatures.

We re-analysed the NGC 2024 integrated dust spectrum and infer that the ambiguity of the decomposition procedure in particular allows a physically consistent fit that matches both the dust and the molecular line data without invoking a massive low-temperature component. These findings and recent reports of stellar activity associated with several of the dense clumps thus question their interpretation as cool protostellar condensations.

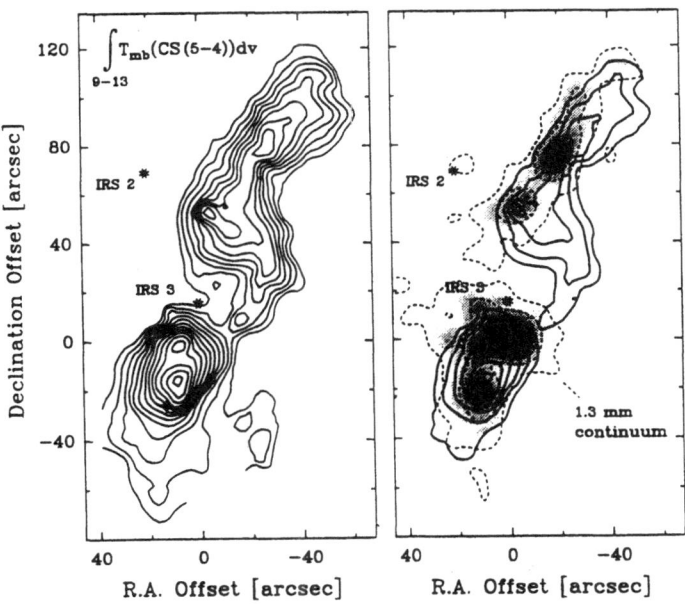

Fig. 1. Velocity-integrated CS(5-4) brightness temperature distribution superimposed on a grey-scale representation of the 1.3 mm dust continuum (Mezger et al. 1988), both taken with 11" angular resolution (IRAM 30-m). Positions of associated IR sources are marked by asterisks, the cross denotes the embedded H_2O maser.

References

Mezger P.G., Chini R., Kreysa E., Wink J.E., Salter C.J. 1988 *Astron. Astrophys.* **191**,44.

Schulz A., Zylka R., Güsten R. 1990 *Astron. Astrophys.* submitted

Thronson H.A., Lada C.J., Schwartz P.R., Smith H.A., Smith J., Glaccum W., Harper D.A., Loewenstein R.F. 1984 *Astrophys. J* **280**,154.

GRAVITATIONAL INSTABILITY INDUCED BY COLLISIONS BETWEEN NON-IDENTICAL CLOUDS

Asao HABE and Kanji OHTA
Department of Physics
Hokkaido University
Sapporo
JAPAN

ABSTRACT. In order to investigate gravitational instability induced by supersonic collisions between small, dense clouds and large, diffuse clouds, we perform the numerical hydrodynamic calculation by SPH code. We show that the large cloud is disrupted by the bow shock induced by the small cloud, and the small cloud is compressed. Final fates of the small cloud are divided into two cases. The first case is that the small cloud is ablated by the ram-pressure of the large cloud and is finally destroyed. The second case is that the central part of the small cloud left from the ablation gravitationally collapses. We discuss the condition dividing these two cases. Our condition show that even if the Mach number of collision velocity is very high, the small cloud can collapse.

1. Introduction

Recently, cloud-cloud collision is proposed to be effective for triggering process of star formation in OB star formation region of our Galaxy (Scoville *et al.* 1986) and in star burst galaxeis (Olson and Kwan 1990).

Cloud-cloud collisions are mainly studied the case of collisions between clouds with same size (e.g. Gilden, 1984, Nagasawa and Miyama, 1987). In this case the gravitational instability condition of colliding clouds is shown. However, since interstellar clouds with various sizes and densities are observed, the case of identical cloud-cloud collision is very rare and collisions between clouds with different sizes and densities are more frequent. The latter case should be investgated. We perform the numerical hydrodynamic calculation of this case.

2. Model and Numerical Results

We assume two clouds with different size, R, and density, ρ. These values are chosen according to the empirical relation obtained by Larson (1981) from the observational results of many molecular clouds. His relation is $\rho \propto R^{-1}$ and $\sigma^2 \propto R$, where σ is the velocity dispersion of gas in clouds. We assume that initial clouds are stable in the meaning of the gravitational stability. These clouds are assumed to collide supersonically each other in head on. Since the shock structures in colliding clouds are well approximated as the isothermal shock, we adopt the isothermal state equations with different temperature for gases of these clouds. We use the smoothed particles hydrodynamic code (SPH) for calculation of self-gravitating gas motion in the axially-symmetric 2-dimension. Since our SPH code is the axially-symmetric 2-dimension, the particle number of 4300 is well enough to calculate the strong isothermal shock with high Mach numbers.

The model parameters and numerical results are shown in Fig.1. The mass ratio of the small cloud and the large cloud is 1:4. The large cloud is disrupted by the strong bow shock induced by the small cloud. We find that our numerical results are divided into two cases for the final fates of small clouds. The first case is that, although the strong shock wave compresses the small cloud and a dense gas layer is formed, the small cloud is finally dispersed by ablation process by the ram-pressure of large cloud. The second case is that, the strong shock wave compresses the small cloud, and finally the central part of the small cloud left from the ablation becomes gravitationally unstable, after the shock wave has passed through the small cloud. As a result it gravitationally collapses and its central density becomes higher than hundred thousand of initial cloud density.

3. Discussion

Numerical results indicate that the residual part of small cloud from the ablatin can collapse. We estimate this mass, m_{com}, and size of this region by a similar discussion to Gilden (1984) and give the gravitational condition of this part of cloud, $m_{com}/M_{com,J} > 1$, as shown in the line in Fig. 1, where $M_{com,J}$ is the Jeanse mass of this part. Figure 1 shows that this criterion well agrees with our numerical results.

We show that even if the Mach number of collision velocity is very high, the small cloud can collapse in the non-identical cloud-cloud collisions. Olsen and Kwan (1990) insisted that in order to explain the high infrared luminosity of interacting galaxies, disruptive collisions must lead to star formation. In this paper, we have shown the case that the disruptive collision can lead to the gravitational collapse of clouds.

References

Gilden,D.L.,1984,$Ap.J$, **279**, 335.
Larson,R.B., 1981, $M.N.R.A.S.$, **194**, 809.
Nagasawa,M. and Miyama,S.M., 1987, $Prog.Theoret.Phys.$, **78**, 1250.
Olson.K.M. and Kwan,J., 1990, $Ap.J$, **349**,480.
Scoville,N.Z., Sanders,D.B. and Clemens,D.P., 1986, $Ap.J.(Letter)$, **310**, L77.

Figure 1. The model parameter and numerical results. Crosses denote the collapse case of small clouds. Filled circles denote the dispersed case of small clouds. The curve denotes the condition given in the text.

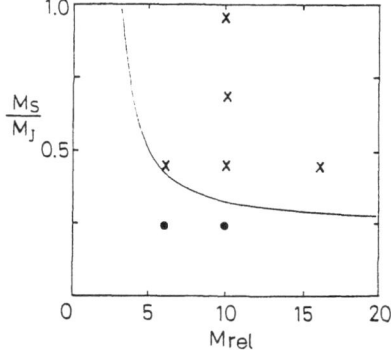

CO MULTITRANSITION OBSERVATIONS OF L 1228 :
A MOLECULAR OUTFLOW
DESTRUCTING ITS PARENT CLOUD?

L.K. Haikala[1,2] and H. Weikard[1,3]

1. I. Physikalisches Institut, Universität zu Köln
2. Observatory and Astrophysics Laboratory, University of Helsinki
3. Groupe d'Astrophysique de l'Observatoire, Grenoble

Abstract. CO, J=3→2, J=2→1 and J=1→0 observations of the extended molecular outflow in the high latitude cloud L 1228 are presented. The ^{12}CO, J=3→2 and J=2→1 observations reveal a well collimated and extended outflow. The ^{12}CO, J=1→0 observations show that the red shifted part of the outflow has blown a large cavity into the parent cloud and has thus dispersed a significant part of the surrounding cloud.

L 1228 is a high latitude (111°.6 , 20°.1) dark cloud. The distance to the cloud is not known but a reasonable estimate is 100 to 200 pc. CO, J=1→0 observations of the cloud are described in Haikala and Laureijs (1989). Strong ^{12}CO self absorption in L 1228 is observed over a one by half a degree area in the cloud.

The outflow has been mapped in the ^{12}CO, J=3→2 (KOSMA 3m telescope) and J=2→1 (POM-2) transitions. The outflow is well collimated (Fig. 1) and nearly identical in both CO transitions. The Onsala 20m telescope was used to make cross scans of the outflow in the CO, J=1→0 transition.

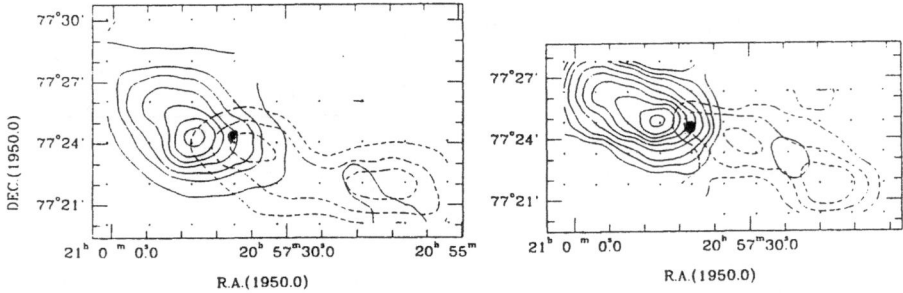

Fig. 1. The blue shifted (continuous line) and red shifted (dashed) ^{12}CO, J=3→2 (KOSMA, on the left) and J=2→1 (POM2, on the right) emission in L 1228. The outflow central source IRAS 20582+7724 is marked in th plot.

In the centre of the blue lobe of the outflow CO emission is totally dominated by the emission coming from the outflowing gas (Fig. 2c). Due to different beam sizes the individual spectra of different transitions cannot be compared directly.

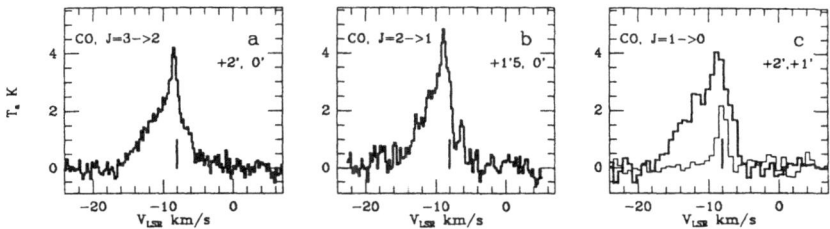

Fig. 2. ^{12}CO, J=3→2 (beam 1'3), J=2→1 (2'5) and J=1→0 (33") spectra in the centre of the blue lobe. The offsets are given from IRAS 20582+7724. ^{13}CO is shown in c (thin line). The tick line is at velocity -8.0 $km\ s^{-1}$

Contour map of the red shifted (-5.5 $km\ s^{-1}$ to 0 $km\ s^{-1}$) emission in L 1228 is shown in Fig. 4a. A large cone like feature extends to the west from the outflow central source. No such feature is seen in the blue shifted emission. This extended red feature is better seen at velocity interval -5.5 to -4.6 $km\ s^{-1}$ (Fig. 4b). The angular size of the cone is half a degree.

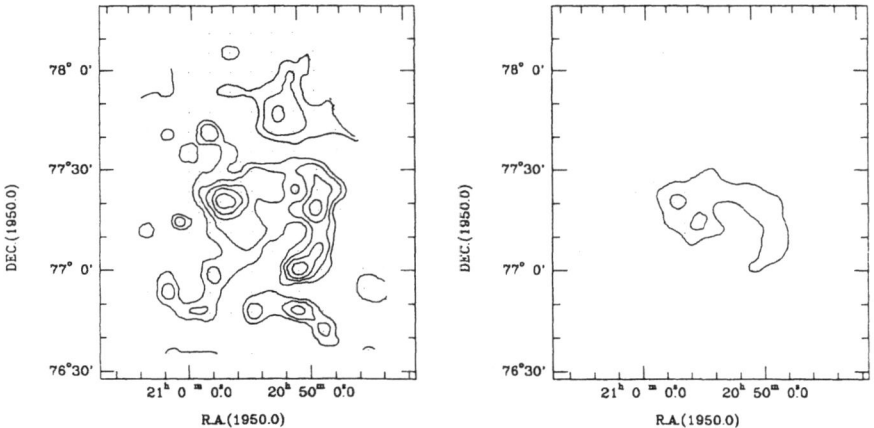

Fig. 3. Red shifted ^{12}CO emission in L 1228. a -5.5 $km\ s^{-1}$ to 0 $km\ s^{-1}$ and b -5.5 $km\ s^{-1}$ to -4.6 $km\ s^{-1}$

The optical extinction to the west of outflow is low. The northern rim of the red cone feature falls into this low opacity hole. The high opacity part of the cloud north of the cone is traced by the slightly blue shifted emission. The cone feature suggests that the outflow has blown a large hole into the parent cloud and has thus destructed at least a part of its parent cloud.

References

Haikala, L.K. and Laureijs, R. : 1989, *Astron. Astrophys.*, **223**,287

AMMONIA CLUMPS IN THE ORION AND CEPHEUS CLOUDS

J. HARJU[1], C.M. WALMSLEY[2] AND J.G.A. WOUTERLOOT[3]

[1] *Observatory and Astrophysics Laboratory, University of Helsinki*
[2] *Max-Planck-Institut für Radioastronomie, Bonn*
[3] *I. Physikalisches Institut, Universität zu Köln*

Abstract. We present statistics of clump properties in the Orion and Cepheus cloud complexes based on ammonia mapping observations. Surroundings of about 50 IRAS sources earlier found to have associated molecular line emission (Wouterloot, Walmsley and Henkel, 1988) were mapped in NH3(1,1) and (2,2) with the Effelsberg 100-m telescope. Our main interest has been in determining the clump sizes and masses on the basis of the ammonia column density distribution, which together with the observed velocity dispersion lead to a rough estimate of the dynamical state. We also have studied the star-clump separations which should give us estimates of the source ages. Special attention has been paid to comparison of our Orion data with the Benson and Myers (1989, hereafter BM89) results in Taurus because the linear resolutions in the two studies are similar.

Molecular clouds contain cool, high density cores, which are often associated with newly born stars detected in infrared surveys - above all IRAS. This has become evident through the work of Myers, Benson and collaborators in nearby molecular cores (see e.g. Beichman *et al.* 1986, Myers *et al.* 1987, BM89), and the surveys by Wouterloot, Walmsley and Henkel in H_2O maser, NH_3 and CO lines towards IRAS point sources in Orion and Cepheus (Wouterloot *et al.* 1986, 1988, 1989).

We have studied the clump properties, such as sizes, masses, temperatures and densities in the Orion and Cepheus giant molecular cloud complexes towards sources detected by Wouterloot et al. (1988). The observations were made with the 100-m telescope in Effelsberg by measuring the $NH_3(1,1)$ and $(2,2)$ transitions. Our Orion sample (L1630 and L1641) consists of 24 separate clumps in the neighbourhood of 16 IRAS point sources. In the Cepheus region, the surroundings of 29 IRAS sources were mapped. A summary of the statistical results of this study is given in the attached table.

The Orion clumps are larger, warmer, more massive and have broader lines than the Taurus clumps of BM89. Despite their relatively large linewidths, most Orion clumps seem to be gravitationally bound. This fits the concept of gravitationally dominated collapse in high-mass star forming regions (Shu *et al.* 1987), and is consistent with the NIR observations of Strom *et al.* (1989a,b,c) towards L1641, where they found a large number of sources embedded in high-opacity cores. 8 of their sources are located inside clumps mapped by us.

In Cepheus, the derived densities are lower than in Orion or Taurus. Especially in the direction of the Cepheus OB3 cloud (adopted distance 730 pc), the cores seem to lack mass to balance their "turbulent" motions. The observed densities are, however, likely to be affected by beam dilution. Furthermore, the uncertainty of the distance makes the mass estimates less reliable.

One distinctive property of Orion is that the embedded stars usually lie very close to the clump centers, in a small range below 0.1 pc which is about a typical clump radius, while in the cases of the other complexes the distribution is broader. If we interpret this star-clump separation as measuring the age of the star, the implication is that most embedded stars detected as IRAS point sources in Orion were born during a short period about 10^5 years ago.

Table. Average clump properties in Orion, Cepheus and Taurus. Columns : 1) infrared luminosity of the IRAS source ($L_\odot$); 2) clump mass ($M_\odot$); 3) clump half power diameter (pc); 4) clump-star distance (pc); 5) $NH_3(1,1)$ line width (kms^{-1}); 6) kinetic temperature (K); 7) maximum excitation temperature (K); 8) maximum H_2 density (cm^{-3}); 9) maximum NH_3 column density (cm^{-2}); 10) number of separate clumps included in the statistics. *Notes : 1) STD means standard deviation. 2) The densities are calculated using the two-level approximation formula of Ho and Townes (1983) on the basis of the derived kinetic and excitation temperatures. For comparison, the densities derived by BM89 based on an LVG analysis are tabulated with emphasized numbers.*

		L_{IR} ($L_\odot$)	M ($M_\odot$)	diam (pc)	d* (pc)	$< \Delta v >$ (kms^{-1})	$< T_{kin} >$ (K)	max(T_{ex}) (K)	max(n_{H_2}) ($\cdot 10^6$cm^{-3})	max(N)) ($\cdot 10^{14}$cm^{-2})	sample
Orion	average	50	26	0.18	0.08	0.81	15.7	9.2	10.0	7.7	24
dist ~ 500 pc	STD	(52)	(22)	(0.06)	(0.05)	(0.28)	(4.6)	(2.9)	(9.4)	(4.7)	
	median	28	16	0.17	0.08	0.75	14.7	8.8	9.2	6.6	
Cepheus (near)	average	110	16	0.18	0.11	1.0	16.7	5.9	2.1	4.6	11
dist ~ 730 pc	STD	(130)	(11)	(0.10)	(0.10)	(0.2)	(1.8)	(1.6)	(1.4)	(3.0)	
	median	55	15	0.17	0.07	1.0	17.0	6.0	2.1	3.3	
Cepheus (far)	average	2.5 (4)	590	0.96	0.53	1.5	20.4	5.7	1.9	4.7	15
dist ~ 3500 pc	STD	(3.0 (4))	(910)	(0.38)	(0.23)	(0.4)	(3.7)	(1.7)	(1.4)	(5.3)	
	median	1.0 (4)	140	0.89	0.56	1.5	19.7	5.7	1.7	2.2	
Taurus	average	1.1	9	0.10	0.13	0.30	10.0	7.6	12.4 *2.8*	9.6	10
dist ~ 140 pc	STD	(1.4)	(11)	(0.04)	(0.10)	(0.07)	(2.1)	(1.6)	(8.4) *(1.8)*	(4.5)	
(BM89)	median	0.6	4	0.08	0.12	0.28	9.6	7.6	11.1 *2.5*	10.0	
notes	*1)*									*2)*	

References

C.A. Beichman, P.C. Myers, J.P. Emerson, S. Harris, R. Mathieu, P.J. Benson, R.E. Jennings : 1986 *Astrophys. J.*, **307**, 337

P.J. Benson, P.C. Myers : 1989 *Astrophys. J. Suppl.*, **71**, 89 (BM89)

Y. Fukui : 1989 in "Low Mass Star Formation and Pre-Main-Sequence Objects" (ESO Workshop, Garching 11-13 July 1989, Proceedings edited by B. Reipurth)

P.T.P Ho, C.H. Townes : 1983 *Ann. Rev. Astron. Astrophys.*, **21**, 239

P.C. Myers, G.A. Fuller, R.D. Mathieu, C.A. Beichman, P.J. Benson, R.E. Schild, J.P. Emerson : 1987 *Astrophys. J.*, **319**, 340

F.H. Shu, F.C. Adams, S. Lizano : 1987 *Ann. Rev. Astron. Astrophys.*, **25**, 23

K.M. Strom, G. Newton, S.E. Strom, R.L. Seaman, L. Carrasco, I. Cruz-Gonzales, A. Serrano, G.L. Grasdalen : 1989a *Astrophys. J. Suppl.*, **71**, 183

K.M. Strom, M. Margulis, S.E. Strom : 1989b *Astrophys. J.*, **345**, L79

K.M. Strom, M. Margulis, S.E. Strom : 1989c *Astrophys. J.*, **346**, L33

J.G.A. Wouterloot, C.M. Walmsley : 1986 *Astron. Astrophys.*, **168**, 237

J.G.A. Wouterloot, C.M. Walmsley, C. Henkel : 1988 *Astron. Astrophys.*, **203**, 367

J.G.A. Wouterloot, C. Henkel , C.M. Walmsley : 1989 *Astron. Astrophys.*, **215**, 131

OBSERVATIONAL CONSTRAINTS ON ANGULAR MOMENTUM TRANSFER DURING GRAVITATIONAL COLLAPSE*

ERIC KETO

Institute of Geophysics and Planetary Physics, L-413,
L.L.N.L., P.O. Box 808, Livermore, Ca., 94550, USA.

ABSTRACT. A simple calculation of the expected spectral signatures of model protostellar accretion flows suggests how the rotation curve of the accretion disk may be deduced from radio frequency molecular line observations. We compare synthetic observations with actual data to derive rotation curves, braking torques, and minimum magnetic field energies required to effect the braking.

1. Introduction

Similar to the way in which galactic rotation curves may be deduced from HI position-velocity of l, v plots, the rotation curves of accretion flows may be deduced from molecular line observations. We illustrate this with synthetic NH_3 spectra computed for 3 model accretion flows.

In th case solid body rotation, $\omega \propto r^0$, the projected velocity is constant along any line of site resulting in a linear dependence of the emission velocity with position across the disk. In the case of angular momentum conservation, $\omega \propto r^{-2}$. The faster rotation in the center of the disk spreads the emission over a larger range in velocity. Figure 1 illustrates the appearance of the intermediate case indicative of partial loss of angular momentum or braking. (The repeating pattern in velocity is caused by the electric quadrupole splitting of the NH_3 line.) A summary of the parameters of several accretion flows from data published elsewhere is shown in Table 1.

Given a rotation curve and infall velocity, the minimum magnetic field energy density required to effect the braking may be estimated from the spin-down torque as,

$$DL/Dt = (DL/Dr)(Dr/Dt) = (4\pi r^2 \rho(r))(r^2 \Omega(r))v_r$$

The results for two flows are presented in Table 2.

---

* Work performed under the auspices of the U.S. Department of Energy by the Lawrence Livermore National Laboratory under contract No. W-7405-ENG-48.

Table 1

Source	Radius	$\omega(r)^{-1}$ (yrs)	$v_r(r)$ (km s^{-1})	$\rho(H_2)$ (cm^{-3})
G10.6C[1] 500M$_\odot$ *	0.35pc	4×10^5 $\propto r^{-2.5}$	2.7 $\propto r^{-0.8}$	2×10^4 $\propto r^{-0.35}$
G10.6B[2] 200M$_\odot$	0.2pc	$\sim 7 \times 10^5$ $r^0 < \omega < r^{-2}$	?	4×10^3
G10.6A[2] 100M$_\odot$	0.2pc	$\sim 7 \times 10^5$ $r^0 < \omega < r^{-2}$	?	1.5×10^4
G34.3[3] 4M$_\odot$	100au	525 ± 20 $\propto r^{-0.5\pm0.05}$	3.6 ± 0.25 $\propto r^{-0.3\pm0.04}$	$4 \times 10^{9\pm.3}$ $\propto r^{-0.4\pm0.1}$

* Masses indicate the total core mass, the radius refers not to the "boundary" of the core, but the radius within the core for which the properties ω, ρ, v are tabulated.

[1] Keto, E. 1990, *Ap. J.*, **355**, 190.
[2] Cool, A., and Ho, P. 1990, *in preparation*. [3] Keto, E., Jeffrey, W., and Proctor, D. 1990 *in preparation*.

Table 2

Source	Radius	DL/Dt (ergs)	Required B (mG)	$\rho(H_2)$ cm^{-3}
G10.6C	0.35pc	2×10^{46}	0.3	2×10^4
	0.025pc	9×10^{45}	10	1.5×10^7
G34.3	500au	4×10^{44}	75	2×10^9
	100au	2×10^{44}	200	4×10^9

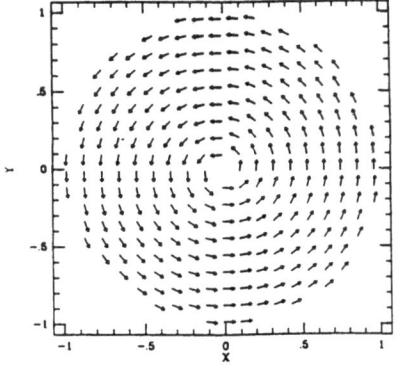

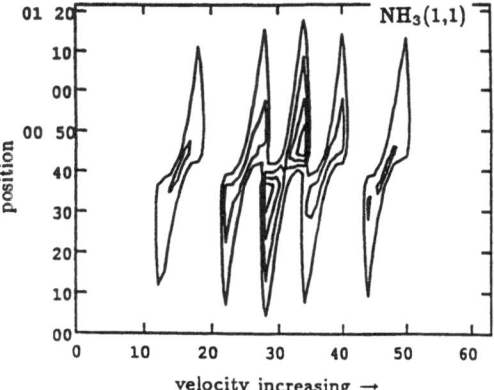

velocity increasing →

MEGAMASERS AS PROBES OF GALAXY
MASS SPECTRUM EVOLUTION

V.K. KHERSONSKII[1] AND N.V. VOSHCHINNIKOV[2]

[1] *Leningrad Department of Special Astrophysical Observatory,*
196140, Leningrad, Pulkovo, USSR

[2] *Astronomical Observatory of Leningrad University,*
198904, Leningrad, Petrodvoretz, USSR

OH megamasers having very high luminosities in the spectral line can be effectively used for the probing of the evolutionary properties of the galaxies in the earliest cosmological epochs. The frequency shift of the emission line uniqually determines the redshift z, which tells about the epoch of emission. One of the important cosmological problems is the investigation of the galaxy mass spectrum in the expanding Universe. There is the empirical relation between the OH and far-infrared luminosities of galaxies. Therefore, if in the earliest cosmological epochs, there were galaxies with sufficient powerful infrared excesses and containing molecular material, they can be detected using the observations of their OH maser emission. The interacting and merging galaxies can be considered as the best candidates for such objects.

We determine the expected number of maser sources in the epoch at redshift z, using the model of galaxy mass spectrum (GMS) evolution in the expanding Universe. In this model, it is assumed that the GMS formation is connected with the collisions and mergers of galaxies and that these interactions are playing the dominant role in the initiation of the star formation process (Khersonskii and Voshchinnikov, 1991, to be published).

We estimate the number of expected OH maser sources in 1Mpc^3 emitting in galaxies at the epoch of redshift z, $N_{OH}(z)$, for a given sensitivity of the radio observations. For the Arecibo radiotelescope, the ratio of this number to the total number density of galaxies at the epoch of redshift z, $S_{OH}(z) = N_{OH}(z)/n_1(t(z))$ increases with z for values of the redshift smaller than 2 to 3. Such a dependence of $S_{OH}(z)$ on z is related to the general growth of the number of starbust galaxies. However, beyond some redshift z_1, $S_{OH}(z)$ decreases rapidly because of the absence of sufficiently massive galaxies in GMS.

Magnetic Fields in the Solar Nebula and the Angular Momentum Transfer

M. Kiguchi[1], S. Narita[2], T. Terasawa[3] and C. Hayashi[4]

[1] Research Institute for Science and Technology, Kinki University, Osaka, 577, Japan
[2] Department of Electronics, Doshisha University, Kyoto, 602, Japan
[3] Department of Geophysics, Kyoto University, Kyoto, 606, Japan
[4] Kyoto University, Kyoto, 606, Japan

1. Introduction

We are now carrying out the project to investigate how the primodial solar disk is formed and evolves. One of the central problems of this project is to study the angular momentum transfer in the nebula caused by the magnetic braking or the turbulent mixing in the boundary layer between the surface of the disk and its high temperature corona envelope.

Here, we summarize the result of analysis for the magnetic braking in 1-dimensional approximation, which will make the basis to interpret the full numerical calculations which we are now carrying out.

2. Angular Momentum Transport in the Radial Direction

As pointed out by Hayashi (1981), when seed magnetic fields which are generated by turbulences have radial component, azimuthal component grows by rotation and decays by Joule loss. For this process, the basic equation is given by

$$\{\frac{\partial}{\partial t} + \Omega_0[r_0 - \frac{3}{2}(r - r_0)]\frac{\partial}{r_0\partial\phi} - \frac{c^2}{4\pi\sigma_e}\nabla\} \begin{pmatrix} H_r \\ H_\phi \\ H_z \end{pmatrix} = \begin{pmatrix} 0 \\ -\frac{3}{2}\Omega_0 H_r \\ 0 \end{pmatrix}. \tag{1}$$

The azimuthal component H_ϕ takes the maximum value at $t \sim t_{\text{decay}}$, $H_\phi^{\text{max}} \simeq (t_{\text{decay}}/t_{\text{Kepler}})H_r(0)$, where t_{decay} is the decay time $t_{\text{decay}} = 4\pi\sigma_e z_0^2/c^2$, z_0 is the scale height of the disk, and t_{Kepler} is the Kepler time. The amplification factor $t_{\text{decay}}/t_{\text{Kepler}}$ takes the value about $10^{-1} - 10^1$ at Jupiter. The angular momentum is transferred by the r-ϕ component of the Maxwell stress tensor along the r-direction. The dynamical viscosity coefficient in this case is given by $\nu \sim (t_{\text{decay}}/t_{\text{Kepler}})c_s z_0$.

3. Angular Momentum Transport by Alfvén Wave

Recently, Terasawa et al. (unpublished) has pointed out the following mechanism: When the magnetic fields are uniform in the z-direction, the ϕ-component of magnetic fields which are generated by rotation is transmitted as the Alfvén wave in the z-direction and the the angular momentum escapes along the z-direction.

In 1-dimensional approximation, it is assumed that $\partial/\partial r = 0$, $v_z = 0$, $H_z = \text{const.}$, where v_z is the z-component of the fluid velocity. The basic equation in this approximation is given by

$$\frac{\partial h}{\partial t} = \frac{\partial}{\partial t}\left(v_\phi + D\frac{\partial h}{\partial z}\right) \qquad \frac{\partial v_\phi}{\partial t} = v_A^2 \frac{\partial h}{\partial z}, \tag{2}$$

where $h = H_\phi/H_z$ is the ϕ-component of magnetic fields normalized by the uniform fields, $D = c^2/4\pi\sigma_e$ is the magnetic diffusion coefficient, σ_e is the electric conductivity, and $v_A = \sqrt{H_z^2/4\pi\rho}$ is the Alfvén velocity. The boundary conditions are given as, at $z = 0$, $h = 0$, $\partial v_\phi/\partial z = 0$, and at $z \to \infty$, where the Alfvén velocity is constant and the diffusion coefficient vanishes, $v_\phi/v_A = -h$ (out-going wave condition).

The Alfvén velocity v_A and the diffusion coefficient D are different in the disk and in the envelope. In the case of thin transition layer, i.e., $D = D_0$, $v_A = v_{A,\text{disk}}$ for $z < z_0$, and $D = 0$, $v_A = v_{A,\text{ext}}$ for $z > z_0$, we can easily find a solution implicitly which has the functional form (compare with the solution by Mouschovias and Paleologou) of $h = -h_0 e^{-\lambda t}\sinh(kz)$, $v_\phi = -v_0 e^{-\lambda t}\cosh(kz)$ in the disk. From the differential equation, we get $\lambda/k^2 = v_{A,\text{disk}}^2/\lambda - D_0$ and $h_0/v_0 = k/\lambda$. From the boundary condition, we get $\tanh(kz_0) = v_0/h_0 v_{A,\text{ext}} = \lambda/k v_{A,\text{disk}}$. If the density contrast between disk and envelope is very large, so that $v_{A,\text{disk}}/v_{A,\text{ext}} \ll 1$, the solution becomes

$$\lambda = \frac{v_{A,\text{disk}}^2}{z_0 v_{A,\text{ext}} + D}, \quad k^2 = \frac{v_{A,\text{disk}}^2}{z_0 v_{A,\text{ext}}(z_0 v_{A,\text{ext}} + D)}, \quad \frac{v_\phi(z = z_0 + \epsilon)}{v_\phi(z = z - \epsilon)} = 1 - \frac{\lambda D}{v_{A,\text{disk}}^2} = \frac{z_0 V_{A,\text{ext}}}{z_0 V_{A,\text{ext}} + D}. \tag{3}$$

The solution for $z > z_0$ is given by $h = h_0 k z_0 e^{-\lambda(t - z/v_{A,\text{ext}} - z_0/v_{A,\text{ext}})}\theta(-z + v_A t + z_f)$, $v_\phi = v_0 e^{-\lambda(t - z/v_{A,\text{ext}} - z_0/v_{A,\text{ext}})}\theta(-z + v_A t + z_f)$, where θ is the step function and z_f is the position of a wave front at $t = 0$. From these results, we can express the in-fall velocity of the gas in Kepler disks as

$$t_r = -\frac{r}{v_r} = \frac{D + v_{A,\text{ext}} z_0}{2 v_{A,\text{disk}}}. \tag{4}$$

Narita et al. (in preparation) has studied the characteristic features of this mechanism in the proto-solar disk. In actual disks, the surface part is ionized by cosmic rays but the inner part is less ionized, so that the electric conductivity σ_e depends on z. The density ρ in a gravitationally equilibrium disk varies exponentially, so that the Alfvén velocity depends on z. In this case, it is not clear how we can modify eq.(4), nor it dose not depend on time. Narita et al. studied these problems numerically.

Numerical results show that

$$\alpha(z) = \frac{1}{1 + \frac{D}{z v_{A,\text{ext}}}}, \tag{5}$$

which describes a velocity gap $v_{\phi,\text{ext}}/v_{\phi,0}$, varies timely. If the magnetic fields are scarcely frozen ($\alpha \ll 1$), the damping time scale for $v_\phi(z)$ is given locally by

$$\tau(z) = \frac{c^2 \rho(z)}{H_z^2 \sigma_e(z)}. \tag{6}$$

This means that when $\alpha \ll 1$, not only the magnetic braking dose not depend on the external density, but also it does not depend global structure of the disk. From this equation, it is expected that the magnetic braking is effective in the transition region between disk and envelope. This means that the in-fall flow of the gas is fast at the surface region of the disk.

References

Hayashi, C., 1981, Prog. Theor. Phys., suppl. **70**, 35.

Mouschovias, T.C., Paleologou, E.V., 1979, Ap. J. **230**, 204.

Physical conditions of star forming sites in the S247/252 molecular complex

C . Koempe[1,2], G. Joncas[3], J.G.A. Wouterloot[4], and H. Meyerdierks[5]

[1] IRAM, Granada, Spain
[2] Max–Planck–Institut für Radioastronomie, Bonn, Fed. Rep. of Germany
[3] Dépt. de Physique, Université Laval, Québec, Canada
[4] I. Physikalisches Institut, Universität zu Köln, Köln, Fed. Rep. of Germany
[5] Radioastronomisches Institut, Universität Bonn, Bonn, Fed. Rep. of Germany

1. Introduction

By now, it is well established that massive stars form in giant molecular clouds. Numerous studies have shown that star formation, instead of being spread uniformly throughout molecular clouds, occurs in dense condensations located within these clouds. The physical conditions in these condensations are therefore critical input parameters for any theory of star formation.

In the following, we present the results of an ongoing case study of the molecular cloud complex associated with the HII regions S247 and S252. The fundamental idea is to completely map the molecular complex with low spatial resolution ($\sim 4'$) in a ^{13}CO line, then identify likely sites of recent star formation, and finally study these sites with high spatial resolution. This approach reveals the distribution of star forming sites throughout the molecular cloud and their physical conditions. The comparison of several studies of this kind will hopefully improve our understanding of the star formation process in galactic molecular clouds.

2. Observations

We used the 2.5 m antenna of the Observatoire de Bordeaux to map the molecular complex S247/252 (distance 2.2 kpc) completely in the ^{13}CO (J=1–0) line. Additional ^{12}CO (J=1–0), CS (J=2–1) and HCO^+ (J=1–0) data were obtained towards S247. A search for OH and H_2O masers was carried out using the quasi-meridian radiotelescope at Nançay, France, and the 100 m antenna at Effelsberg, Germany, respectively. We also obtained higher resolution (40") data towards the four IRAS point sources 06061+2151, 06058+2138, 0607+2138, and 06069+2142 located in the S247 molecular cloud in the NH_3 (1,1) and (2,2) lines using the Effelsberg antenna. The first two sources were mapped, while the line intensities of the other two sources were only measured towards their central position.

3. Data analysis

In the ^{13}CO (J=1–0) line, two main molecular clouds are seen: a northern cloud close to the S247 HII region, and a southern cloud associated with the S252 optical nebula. They appear to be linked by a bridge of molecular gas. Several fragments can be distinguished in the S247 and the S252 cloud. Towards S247, the same fragments are seen in our CS and HCO^+ data indicating a minimum H_2 density of 10^4 cm^{-3}. The same spatial structure as that seen in the ^{13}CO data is revealed by the IRAS data. Combining our ^{13}CO and ^{12}CO data, we found the total mass of the S247 cloud to be 37,000 $M_\odot$ which is similar to that of the S252 cloud (25,000 $M_\odot$) determined by Lada and Wooden (1979). We searched the IRAS point source catalog for objects that are located inside the ^{13}CO boundary of the molecular complex: 47 sources were found. Those sources that are located close to high density fragments are likely to be young stellar objects. Water masers were found towards five IRAS sources, and two OH masers in the vicinity of the S247 optical nebula. A detailed account of our mm, IR, and maser emission data can be found in Koempe et al. (1989). Contour maps of NH_3 (1,1) obtained towards IRAS 06061+2151 and IRAS 06058+2138, displayed in Fig. 1, show a maximum close to the position of the IR source. The kinetic temperatures T_K derived from the ammonia data are also shown. In both cases, the maximum temperature is found close to the position of the IR source but decreases fast as one moves away from the source. This behavior is expected if the IR source is heating the surrounding gas from the inside. The kinetic temperatures derived for IRAS 0607+2138 and IRAS 06069+2142 are 18 K and 24 K, respectively. A minimum H_2 density of 10^4 cm^{-3} was derived towards all four IR sources on the basis of the NH_3 observations.

4. Conclusion

The S247/252 molecular complex is composed of two well separated clouds that are linked by a bridge of molecular material. Both clouds show a number of fragments. It should be noted that the masses of these fragments lie in the range of small molecular clouds. In the S247 cloud each fragment is associated with at least one IRAS point source. Towards five IR sources, H_2O masers were detected indicating that these sources are indeed young stellar objects. Four IR sources are associated with compact HII regions (S252 A, S252 B, S252 C, and S252 E). It is thus likely that the IRAS sources, distributed throughout the S247/S252 gas complex, represent young stellar objects at different evolutionary stages. The mechanism(s) that triggered star formation in this particular complex remain(s) still to be discovered.

5. References

Koempe C., Joncas, G., Baudry, A., Wouterloot, J.G.A., *Astron. Astrophys.*, **221**, (1989)
Lada, C.L., Wooden, D., *Astrophys. J.*, **232**, 158, (1979)

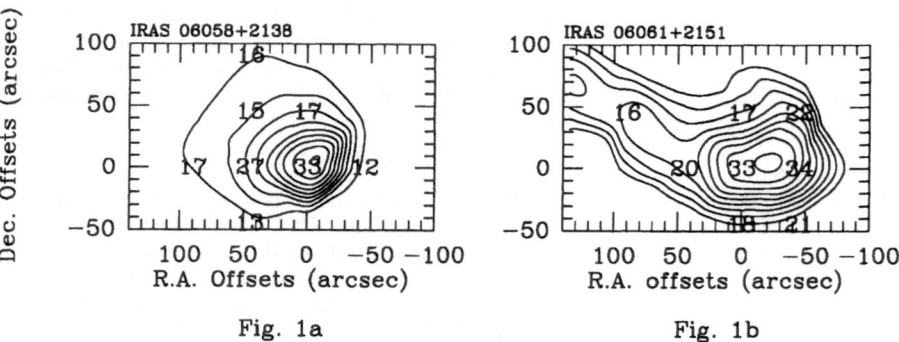

Fig. 1a   Fig. 1b

Fig. 1a: Integrated NH_3 (1,1) contour plot of IRAS 06058+2138. The lowest contour level is 0.6 K km s^{-1} and the spacing between levels is 0.2 K km s^{-1}.

Fig. 1b: Integrated NH_3 (1,1) contour plot of IRAS 06061+2151. The lowest contour level is 0.5 K km s^{-1} and the spacing between levels is 0.15 K km s^{-1}.
In both maps, the IR source is located at the (0,0) position. The bold face numbers give the kinetic temperature at different offset positions.

HIERARCHICAL FRAGMENTED STRUCTURE OF MOLECULAR CLOUDS PRODUCED BY SUPERSONIC TURBULENCE

IGOR G. KOLESNIK and YAROSLAV YU. OHUL'CHANSKY
Main Astronomical Observatory
Ukrainian Academy of Sciences
252127 Kiev
USSR

For GMC's the two prominent properties are typical. They consist of dense molecular gas clumps concentrating to the GMC centre and filling only a few percent of a total volume. And these clumps participate in chaotic motions with velocities v_t exceeding as a rule the sound velocity c_o at the temperature of molecular gas. This phenomenon is considered as a supersonic molecular cloud's turbulence. The compressibility of turbulent matter becomes very important with such velocities. Thus in application to GMC it is necessary to develop the theory of turbulence and fragmentation under transsonic and supersonic random motions. The hydrodynamic flow velocity field can be divided into the potential and vortical components. When transsonic or supersonic motions prevail the potential component is become more important that stimulates the shock wave's stochastic field development. Ohul'chansky (1988,Kinematics and Physics of Celestial Bodies 4,3) has described this process on the base of Burgers' equation treatment. In this paper we apply this approach for conditions of GMCs that permit the supersonic turbulence' spectrum evolution, the large density fluctuations development, and clumps formation to consider.
As was shown by Kolesnik (1987,Kinematics and Physics of Celestial Bodies 3, 47) the supersonic motions of molecular gas can be developed from subsonic turbulence of the warm gas when its temperature rapidly drops down to about 100 K.
The evolution of initial Kolmogorov's spectrum on the base of three-dimentional Burgers' equation was considered by Ohul'chansky(1988). Qualitevely the spectrum evolves from Kolmogorov's one with exponent $-5/3$ to more steeper ones,the exponent being appreciated from -2 to -3. The characteristic time of evolution depends on the wave number k.
The further evolution of turbulent medium is following.The range of the shock wave forming harmonics is moved into the smaller wave numbers k gradually. The interacting shock wave's ensemble is formed. Because of isothermal conditions the magnitude of density fluctuations can rise to the large values. Most dense clumps are formed when shock waves collide in head-on. This promotes to the filamentary structures formation in molecular clouds. On the other hand the oblique shock waves interac-

tion practically doesn't increase the density fluctuation. It leads only to the decreasing of the angle between interacting oblique shock fronts. This also provokes the plane dense structure formation. Therefore in the evolving supersonic turbulent medium the flat dense structures will be pronounced more and more by the stochastic shock fronts interactions. The statistics of density peaks is determined by the totality of all multy-point joint probabilities of physical quantity and its derivatives. In Gauss case it is possible to derive the size distribution of fragments

$$(dN/d\lambda) \propto \lambda^{-2m-3}, \quad \lambda \geq \sqrt{<\lambda^2>} \; ; \quad (dN/d\lambda) \propto \lambda, \quad \lambda < \sqrt{<\lambda^2>} \; .$$

Here λ is the clump's length scale. As it is seen the large scale clumps distribution is sensitive to the geometry of fragments. Received properties of the size distribution function can be used for conclusions what type of fragmentation processes is responsible for the clumps formation in the observing objects. The typical density in the centre of clumps may be appreciated in the following way. When accepting that filamentary flattened stuctures mainly are the consequence of collisions of shock waves, then it follows that for such structures $\Delta\rho/\rho_0 \propto <Ma>^4$

($<Ma>$ is the typical Mach number of shock waves), because in GMC the passing of one shock wave gives the velocity jump Ma^2. The quantity $<Ma>$ is proportional to v_t/c_0 , where v_t is the typical velocity of turbulent motions. The quantity $<\lambda>$ is determined by the scale l_c (the scale l_c is the scale on which the turbulent velocity difference equals c_0) and by other space parameters of physical medium. These ones may be the typical size of eddies in primary stage of evolution of molecular cloud and also the typical scale of inhomogenity of cloud.

During the evolution of molecular cloud the following situation is possible. In the distinguished dense core of GMC the turbulence has lost the memory about the primordial hierarchy of eddies yet, and in the outer layers of GMC the system keeps the memory and the typical size of clumps corresponds to the size of eddies. Thus, in the outer layers the size of clumps is much greater than in the core. Besides, the interaction between the turbulent motions in the outer layers of cloud and inner dense core results to the strong compression of matter in the intermediate layer which may cause the violent star formation.

THE BIPOLAR-FLOW PHENOMENON

Wolfgang Kundt,
Institut für Astrophysik der Universität Bonn
Auf dem Hügel 71
53 Bonn 1
Federal Republic of Germany

ABSTRACT: Very young stars, known as 'pre-T-Tauri stars' (PTTS) or 'young stellar objects' (YSO), tend to be surrounded by elongated outflow regions involving 'Herbig-Haro objects' (HH). Such 'bipolar flows' (BF) are reminiscent of the extragalactic radio sources for which a consensus has formed in 1986 that their jets consist of extremely relativistic pair plasma, of typical Lorentz factor $10^{4\pm1}$, generated by an 'active galactic nucleus' (AGN); [refs 6,7]. Here I collect new circumstantial evidence for the relativistic nature of BFs.

In fact, the relativistic nature of the jet plasma of stellar BFs has been derived, in ref 2, from both particle number and momentum conservation in the head of the flow where the jet turns subsonic across a terminal shock. A most plausible way of generating a relativistic flow is through localized magnetospheric discharges near an extremely spinning, strongly magnetized object; the pair plasma results via collisions of the charges with stellar photons once the discharge energy exceeds threshold (of order 10^{12} eV). This criterion applies equally to all four families of bipolar flows, viz. to YSOs, AGN, young binary neutron stars, and young binary white dwarfs (planetary nebulae) [ref 6]. The magnetospheric pair plasma is pumped centrifugally into a surrounding overpressure bubble [ref 3] which discharges in the form of two narrow, antipodal jets; fig. 1. In this way, the central engine combines the properties of an ordinary star with those of a fast, magnetized rotator.

The predicted relativistic nature of the BFs driven by YSOs has not easily appealed to the scientific community because of the lack of evidence: One observes low-ionization emission lines from 'slowly' moving HHs ($v \le 10^{-3}$ c). On the other hand, freely propagating pair-plasma jets are thought

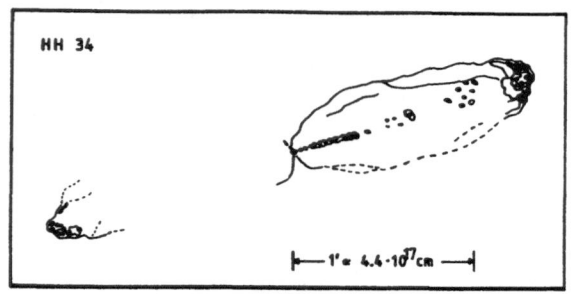

Fig.1:Optical
CCD contour
map (drawn)
of HH 34 ;
from Bührke
et al: A & A
200,99(1988).

to be loss-free; most of the observed radiation comes from
the impacted, heavy channel-wall and bowshock material which
has absorbed part of the jet momentum. Nevertheless, the
radio jet of L1551/IRS 5 shares a spectral index of α = 0.3
with the synchrotron sources in several AGN [refs 1; 6,7],and
so do the (jet) sources HH 1-2, S 68 (Serpens), NGC 2071,
G 192.58-0.04, and (probably) the polarized radio double S
187 [refs 8,9] as well as an HH-like 'streamer' near V 571
Ori [ref 10]. Even more striking is the finding, in ref 5, of
strongly non-thermal radio emission from three T-Tauri stars
whose size is a few times stellar: TTSs are understood to
descend from PTTSs, hence to be spun down and less magnetic
compared with the former!

Another observation which proves the occasional
narrowness of the outflows from YSOs is the UV variability of
HH 29, the moving emission knot at a distance of $10^{17.6}$ cm
from L1551/IRS 5. According to ref 4, HH 29 has varied by a
factor of 4 within five months. The authors conclude at a
size of the emission region of $\leq$ 10 AU, some $10^{-3.5}$ times
the distance from the central engine! Clearly, such extreme
focussing asks for an essentially weightless jet substance.

References:

1. Bieging, J.H. and Cohen, M.: Ap.J. 289, L5 (1985)
2. Blome, H.J. and Kundt, W.: Astroph. Sp.Sci. 148, 34 (1988)
3. Camenzind, M., in: Neutron Stars and their Birth Events,
     NATO ASI C 300, ed. W. Kundt, Kluwer 1990.
4. Cameron, M. and Liseau, R.: UV-observs. of HH obj. associ-
     ated with bipolar molecular outflows: A & A, in press
5. Feigelson, E.D., Lonsdale, C.L. and Phillips, R.B.: AAS
     20.07, Jan. 1990
6. Kundt, W.: Astrophysical Jets and their Engines, NATO ASI
     C 208, Reidel 1987.
7. Kundt, W.: The Galactic Center, Astroph. Sp.Sci., in press
8. Rodríguez, L.F., et al: Ap.J. 346, L85 (1989)
9. Snell, R.N. and Bally, J.: Ap.J. 303, 683 (1986)
10.Yusef-Zadeh, F., et al.:Ap.J. 348, L 61 (1990).

A HYDRODYNAMICAL MODEL FOR THE FRAGMENTATION OF THE W49A STAR-FORMING REGION*

JOHN LATTANZIO[1,2], ERIC KETO[2] and JOE MONAGHAN[1]

[1]Department of Mathematics, Monash University,
Clayton, Victoria, 3168, AUSTRALIA.

[2]Institute of Geophysics and Planetary Physics, L-413,
L.L.N.L., P.O. Box 808, Livermore, Ca., 94550, USA.

ABSTRACT. We present a 3-D hydrodynamical and radiative transfer simulation which suggests that the circular "necklace" of massive star formation in W49A may result from fragmentation via a ring instability during the collapse of a rotating cooling molecular cloud.

1. Introduction

The star-forming region W49A is characterized by a rotating ring of HII regions (Welch et al. 1987). Spectral lines toward several of the HII regions show red and blue shifted absorption and emission components split in an inverse P-Cygni profile characteristic of infall and gravitational collapse (Keto 1990a). Welch et al. interpret this as evidence for uniform collapse of a several pc scale molecular envelope onto the ring. They postulate a central mass of $5 \times 10^4 M_\odot$ to keep the orbiting HII regions in centrifugal equilibrium, and motivate the infall of the cloud envelope. We present here an alternative model.

2. The Calculations

We begin with the cooling spherical cloud described in Monaghan and Lattanzio (this volume, and 1990). The gas rapidly collapses down the rotation axis and forms a thin disc. After 2 initial free fall times (t_{ff}) a ring forms, with material in the centre cleared due to the centrifugal force. The ring then fragments into 5 major condensations (see figure 1). These cores have masses of about $1000 M_\odot$ and radii about 0.6pc.

We expect stars to form within these cores, on a scale below our resolution. We have thus assumed that in each core there is an HII region of diameter 0.1 pc. The model HII

---

* Work performed under the auspices of the U.S. Department of Energy by the Lawrence Livermore National Laboratory under contract No. W-7405-ENG-48.

regions serve 2 purposes. They heat the surrounding molecular gas (Scoville and Kwan, 1976) and provide a background source of continuum radiation against which the foreground gas may be seen in absorption. Synthetic HCO^+ spectra calculated with the code described by Keto (1990b) are shown in figure 1. The velocity field in the model shows that the splitting of the emission and absorption lines arises from small scale rotation and collapse within the individual cores. The magnitude of the splitting is about the same in each of the different cores because the fragmentation process results in cores of approximately equal mass and radius. Whether the splitting is seen in emission or absorption is dependent on the assumed radiation temperature of the continuum relative to the lines.

3. The New Model

In this model we propose that the ring of HII regions results from the fragmentation, through a ring-mode, of a collapsing rotating cloud of gas. There is no central mass to keep the fragments in stable orbits, and further hydrodynamical evolution shows the ring to be a transient phenomenon with a lifetime of $< t_{ff}$. Gravitational collapse at the time of formation of the cores is entirely localized about the individual HII regions while the surrounding disc material is only weakly bound to the structure. Details will be published elsewhere (Keto et al. 1990).

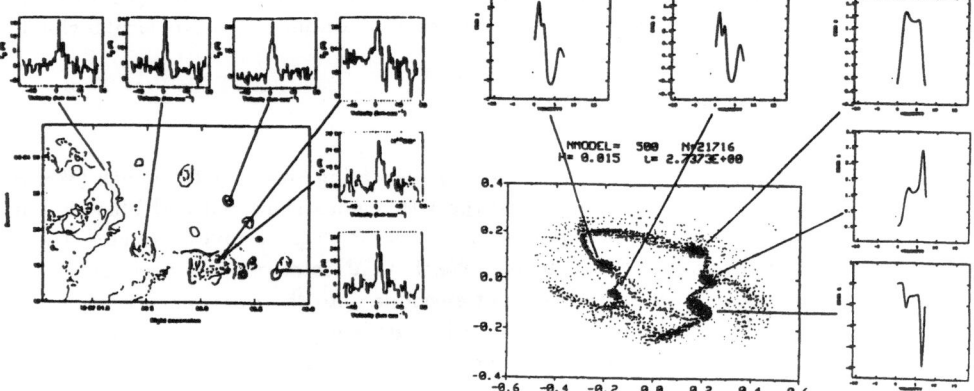

Figure 1: W49A (left) from Welch et al. (1987) and our model (right), at a viewing perspective suggestive of W49A.

References

Keto, E. R., 1990a, *Astrophys. J.*, **355**, 190.

Keto, E. R., 1990b, *Astrophys. J.*, in press.

Keto, E. R., Lattanzio, J. C., and Monaghan, J. J., 1990, in preparation.

Monaghan, J. J., and Lattanzio, J. C., 1990, *Astrophys. J.*, submitted.

Scoville, N., and Kwan, J., 1976, *Astrophys. J.*, **206**, 718.

Welch, W. J., Dreher, J. W., Jackson, J. M., Tereby, S., and Vogel, S. N., 1987, *Science*, **238**, 1550.

"CO LINE BROADENING BY SATURATION EFFECT IN MOLECULAR CLOUDS"

Jacques R.D. Lépine
Instituto Astronômico e Geofísico
Universidade de São Paulo, Brazil

Abstract

A new discussion of the evidences for turbulences in molecular clouds, taking into account line broadenning by saturation at large optical depths, shows that the generally accepted power law dependence of turbulent velocity with cloud size must be revised.

I. Introduction

After the publication of a paper by Larson (1981), several authors have strengthened the conclusion that molecular clouds exhibit a turbulent behaviour with a power law dependence of the turbulent velocity σ on the size L of the cloud (eg. Leung et al., 1982, Myers, 1983; Henriksen and Turner, 1984). According to Larson, the relation is:

$$\sigma \, (km^{-1}) = 1.1 \, L^{0.38} \; (pc) \tag{1}$$

We show that although the CO line width data can be fitted by the equation above, this result should not be interpreted as a Kolmogoroff spectrum of turbulence, since line broadenning by saturation must be taken into account.

II. The Model

We consider molecular clouds with constant density and temperature, and a beam filling factor equal to 1. In LTE the profile of a CO line expressed as a function of velocity is given by:

$$T_a(v) = (T - 2.7) \, (1 - \exp(-\tau(v))) \tag{2}$$

where T is the kinetic temperature and the optical depth $\tau(v)$ is proportional to the CO column density N:

$$\int \tau(v) \, d_v = \frac{8 \, \pi^3 \, v \, \mu^2}{3 \, h \, c} \; \frac{[1 - e^{-hv/kT}]}{Q} \; N \tag{3}$$

where Q is the partition function, and μ the electric dipole matrix element of the transition. Adopting a gaussian profile, we have:

$$\tau(v) = \tau_0 \, e^{-(v/\Delta v)^2} \tag{4}$$

451

and

$$\int \tau(\nu) \; d\nu = \frac{\nu}{c} \int \tau(v) \; dv = \frac{\nu}{c} \; \pi^{1/2} \; \Delta v \; \tau_0 \tag{5}$$

where τ_0 is the optical depth at the center of the line and Δv is the Doppler width, which includes the thermal width and microturbulence:

$$\Delta v = (\frac{2 \; kT}{m} + v_t^2)^{1/2} \tag{6}$$

One must pay attention to the distinction between the Doppler width of the absorption coefficient, Δv, and half-linewidth at half-maximum of the line, σ, which includes broadenning by saturation.

In figure 1 we show $\sigma(C^{13}O, j=1-0)$ as a function of optical depth. The data points are the same data collected in the literature by Larson (1981), except that we only keep the $C^{13}O$, data, since different species have different optical depths and thus do not show the same amount of broadenning. We used a density of $5 \; 10^3 \; H_2 \; cm^{-3}$ in order to scale cloud sizes to optical depths. A single value of the Doppler width does not fit correctly the data (not shown). However if we use a Δv law which first increases linearly with τ and then saturates, such as:

$$\Delta v \; (kms^{-1}) = 2.5 \; (1 - e^{-\tau_0}) \tag{7}$$

then a good fit can be obtained (curve 1). For comparison we show (curve 2) a fit of the data with a power law of index 0.38.

Figure 1

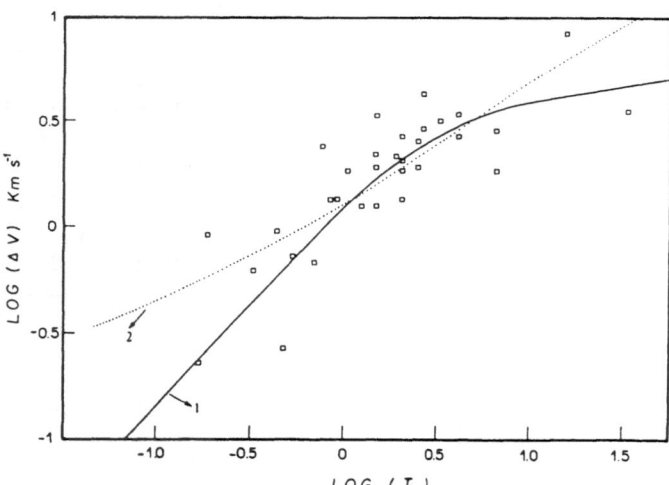

III. Discussion

We conclude that line broadenning due to saturation cannot explain alone the observed increase of the $C^{13}O$ (J = 1-0) linewidth with the size of the clouds; however, if this broadenning is taken into account, and if for consistency only the $C^{13}O$ data is examined, the turbulence is better described by an expression like equation (7) than a power law. In other words, the Doppler width seems to reach a constant value for large clouds. This is not an unexpected result, if, for instance, the turbulence is produced by localized sources like the winds from embedded young stellar objects. Our result is in better agreement with the regular pattern of the magnetic field often observed in molecular clouds than the Kolmogoroff spectrum interpretation.

References

Henriksen, R.N., Turner, B.E.: 1984, Astrophys. J. <u>287</u>, 200

Larson, R.B.: 1981, Mon. Not. R. astron. Soc. <u>194</u>, 809

Leung, C.M., Kutner, M.L., Mead, K.N.: 1982, Astrophys. J. <u>262</u>, 583

Myers, P.C.: 1983, Astrophys. J. <u>270</u>, 105

Sanders, D.B., Solomon, P.M., Scoville, N.Z.: 1984, Astrophys. J. <u>276</u>,182

THE HIGH LATITUDE CLOUD LYNDS 1642 IS NOT BREAKING UP

T. Liljeström

Helsinki University Observatory
Tähtitorninmäki
SF-00130 Helsinki, Finland

ABSTRACT. CO, HCO$^+$ and NH$_3$ observations have been carried out towards the high latitude cloud L1642 using the 1.2-m GISS, the 11-m NRAO, the 14-m Metsähovi and the 100-m Effelsberg radio telescopes. The velocity field of the CO gas indicates a core-halo structure. The core component has a constant radial velocity, whereas the halo gas is slightly (< 1 km/s) redshifted, as compared to the radial velocity of the core, and shows velocity gradients towards the cloud edges in a similar way as the HI gas associated with L1642. Within the border of the ^{13}CO emission the mass of the cloud is estimated to be some 76M$_\odot$. In contradiction to Magnani et al.(1985), who claimed that L1642 belongs to a population of very young high latitude clouds which are breaking up, this study supports the view that L1642 is in virial equilibrium and significantly older than 10^6 yr. The virial equilibrium of L1642 enables a distance determination of ~ 190 pc to the cloud core.

1. SIGNS OF SHOCK INDUCED CLOUD IMPLOSION IN L1642

The Orion-Eridanus region contains large arc-like features, which clearly imply that supernova explosions have disturbed the interstellar medium. Due to the lower density of the medium at high galactic latitudes, disturbances can propagate to much larger distances as compared to the galactic disk. Thus it is natural to expect that shock fronts from supernovae of the Orion I OB association have passed also the 190 pc distant L1642 flattening the cloud in the direction of the arriving shock front and initiating a flow in the cloud envelope. It is also noteworthy that the extinction map of L1642 (Liljeström et al., 1988) shows pronounced dust "tongues" penetrating outwards from the cloud, a characteristic sign of Rayleigh-Taylor and Kelvin-Helmholtz instabilities.

These similarities with Woodward's model(1976) support an interpretation of a shock induced cloud implosion. Woodward's simulations predict also the formation of cloud condensations, some of which may be dense enough to undergo gravitational collapse. L1642 has produced two low-mass double stars. The more embedded one, associated with a CO outflow, is located in the outer part of the core region in a "tongue" structure.

This is the predicted geometry and location for a new star in the model of Woodward (1976). The time for a shock driven low mass star formation is of the order 10^7 yr (Woodward, 1976) which is an order of magnitude longer than the dynamical time scale derived for high latitude clouds (Magnani et al., 1985).

2. DYNAMICAL STATE OF L1642 AND CORE DISTANCE

The virial theorem for a stationary cloud is $M(M_\odot) = C \, \sigma^2 \, R_{pc}$, where M is the cloud mass, R the effective cloud radius, C a constant depending on cloud geometry and density structure, and σ the 3-dimensional velocity dispersion. The best distance estimate to L1642, based on uvby and H_β photometry (Franco, 1989), is 114< r <230pc. Adopting a mean value, r = 170 pc, the virial masses of L1642 range from 50 to 97 $M_\odot$ when 4 different cloud models are considered. Comparing these with the observed M_{tot} = 75 $M_\odot$ (for r=170pc) it is obvious that L1642 is in virial equilibrium.

The virial equilibrium of L1642 enables a distance determination to the cloud core. From the virial theorem one obtaines

$$\sigma_{vir}(3-dim) = \sqrt{M(M_\odot)/C \, R_{pc}} = \sqrt{Ar_{pc}^2/C \, Dr_{pc}} \leftrightarrow r_{pc} = \sigma^2 \, C \, D/A \qquad (1)$$

where A and D are the numerical coefficients of r^2 and r_{pc} in the observed cloud mass and effective cloud radius, respectively, and C the constant in the virial theorem. Substituting the numerical values obtained from the observations the most probable core distance of 190 pc is obtained, which is in accordance with the results of Franco (1989).

The line widths of ^{13}CO and even ^{12}CO have been commonly used to derive the velocity dispersion of a cloud. However, in addition that these line widths have not been corrected for opacity broadening, their use causes an observational selection bias towards the higher velocity dispersions of the envelope gas, which do not necessarily reflect the gas motions which oppose the gravitational collapse of a cloud. Especially it should be stressed that if systematic gas motions are included into the kinetic energy of a cloud (as e.g. Magnani et al. (1985) in L1642), the virial theorem of a bound and stationary system must be changed to

$$(1/t) \int_0^t (d/dt) \sum \vec{p}_j \cdot \vec{r}_j = 2 < E_{kin}> + < \sum \vec{F}_j \cdot \vec{r}_j> \qquad (2)$$

where $\vec{p}$ is the momentum vector, t the interaction time, and < > the symbol for time average. However, in this case the relation $<v^2> = \overline{v^2}$ is not valid any more. Therefore, the velocity component of the halo gas should be examined carefully before it is included into virial theorem considerations.

REFERENCES

Franco, G.A.P. 1989, Astron. Astrophys. 223, 313
Liljeström, T. and Mattila, K. 1988, Astron. Astrophys. 196, 243
Magnani, L., Blitz, L. and Mundy, L. 1985, Astrophys. J. 295, 402
Woodward, P. 1976, Astrophys. J. 207, 484

LARGE SCALE INTERACTION OF THE OUTFLOW AND QUIESCENT GAS IN ORION

J. MARTIN-PINTADO, A. RODRIGUEZ-FRANCO AND R. BACHILLER

1) Centro Astronómico de Yebes, Apartado 148, 19080 Guadalajara (Spain)

The IRAM 30-m radiotelescope have been used to obtain, with high angular resolution, the spatial distribution and the physical conditions of the quiescent gas in Orion A, and to search for high velocity molecular gas far away from the well known molecular outflow around IRc2. To study the quiescent gas we mapped a region of 200"x300" around IRc2 in the J=12-11 and J=16-15 lines of HC3N with angular resolutions of 22" and 17" respectively. The left panel of Fig.1 shows the spatial distribution of the high density quiescent gas around IRc2 for different radial velocities. Beside the already known molecular ridge north of IRc2 (see e. g. Bartla et al. 1983), we find four very thin (nearly unresolved) and long filaments, like "fingers", stretching from IRc2 to the north and west. The deconvolved size of the longest fingers is $\approx180"x15"$. From a multi-transition analysis of the HC3N emission we derive H2 densities of 1-8 10^5 cm^{-3}, kinetic temperatures larger than 40 K and masses of ≈10 Mo. Our high sensitivity observations of the J=2-1 line of CO at selected positions (see right panel ib Fig. 1) show widespread molecular gas with high velocities wings over the region where the molecular fingers and the HH objects are observed (see Fig.1). The high velocity emission occurs over a range of ±40 km s^{-1}. This high velocity gas is more extended (up to 150" from IRc2) than the very compact (40") and well studied molecular outflow around IRc2 (see e.g. Wilson et al. 1986). The terminal velocities of the CO wings decrease from 100 km s^{-1} (corresponding to the very fast molecular flow) to the typical terminal velocities of the extended high velocity gas when the distance to IRc2 changes from 40" to 60". The origin of the large scale high velocity gas is unknown, but it is very likely the link between the very compact (40") and fast (±100 km s^{-1}) molecular outflow around IRc2 and the ionized high velocity gas and the HH objects (Martín-Pintado et al. 1990). The mass, momentum and energy of the extended high velocity gas are crudely estimated to be ≈1 Mo, ≈20 Mo km s^{-1} and ≈2 10^{45} erg respectively (i.e. a factor of ≈10 smaller than those of the fast molecular outflow). The location, at the edges of the molecular fingers, and the proper motions of the HH objects (see Fig. 1) suggest the stellar wind is interacting with the molecular fingers. If this interpretation is correct, the influence of the molecular outflow in Orion on the surrounding molecular clouds must be revised.

Acknowledgments: This work has been partially supported by the Spanish CICYT under grant number PB88-0453.

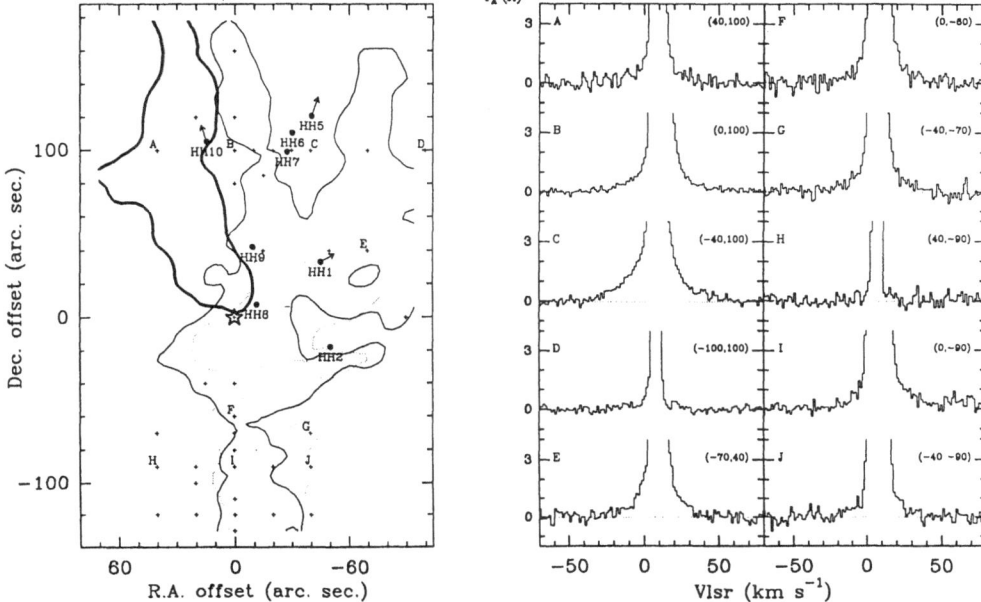

Figure 1 .- The left panel shows a sketch of the most outstanding features (the molecular fingers and the molecular ridge) as observed in the J=12-11 line of HC3N. The molecular ridge is represented by a solid thick contour corresponding at a level of 2 K km s⁻¹ of the HC3N integrated intensity between 9.5 and 10 km s⁻¹. The molecular fingers are delineated by thin and dotted contours taken at levels of 0.7 (HC3N integrated intensity between 8 and 8.5 km s⁻¹) and 1 K km s⁻¹ (HC3N integrated intensity between 7 and 7.5 km s⁻¹) respectively. The position of IRc2 is shown by a star. The HH objects are shown as filled circles and the direction of their proper motions is indicated by arrows (Axon and Taylor 1984; and Jones and Walker 1985). The HH objects are located close to the edges of the molecular structures and they seem to move in the direction of the fingers. The crosses show the positions where the J=2-1 CO spectra have been measured. The capital letters close to some crosses indicate the position of the CO spectra displayed on the right panel.

The right panel shows a sample of spectra in the J=2-1 line of CO taken in the Orion A molecular cloud toward positions far away from IRc2. The positions are indicated by their offsets, in arcseconds, with respect to IRc 2 (upper right corner on every spectra) and by a capital letter (upper left corner) which refer to the left panel.

References

Axon, D. J., Taylor, K. 1984, M.N.R.A.S., **207**, 241.
Bartla, W., Wilson, T. L., Bastien, P, Ruf, K. 1983, Astr. Ap., **128,** 223.
Martín-Pintado, J., Rodriguez-Franco, A., Bachiller, R. 1990, Ap. J. (Letters), **357,** L47.
Jones, B.F., Walker, M.F. 1985, A.J., **90**, 1320.
Wilson T. L., Serabyn, E., Henkel, C. 1986, Astr. Ap. (Letters), **167**, L17.

Physical parameters in extragalactic star forming regions

R. MAUERSBERGER, C. HENKEL, AND L.J. SAGE

MPIfR, Auf dem Hügel 69, D-5300 Bonn, F.R.G.

To use a molecule as a diagnostic of a single star forming region, one has to observe it in different transitions. That is especially true for galaxies, where the beam filling factor is small and unknown. From studies of Galactic objects, the CS molecule is a good tracer of dense gas. Toward the IR galaxies NGC 253, IC 342 and M 82, Mauersberger and Henkel (1989) observed the J=2—1, 3—2 and 5—4 transitions of CS at 3, 2 and 1.3 mm wavelength, repectively. They also measured emission of the isotope $C^{34}S$, which has an abundance $\sim 1/23$ that of the main isotope. This study revealed that the H_2 densities in the circumnuclear gas exceed 10^4, and, in the case of NGC 253, 10^5 cm^{-3}. In all three cases, the distribution of the dense gas traced out by CS follows the CO distribution.

In order to confirm this high density for NGC 253, we (Mauersberger et al., 1990) conducted a multi-level study of HC_3N (cyanoacetylene) toward NGC 253. We measured six transitions of this heaviest molecule detected outside our Galaxy. Comparing the relative intensities of the J components (Fig. 1) we reached the following conclusions:

1. The 3mm transitions can only be fitted if the bulk of the gas seen in HC_3N has a density of $\sim 10^4$ cm^{-3} and a temperature of ~ 30 K. The filling factor should be comparable to CO (i.e. $> 10\%$).

2. This component cannot reproduce the observed emission of the 2 mm and 1 mm lines of HC_3N. These lines require a gas component with a much higher excitation. The high excitation model applied to fit the high J tail is a conservative guess and has as parameters $n(H_2) = 10^5$cm^{-3} and $T_{kin} = 150$ K. The filling factor should be low (0.1%).

3. The existence of two components is very similar to our own Galactic center, where Walmsley et al. (1986) have detected two components, the denser one being less excited than the dense component toward NGC 253.

Besides the studies summarized above, there are a number of other methods of determining the state of the densest molecular gas in external galaxies:

H_2CO. In our Galaxy, the 2 cm line of formaldehyde is normally seen in absorption. Only toward a handfull of Galactic sources, thermal emission has been observed. The explanation of the emission requires densities exceeding 10^5 cm^{-3}. 2-cm emission toward M 82 was detected by Baan et al. (1990).

N_2H^+. This molecular ion, the third detected outside the Galaxy, has been measured toward five galaxies and has been mapped toward NGC 253 and M 82. Unlike other molecules, N2H$^+$ seems to be centrally peaked toward M 82.

CH_3CCH. Methyl acetylene is the most complex molecule detected outside our galaxy, namely toward M82. It is, however, difficult to disentangle its K-lines from the velocity structure of Galaxies.

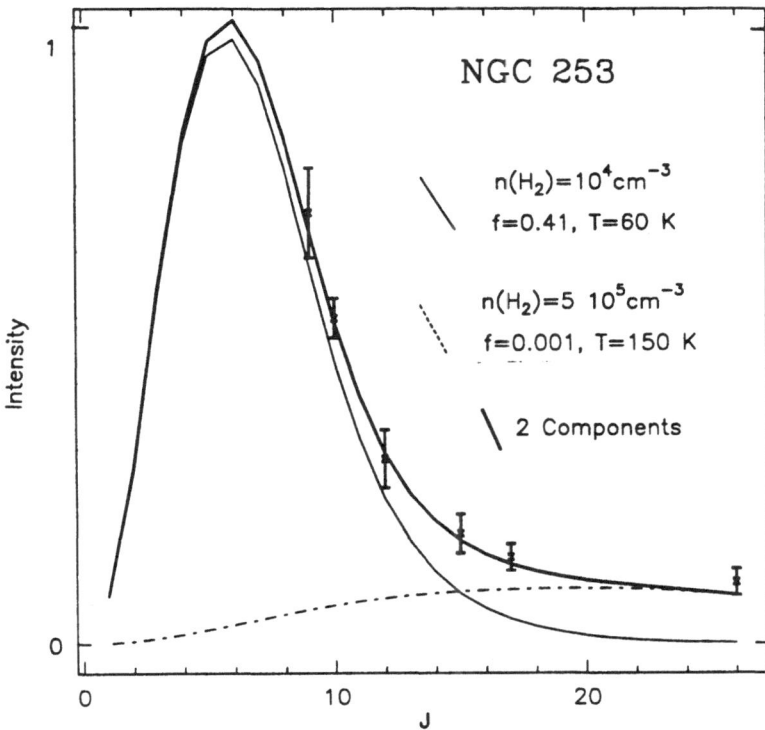

Fig. 1: Fits to our data with a two component model. Intensities are scaled with respect to the maximum emission. The thick line represents a two component model fitting the measured data. The thin and the dashed lines show the contribution of each of the two components. The parameter for the relative HC_3N abundance per velocity gradient, $X(CO)/(dv/dr)$, has been chosen to be $5\,10^{-10}/(km\,s^{-1}/pc)$.

Unlike in the center of our Galaxy, CH_3CCH emission is stronger than that of CH_3CN.

SiO. SiO is mainly observed toward regions with a very elevated temperature and in shocked gas. It could therefore be used as a 'chemical thermometer'. We have detected its $J=2$—1 line of the vibrational ground state (v=0) toward the nucleus of NGC 253 but not toward M 82 and IC 342. It may be that temperatures in NGC 253 are higher than in the other two galaxies.

References

Baan, W., Henkel, C., Schilke, P., Mauersberger, R., Güsten, R.: 1990, *Astrophys. J.* **353**, 132
Henkel, C., Mauersberger, R., Schilke, P.: 1988, *Astron. Astrophys.* **201**, L23
Mauersberger, R., Henkel, C.: 1989, *Astron. Astrophys.* **223**, 79.
Mauersberger, R., Henkel, C., Sage, L.J.: 1990, *Astron. Astrophys.*, in press
Walmsley, C.M., Güsten, R., Angerhofer, P., Churchwell, E.: 1986, *Astron. Astrophys.* **155**, 129

An 8″ resolution CO (J=3–2) map of IC342

R. MAUERSBERGER[1], A. SCHULZ[1], J.W.M. BAARS[1], H. STEPPE[2]

1. MPIfR, Auf dem Hügel 69, D-5300 Bonn, F.R.G.
2. IRAM, Avda. Divina Pastora 9, E-18012 Granada, Spain

Introduction

IC342 (Distance 4 Mpc) is one of the most suitable sources for extragalactic molecular line studies. Toward its nucleus, a great number of molecular species have been found (see Henkel and Mauersberger, 1990); it is also one of the few galaxies investigated in molecular multi-level studies (Mauersberger and Henkel, 1989). In particular, CO shows strong emission: A 7″ resolution interferometric map of the central parts of this galaxy in the $^{12}CO(1-0)$ transition by Lo et al. (1984) reveals that the circumnuclear molecular gas is distributed in a bar (size 15″ × 70″) (330×1500 pc) extending from the nucleus towards the spiral arms. An interferometric map of the 1—0 line of CO by Ishizuki et al. (1990) shows that the inner part of the bar forms a molecular ring of diameter 110 pc. This inner ring also emits 2 and 6 cm continuum radiation (Turner and Ho, 1983). The kinetic temperature of the denser molecular gas is > 50 K (Martin and Ho, 1986). The H_2 density of the gas component seen in CO (Eckart et al., 1990) and CS (Mauersberger and Henkel, 1989) is ∼ 10^4 cm^{-3}.

Instrumentation, observations and results

The observations are the result of first measurements using a 345 GHz Schottky receiver at the IRAM 30-m telescope. The CO (J=3-2) spectra are shown in Figure 1. These measurements have demonstrated that the 30-m telescope is suitable for 0.87 mm wavelength observations with respect to a) atmospheric conditions, b) surface accuracy and c) pointing stability. Averaged over a year, for more than 10% of the observing time, the amount of precipitable water vapour in the atmosphere is expected to be <2 mm. The beam width nearly gets down to that of conventional millimeter range synthesis telescopes. With the availability of broader spectrometers, the usable instantaneous bandwidth could be 1 GHz and, hence, extragalactic objects with broader line emission could be observed.

Interpretation

In Fig. 2, the features observed toward IC 342 are sketched. The shaded area marks the molecular bar seen in the CO (1—0) emission. The thick lines show where we have seen the CO (J=3—2) hotspots. The thin lines denote the cm radio emission (Turner and Ho) and the cross the 2μm IR source.

The radio emission traces regions of newly formed young stars. They are concentrated in a ring of ∼ 100 pc diameter located in the center of a bar-like structure seen in the J=1—0 line of CO. It could be inferred that the CO bar supplies the molecular gas needed for ongoing star

461

formation. At the junction of the molecular bar and the 100 pc-ring, the molecular gas could be heated to the high temperatures derived from ammonia and compressed to high densities. Possible interpretations include the dissipation of flow motion of the molecular gas through the bar, the interaction between the molecular gas and the young stellar population or the fission of a nuclear molecular cloud (Fujimoto, these proceedings)

Presumably, the gas seen in our high resolution CO (3—2) map is being compressed to densities critical for star formation, and the next generation of stars will form in the two molecular hotspots.

References

Eckart, A., Downes, D., Genzel, R., Harris, A.I., Jaffe, D.T., Wild, W.: 1990, Astrophys. J. **236**, 441

Henkel, C., Mauersberger, R.: 1990 in: "Dynamics of Galaxies and Molecular Clouds Distribution", IAU Symp. **147**, eds.: F. Combes and F. Casoli, in press

Ishizuki, S., Kawabe, R., Ishiguru, M., Okumura, S.K., Morita, K., Chikada, Y., Kasuga, T.: 1990, *Nature* **344**, 224

Lo, K.Y. et al.: 1984, *Astrophys. J.* **282**, L59

Martin, R.N., Ho, P.T.P.: 1986, *Astrophys. J.*, **308**, L7

Mauersberger, R., Henkel, C.: 1989, *Astron. Astrophys.* **223**, 79).

Turner, J.L., Ho, P.T.P.: 1983, *Astrophys. J.* **268**, L79

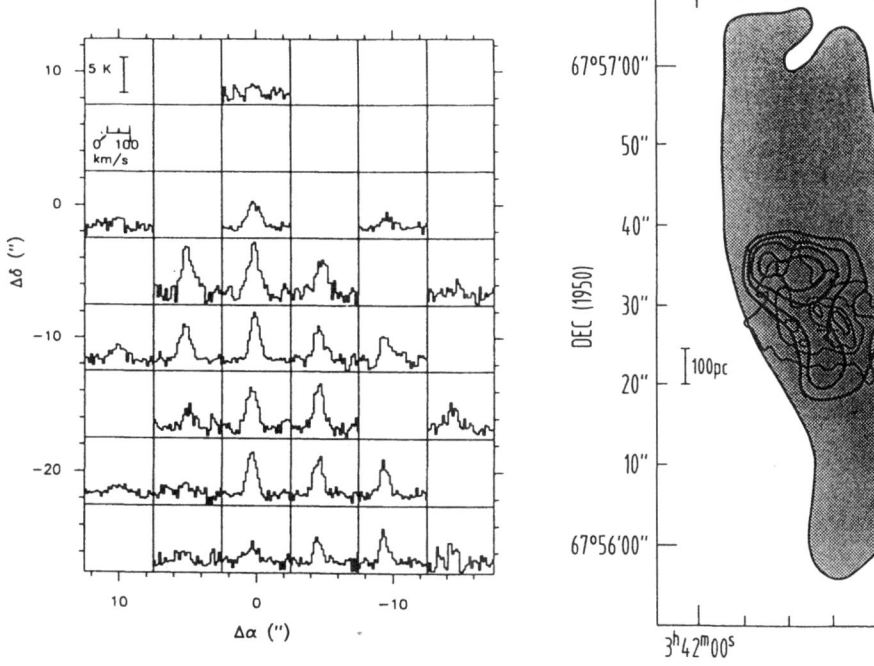

Fig. 1.:CO (3-2) profiles toward IC 342.　　*Fig. 2.: The nuclear region of IC 342 (see text).*

A Large Scale Survey of Dense Cores
and Molecular Outflows in Ophiuchus

Akira Mizuno*, Satonori Nozawa†, Takahiro Iwata*, and Yasuo Fukui*

*Department of Astrophysics , Nagoya University, Nagoya 464-01, Japan
†Research Institute of Atmospherics, Nagoya University , Toyokawa 442, Japan

We have been surveying dense molecular cores in Ophiuchus region including ρ Oph, L234, and L43 with the 4m radio telescope at Nagoya University since 1985. We have already mapped ~18° × 12° area with 2′ or 4′ grid spacing in ^{13}CO (J=1-0) spectra. We have identified ~50 dense cores (we call "^{13}CO cores"). Typical mass, density, and size of the ^{13}CO cores are ~20 $M_\odot$, ~3 × 10^3 cm^{-3}, and ~0.3 pc, respectively (Nozawa et al. 1990). We also surveyed molecular outflows in ^{12}CO (J=1-0) spectra toward 13 *IRAS* point sources associated with ^{13}CO cores in Ophiuchus. As a result of the survey, we have found 5 molecular outflows in the filamentary dark clouds and 5 regions exhibiting high velocity wings in the ρ Oph main body.

1. ρ Oph-East

ρ Oph-East (Fukui *et al.* 1986, Mizuno *et al.* 1990a) is the most spectacular one which is associated with IRAS16293-2422, being discovered by Wootten and Loren (1987) independently. It consists of five distinct outflow lobes. Four of them are compact (≲ 3′) and apparently form two pairs of bipolar outflows, and the fifth lobe is an extended (~10′) monopolar blue-shifted lobe. By NH_3 observations with the 100m telescope at Effelsberg, we found a dense core just toward the eastern edge of the compact blue eastern lobe. The velocity of the dense core is blue-shifted by ~ 0.5 km s^{-1} from the rest of the NH_3 cloud. Calculated momentum of the CO lobe is large enough for causing such a velocity shift if a significant portion of the outflow momentum is transferred to the NH_3 core. This provides the first direct evidence for an outflow to accelerate interstellar molecular gas, strongly suggesting the dynamical importance of outflow in cloud cores where stars are formed.

A high resolution ^{12}CO (J=1-0) map (Figure 1) made with the Nobeyama Millimeter Array (Mizuno et al. 1990b in preparation) reveals the distribution of the high velocity gas near the driving source. The blue-shifted gas is located on the east side and the red-shifted gas on the west side, suggesting that the axis of the bipolar outflow is oriented in the E-W direction. In the vicinity of the IRAS source, the NE-SW bipolar flow is not seen. Taking this result into consideration, we suggest that the NE-SW bipolar flow is formed by some secondary effect such as a dynamical interaction between the high-velocity outflow and the dense ambient cloud.

2. IRAS16285-2356

This outflow was discovered in the ρ Oph northern streamer with the Nobeyama

45m radio telescope. It has a very short dynamical timescale, ~1×10^4 yr. The outflow is apparently associated with a cirrus type IRAS point source, 16285-2356, detected only in the 100μm band. About 90" north of the IRAS source, another IRAS point source, 16285-2355, is located. 16285-2355 is detected in the 12, 25, and 60μm bands and has a cooler color indices than those of T Tauri stars. Levreault *et al.* (unpublished data) suggested that there is a molecular outflow associated with 16285-2355. However, the high velocity lobes are localized just toward 16285-2356 in our high resolution map with 17" beam and 15" sampling, suggesting that 16235-2356 is preferable to 16285-2355 as a driving source. We think it probable that IRAS 16285-2356 is a young and very low-mass (ie. less luminous) protostar, so that the fluxes of 12, 25, and 60μm may be less than the IRAS sensitivity limits.

References

Fukui, Y., Sugitani, K., Takaba, H., Iwata, T., Mizuno, A., Ogawa, H., and
    Kawabata, K. 1986, *Ap. J. (Letters)*, **325**, L13.
Mizuno, A., Fukui, Y., Iwata, T., Nozawa, S., and Takano, T. 1990, *Ap. J.*,
    356, 184.
Mundy, L. G., Wilking, B. A., and Myers, S. T. 1986, *Ap. J. (Letters)*, **311**, L75
Nozawa, S., Mizuno, A., Teshima, Y., Ogawa, H., and Fukui, Y. 1990, (submitted to
    Ap. J. Suppl.).
Wilking, B. A., Lada, C. J., and Young, E. T. 1989, *Ap. J.*, **340**, 823
Wilking, B. A. and Lada, C. J. 1983, *Ap. J.*, **274**, 698
Wootten, A. and Loren, R. B. 1987, *Ap. J.*, **317**, 220

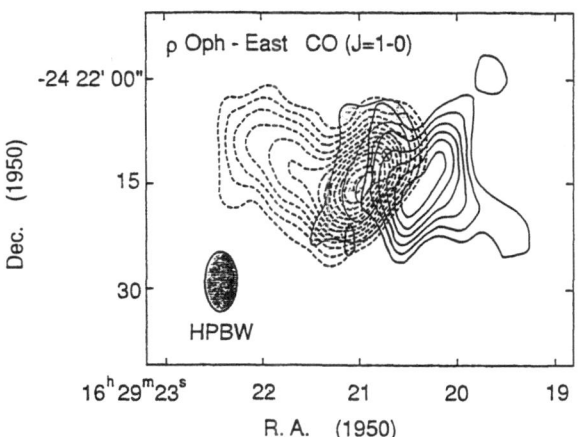

Figure 1   Interferometric map of ρ Oph-East taken with the Nobeyama millimeter array is superposed on the 2.7 mm continuum map (Mundy, Wilking, and Myers 1986). The velocity intervals are 0.4 to 3.2 km s^{-1} (low velocity blue wing) and 8.5 to 12.5 km s^{-1} (high velocity red wing). Contours extend from 2σ rms noise with a 1σ step. 1σ rms noises are 70 mJy beam^{-1}.

A HYDRODYNAMICAL STUDY OF FRAGMENTING GAS CLOUDS*

JOE MONAGHAN[1] and JOHN LATTANZIO[1,2]

[1]Department of Mathematics, Monash University,
Clayton, Victoria, 3168, AUSTRALIA.

[2]Institute of Geophysics and Planetary Physics, L-413,
L.L.N.L., P.O. Box 808, Livermore, Ca., 94550, USA.

ABSTRACT. We present 3-D hydrodynamical calculations of collapsing rotating gas clouds, with molecular cooling. We find that cooling significantly inreases the number of fragments.

1. The Fragmentation Calculations

The details of fragmentation are not well understood. We have used SPH to study the collapse of rotating gas clouds. We have recently modified our code to accurately solve for the gravitational potential in disk systems, and have added cooling from H_2 and CO molecules (taken from Hollenbach and McKee 1979). Details may be found in Monaghan and Lattanzio (1990).

We study the collapse of a $10^4 M_\odot$ gas spheres, with initial radius 12.6 pc and temperature 70K. The density is uniform on average, with perturbations of standard deviation 14%. The cloud has $\alpha = E_{th}/|E_g| = 0.30$. Uniform rotation is added, with $\beta = E_{rot}/|E_g| = 0.47$. We present two calculations, one isothermal and one with molecular cooling. Both have the same field of density perturbations.

Figure 1 shows the particles projected onto the equatorial plane for the isothermal evolution. Initially the gas rapidly collapses down the rotation axis and forms a thin disc. We see the development of an asymmetric mass distribution which finally grows into a binary. We note that this is very similar to the SLING amplification mechanism proposed by Adams et al. (1989).

Figure 2 shows the effect of molecular cooling. A ring forms which then fragments into 5 major condensations. This ring bears a remarkable similarity to the W49A star formation region (Welch et al. 1987). For further details see Lattanzio et al. (in this volume).

* Work performed under the auspices of the U.S. Department of Energy by the Lawrence Livermore National Laboratory under contract No. W-7405-ENG-48.

2. Summary

The fragmentation of clouds depends on the equation of state. Isothermal clouds, if they fragment, do so into fewer lumps than when cooling is included. The number of fragments does not scale with temperature in the same way as the Jeans mass.

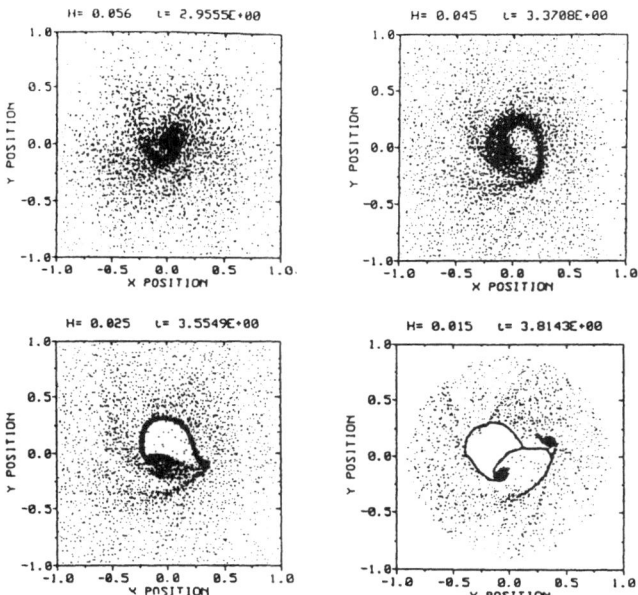

Figure 1: Evolution of the isothermal sequence. Times are in units of the initial free-fall time.

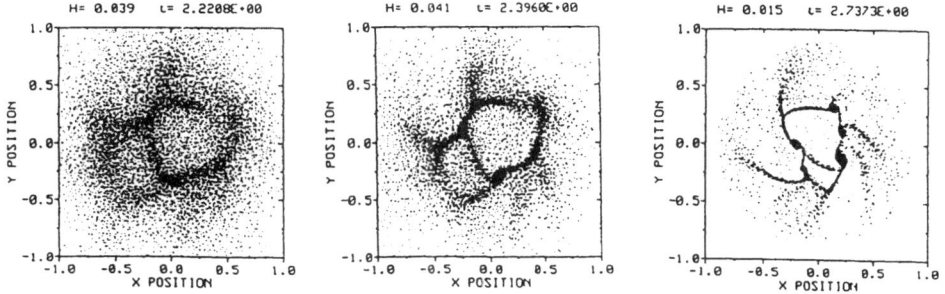

Figure 2: Evolution of the cooling sequence. Times are in units of the initial free-fall time.

References

Hollenbach, D. and McKee, C., 1979, *Astrophys. J. Suppl. Ser.*, **41**, 555.
Monaghan, J. J., and Lattanzio, J. C., 1990, *Astrophys. J.*, submitted.
Adams, F. C., Ruden, S. P., and Shu, F. H., 1989, *Astrophys. J.*, **347**, 959.
Welch, W. J., et al,1987, *Science*, **238**, 1550.

NEW SAMPLE OF YOUNG STELLAR OBJECTS

S.M.MOY, G.H.MACDONALD & R.J.HABING
University of Kent
Canterbury
Kent CT2 7NT.
United Kingdom.

ABSTRACT. In recent years there has been much interest in the study of large samples of molecular cloud cores and related infrared sources in an attempt to observe true protostars - objects in transition between a molecular cloud core and a young stellar object (YSO). We present here a survey of 48 possible protostellar objects chosen initially by their IRAS colours and subsequently observed in (1–0) HCO$^+$ emission at Onsala in Feb 1990. Future observations in (3–2)^{13}CO & (3–2)^{12}CO will be made with the JCMT and in (1,1) and (2,2) NH$_3$ emission with the Bonn 100m telescope.

1. Introduction

A search for possible protostars can be undertaken in one of two ways: candidate objects can be chosen in regions of high visual obscuration containing known dense molecular cores (Myers & Benson, 1983), or, more commonly, objects can be selected by their infrared (IRAS) colours. (Beichman 1986, Myers 1987, Heyer 1987 and Scalise 1989). In both cases follow-up observations can be made using molecular line emission and/or further mid- to near- infrared photometry.
A protostar can be defined as a region within a molecular cloud where the star forming efficiency approaches 100%; thus all the gas and dust present within that region will collapse to form a star. This proceeds from the inside out, increasing the temperature and pressure of the protostar until the initiation of deuterium 'burning'. To date observations have failed to detect such an object, with most detections revealing outflow material with little or no sign of collapse. No object has been detected with simply inflow alone: this may be because such activity occurs over a very short time span compared with the outflow liftime, requiring a large sample of objects to be observed. A further reason may be that œOxisolated inflow occurs over a region that cannot be properly resolved, thus 'diluting' any otherwise detectable velocity shifts due to collapse.

2. The IRAS Sample

The core of a protostellar object is shrouded by a thick optically opaque envelope of gas and dust which is heated initially by the release of gravitational energy and then later by the radiation energy emitted by deuterium 'burning'. The dust within the envelope re-radiates this energy at infrared wavelengths.

One can define a spectral index α for a given spectral energy distribution (SED), (Lada, 1987), as

$$\alpha = \frac{d\log\lambda F_\lambda}{d\log\lambda} .$$

Lada classifies young stellar objects (YSO's) according to their spectral energy distribution into three classes:-

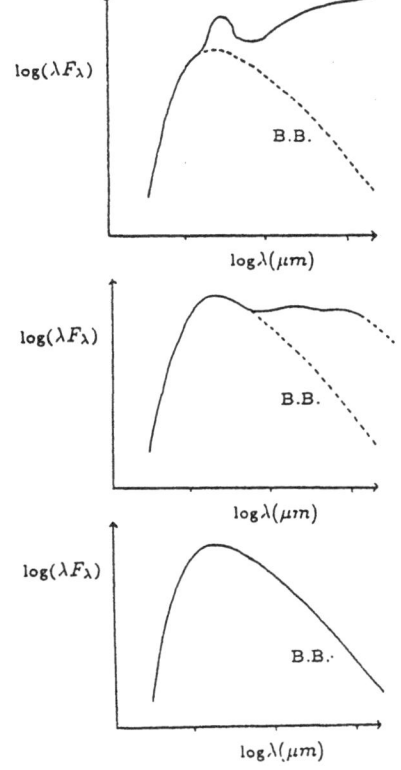

Class I.
These have a SED broader than for a single blackbody, with a positive spectral index between 12 to 100μm. This can be attributed to large amounts of circumstellar dust.

Class II.
These have negative spectral indices from 12 to 100μm. Again these have SED's broader than for single blackbodies. Their spectral index is due to some circumstellar dust. These objects are usually found to be embedded T-Tauri stars.

Class III.
Again these have a negative spectral index but SED's that correspond to a single black body which is only slightly reddened.

It is assumed that these classes represent an evolutionary sequence starting at the young, heavily shrouded class I objects and proceeding through class II to the older class III sources. Real sources are found to have a whole range of SED's that vary continuously in shape from class I to class III. Starting from this hypothesis, we selected all sources in the IRAS point source catalogue with positive spectral indices between all four wavelengths, not associated with possible extragalatic objects, and with all flux measurements being of at least moderate quality. These selection criteria produced ~4,500 sources, the vast majority of which had never been previously observed in molecular line emission.

Two further selection criteria were imposed to produce a smaller, yet complete, sample of 48 objects, these were:-

　　i) All spectral indices had to be greater than 2.

　　　$(\alpha_{12,25} > 2,\ \alpha_{25,60} > 2,\ \alpha_{60,100} > 2)$ and

　　ii) Spectral indices increased with increasing wavelength.

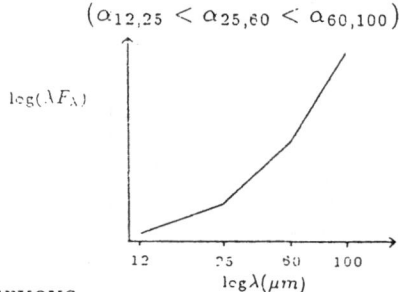

3. Previous Surveys

By bringing together many of the IRAS sources included in other surveys and plotting them on colour-colour diagrams, we investigated whether particular types of objects were confined to given regions in colour space. Caution must be exercised when combining sources from so many surveys - not only do they lie in different regions of space but they will have been selected in different ways , although the selection criteria will have been chosen to select possible YSO's and protostars and therefore be similar.
Looking at figs 1 & 2 it is apparent that certain phenomena are confined to areas of colour space; all the sources except those of Scalise (1989) have IRAS luminosities low to intermediate (~0.1 to ~200 L$_\odot$), although most are less than ~20 L$_\odot$.
Sources with maser and outflow activity predominately are found in similar regions: Wouterloot (1989) concluded that the two are correlated, and that sources with a higher far-infrared luminosity have generally higher H_2O maser luminosity. There appear to be close correlations between IRAS colours and near infrared and optical fluxes. Myers et al. (1987) found that objects that had a high spectral index s between 2 and 25μm were more likely to be optically invisible and near molecular cores than those with a low value of s, whilst extinction reaches a maximum when $s=2$ to 3, and implies a gas and dust density much greater than that deduced from molecular line observations, suggesting that the gas within the cores may be clumped,or have a non-spherical geometry. Similar results were obtained by Beichman (1986) who found a correlation between the presence of an optical counterpart and distance from the associated cloud core. The IRAS colours at 12 and 25μm

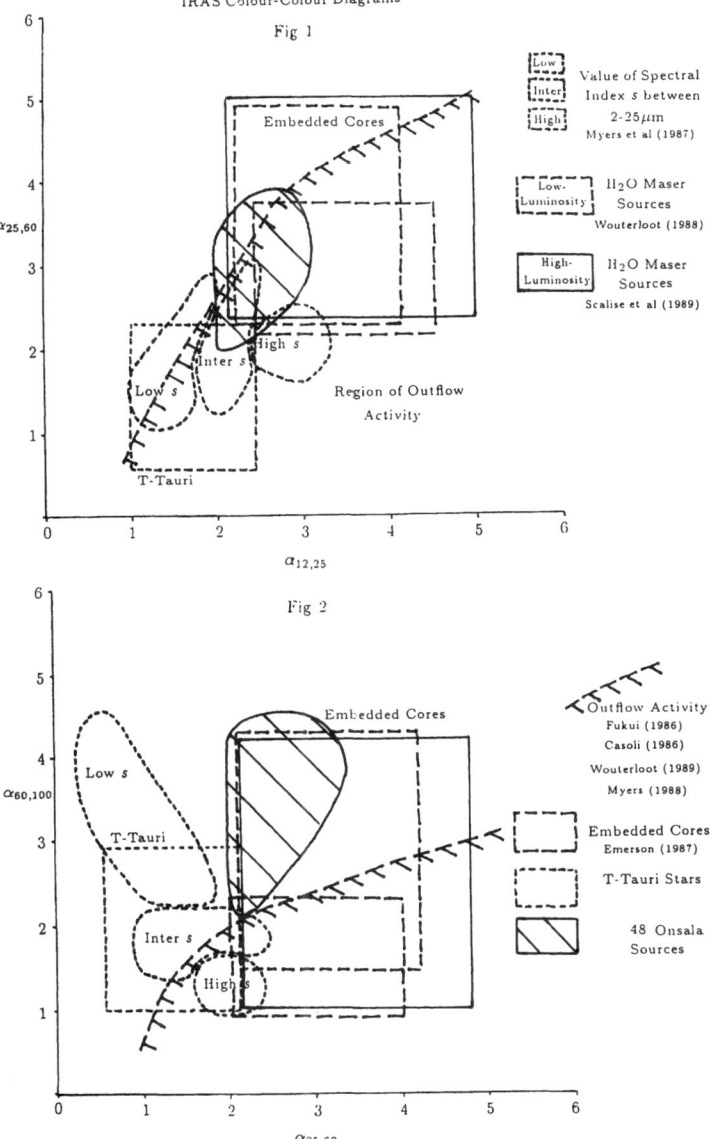

IRAS Colour-Colour Diagrams

Fig 1

Fig 2

tend to show that the more obscured of these objects lie in the region of embedded sources defined by Emerson (1987), whilst those having optical counterparts tend to lie within the T-Tauri region. In figure 2, in which a third dimension is added to colour- colour space with $\alpha_{60,100}$, the simple evolutionary correlation, seen in fig 1, of decreasing s between embedded cores and T-Tauris is weaker. However, since IRAS fluxes at 100μm are subject to unknown amounts of 'cirrus' contamination, this evidence does not necessarily contradict the conclusions of Myers et al.

The mid-infrared flux between 2 to 25μm is dependent, according to theoretical calculations by Adams & Shu (1985), on the presence of material within a shell, or possibly a disk. This governs emission between 5 to 30μm, intercepting substantial amounts of radiation and allowing more distant grains to be warmed with a cooler distribution of photons than that from the star. The presence of such a structure would explain the cooler grain distribution required to fit far infrared observations. Thus the value of s (between 2 to 25 μm) may be indicative of the presence of the circumstellar disk; it should be noted that high and intermediate s value sources lie in a region associated with outflows and adjacent to one where masers occur. Using the same sample, a CO survey by Myers (1989), showed that outflows were only associated with IRAS sources near molecular cores.

From the IRAS colour plots no clearcut picture emerges. Although outflow sources predominately are found in the region enclosing embedded cores (Emerson, 1987), and that these sources border that enclosing T-Tauri stars, it is still not clear how a protostar moves in colour space as it evolves. Outflow is one of the easiest phenomena to detect, with adequate mapping, and is believed to last approximately half the life time of a protostar ($\sim 2 \times 10^5$yr, Snell et al.1988).

The accretion model of Adams and Shu (1986) predicts a protostar gaining mass and becoming warmer as it moves towards the region occupied by T-Tauri stars. From theoretical work by Stahler Shu & Taam (1980) accretion luminosity is proportional to the protostellar mass, thus luminosity and temperature should increase with time as the core assembles more material. Berrilli et al (1989), taking a large sample of known CO sources (outflows) and Herbig Haro excitation sources, found that there was no correlation between their luminosity and IRAS colours; they were evidently not becoming warmer as their luminosity increased. If the model of Adams & Shu is correct then such exciting sources cannot be described by a purely accretional model and are thus more evolved objects than true protostars.

4. Sample of 48 IRAS sources

4.1. CASE FOR PROTOSTARS

The sample of 48 IRAS sources, (Table 1), was found to occupy a very small area of colour space. Almost all lay outside the region predominately occupied by outflow and maser sources, and all have colour temperatures $T(\frac{60}{100}) \sim 20$ to 25K. They all have spectral energy distributions which peak longwards of 100μm, corresponding to temperatures less than 29K.

Table 1. 48 IRAS Source Fluxes and Spectral Indices

NAME	F_{12}	F_{25}	F_{60}	F_{100}	$\alpha_{12,25}$	$\alpha_{25,60}$	$\alpha_{60,100}$
00040+6742	0.55	1.33	9.77	47.66	2.20	3.27	4.10
00322+6315	0.29	0.65	3.51	12.50	2.10	2.93	3.49
00361+5911	2.64	5.68	14.20	33.48	2.04	2.05	2.70
00412+4132	0.37	0.87	6.42	21.56	2.20	3.28	3.37
00544+5609	0.49	1.26	6.65	28.12	2.29	2.90	3.82
02407+6029	1.20	4.51	29.07	90.17	2.80	3.13	3.22
02500+6905	0.52	1.10	7.66	29.12	2.02	3.22	3.61
03111+5938	0.38	1.27	5.73	17.17	2.64	2.72	3.15
03429+2423	0.42	1.14	4.68	10.91	2.36	2.61	2.66
04033+5103	0.28	0.76	4.58	14.67	2.36	3.05	3.28
04362+4913	0.42	1.20	5.03	17.79	2.43	2.64	3.47
04480+4530	0.93	2.23	14.93	46.96	2.19	3.17	3.24
04487+3942	0.38	0.84	2.45	8.77	2.10	2.20	3.50
05177+3636	0.74	1.59	4.29	15.90	2.00	2.10	3.60
05286+1203	0.78	1.19	9.35	70.37	2.22	2.81	4.95
05375−0040	0.73	1.72	5.00	27.98	2.17	2.22	4.37
05382−0324	0.25	0.53	5.66	35.46	2.02	3.71	4.59
05393+2248	1.10	2.94	11.25	25.09	2.34	2.53	2.57
05435−0015	1.93	5.13	20.87	67.93	2.33	2.60	3.31
05590+2008	0.41	0.90	5.72	23.03	2.07	3.11	3.73
06405−0356	0.31	0.65	4.38	13.75	2.01	3.18	3.24
06522−0350	1.45	3.56	23.12	69.62	2.22	3.14	3.16
18216+6418	0.19	0.40	1.13	2.16	2.01	2.19	2.27
18437−0216	3.87	14.86	146.81	640.72	2.83	3.62	3.88
18533+0750	2.56	5.55	25.84	363.55	2.05	2.76	6.18
19009+0204	0.61	1.42	8.09	24.10	2.15	2.99	3.14
19116+1155	2.69	6.28	43.47	370.44	2.16	3.21	5.19
19188+1400	1.52	6.20	71.65	443.40	2.90	3.79	4.56
19253+1748	3.33	11.30	52.49	136.06	2.66	2.75	2.86
19262+1924	3.86	12.39	70.63	468.50	2.58	2.98	4.70
19287+1816	1.21	3.80	21.22	99.11	2.56	2.96	4.02
19459+2558	0.36	1.34	6.93	24.86	2.79	2.88	3.50
19474+2637	4.21	14.75	92.02	319.28	2.71	3.09	3.44
19509+2725	0.48	1.06	3.50	16.76	2.08	2.36	4.07
20049+3326	0.97	2.26	6.74	24.18	2.15	2.25	3.50
20072+2720	0.78	2.19	14.30	44.87	2.41	3.14	3.24
20136+4025	0.77	1.90	13.34	50.64	2.23	3.23	3.61
21026+4932	1.38	3.98	15.29	37.51	2.44	2.54	2.76
21098+5358	0.40	0.88	7.05	28.79	2.07	3.37	3.75
21143+5159	0.38	0.85	5.07	19.20	2.10	3.04	3.61
21306+5733	0.65	1.44	9.01	38.47	2.08	3.09	3.84
21350+4943	0.32	0.74	5.48	18.44	2.14	3.29	3.38
22174+6042	0.46	1.67	9.27	35.81	2.75	2.96	3.65
22178+6317	5.16	14.37	61.59	212.67	2.39	2.66	3.40
22452+5835	0.31	0.66	3.40	23.94	2.00	2.80	4.80
22529+5704	0.34	0.83	2.51	10.53	2.22	2.26	3.81
23091+6211	0.33	0.88	3.16	18.85	2.38	2.46	4.50

We believe that this sample may represent a selection of extremely cool and still heavily enshrouded protostars. The SED's of our sample are very similar to IRS2 found in the dark cloud Barnard 5. This object has a colour temperature of $\sim$25K and a mass which exceeds the calculated Jeans mass for this cloud. It is unlikely to be a density enhancement within the cloud heated by the interstellar radiation field (Beichman, 1984). A similar very cool object lies in the centre of B335, this also peaks longward of 100μm. (Keene et al, 1983).

From the NH_3 studies by Myers & Benson (1983) and Wouterloot & Walmsley (1988), almost all dense molecular cores associated with possible protostellar IRAS sources had kinetic temperatures less than 25K. Further, the study by Myers & Benson showed that even under the two extremes for supporting motion, thermal and Doppler, the cloud cores would at best be unstable and at worst, be undergoing collapse.

4.2. LUMINOSITY

Luminosity has been calculated where possible using the distance obtained from distance modulii of associated HII & OB clusters, using the expression for the infrared flux of Boulanger (Casoli, 1986). Most of the sources appear to be of intermediate luminosity, much lower than the outflow sources of Snell (1988) $\sim 10^4 L_\odot$, yet slightly greater than those of Beichman et al (1986) $\sim$2-3$L_\odot$, (Table 3).

4.3. MASS

TABLE 2. Source mass comparisons

	Wavelength Range(μm)	Mass ($M_\odot$)	Number in Sample	
Sources with	12–25	$1 \pm 0.4 \times 10^{-6}$	(16)	
Optical	25–60	$4 \pm 2 \times 10^{-4}$	(9)	Beichman
Counterparts	60–100	$3 \pm 2 \times 10^{-5}$	(5)	1986
Sources without	12–25	$1 \pm 0.5 \times 10^{-5}$	(14)	
Optical	25–60	$4 \pm 1 \times 10^{-3}$	(22)	Beichman
Counterparts	60–100	$4 \pm 1 \times 10^{-1}$	(23)	1986
	12–25	1.4×10^{-5}	(16)	
	25–60	5.0×10^{-3}	(22)	Our Sample
	60–100	7.7	(22)	

Following Beichman (1986) the mass of gas & dust in an optically thin region, emitting at a particular wavelength, may be calculated, assuming grain properties and temperatures of Hildebrand (1983).

TABLE 3. Source parameters and associated OB clusters

NAME	Assoc	D(kpc)	$L(L_\odot)$	$M_{12,25}$	$M_{25,60}$	$M_{60,100}$
05177	Aur OB1	1.32	42	5.7×10^{-6}	9.5×10^{-4}	0.21
00322	Cas OB14	1.10	19	2.8×10^{-6}	7.8×10^{-4}	1.91
	Cas OB4	2.88	128	1.9×10^{-5}	5.2×10^{-3}	12.9
00544	S184	2.20	154	2.8×10^{-5}	7.7×10^{-3}	25.7
02500	$\sim$CAM OB1	1.00	33	3.0×10^{-6}	9.7×10^{-4}	3.5
03111	CAM OB1	1.00	24	6.7×10^{-6}	2.3×10^{-3}	1.2
	S202	0.80	15	4.3×10^{-6}	1.5×10^{-3}	0.77
03429	$\sim$PER OB2	0.40	2.88	7.6×10^{-7}	3.0×10^{-4}	0.08
04033	CAM OB1	1.00	18.6	4.0×10^{-6}	1.1×10^{-3}	0.96
05375	ORI OB1	0.50	8.1	1.6×10^{-6}	4.3×10^{-4}	2.35
05435	ORI OB1	0.50	23	1.1×10^{-6}	1.6×10^{-4}	0.016
19459	$\sim$VUL OB1	2.00	117	3.6×10^{-5}	1.3×10^{-2}	12.6
20049	$\sim$CYG OB3	2.29	190	3.8×10^{-5}	1.0×10^{-2}	14
20136	CYG OB8	2.29	288	3.4×10^{-5}	1.1×10^{-2}	27.6
20555	CYG OB7	0.83	15.5	3.9×10^{-6}	1.1×10^{-3}	2.1
21026	CYG OB7	0.83	42.6	1.6×10^{-5}	4.5×10^{-3}	1.57
21143	CYG OB7	0.83	15.8	2.1×10^{-6}	5.8×10^{-4}	1.58
21306	CEP OB2	0.83	28.8	3.4×10^{-6}	9.9×10^{-3}	4.3
21350	CYG OB7	0.83	15.5	1.9×10^{-6}	5.7×10^{-4}	0.8
22178	CEP OB2	0.83	195	4.2×10^{-5}	1.6×10^{-2}	12.8
22529	CEP OB1	3.47	263	3.8×10^{-5}	1.2×10^{-2}	21.1
23091	CEP OB3	0.87	15	3.0×10^{-6}	1.0×10^{-3}	4.51
	CAS OB2	2.63	134	2.8×10^{-5}	9.9×10^{-3}	41.3
00040	$\sim$CEP OB4	0.84	34	3.4×10^{-6}	1.2×10^{-3}	7.7
02407	CAS OB6	2.19	501	1.4×10^{-4}	5.2×10^{-2}	29
	CAM OB1	1.0	112	6.3×10^{-5}	2.5×10^{-2}	6.1
18437	S66	3.2	6760	9.6×10^{-4}	3.8×10^{-1}	1045
21098	CYG OB7	0.83	21	2.1×10^{-6}	5.9×10^{-4}	2.2
22174	CEP OB2	0.83	29	7.1×10^{-6}	2.8×10^{-3}	2.9
22452	CEP OB2	0.83	15	2.2×10^{-6}	5.4×10^{-4}	0.32
	CEP OB1	3.47	263	3.9×10^{-5}	9.0×10^{-3}	5.6
05382	ORI OB1	0.5	7.5	4.7×10^{-7}	1.1×10^{-4}	2.7

$\sim$ Source just outside OB cluster boundary

Taking an average mass for each wavelength band, (neglecting 18437$-$0216 and all sources with ambiguous distances), gives results not too dissimilar from those

474

obtained by Beichman for sources without optical counterparts, i.e. younger more embedded objects. Our 48 sources have around ten times more cool material emitting between 60 and 100μm; this is not particularly surprising judging from their enhanced SED's at these wavelengths, (Table 2).

4.4. HCO$^+$ OBSERVATIONS

HCO$^+$ (1–0) observations were made towards all 48 sources with the Onsala 20m telescope in Feb 1990, to determine if they were associated with molecular material. We obtained detections in 13 with T_A^* ranging from 0.2 to 0.8K with an average of 0.6K. Although this is more probably a distance effect rather than a reflection of the intrinsic source properties as non detections ocurred in nearby high, as well as low, IR luminosity sources.
Six of the stronger detections were then mapped on a 9 point grid , 5 of these proved to be extended with peaks not coinciding with the IRAS position. None showed high velocity wings, a sign associated with possible outflow activity and, according to Berrilli, associated with more developed YSO's.

5. Conclusions

We believe that these 48 sources represent a complete sample of protostellar objects, deeply embedded in their molecular clouds and still too young to be undergoing any large scale outflow.
NH$_3$ (1,1) & (2,2) observations will be made in December towards all 48 sources, enabling calculations similar to those of Myers & Benson (1983), to be made, determining if collapse is occuring in molecular cores. With sources of known distance an independent estimate can be placed on core masses allowing us to verify if the IR luminosities are proportional to their accreted mass, a sign that the object is still in a purely accretional, protostellar phase.
Time has been granted at the JCMT to observe all sources at high resolution ^{12}CO(3–2) & ^{13}CO(3–2), in an attempt to observe directly any infalling material, and to do far infrared photometry at 450 & 800μm, to fix the frequency range containing the emission peak.

6. Acknowledgments

S.M.M gratefully acknowledges financial support from the U.K. Science and Engineering Research Council.

7. References

Adams,F.C.,Shu,F.J.,1985.*Astrophys. J.* **296**,655

Adams,F.C.,Shu,F.J.,1986.*Astrophys. J.* **308**,836

Beichman,C.A.,et al.,1984.*Astrophys. J. Lett.* **278**,L45

Beichman,C.A.,Myers,P.C.,Emerson,J.P.,Harris,S.,Mathieu,R., Benson,P.J.,
Jennings,R.E.,1986.*Astrophys. J.* **307**,337

Berrilli,F.,Ceccarelli,C.,Liseau,R.,Lorenzetti,D.,
Saraceno,P.,Spinoglio,L.,1989.*Monthly Notices Roy. Astr. Soc.* **237**,1

Casoli,F.,Dupraz,C.,Gerin,M.,Combes,F.,Boulanger,F.,1986.*Astron. Astrophys.*
169,281

Emerson,J.P.,1987 IAU Symposium No.115,"Star Forming Regions". Reidel

Fukui,Y.,Sugitani,K.,Takaba,H.,Iwata,T.,Mizuno,A.,Ogawa,H.,
Kawabata,K.,1986.*Astrophys. J.* **311**,L85

Heyer,M.H.,Snell,R.L.,Goldsmith,P.F.,Myers,P.C.,1986.*Astrophys. J.* **321**,370

Hildebrand,R.H.,1983. Q.Jl R.astr.Soc **24**,267

Keene,J.,Davidson,J.A.,Harper,D.A.,Hildebrand,R.H.,Jaffe,D.T.,
Loewenstein,R.F.,Low,F.J.,Pernie,R.,1983.*Astrophys. J.* **274**,L43

Lada,C.J.,1987 IAU Symbosium No.115,"Star Forming Regions". Reidel

Myers,P.C.,Benson,P.J.,1983.*Astrophys. J.* **266**,309

Myers,P.C.,Fuller,G.A.,Mathieu,R.,Beichman,C.A.,Benson,P.J.,
Schild,R.E.,Emerson,J.P.,1987.*Astrophys. J.* **319**,340

Myers,P.C.,Heyer,M.,Snell,R.L.,Goldsmith,P.F.,1988.*Astrophys. J.* **324**,907

Scalise,E.,Rodriguez,L.F.,Mendoza-Torres,E.,1989.*Astron. Astrophys.* **221**,105

Snell,R.L.,Huang,Y.L.,Dickman,R.L.,Claussen,M.J.,1988.*Astrophys. J.* **325**,853

Stahler,S.W.,Shu,F.H.,Taam,R.E.,1980.*Astrophys. J.* **241**,637

Wouterloot,J.G.A.,Walmsley,C.M.,1986.*Astron. Astrophys.* **168**,237
Wouterloot,J.G.A.,Walmsley,C.M.,Henkel,C.,1988.*Astron. Astrophys.* **203**,367

Wouterloot,J.G.A.,Walmsley,C.M.,Henkel,C.,1989.*Astron. Astrophys.* **215**,131

COMPARISON OF TURBULENCE IN HII REGIONS AND MOLECULAR CLOUDS

C. R. O'DELL
Department of Space Physics and Astronomy
Rice University
P. O. Box 1892
Houston, Texas, 77251 USA

ABSTRACT. Both the HII Regions and the Molecular Clouds show broadening of their emission lines beyond that expected from thermal motion and this is ascribed to turbulence. Turbulence in molecular clouds generally agrees with a model where the velocity of motion is determined by the Alfvén velocity.

Turbulence in Galactic HII Regions and Giant Extragalactic HII Regions can also be studied by the width of the emission lines. The magnitude of the turbulent velocities in these regions are characteristically about 10 km/s. There is a general increase in turbulent velocity with the size of the HII Region, and this relation is close to but different from the one third power dependence expected from the most naive application of Kolmogorov theory. When a detailed study is conducted of each Galactic HII Region by means of the structure function, one finds that there is not agreement with Kolmogorov theory.

The Size-Turbulent versus Velocity relation for Galactic HII Regions differs slightly from the better defined velocity relation for Giant Extragalactic HII Regions. This difference is probably due to the fact that the larger extragalactic objects are probably complexes of multiple individual HII Regions. There is no evidence that broadening of extragalactic HII Regions is due to motion about a common center of mass.

1. TURBULENCE IN MOLECULAR CLOUDS

Broadening of radio molecular emission lines arising from these cold regions generally is greater than expected from thermal broadening. This extra broadening is small, but certainly real, in the Cores of these clouds (e.g., Meyers 1983) and is very pronounced in studies of the full Molecular Complexes (e.g., Dame et al. 1986). The results of these studies are shown in Figure 1, where the data is grouped by intervals in the logarithm in the diameters.

There is a rudimentary theory that can explain the increase in turbulent velocity with size that is seen in the Molecular Clouds. As shown in the reviews by Meyers (1987) and Falgarone and Perault (1987) there is a general agreement with the one half power law dependence that would apply if the turbulent velocities were limited by the Alfvén wave velocities. We show in the figure that the slope of the relation determined from the cores and the complexes is very similar and taken together has a slope of 0.52, which is remarkably close to the value of one half expected from the magnetic model.

2. LINE BROADENING IN EXTRAGALACTIC HII REGIONS

The existence of a general relationship between line width and size of Giant Extragalactic HII Regions (GEHR) was first suggested by Melnick (1977), who argued that the large velocities were due to large scale kinematic motion about a center of mass. This interpretation did not recognize that similarly large line widths were already seen in the Galactic HII Regions. Roy et al. (1986) have determined the size-velocity data set with the greatest accuracy and the figure also shows their grouped results. These data are corrected for thermal broadening.

3. LINE BROADENING IN GALACTIC HII REGIONS

In many ways it is more difficult to get a measure of the total turbulent velocity in the Galactic HII Regions (GHIIR) because of their large angular sizes. My colleagues and I have been mapping velocities across the faces of GHIIR for over a decade in order to accurately model the evolution of these objects (c.f. references in O'Dell and Castañeda 1987). In a given line of sight we see a particular value of the line width and we also see random variations in the radial velocity from point to point across the face of the nebulae. The dispersion of values of the radial velocities can be used to correct the average observed line width β through the equation

$$\beta = (\beta_{obs}^2 + 2 \times SD^2)^{1/2}$$

where SD is the standard deviation. Figure 1 also shows the results for the turbulence in GHIIRs, with the values representing the entire nebulae in the same fashion as the GEHRs and the Molecular Clouds.

Examination of the figure shows that one can fit a simple curve of the form log $\beta = 0.78 + .25 \log D$ to all of the HII region data, both Galactic and extragalactic. This slope is close but demonstrably different from the expectation of 1/3 that would result if all HII regions were similar and naively behaved according to Kolmogorov theory for an incompressible gas. Although this theory has many inapplicable assumptions, it agrees with laboratory and measurements made in nature for a wide variety of gases and fluids. As such, it is not surprising that there would be general agreement in this astronomical application of the theory.

The data indicate that there may be a slightly different relation of size and velocity for the GHIIR and the GEHR. This would not be unexpected, as the GHIIR are clearly single objects while the GEHR are probably multiple HII Regions that are not resolved spatially because of their great distances.

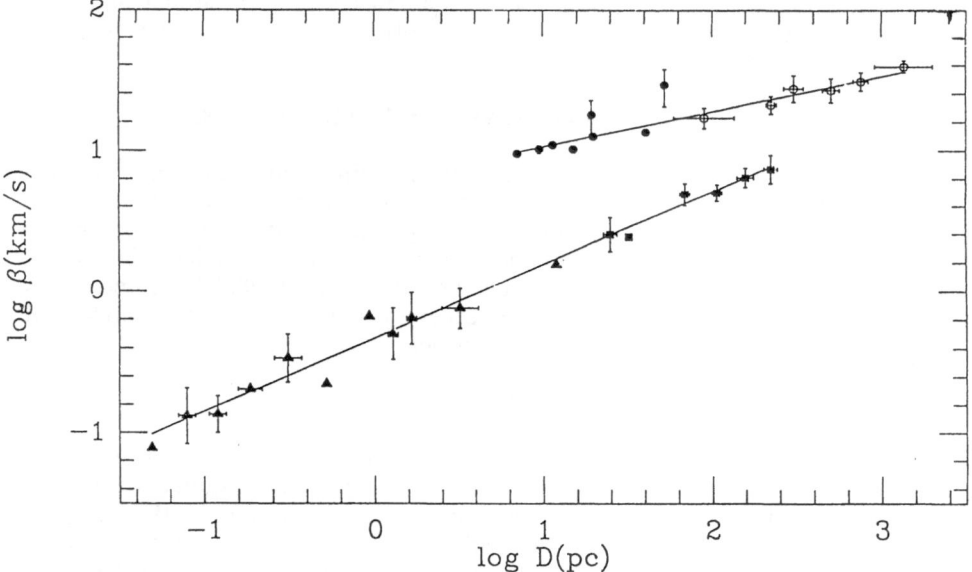

Figure 1. The relation between size and e^{-1} velocity width is shown for the Cores (▲), Complexes (■), GEHRs (o), and GHIIRs (●). The solid lines are the corresponding least square fits for each kind of object. The Core's data are for the turbulent component of motion.

4. DETAILED TESTS OF KOLMOGOROV THEORY IN GALACTIC HII REGIONS.

It is possible to use the detailed study of random radial velocity variations across the face of the GHIIR's to test the applicability of Kolmogorov theory. The theoretical framework for this was established by von Hoerner (1951) who showed that the most useful statistical measure of the random velocities is the structure function

$$B(r)= \quad <|V(r')-V(r'')|^2>$$

This function should have a slope of 5/3 in the inner region of a nebula, begin to flatten when the average separation is comparable to the depth of the nebula, and would be 2/3 if one ever reached distances much larger than the depth of the nebula. The results from the study of a number of GHIIR (O'Dell and Castañeda 1987) demonstrates that when studied in detail the Galactic HII Regions do not agree with Kolgomorov theory. This means that general relationship between size and turbulent velocity inferred from the line broadening must reflect some mechanism different from Kolgomorov
theory.

5. CONCLUSIONS

Our summary of the combined data for cores and entire molecular cloud complexes shows that the theory of random motions of material being limited by the Alfvén wave velocities can explain the observed line broadening over a size range of more than three orders of magnitude.

The internal velocities in Galactic HII Regions and Giant Extragalactic HII Regions shows a well defined correlation with size over a size range of a factor over two orders of magnitude but with a very different slope than that for molecular clouds. The continuity between Galactic and extragalactic objects argues against the latter being broadened by large scale kinematic motions.

Detailed determination of the turbulent velocities within Galactic HII Regions from studies of the radial velocities along various lines-of-sight indicates that there is not any agreement between the observations and the simplest application of Kolmogorov theory to these nebulae.

REFERENCES

Dame, T. M., Elmegreen, B. G., Cohen, R. S., and Thaddeus, P. (1986) Ap.J., 305, 892.

Falgarone, E. and M. Perault (1987) in Physical Processes in Interstellar Clouds, ed. G.E. Morfill and M. Scholer (Dordrecht: Reidel) p. 59.

Melnick, J. (1977) Ap. J., 213, 15.

Meyers, P. C. (1983) Ap. J., 270, 105.

Meyers, P. C. (1987) in Interstellar Processes, ed. D. J. Hollenbach and H. A Thronson, Jr. (Dordrecht: Heidel) p. 71.

O'Dell, C. R. and Castañeda, H. O. (1987) Ap. J., 317, 686.

Roy, J. R., Arsenault, R. and Joncas, G. (1986) Ap.J., 300, 626.

von Hoerner, S. (1951) Zs.Ap., 30, 17.

FRAGMENTARY STRUCTURE IN THE L1551 MOLECULAR OUTFLOW

NICHOLAS D. PARKER and GLENN J. WHITE
Queen Mary & Westfield College
Mile End Road
London E1 4NS

ABSTRACT. We have obtained a high resolution map of the ^{12}CO J=3–2 emission in the vicinity of HH102 in the L1551 molecular outflow. The data reveal the presence of several bright clumps within the thin shell of low-velocity blue-shifted outflowing gas. There is evidence for further fragmentation in the clumps, as signified by the high derived excitation temperatures and low beam-filling factors. The region of peculiarly energetic activity associated with HH102 coincides with the projected location of impact of a radio jet from IRS5 with the dense ambient gas surrounding the outflow cavity.

^{12}CO J=3–2 spectra were obtained using the James Clerk Maxwell Telescope (JCMT). At 345 GHz the beamwidth was $\Theta_{\text{beam}} \sim 15$ arcsec. Figure 1 shows the map of integrated intensity (T_A^*) in the low-velocity blue-shifted gas ($v_{\text{lsr}} = 5 - 7$ km s^{-1}, the ambient cloud line-centroid velocity being $v_{\text{lsr}} \sim 7$ km s^{-1}). The thin shell, running roughly northeast – southwest is clearly defined, although this is not the case at larger blue-shifted velocities. The shell separates into three clumps, labelled A, B and C, which lie along the locus of peak intensity. A fourth clump, labelled D, is found interior to the arc of the shell. These clumps have size $\sim$ 20 arcsec and a roughly regular spacing of $\sim$ 55 arcsec. Thus, since $\Theta_{\text{beam}} \simeq 15$ arcsec, the shell has width of order 20 – 25 arcsec over much of its extent. In some places the shell appears narrower than this, perhaps being unresolved, hence $\Theta_{\text{shell}} \lesssim 15$ arcsec. The features seen in these data are consistent with recent high resolution maps obtained using Maximum Entropy image reconstruction (Moriarty-Schieven *et al.* 1987).

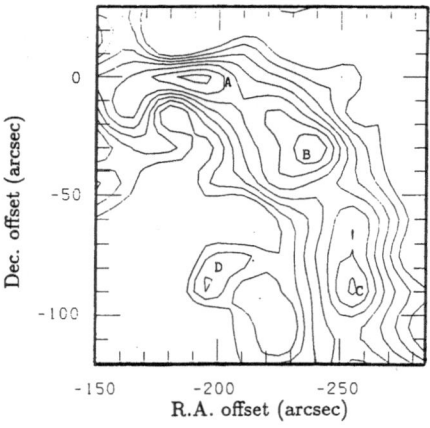

Figure 1. Integrated intensity in the low velocity blue-shifted CO J=3–2 emission in the vicinity of HH102.

Uchida *et al.* (1987) obtained ^{12}CO J=1–0 maps of L1551 with resolution equal to that of the observations above ($\Theta_{\text{beam}} \simeq 16$ arcsec for the NRO 45-m telescope at 115 GHz). Comparing the line ratios for the clumps, we find the values of $T_R^*(3 - 2)/T_R^*(1 - 0)$ are an order of magnitude too low to be consistent with those expected for the hot, optically-thin case. More likely is the optically-thick case, in which (assuming LTE) the excitation temperature , T_{ex}, is obtained from the following expression;

$$\frac{T_R^*(3-2)}{T_R^*(1-0)} \simeq 3\frac{e^{h\nu_{10}/kT_{ex}}-1}{e^{3h\nu_{10}/kT_{ex}}-1} \quad .$$

Values of T_{ex} are found to range from ~ 25 K for clump C, up to ~ 40 K for clump B. Corresponding beam-filling factors are all considerably less than 1, with typical values $f \sim 0.2 - 0.3$. Volume filling factors are smaller again since they scale as $f^{\frac{3}{2}}$. The emergent picture of the blue-shifted outflow at low velocities is one of a thin shell comprising a moderately hot clumpy structure on scales smaller than ~ 0.01 pc (assuming $D \sim 140$ pc to L1551).

Narrow-band Hα CCD images of the region (Snell et al. 1985) show that the shock-excited emission originates in clumps generally lying interior to the CO shell (i.e. closer, in projection, to IRS5). If the HH102 region is delineating the cavity – ambient cloud interface then this geometry can be explained as a high-velocity wind being shocked immediately interior to this interface, with the swept up shell lying just behind the shocked region. The HH102 region exhibits velocity dispersions, apparent density enhancements and shock-excited emission signifying a particularly strong interaction of a wind or jet with the surrounding dense molecular material. Figure 2 shows the distribution of high-velocity CS emission (Snell & Schloerb 1985) together with a VLA map of the 2 cm radio continuum (Rodriguez 1987).

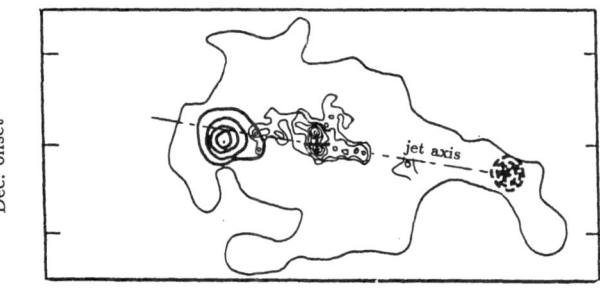

R.A. offset

Figure 2. Contours of 2 cm radio continuum emission (thin solid contours) overlaid on a map showing the high-velocity CS emission. Thick solid contours are the red-shifted CS integrated intensity and dashed contours are blue-shifted. The maps are on different scales but their relative orientations and (0,0) positions are correct.

Two faint radio 'jets' emanate from IRS5. They are not aligned with the symmetry axis of the outflow, but instead project directly towards the high-velocity CS peaks. We suggest that these jets are responsible for the activity in the HH102 region, in which the jet material suffers an oblique impact at the boundary of the cavity. Subsequent 'splaying-out' of the flow along the walls of the cavity, coupled with viscous entrainment (Dyson et al. 1988), could then explain the presence of both red- and blue-shifted CO gas at this location and its equivalent diametrically opposite IRS5. In this case it would not be necessary to invoke the presence of slow rotation of the outflow about its axis (e.g. Uchida et al. 1987) to explain the CO data.

Dyson et al. , 1988. *Mass Outflows from Stars and Galactic Nuclei*. Kluwer Academic.
Moriarty-Schieven et al. , 1987. *Astrophys. J.*, **319**, 742.
Rodriguez, 1987. *IAU symp. no. 115, Star Forming Regions*. Dordrecht:Reidel.
Snell et al. , 1985. *Astrophys. J.*, **290**, 587.
Snell & Schloerb, 1985. *Astrophys. J.*, **295**, 490.
Uchida et al. , 1987. *IAU symp. no. 115, Star Forming Regions*. Dordrecht:Reidel.

OBSERVATIONS OF VARIABILITY OF H_2O MASER SOURCES ASSOCIATED WITH STAR FORMATION REGIONS

M. I. PASHCHENKO, E. E. LEKHT
Sternberg State Astronomical Institute
13 Universitetskij prospect
Moscow V-234, 119899 USSR

I. I. BERULIS, R. L. SOROCHENKO
Lebedev Physics Institute,
Academy of Sciences, Moscow

An outstanding feature of the water maser emission from star formation regions is the strong time variability, which was recognised soon after the discovery of interstellar H_2O masers. It is likely that the observed considerable variations of the H_2O line profiles as well as of the flux density reflect the disturbancies in the protostellar nebule probably associated with the outflow from the newly formed stars.
The conventional viewpoint is that H_2O masers represent unstayable systems of infiniteley moving clouds assosiated with very young stars that are losing their mass. Our observations of H_2O sources during several years confirm the point: the majority of the observed H_2O masers associated with star formation regions by all their features (broad unstayable spectra, variable intensity, frequently changing flashy character) can be identified as objects of that type. Yet the recent years made it more obvious that some H_2O sources evidently do not match the conventional model of the maser sources.
The observations. of H_2O maser sources were carried out from November 1979 till 1990 using the 22-meter fully steerable, parabolic reflector (RT-22) at the Pushchino Radio-Astronomical Station of the Lebedev Physical Institute. The receiver was equiped with a maser amplifier at its input and a 96-channel filter bank spectrum analyser. The

resolution in radial velocity was 101 m/sec. We carried
out for a decade monitoring observations of 20 different
types of sources associated with star formation regions,
with a mean sample interval of 1-2 months. The examples of
"classical" and "peculiar" H_2O masers are presented.

During the period from 1980 to 1990 several bursts of
maser emission quite different in character were detected
for W75N.
In the H_2O spectra of Cep. A one could distinguish at

different epochs 5 groups of emission features at
different radial velocities. Neither clear correlation
betweeen the feature intensities in different groups, nor
any periodicity in their appearence were detected. Very
narrow emission picks are sometimes observed in the spectrum. The analysis shows that these picks could be produced by dense condensations immersed in a less dense turbulent medium.
In the H_2O spectrum of S269 two single (nonblended)

slowly and monotonously changing in intensity emission
features have been constantly observed during the last 10
years. The radial velocity of them keeped constant during
the whole period of observations.
The extreme variability of the H_2O emission intensity

has been considered to be one of the fundamental features
of H_2O sources. So particularly interesting were the cases

of surprising constancy of some emission features of the
H_2O spectrum (S269, W44C). In these cases we obviously deal

with stayable circumstellar structures.
The spectra of some masers have a triple line profile. We regularly observe several sources of such type, e.g.
S140, S255. As it was first shown by Elmegreen and Morris
(1979) some H_2O sources could rather be associated with

stable circumstellar formations (disks). Our observations
indicate that spectra from several sources are with a high
probability associated with such structures. The stayable
triplet structure is particularly typical for such
formations.
Some groups of radioastronomers, basing on observations of H_2O masers in star formation regions, concluded

that there is a common dependence of the width of the
bright lines on their intensity: $\Delta\nu \sim F^{-0.5}$ We would
like to confirm the universal character of this depen-

dence: besides the masers in regions of star formation we have found the similar dependence for circumstellar masers of old stars. For instance for the semiregular variable RT Vir it was: $\Delta\nu^{-2} \sim \ln F$ (Berulis et al. 1987). We feel that the mentioned universal dependence still waits for its explanation, though several attempts to explain it already existe. However, in some cases the dependence is more complicated.

Thus, our observations show a multitude of H_2O maser types. It is clear now that H_2O masers in regions of star formation are connected not only with the infinitely moving clumps of strong stellar wind, but also with stable circumstellar structures, most probably - with disc-like structures.

REFERENCES

Elmegreen B.G. and Morris M. (1979) Astrophys. J. 229,593.
Berulis I.I., Lekht E.E., Pashchenko M.I. (1987) Sov. Astron. Lett. 13(2), 124.

A MODEL OF THE DISTRIBUTION OF IONISED CARBON IN M17SW.

Simon Pinnock and Tania S. Monteiro.
Queen Mary & Westfield College,
Mile End Road,
London, E1 4NS.

ABSTRACT. The observed distribution of ionised carbon with depth into the edge-on photodissociation region (PDR) in M17SW is not consistent with results from chemical models which assume a constant or smoothly varying density. These observations have been interpreted to indicate that the PDR actually consists of many clumps embedded in a low density interclump medium (1). A chemical model is described which is designed to simulate such a clumpy PDR.

1. Introduction.

When a high mass star forms inside a molecular cloud it ionises the material around it to form an HII region. Between this and the cool molecular cloud exists a warm layer consisting of material which is partly ionised and photodissociated by far ultraviolet radiation (FUV) from the star. These photodissociation regions (PDR) are of great interest since they may be places where star formation is induced by the interaction of previously formed young stars with their parent molecular cloud. To interpret observations of these systems in terms of their physical characteristics, chemical models are used to predict the abundance, and to varying degrees the excitation of the observed species. When applied to observations of an edge-on system like the HII region/molecular cloud interface in M17SW, it is found that the models cannot reproduce the observed distribution of species. Scans of [CII]158μm emission from M17SW (1,2) show that the ionised carbon abundance decreases much less rapidly with distance from the cloud surface than expected for the cloud density derived from molecular line observations (3). To account for this it has been suggested that, at least to a depth of a few parsecs, the molecular cloud has a clumpy or fragmented structure, with the interclump material having a low density and a low FUV optical depth (1). Thus some of the FUV passes through the gaps between the clumps to ionise the surfaces of fragments deep within the cloud, so increasing the depth to which [CII] is detected.

2. The Model.

To simulate observations of an edge-on, fragmented PDR, a model is constructed from a spatially random distribution of cloud fragments. Each spherical clump consists of an optically thick cool dense core, and a relatively less dense warm envelope. For each clump, its probability of being illuminated by radiation from the ionising source, being in the shadow of foreground cores, or being prescreened by some amount of foreground envelope material, is calculated from the survival probability of a ray of light travelling from the FUV source to the centre of the clump. Having, calculated the distribution of conditions under which clumps exist, a one dimensional chemical model (4) is run for a representative population. This gives the average densities of say, C$^+$, for each of the envelopes and cores. Using these values the column density along each line of sight is calculated.

This is converted to an intensity via the transfer equation assuming a constant source function and using the relations of Crawford et al. (5) for the optical depth. This emission is then convolved with a gaussian beam pattern to simulate the [CII]158μm observations.

3. Results.

Observations of the PDR in M17SW (1,6) have indicated the presence of warm dense gas T$\simeq$250K, and $n_{H_2} \simeq 10^{4.5}$cm^{-3}, coexistent with cooler less dense material, T$\simeq$50K, and $n_{H_2} \simeq 10^{5.5}$cm^{-3}.

For the FUV radiation to penetrate two or three parsecs into the cloud, and still cause ionisation of carbon on the surfaces of clumps there, then the density of the interclump medium must be less than $n_{H_2} \sim 500$cm^{-3}, assuming it is not dust depleted. Many high resolution single dish and interferometer observations have revealed density enhancements of a few tenths of a parsec across, or less. These values are used for the standard clump density and temperature structure. With a clump diameter of 0.4pc, an extent along the line of sight of 2pc, and a volume filling factor of 0.2, the model gives the intensity distribution as shown in Fig. 1. This corresponds to two to three clumps along a line of sight through the cloud.

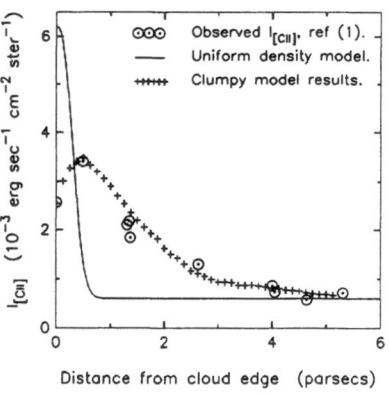

Figure 1. [CII]158μm emission in M17SW.

The simulated $I_{[CII]}$ scan matches the observed one to within the accuracy of the observations. The [CII] intensity curve convolved with a 50'' beam, predicted for a PDR of uniform density $n_{H_2} = 10^{4.5}$cm^{-3} is also plotted. The extended low intensity emission observed by Stutzki et al. (1) is supposedly due to embedded B-type stars, so one does not expect the model to reproduce this characteristic. Thus a constant C$^+$ column density corresponding to this emission has been added on to both model results before calculating the intensity. It is evident that the results of the clumpy model are a much better fit to the observations than the uniform density model.

4. Conclusions.

By using a model that attempts to mimic both the chemical and physical characteristics of a PDR, it has been possible to add weight to the interpretation of the observations in terms of a clumpy or fragmented PDR. If the M17SW PDR is not as closely edge on as has been thought, then this conclusion is more tentative. A more complicated interface geometry may produce a similar [CII] distribution without clumping. Currently work is underway to include observations of CO and other molecular species to better constrain this model.

5. References.

(1) Stutzki, J., Stacey, G.J., Genzel, R., Harris, A.I., Jaffe, D.T., Lugten, J.B., 1988, Ap.J., **332**, 379.

(2) Matsuhara, H., Takao, N., Shibai, H., Okuda, H., Mizutani, K., Maihara, T., Kobayashi, Y., Hiromoto, N., Nishimura, T., and Low, F.J., 1989, Ap.J.Lett., **339**, L67.

(3) Snell, R.L., Mundy, L.G., Goldsmith, P.F., Evans, N.J., Erickson, N.R., 1984, Ap.J., **276**, 625.

(4) Monteiro, T.S., and Pinnock, S., 1990, in prep.

(5) Crawford, M.J., Genzel, R., Townes, C.H., and Watson, D.M., 1985, Ap.J., **291**, 755.

(6) Harris, A.I., Stutzki, J., Genzel, R., Lugten, J.B., Stacey, G.J., Jaffe, D.T., 1987, Ap.J.Lett., **322**, L49

ESTIMATES OF MAGNETIC FIELDS IN INTERSTELLAR GAS CLOUDS FROM 18-CM OH SATELLITE LINES

G. M. Rudnitskij
Sternberg Astronomical Institute
Moscow State University
Moscow V-234, 119899
U.S.S.R.

In interstellar clouds, where "thermal" /nonmaser/ OH emission in the 18-cm lines is observed, prominent anomalies of the intensities of the two satellite lines, 1612 and 1720 MHz, are often seen. If a cloud is observed against a background source of radio continuum, then one satellite may be in strong absorption, while the other one is in emission almost as strong as that in the main lines, 1665 and 1667 MHz. In those clouds, for which several positions were observed in the OH lines, the roles of the satellites are sometimes interchanged with the displacement along the cloud. At the same time, the main lines usually have the intensity ratio close to the equilibrium one. This behaviour indicates to deviations of the level populations of OH molecules from the Boltzmann ones.

Such extended clouds with the anomalous OH satellites were classified by Caswell and Haynes /1975/ as Class IIc OH sources. Many examples can be found in Caswell and Haynes /1975/, Haynes and Caswell /1977/, Pashchenko /1979/, Lucas et al. /1979/ and in other works.

These satellite anomalies can be explained by the model of spin alignment of OH molecules in an external magnetic field, by the action of the infrared radiation of an embedded /protostellar or young stellar/ object at the wavelengths about 100 mcm /Burdyuzha and Varshalovich, 1973/. Thereby, the satellites' intensities are affected first of all. A consideration of the IRAS Point Source Catalog shows that many OH clouds with satellites' anomalies really contain strong far-infrared emitters.

The geometry of the model is presented on Figure 1. On Figure 2, "maps" of satellites' anomalies in a cloud, as viewed at different angles ϑ, are shown. These "maps", when compared with the observed ones, allow in principle to estimate the direction of the magnetic field with respect to the line of sight. Earlier measurements of

488

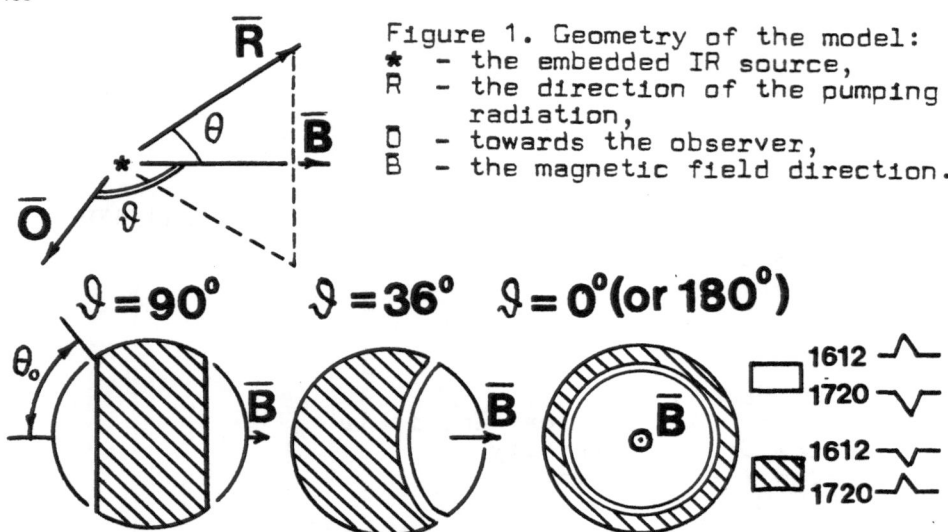

Figure 1. Geometry of the model:
* - the embedded IR source,
R - the direction of the pumping radiation,
O - towards the observer,
B - the magnetic field direction.

$\vartheta = 90°$ $\vartheta = 36°$ $\vartheta = 0°$ (or $180°$)

| | 1612 |
| | 1720 |

| | 1612 |
| | 1720 |

Figure 2. Simplified model "maps" of a spherical OH cloud with an IR source in its centre and with a homogeneous magnetic field; $\theta_o = 54°$ is the critical angle in the model of Burdyuzha and Varshalovich /1973/.

interstellar magnetic fields by means of Zeeman doublets with components of opposite circular polarization /e.g., Kazès et al. /1988/, Goodman et al. /1989// yield but the longitudinal component of the field /+ or -/. Our "maps" in combination with Zeeman results can give the true vector of the magnetic field intensity.

However, the situation in a real cloud may be more complicated, with the regions of emission and absorption in the same satellite overlapping; this may result in a "P Cyg" /or "inverse P Cyg"/ type profile, which is sometimes really observed /Pashchenko, 1979/.

A detailed publication on this subject is now in preparation.

References

Burdyuzha, V.V., Varshalovich, D.A.: 1973, Soviet Astron. 16, 597.
Caswell, J.L., Haynes, R.F.: 1975, M.N.R.A.S. 173, 649.
Goodman, A.A., Crutcher, R.M., Heiles, C., Myers, P.C., Troland, T.H.: 1989, Astrophys. J. Letters 338, L61.
Haynes, R.F., Caswell, J.L.: 1977, M.N.R.A.S. 178, 219.
Kazès, I., Troland, T.H., Crutcher, R.M., Heiles, C.: 1988, Astrophys. J. 335, 263.
Lucas, R., Le Squeren, A.M., Kazès, I., Encrenaz, P.J.: 1978, Astron. Astrophys. 66, 156.
Pashchenko, M.I.: 1979, Soviet Astron. Letters 5, 326.

^{15}NH$_3$ MILLIMASERS TOWARD NGC 7538–IRS 1

PETER SCHILKE

MALCOLM WALMSLEY

Max-Planck-Institut für Radioastronomie, Auf dem Hügel 69, D-5300 Bonn 1, RFA

1. Observations and Results

Ammonia towards the UCHII–region NGC 7538–IRS 1 shows an extremely hot component observed in absorption in several transitions and a so far unexplained (9,6) maser (see Fig. 1). This source is the only known source in the Galaxy where the 141 GHz line (1450 K above ground) of vibrationally excited Ammonia is seen in absorption (Schilke et al. 1990). An additional unique feature is the ^{15}NH$_3$(3,3) maser observed by Mauersberger et al. (1986) in Effelsberg and by Johnston et al. (1989) with the VLA. There are theoretical models to explain such a maser (see e.g. Flower, Offer and Schilke 1990) in low density regions where ortho-H$_2$ is underabundant. In this study, our aim was to investigate ^{15}NH$_3$ in more detail. We succeeded in observing the (1,1) and (2,2) lines of ^{15}NH$_3$ in absorption and, surprisingly, the (4,3) and (4,4) lines in emission (see Fig. 1).

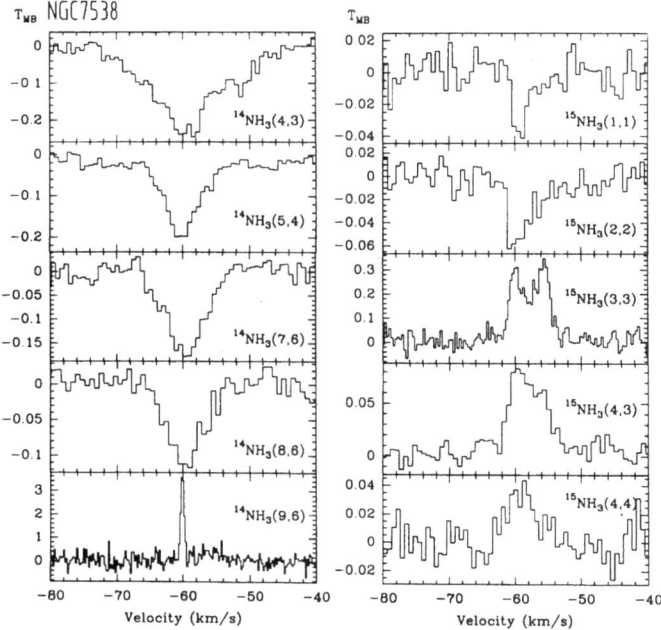

Fig. 1.   Spectra of ^{14}NH$_3$ (left) and ^{15}NH$_3$ (right)

We conclude that the (4,3) and (4,4) lines are weak masers amplifying the continuum background of NGC 7538–IRS 1. The ^{14}NH$_3$ lines show no pecularities. In a Boltzmann–plot for the ^{14}NH$_3$ absorption lines, one finds a rotation temperature of 150 K. The total ^{14}NH$_3$ column density is $5 \cdot 10^{18}$ cm^{-2}. The linear shape of the plot suggests thermalization and therefore high H$_2$–densities ($> 10^6$ cm^{-3}).

2. Theoretical Model

The current model of Flower, Offer and Schilke (1990) fails to explain the (4,3) and (4,4) emission. Hence, we decided to look at other possibilities. The detection of the vibrationally excited line (Schilke et al. 1990) gave us a hint that the first vibrationally excited state may be involved in the pumping process. And indeed, a comparison of the transition probabilities shows that transitions via the vibrationally excited state lead to inversion. The molecules are pumped to the $v_2 = 1$ state by absorption of 10μm IR photons emitted by hot, local dust. A critical dust temperature for this process can be derived based on the approach of Carroll and Goldsmith (1981). One compares the probabilities for rotational transitions with these for vibrational transitions. It turns out that the critical dust temperature is in the order of 130 K.

Detailed calculations were performed using a 260 level LVG statistical equilibrium program. The results agree qualitatively with the simple estimates. Trapping in the highly optically thick transitions of ^{14}NH$_3$ seems to quench the inversion of this species, as seen in calculations with a 300 times higher NH$_3$ abundance. The absence of the ^{14}NH$_3$ counterpart of the ^{15}NH$_3$ masers can therefore be explained in the frame of the model.

References

Carroll T.J., Goldsmith P.F. : 1981 *Astrophys. J.* , **245**, 891

Flower D.R., Offer A., Schilke P. (1990) *Monthly Notices Roy. Astron. Soc.* , **244**, 4p

Johnston K.J., Stolovy S.R., Wilson T.L., Henkel C., Mauersberger R. : 1989 *Astrophys. J.* , **343**, L41

Mauersberger R., Wilson T.L., Henkel C. : 1986 *Astron. Astrophys.* , **160**, L13

Schilke P., Mauersberger R., Walmsley C.M., Wilson T.L. : 1990 *Astron. Astrophys.* **227**, 220

A VERY STRAIGHT AND COLLIMATED OUTFLOW IN THE CORE OF OMC-1

J. SCHMID-BURGK, R. GÜSTEN, R. MAUERSBERGER, A. SCHULZ and T. L. WILSON
Max-Planck-Institut für Radioastronomie, Auf dem Hügel 69, D-5300 Bonn 1, RFA

Summary

We have recently discovered a large-scale (200″) outflow system in the core of OMC-1 (fig. 1), centered about 100″ South of IRc2 and extending over some 120″ (red lobe) resp. 60″ (blue) along a position angle of −31° (Schmid-Burgk et al. 1990). The blue lobe which might actually *protrude into the HII region M42* is poorly defined in CO 2-1, but the red lobe reveals a number of remarkable properties which we summarize here:

1. The outflow is very straight and smooth. Over the full length of 120″, the center of any cross scan deviates by not more than about 1″ from a straight line. This line passes to within 2″ the peak of the submm source FIR4 of OMC-1 (Mezger, Wink and Zylka 1990) and the mm continuum peak CS3 (Mundy et al. 1986); it also cuts across the red and blue SiO-outflow lobes recently discovered some 5-10″ to either side of FIR4 (Ziurys, Wilson and Mauersberger 1990). It thus seems that the "base" of our large-scale CO jet can be seen as well.

2. The collimation (length to width ≥ 15) is among the highest ones known. The diameters of the most redshifted velocity components are unresolved at 230 GHz, i.e. smaller than 4″, over the full extent of the jet.

3. Where measurable, the lobe diameter depends on velocity in a systematic way at any given distance from the origin of the flow, the most redshifted velocities being the narrowest. All velocity components are centered on the same axis. The fastest motions thus are embedded in a wider envelope of slow flow. The total V_{LSR} range of the redshifted wing extends over ≈ 17 km s^{-1}.

4. For any velocity, the signal strength increases with distance from FIR4, culminating 60″ away before a steep decline sets in. The decline zone is unresolved by our 12.5″ beam. At $\approx 65''$, the first stage of the outflow must thus come to an *abrupt halt*. Further on, the flow continues to be seen at lower levels, eventually building up to a second peak near 110″. As before, buildup of signal strength proceeds sequentially from low velocities to high. This results in a linear increase of flow speeds with distance, hence an *acceleration approaching exponential behaviour* in time. The characteristic time scale is 17000 years, to be reduced by projection factors which may be considerable.

5. Where measurable, the flow's brightness temperature is invariably at a constant value near 45 K, similar to that of the ambient gas (40...70 K). No significant heating or cooling seems connected with the flow, *not even at the position of abrupt decline (the "obstacle").*

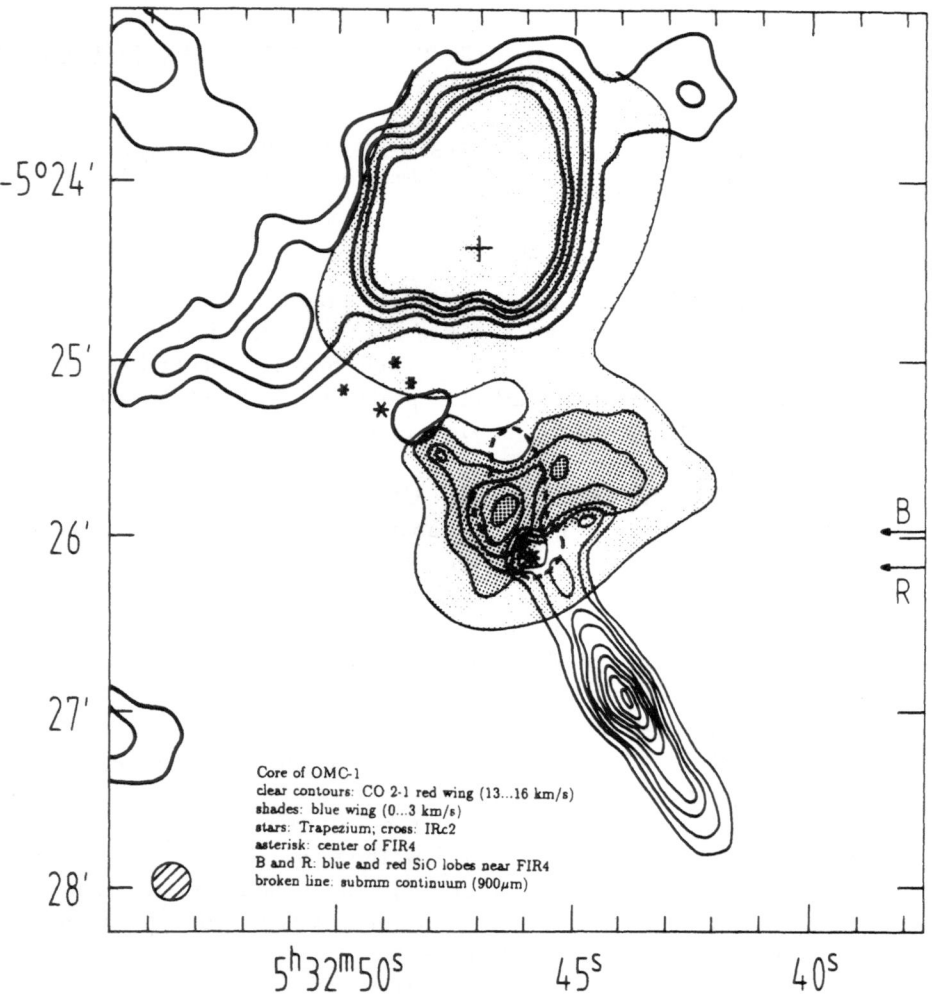

Fig.1. The newly discovered outflow and its surroundings in OMC-1. The four stars are the Trapezium, the cross is IRc2, and FIR4, the suspected origin of the bipolar flow system, is indicated by the asterisk some 80″ South-West of the Trapezium stars.

6. The flow is optically thick in CO 2-1 everywhere except at the very tip. The densities must thus be at least a few 10^5 cm^{-3}, and the total mass loss rate could reach several $10^{-4}M_\odot$ per year. At $\approx 75''$ from FIR4, the ^{12}CO/^{13}CO 2-1 intensity ratio is 24, indicating $n_{H_2} \approx 3 \cdot 10^5$ cm^{-3}. *Densities probably exceed the ambient gas values.*

7. A "halo" of what nearly seems like *countercurrents*, of blueshift a few km s^{-1} relative to the ambient molecular gas, can be seen to a distance of $10''$ from the axis of the redhifted jet. This might be ambient gas somewhat compressed (but not heated) by the action of the flow, or turbulent matter transported from the vicinity of FIR4, or else rotation of the jet material around its length axis.

8. Near FIR4 itself, a structure perpendicular to the outflow axis is clearly seen in the blue wing of CO 2-1 at $V_{LSR} \approx 0$ to 5 km s^{-1} to the West and, much less clearly, in the red to the East of the lobes. This invites speculations on a *dynamical structure* from which the flow might emanate.

References

Mezger P. G., Wink J. E., Zylka R. 1990 *Astron. Astrophys.* **228**, 95.

Mundy L. G., Scoville N. Z., Baath L. B., Masson C. R., Woody D. P. 1986 *Astrophys. J. Lett.* **304**, L51.

Schmid-Burgk J., Güsten R., Mauersberger R., Schulz A., Wilson T. L. 1990 to appear in *Astrophys. J. Letters.*

Ziurys L. M., Wilson T. L., Mauersberger R. 1990 *Astrophys. J. Letters* **356**, L25.

Density and kinematics of the W49A cloud core

E. Serabyn
Division of Physics, Mathematics and Astronomy
California Institute of Technology, 320-47
Pasadena, CA 91125

R. Güsten
Max-Planck-Institut für Radioastronomie
Auf dem Hügel 69, 5300 Bonn 1, FRG

The dense core of the W49A molecular cloud (Miyawaki *et al.* 1986, Schloerb *et al.* 1987) has been mapped in 5 different transitions of CS and $C^{34}S$, in order to determine its density structure. The three lower frequency transitions (CS J=3-2, CS J=5-4 and $C^{34}S$ J=5-4) were observed with the IRAM 30m telescope, and the two highest frequency transitions (CS J=7-6 and $C^{34}S$ J=7-6) with the Caltech Submillimeter Observatory. The beamsizes were in the range $12''$ to $20''$. As a calibration check, the CS J=7-6 line was observed with both telescopes, and was found to give a consistent temperature scale. The spectra at the peak of the emission are shown in Fig. 1.

A map of the dense core in the optically thin $C^{34}S$ J=5-4 line is shown in Fig. 2. The cloud core is seen to extend in a NE-SW direction, and probably consists of two main sub-components displaced from each other in this direction. The brightest emission is well centered in the ring of compact HII regions seen in the radio continuum (Welch *et al.* 1987), and shows a large velocity gradient across its body. A position-velocity plot indicates that the southwestern cloud is likely rotating. Surprisingly, the direction of rotation is *opposite* to that inferred from the compact HII regions. This may be due to optical depth effects in the lower frequency radio recombination lines (Welch, priv. comm).

Statistical equilibrium calculations under the large velocity gradient asumption yield a density of about 7×10^6 cm^{-3} at the center of the cloud, much higher than H_2CO measurements indicate (Dickel and Goss 1990). Assuming a CS abundance of 2.5×10^{-9}, the inferred molecular hydrogen column density is 1.4×10^{24} cm^{-2}. The area and volume filling factors of this dense gas are ~ 0.4 and 0.1, respectively, suggesting that only a few unresolved dense knots are present in our beam. We speculate that these knots would most likely be associated with the compact HII regions seen by Welch *et al.* (1987). The molecular mass in the region is estimated in two ways: both the molecular excitation model and the virial theorem yield a mass of about 10^4 $M_\odot$.

References:
Dickel, H.R. and Goss, W.M.: 1990, *Astrophys. J.*, **351**, 189.
Miyawaki, R., Hayashi, M. and Hasegawa, T.: 1986, *Astrophys. J.*, **305**, 353.
Schloerb, F.P., Snell, R.L. and Schwartz, P.R.: 1987, *Astrophys. J.*, **319**, 427.
Welch, W.J., Dreher, J.W., Jackson, J.M., Terebey, S. and Vogel, S.N.: 1987, *Science*, **238**, 1550.

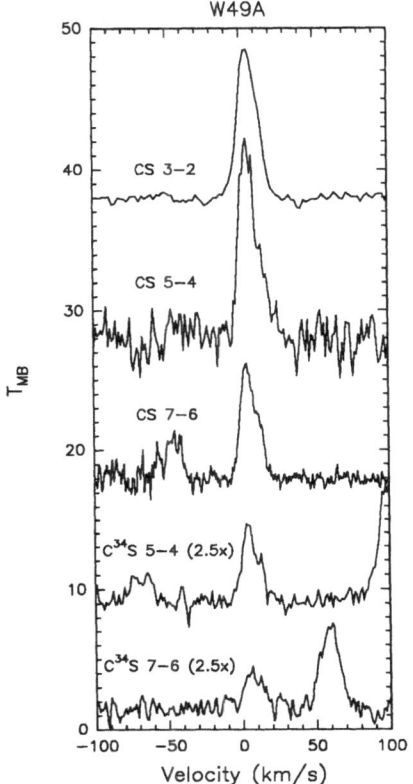

Fig. 1. Spectra of the 5 CS and $C^{34}S$ transitions observed toward $(\alpha, \delta) =$ (19:07:50.0, 09:01:20). The lines all appear at an LSR velocity of 4 km s^{-1}.

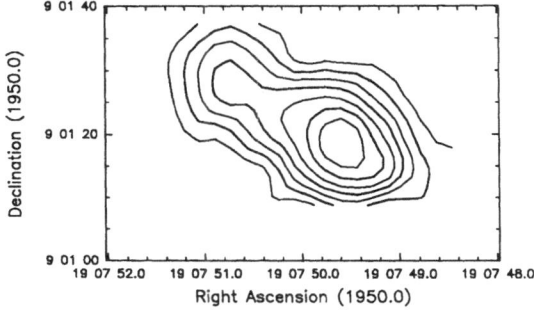

Fig. 2. Integrated intensity contour map of the $C^{34}S$ J=5-4 emission.

Synthetic Linemaps for Hierarchial Clouds

L. G. STENHOLM
Astronomical Observatory
Uppsala University
Box 515
S 751 20 Uppsala
Sweden
and
Swedish Defence Research Establishment
S 172 90 Sundbyberg
Sweden

Molecular clouds are characterized by large density contrasts and pronounced internal motions. Some information on their internal physics has been acquired through relatively simple data analysis and through theoretical modelling, mainly hydrodynamical calculations. More detailed radiative transfer models have been limited to spherical cloud models.

A related but less recognized problem is the classification of molecular clouds, or otherwise stated, can we compress the information in the observed two-dimensional maps for various line parameters in such a way that the physical differences are retained, while the random component has been removed? This is a more fundamental formulation of the special problem of comparing the appearance of two molecular clouds, or of the problems of quantifying the velocity correlation within the clouds.

This paper addresses the question of radiative transfer in fully three dimensional cloud models. the problem is solved under the limiting condition of LTE. We have here concentrated our effort on the ^{13}CO J=1-0 line which is formed in LTE in the clouds.

The basic idea is to divide the cloud region in small cubic cells with constant physical conditions. The smallest possible scale-length of a fluctuation which can be represented by our model is thus the side length of such a cell. The largest possible scale-length is naturally the size of the cloud region. There are no limitations on the variations of the physical conditions between two neighbouring cell.

The cloud is organised in a number of generations of fragments, with identical physical conditions for all fragments of the same order. The velocity field is divided in two parts, the microscopic velocity field and the macroscopic velocity field. A number of different velocity fields and their combinations have been tried;

1. Each fragment moves through space with the velocity specified by eq. 2.
2. The fragments rotates.
3. Radial expansion or contraction is assumed.
4. The velocity field is formed by a spectrum of Alfven waves.

The line parameters is defined in the standard way but note that no separation in line components is done. Each model produces of the order of 10000 spectra. The number of independent units of information in the model spectra is thus around 25 (spectral range covered divided by the width of the thermal profile). The total information content per model is thus typically 250000 information units.

The data analysis therefore rests on two postulates:

1. The total information pool contains true information on the physical processes, as well as "noise" due to physical causes which are well represented by the chaos concept.

2. It is possible to reduce the data, without prior information on the physics, in such a way that the essence of true information is retained in a few parameters, and to eliminate the noise in the process.

The first reduction step is to compress the frequency information by the introduction of the standard line parameters. The line parameters used (T_B, V and Dv) can be regarded as the lowest order moments of the intensity distribution in a spectra. The integrated intensity is a measure of the total number of photons (energy) in a spectrum.

The maximum amount of information retained after this step is 40000 units.

We can approach the spatial compression by using an analogy with the frequency case. The spatial information in a line parameter map is strongly influenced by chance, relative distances of fragments will depend for ex. on line of sight and starting conditions for the evolution. The spatial correlation information is thus ignored and one retains only the statistical distribution (i.e. number of elements per line parameter interval). The statistical moments for the distribution of each of line parameters can now be determined.

We have now reduced the information content to 12 parameters for each model (three moments for each line parameter). The cluster analysis is now used in the final step of the analysis to determine the multidimensional distance between any two models and to group similar models together.

If the method is successful each cluster should contain clouds which are in some way physically similar.

The separation according to important physical parameters is found to be very good, though not totally efficient.The calculations presented here indicates that the method is most sensitive to changes in the velocity field and least sensitive to the details of how the fragments are defined. These results can probably be explained by noting that changes in the velocity field influences 3 of the four line parameters (T, V and Dv) in contrast to density and abundance structures which mainly influences the peak intensity and the integrated intensity. The specifications of fragment distribution finally influences all line parameters but only indirectly through the velocity and density variations connected with the various fragment orders. The method presented is thus obviously capable of compressing substantial physical information on the cloud into a few characteristics and of eliminating most of the random component, but its sensitivity probably varies for the various physical parameters determining the cloud physics.

The cluster analysis show that *models 47, 48 and 52 are closest to the observed B5 cloud (e.g. the three models and B5 belong to the same cluster). All three model clouds are basically wave cloud models but models 48 and 52 have a small translational component in the velocity field.* The multidimensional distance to B5 is larger for model 47 than for the other two.

Models 48 and 52 are characterized by a relatively small density gradient and by a whole spectrum of interfering "chaotic" waves, the amplitudes of the waves are independent of frequency (for simplicity) and the velocity does not depend on density. It was impossible to find a way to modify the hierarchical models so that they could be brought in closer agreement with the B5 characteristics. The major stumbling block is the line width. All hierarchical models tried have two problems, large regions with small Δv and very chaotic line profiles in contradiction to the observed line profiles. *It may be possible to bring the hierarchical models in reasonable agreement with B5 if one postulates that all mass is contained in the highest order fragments. An example is model 6, which has a multi-dimesional distance to the B5 cloud which is comparable to the distances of the most successful models.*

The models computed cover a relatively small part of the possible range of parameters and this work is in that sense rather exploratory, even though the modelling were quite extensive. Any conclusions on the physics of the B5 cloud can thus be premature, but the conclusions given above seems to be rather independent of the model details.

BRIGHT-RIMMED CLOUDS WITH IRAS POINT SOURCES: CANDIDATES FOR STAR FORMATION BY RADIATION-DRIVEN IMPLOSION

Koji Sugitani[1,2], Yasuo Fukui[3], And Katsuo Ogura[4]
[1]I. Physikalisches Institut, Universität zu Köln
[2]College of General Education, Nagoya City University
[3]Department of Astrophysics, Nagoya University
[4]Kokugakuin University

We present preliminary results of a survey of bright-rimmed clouds associated with IRAS point sources.

Bright-rimmed globules associated with old HII regions have long been suspected as a potential site for star formation. Physical conditions of such clouds seem to well match to models of radiation-driven implosion, which have been studied as an effective process for induced star formation (e.g. Klein et al. 1985, Bertoldi 1989). Bright-rimmed globules associated with IRAS point sources are good candidates for the sites of induced star formation. Three well-established cases of radiation-driven implosions in bright-rimmed globules (Ori I-2, IC1396-n, and L1206) were reported (Sugitani et al. 1989). Similar examples were also reported in HH46/47 (Olberg et al. 1989) and in GN21.38.9 (Duvert et al. 1990). However, the samples are still not numerous enough to establish comprehensive understanding of star formation by such implosion process.

We have surveyed for bright-rimmed clouds (cometary globules, and small clouds with curved bright rims) associated with IRAS point sources. The Palomar Sky Survey prints and IRAS Point Source Catalog were used for this survey. The surveyed regions are mainly around HII regions of Sharpless (1959) with texts of ~60' or larger. By using overlay maps showing IRAS point sources for the PSS prints, we selected the clouds having IRAS sources just surrounded by the curved rims. To exclude emission from diffuse dust, we included only the IRAS sources which are detected at 25 μm and, at least, at one more band and have correlation coefficients of F or better at 25 μm.

44 bright-rimmed clouds associated with IRAS sources have been selected from 18 HII regions. The most prominent region is S131(IC1396) having eleven bright-rimmed clouds. The second regions are S190 (IC1805) and S199 (IC1848), both of which have 5 clouds. Nine of the 44 clouds are associated with molecular outflows and one with a HH object, which are good signposts of young stellar objects. Most of their radii, R, determined by optical measurement are $\lesssim 0.5$ pc, similar to those of Bok globules (Fig. 1). We estimated their masses by assuming a density of 3×10^4 cm^{-3}. Most of the clouds have masses of ~1 - 100 $M_\odot$, similar to Bok globules. We calculated IR luminosities of the IRAS sources from the IRAS 25 μm fluxes (F.g. 2). The luminosities of the associated IRAS sources are relatively large, mostly 10-10^3 $L_\odot$, compared to those of the IRAS sources associated with dark globules or dense cores of dark cloud complexes (Beichman et al. 1986). This suggests that intermediate-mass stars are mainly formed in bright-rimmed clouds. On the other hand, the range of the luminosities is wide, ~4 - $10^5 L_\odot$, suggesting that relatively low-mass to high mass stars are formed in these bright-

rimmed clouds under the influence of ionization-shock fronts. This fact appears to be consistent with the prediction of radiation-driven implosion models for star formation (e.g. Bedijin and Tenorio-Tagle 1984, and Klein et al. 1985). We estimated IRAS luminosity to cloud mass ratios following Sugitani et al. (1989). These ratios are significantly greater than that in dark globules or in dense cores of dark cloud complexes, suggesting more effective star formation in the bright-rimmed clouds.

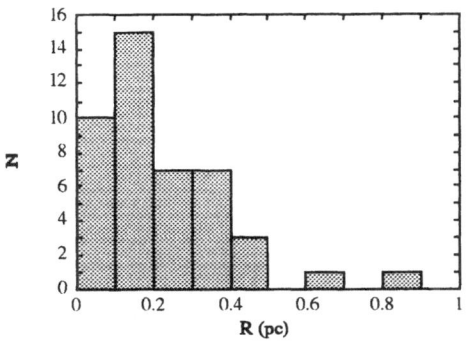

Fig. 1 --- Distribution of the cloud radii.

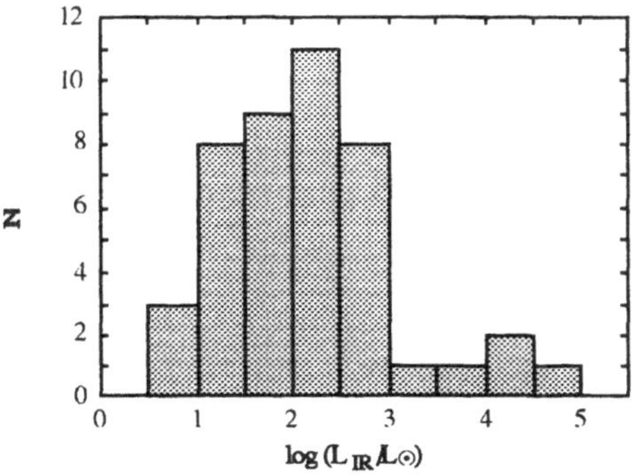

Fig. 2 --- Distribution of the luminosities of the IRAS point sources associated with the bright- rimmed clouds.

A detailed report of this survey will be presented elsewhere with a list and photographs of the bright-rimmed clouds.

REFERENCES

Bedijin, P. J., and Tenorio-Tagle, G. 1984, Astr. Ap., **135**, 81.

Beicheman, C. A., Myers, P. C., Emerson, J. P., Haarris, S., Mathieu, R., Benson, P. J., Tennings, R. E. 1986, Ap. J., **307**, 337.

Bertoldi, F. 1989, Ap. J., **346**, 735.

Duvert, G., Gernicharo, J., Bachiller, R., and Gómez-González, J., Astr. Ap., **49**, 57.

Klein, R. I., Sandford, M. T., and Whitaker, R. W. 1985, in Protostars and Planets II, ed. D. C. Black and M. S. Mathews (Tucson: University of Arizona Press), p. 340.

Olberg, M., Reipurth, B., and Booth, R. S. 1989, in The Physics and Chemistry of Interstellar Molecular Clouds, ed. G. Winnewisser and J. T. Armstrong (Heidelberg: Springer-Verlag).

Sharpless, S. 1959, Ap. J. Suppl., **4**, 257.

Sugitani, K., Fukui, Y., Mizuno, A, and Ohashi, N. 1989, Ap. J. (Letters), **342**, L87.

Density structure of dense cores in the Cepheus molecular cloud

K. Sunada[1], T. Hasegawa[2], M. Hayashi[1], Y. Fukui[3], and K. Sugitani[4]

1. *Department of Astronomy, University of Tokyo*
2. *Institute of Astronomy, University of Tokyo*
3. *Department of Astrophysics, Nagoya University*
4. *College of General Education, Nagoya City University*

SUMMARY   We mapped 5 dense cores in the Cepheus molecular cloud in the optically thin $C^{18}O$ $(J = 1 - 0)$ line using the 45-m telescope at Nobeyama and derived density profiles around those cores assuming spherical symmetry. Cloud cores are selected from the ^{13}CO $(J = 1 - 0)$ map obtained through the unbiased survey program with the 4-m telescope at Nagoya university.

The results are summarized as below:

1. The cores with protostellar *IRAS* sources have the density profile consistent with the r^{-2} distribution, being inferred from the quasi-static isothermal case, while those without *IRAS* sources tend to have less steep density gradient. (Fig. 1, Table 1)

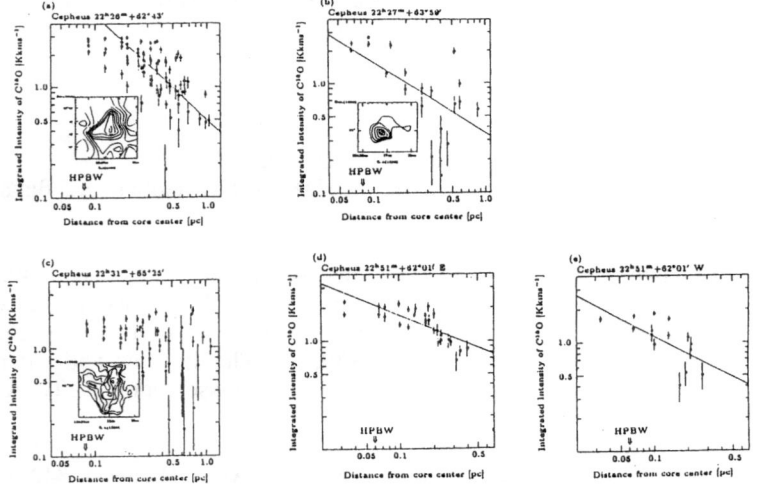

Fig. 1 — The $C^{18}O$ intensity, i.e. the gas column density, is plotted as a function of radius measured from the centers of the observed cores with associated *IRAS* sources (a and b) and without *IRAS* sources (c, d, and e). The $C^{18}O$ intensity for the cores with embedded *IRAS* sources (a and b) significantly decreases with the increasing radius from the core centers, while for the cores without *IRAS* sources (c, d, and e) the $C^{18}O$ intensity remains relatively large even at large distances away from the centers. The straight lines show the result of least square fitting for the $C^{18}O$ intensity. The map obtained with 4-m telescope at Nagoya university is inserted (a, b, and c).

2. All the observed cores are stable in gravitational equilibrium. (Table 1)
3. The cores with steeper density gradient have larger average column density toward their centers. (Fig. 2)

These results suggest that cores with less steep density gradient contract in a quasi-static manner under gravitational equilibrium and evolve into cores with steeper density gradient to form protostars at their centers.

Table 1

[1] Core Name	[2] $r(C^{18}O)$	[3] T_{ex} (K)	[4] α	[5] Size (pc)	[6] $N(H_2)$ ($\times 10^{21} cm^{-2}$)	[7] Velocity gradient ($kms^{-1}pc^{-1}$)	[8] Mass ($M_\odot$)	[9] Virial Mass ($M_\odot$)
with an IRAS point source								
$22^h26^m+62°43'$	0.05	33	-1.0	0.27	16.1	2.6	122	128
$22^h27^m+63°59'^{(1)}$	0.10	13	-0.8	0.20	14.3	2.6	16	74
without an IRAS point source								
$22^h31^m+65°25'$	0.12	15	$0^{(2)}$	0.40	10.5	1.6	83	109
$22^h51^m+62°01'E^{(1)}$	0.12	19	-0.4	0.21	12.5	—	37	92
$22^h51^m+62°01'W^{(1)}$	0.15	16	-0.5	0.15	10.0	3.0	11	56

note) (1) These cores are associated with HII region.

(2) Correlation coefficient is 0.3.

Table 1 — We estimated the power law indices of the column density distribution from the least square fitting (column 4). This result suggests that the density profiles of the cores with IRAS sources are consistent with the r^{-2} distribution under the assumption of spherical symmetry. On the contrary, cores without IRAS sources have less steep density gradient. The mass estimated from the observed column density (column 8) agrees well with the dynamical mass estimated from the observed line width (column 9), suggesting that these cores are in gravitational equilibrium.

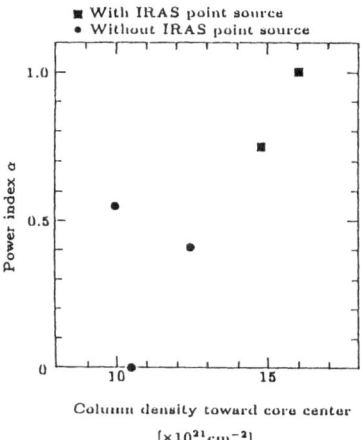

Fig. 2 — This figure shows the power law indices of the column density profiles versus the column density averaged over the radius of 0.15 pc around the core centers. The cores with steeper density gradient have larger average column density toward them.

Turbulence in TMC1-C and ρ-Oph core

K. SUNADA[1], Y. KITAMURA[2], T. HASEGAWA[3], M. HAYASHI[1]

1. Department of Astronomy, University of Tokyo
2. Department of Liberal Arts, School of Allied Medical Sciences, Kagoshima University
3. Institute of Astronomy, University of Tokyo

SUMMARY We made ^{13}CO and $C^{18}O$ (both $J = 1 - 0$) maps of the $4' \times 4'$ (0.2 pc $\times$ 0.2 pc) area toward TMC1-C and ρ-Oph core with the 45-m telescope of the Nobeyama radio Observatory in order to find small scale velocity fluctuation in molecular clouds based on the analysis described by Kleiner and Dickman (1987).

We found followings:

1. The coherent length in TMC1-C is 0.02 pc, which is of order smaller than the value 0.1 pc derived by Kleiner and Dickman. We obtained the smaller coherent length because the large scale systematic velocity gradient corresponding to the size scale of 0.1 pc is subtracted before the analysis. The presence of the new scale means that turbulence is actually hierarchical down to 0.02 pc.

Fig. 1 — These figures show auto-correlation functions (ACF) in TMC1-C for (a) ^{13}CO and (b) $C^{18}O$ measurements. Figure (c) shows the ACF for which large scale systematic velocity gradient corresponding to the size scale of 0.1 pc has not been subtracted. The coherent length, i.e, the e-folding length of ACF, is 0.02 pc for both ^{13}CO and $C^{18}O$, being smaller than the value of 0.1 pc derived by Kleiner and Dickman(1987) because of the subtraction of the large scale systematic velocity gradient shown in (c). Our result means that turbulence has a coherent length as small as 0.02 pc, suggesting that turbulence in molecular clouds is hierarchical down to this small scale length.

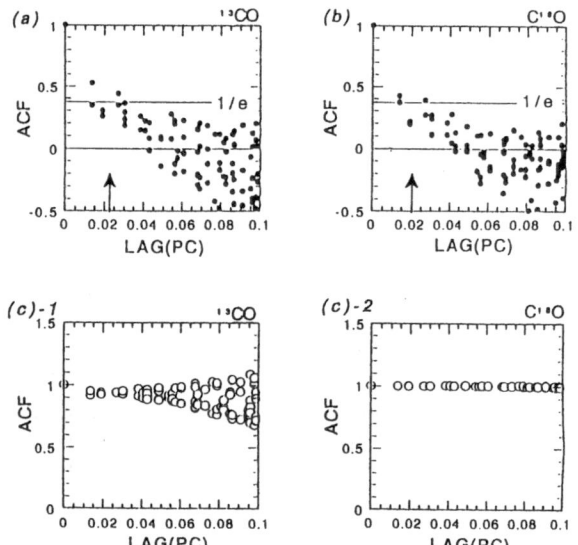

2. The small scale coherent length in TMC1-C (0.02 pc) is significantly smaller than that in ρ-Oph core (0.03 pc measured for $C^{18}O$ and 0.04 pc for ^{13}CO). This may be related to the recent observational result that typical masses of newly formed stars in Taurus are significantly smaller than those in ρ-Oph (Cohen and Kuhi 1979, Rieke et al. 1989), suggesting that turbulence plays an important role in determining stellar masses, or vice versa.

Reference

Cohen, M., and Kuhi, L. V. 1979, *Ap. J. Suppl.*, 41, 743
Kleiner, S. C., and Dickman, R. L. 1984, *Ap. J.*, 286, 255
Kleiner, S. C., and Dickman, R. L. 1985, *Ap. J.*, 295, 466
Kleiner, S. C., and Dickman, R. L. 1985, *Ap. J.*, 295, 479
Kleiner, S. C., and Dickman, R. L. 1987, *Ap. J.*, 312, 837
Rieke, G. H., Ashok, N. M., and Boyle, R. P. 1989, *Ap. J. (Letters)*, 339, L71

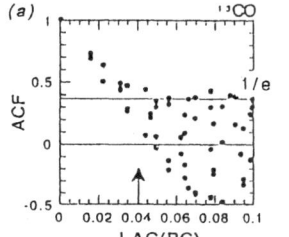 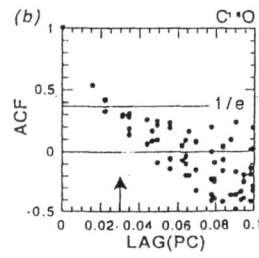

Fig. 2 — ACFs measured in ρ-Oph core are shown for (a) ^{13}CO and (b) $C^{18}O$. The correlation lengths are 0.03 pc for $C^{18}O$ and 0.04 pc for ^{13}CO, respectively. These correlation lengths are significantly larger than the value for TMC1-C (0.02 pc), even when the effect of degradation for coherent length due to noise is taken into account. We conclude that the coherent length of turbulence is different among molecular clouds.

Table 1

	TMC1-C		ρ-Oph core	
Distance(pc)	140		160	
Mapping Area	0.16 pc ×0.16 pc		0.19 pc ×0.19 pc	
$T_{ex}(K)$ [1]	11		53	
	^{13}CO	$C^{18}O$	^{13}CO	$C^{18}O$
r ($C^{18}O$) [1]		0.55		0.15
Coherent length(pc)	0.02	0.02	0.04	0.03
Corrected Factor [2]	1.22	1.20	1.06	1.04
Corrected	0.024 pc	0.024 pc	0.04 pc	0.03 pc
Mass($M_\odot$)		0.3		3.7

note) (1) These values are averaged all over the observed positions. These values at each position are obtained by the assumption of LTE and of the abundance ratio (X(^{13}CO)/X($C^{18}O$)) is 5.5.

(2) This value reveals the degree of the decorrelated due to the ramdom noise. The way of the estimation are followed by Kleiner and Dickman.

Table 1 — This table shows the total mass contained inside the coherent length for TMC1-C and ρ-Oph core. The mass in the coherent length in ρ-Oph core is $\sim$ 10 times larger than that in TMC1-C. This may be related to the recent results that ρ-Oph core tends to produce relatively high mass stars (Rieke et al.1989), while solar-type stars are predominantly formed in TMC (Cohen and Kuhi, 1979), suggesting that turbulence plays an important role in determining stellar masses.

Radiative Transfer of CO through Clumpy Molecular Clouds with External UV Heating

Jan A. Tauber[1] and Paul F. Goldsmith[2]

[1]Radio Astronomy Laboratory, Univ. of California, Berkeley, CA

[2]Five College Radio Astronomy Observatory, Univ. of Massachusetts, Amherst, MA.

We have developed a model which simulates the radiative transfer of molecular line emission through clumpy molecular clouds. The dynamical structure of the model cloud is based on the work of Kwan and Sanders (1986). The model incorporates the existence of an intense source of UV photons at the surface of the cloud. The UV source heats the clumps and creates kinetic temperature and CO abundance gradients within them. The amount of heating depends on the intensity of the UV field, which decreases from the surface to the core of the cloud due to attenuation by dust. We treat in detail the photochemistry and self-shielding properties of CO as a function of UV intensity and gas density in order to obtain the CO line intensities emerging from each clump. The line intensity emerging from the cloud is obtained by integrating the emission from all clumps along the line of sight, weighted by an area covering factor, and attenuated by the opacity of intervening clumps. The effects of the heating are significantly noticeable on the line intensities of CO transitions arising from levels with J between $\sim$ 3 and $\sim$ 7. We apply our model to the case of the Orion A molecular cloud, and in particular to observations of the J=3$\rightarrow$2 ^{12}CO and ^{13}CO lines. The model is in general agreement with the observed enhanced intensity of the ^{12}CO J=3$\rightarrow$2 transition relative to the J=1$\rightarrow$0 transition throughout the central $\sim$ 10' region of Orion. It also produces centrally peaked spectral lines whose intensity is maximum in a shell-like distribution centered on the Trapezium HII region, as is observed.

References

Kwan, J., and Sanders, D.B. 1986, *Ap.J.*, **309**, 783

CHARACTERISATION OF SPATIAL STRUCTURE
IN MOLECULAR CLOUDS

SHOBA VEERARAGHAVAN

Harvard-Smithsonian Center for Astrophysics,
60 Garden Street, Cambridge, MA 02138, U.S.A
E-mail SHOBA@CFA.HARVARD.EDU

GARY A. FULLER

Harvard-Smithsonian Center for Astrophysics,
60 Garden Street, Cambridge, MA 02138, U.S.A
E-mail FULLER@CFA.HARVARD.EDU

Abstract. Two topological tools for studying the global structure of molecular clouds, the genus and the contour-crossing statistic, are discussed. Preliminary results for the Taurus molecular cloud complex are presented.

1. Introduction

Molecular clouds display structure on all observed length scales. If the complexities of this structure can be characterised by a finite number of statistical and/or topological measures, these measures could be used in two ways. Firstly, they might be correlated with cloud properties such as the star-forming efficiency or the local stellar density. Secondly, they could be used to determine which physically motivated models of cloud structure best match with the observations. Realisations of such a model can then be used to investigate other properties of real clouds which are not directly observable, such as the extent to which the interstellar radiation field penetrates into the inner regions of a molecular cloud, or the extent to which the density structure determines the efficiency of energy transport within the molecular cloud.

Several techniques have been investigated to quantify the properties of observed distributions of objects in astrophysics. The autocorrelation function (Peebles 1980) has proved useful in quantifying the large-scale distribution of galaxies, and has been applied to the structure of molecular clouds (Houlahan and Scalo 1990). Fractal analysis has also been used (e.g. Dickman *et al.* 1990). We are currently investigating two topological measures of structure which have been developed to study the galaxy distribution. These are the genus (Gott, Melott and Dickinson 1986) and the contour crossing statistic (Ryden 1988, Ryden *et al.* 1989).

2. Topological Characterisation

We present preliminary results for the topology of the Taurus molecular cloud as mapped in the ^{12}CO J $= 1-0$ transition (Ungerechts and Thaddeus 1985). Here, we use two statistics to describe surfaces (1-dimensional, hence contours) of constant integrated emission. If the emission is optically thin, such surfaces trace surfaces of constant column density.

2. 1. GENUS

The genus, $\mathcal{G}$, of a surface is an invariant measure of its connectedness. Formally, it equals the number of topologically distinct closed curves which can be drawn on the surface without cutting it into two unconnected pieces :

$$\mathcal{G} = 1 + \text{ Number of Holes } - \text{ Number of Disconnected Segments}$$

Thus $\mathcal{G}$ of a surface of constant emission can be used to characterise the connectedness of the surface in terms of either the value on the surface or alternatively, the average emission interior to the surface. A large and negative value of $\mathcal{G}$ implies that the emission is distributed in many isolated clumps. On the other hand, a large positive value of $\mathcal{G}$ suggests that the emission occurs in a medium which permeates the cloud.

For the Taurus data, the run of genus with integrated emission is negative (figure 1), suggesting that the structure consists of isolated clumps at most levels of emission : most of the emission arises from small disconnected regions.

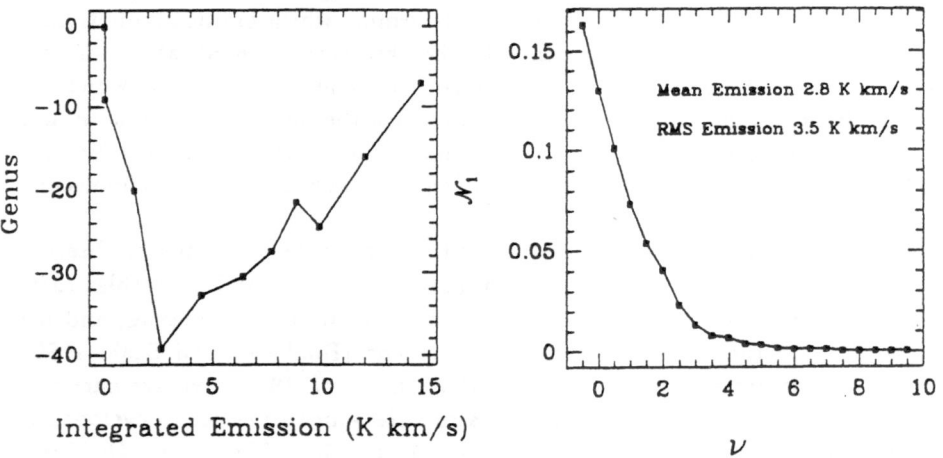

Fig. 1.   The genus (left) and contour crossing statistic, $\mathcal{N}_1$ (right) for the integrated $J = 1 - 0$ ^{12}CO emission from the Taurus molecular cloud.

2. 2. CONTOUR-CROSSING STATISTIC

The genus statistic cannot distinguish between topologically identical but common-sensically distinct objects (such as a banana and an orange). However, physical processes such as the energy or mass flux across a surface enclosing some given volume certainly depend on the shape of the surface as well as its topology. The one dimensional contour-crossing statistic, $\mathcal{N}_1$, is complementary to the genus statistic in this respect, and is defined as the mean number of times a randomly directed line crosses the surface per unit length of the line. In higher dimensions $\mathcal{N}_2$ and $\mathcal{N}_3$ are respectively the contour-length statistic which is the mean length of the intersection of the surface with a plane, and the contour-area statistic which is the mean surface area per unit volume. If the distribution of emission is known, the three statistics are related.

We calculated $\mathcal{N}_1$ for isoemission surfaces by counting the mean number of crossings of the surfaces per pixel within the emitting region (figure 1). If the emission was distributed as a Gaussian random field, we would expect $\ln(\mathcal{N}_1) \propto -\nu^2$, where ν is the deviation of the emission from the mean in units of the root-mean-square emission. Instead we find that $\ln(\mathcal{N}_1) \propto -\nu$. The emission is probably not well modelled by a Gaussian distribution, which is consistent with the results from the genus.

References

Dickman, R. L., Horvath, M. A. and Margulis, M. *et al.* 1990, *preprint*

Gott, J. R., Melott, A. L., and Dickinson, M., 1986, *Astrophy. J.*, **306**, 341.

Houlahan, P., and Scalo, J., 1990, *Astrophy. J. Suppl.*, **72**, 133.

Peebles, J. 1980, *The Large-Scale Structure of the Universe*, Princeton : Princeton University Press

Ryden, B., 1988, *Astrophy. J.*, **333**, L41.

Ryden, B., *et al.* 1989, *Astrophy. J. Suppl.*, **340**, 647.

Ungerechts, H. and Thaddeus, P. 1985, *Astrophys. J. Suppl.*, **63**, 645.

MEASURING THE FRACTAL STRUCTURE OF INTERSTELLAR CLOUDS

M.G.R. VOGELAAR
Kapteyn Laboratory
Rijks Universiteit Groningen
Postbus 800
9700 AV Groningen
The Netherlands

B.P. WAKKER
Department of Astronomy
University of Illinois
1002 W Green St
Urbana IL61801
USA

U.J. SCHWARZ
Kapteyn Laboratory
Rijks Universiteit Groningen
Postbus 800
9700 AV Groningen
The Netherlands

ABSTRACT. To study the structure of interstellar clouds we used the so-called *perimeter-area* relation to estimate fractal dimensions. We studied the reliability of the method by applying it to artificial fractals and discuss some of the problems and pitfalls. Results for two different cloud types (high-velocity clouds (HVCs) and infrared cirrus) are summarized. We find dimensions $1.2 < D < 1.55$, somewhat higher than found in previous, similar studies.

1. The Perimeter-Area Relation

Mandelbrot's relation between area and perimeter of a closed fractal curve, $P(l) = \rho A(l)^{D/2}$, provides an algorithm to estimate the fractal dimension of interstellar clouds. The projection on the sky of the emission gives a brightness distribution and the brightness contours form the cloud perimeters. Within these clouds we define "objects" as sets of connected pixels with brightness above a given level. If clouds are real fractals we expect $1 < D < 2$. The factor ρ is related to the shape of the cloud (Feder 1988) and is called the prefactor.

2. Data

Five HVC fields were measured (21-cm H I WSRT data, see e.g. Wakker & Schwarz 1990). The original data are channel maps at $1'$ resolution. For the measurements all channels were used together, as if they formed one big mosaiced map. We also used $100\mu m$ data from Laureijs (1989): the relatively isolated dark clouds L134, G240−66 and G102+70 and a field named "Ring" which may contain a supernova remnant. Further we used the $100\mu m$ maps of HVCs A III /A IV and M I (Wakker & Boulanger 1985) and an unpublished, $60° \times 60°$ map of the anticentre. The resulting dimensions and prefactors are summarized in Fig. 1.

3. Problems and pitfalls

There are some problems associated with using the *perimeter-area* relation.

–As all available data are on grids, perimeters can only be measured approximately. The value depends on the way corners are treated. Normally, this only influences the value of the prefactor. Furthermore, the gridding implies that a minimum number of pixels is needed to define an object. We found 25 pixels to be a good number.

–A severe problem is the presence of noise. At low levels and at levels close to the map maximum, the noise distorts the contours considerably. So, we excluded contours below 3 sigma and closer than 4 sigma to the peak. A preliminary test on the influence of the signal-to-noise on the derived dimension gave different results for an artificial fractal and a real observation. For an artificial fractal constructed using fractional Brownian motion (fBm) the estimate was much too high for signal-to-noise ratios below 50. Adding noise to a real observation gave a constant estimate for ratios above 15.

–For the HVCs we assumed that the dimension is independent of the velocity interval. To test this, we measured D for different intervals, containing a comparable number of

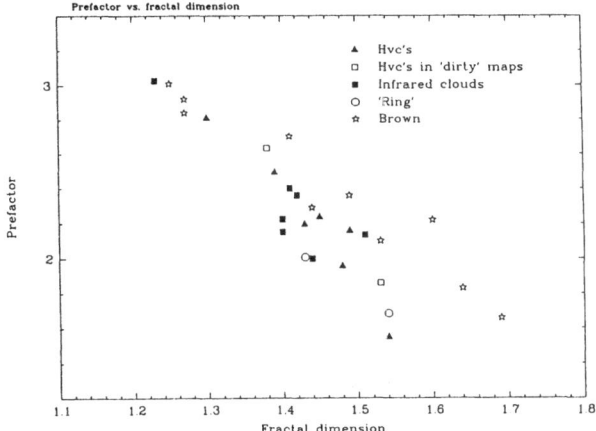

Figure 1. Estimated prefactor ρ versus estimated fractal dimension D for all fields.

objects. Probably because of low signal-to-noise in edge channels, the estimates then seem to depend on the chosen interval. This effect makes estimating dimensions for HVCs more difficult.

–A complication is posed by several projection effects. Firstly, we observe the projection of a three-dimensional cloud on a plane, not a slice through the cloud. It is not mathematically necessary that these two dimensions are related. From a physical point of view the interesting dimension is that of the slice. Some evidence from laboratory studies of turbulence suggests that the dimensions may indeed be different (Méneveau 1989). A second effect is that in one particular field there may be many overlapping clouds at different distances, each of which may have different intrinsic structure. A way to minimize this effect is to use velocity information to separate clouds. Finally, the cloud structure may be multi-fractal, i.e. consisting of intertwining structures with different fractal dimensions. A possible indication of the presence of multi-fractal structure can be found in the plots of area vs perimeter. A large scatter means there is not an easily definable single dimension.

4. Prefactor

All our maps had different gridspacings, so a correction must be applied to compare prefactors. The resulting values should be approached with care. Also, the value depends on the approximation used to construct the perimeter from horizontal, vertical or diagonal (at corners) segments (in the original derivation it was assumed that the ruler could have any orientation).

In principle ρ can have any value for a given D. However, in our measurements, the corrected prefactor and the dimension appear related (Fig. 1). That the same relation is obtained for artificial clouds suggests it is an artifact of the method used.

5. Conclusions

The real reason for applying fractal geometry to the ISM is to understand the physical processes underlying its structure. Clouds subject to the same influences should have the same dimension. The connection with physics is not yet clearly understood, however. Some models of turbulence predict fractal dimensions in specific cases. Méneveau (1989) predicts $D = \frac{4}{3}$ from a model for turbulent flow. Hentschel & Procaccia (1984) calculated from their theory of "relative turbulent diffusion" that $1.37 < D < 1.41$. The dimensions we found are usually higher than, but close to, these theoretical values; dimensions for infrared cirrus are a little closer to the predictions than those for HVCs, possibly because of the lower signal-to-noise ratio.

References

Falgarone E., 1989, *Structure and Dynamics of the Interstellar Medium*, eds. G. Tenorio-Tagle, M. Moles, J. Melnick, IAU Colloquium **120**, Springer Verlag, Berlin, p68

Feder J., 1988, *Fractals,* Plenum Press, New York/London

Hentschel H.G.E., Procaccia I. 1984, *Phys. Rev. A.,* **29**, 1461

Laureijs R., 1989, Ph. D. Thesis, University of Groningen

Méneveau, C., 1989, Ph. D. Thesis, Yale University

Wakker B.P., Boulanger F., 1986, *Astron. Astrophys.* **170**, 84

Wakker B.P., Schwarz U.J., 1990, submitted to A&A

THE W75 – CYGNUS - X IRAS LOOP:- OB - 'BUBBLE' OR SNR ?

D.Ward-Thompson & E.I.Robson

School of Physics & Astronomy, Lancashire Polytechnic, Preston, UK

Abstract. IRAS Calibrated Raw Detector Data (CRDD) are presented of a part of the Cygnus-X region, incorporating W75, DR21 and W75N, and a previously unknown loop of dust emission is observed. This loop is interpreted as a spherical shell-like shock front, and two alternative explanations for its origin are explored – a wind-blown bubble around an OB association, and an old supernova remnant. The arguments for each are outlined, and it is deduced that there are insufficient OB stars old enough to have formed the loop by combined stellar wind action, although a SNR appears consistent with the data.

IRAS Data

The W75 region lies on the north-eastern edge of the Cygnus-X complex, at an approximate distance of 2kpc. It contains the molecular clouds W75 and W75N, and the bipolar outflow source DR21. IRAS CRDD of a 2°-field around W75 were obtained from the UK Rutherford Appleton Laboratory, and a new method of destriping the data was used (Ward-Thompson *et al.* 1989 – Paper I; Ward-Thompson & Robson 1990a – Paper II). Figure 1 a & b show isophotal contour maps of the region at 12 and 100μm. The previously known infrared sources are marked. The spatial correlation between the point sources and extended emission suggests embedded stars within extensive dust clouds.

A ridge of emission stretches from the south-east of the maps and ends in a loop in the centre of the field. The brightest part of the loop is a knot at its western edge, centred roughly at 20h33m +42°15′ and extending some 20 arcmin to both north and south, hereafter called K1. The loop does not correspond spatially with any previously known feature, although the centre of the loop corresponds to a minimum of CO emission (Dame & Thaddeus 1985). All of these observations can be interpreted in terms of a shell of gas and dust, whose edge can be seen by limb-brightening. We therefore suggest that the loop seen in the IRAS raw data in Figure 1 is an expanding shell-like shock front, which is sweeping up material in the arc of K1 (Ward-Thompson & Robson 1990b – Paper III). The mass of dust in a region 35pc across was calculated from the 100-μm flux to be $450M_{\odot}$. If the canonical gas-to-dust mass ratio of 100 is used, the mass of the cloud is $4.5\times10^{4}M_{\odot}$.

(i) OB Association

One explanation for the shell is a wind-blown bubble around an OB association. The only possible location for this association is within W75N, as that is the only place near to the loop centre with

512

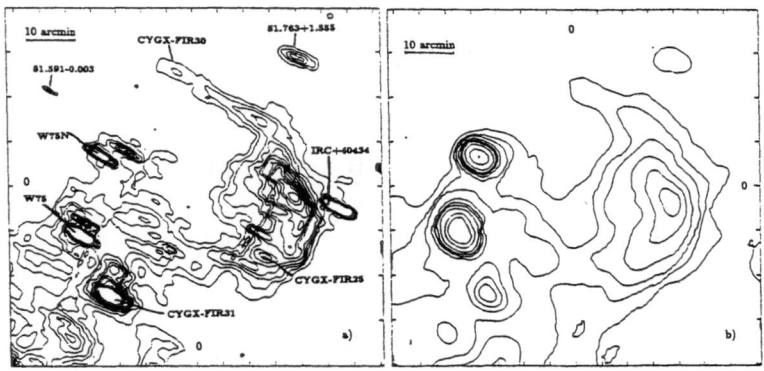

Figure 1: W75-CygX at (a)12μm & (b)100μm, (0,0) is 20h35m+42°20'.

a high enough A_V to mask such an association. The radius of a wind-blown OB shell is given by $R = 27(L_w t^3/n_o)^{1/5}$, where: R is the radius of the loop in pc = 35pc; L_w is the combined stellar wind luminosity (in units of 10^{36}erg/s); t is the age of the loop (in units of 10^6year); and n_o is the mean ambient cloud density = 50cm^{-3}.

This gives the total combined stellar wind luminosity, L_w as 2×10^{35}erg/s, and the age of the association, t as 10^7years. This stellar wind corresponds to 2 stars of type B0 or earlier, whose combined stellar winds must act on the surrounding interstellar medium for $\sim 10^7$years. W75N contains 3 stars of type B0.5, mass $\sim 15 M_\odot$ and age $\leq 10^6$years. In fact one of the stars may be $\leq 10^4$years old (Moore *et al.* 1988). After 5×10^6years each one should typically become a supernova. So there is **not enough time** for the bubble to have formed. If the required age of the bubble is taken in the above equation to be only 10^6years, then L_w must be 2×10^{38}erg/s, or the equivalent of 200 O7III stars, which is clearly impossible.

(ii) Supernova remnant

Another possible origin of the shell is an old supernova remnant. If this were the cause of the shell, any optical remnant in the centre of the loop would be extinguished by the large A_V. The age of the SNR can be calculated from the equation: $R = (2Et^2/\rho_o)^{1/5}$, where: R is the radius of the loop = 17pc; E is the energy of the supernova explosion = 10^{51}erg; t is the age of the loop; and ρ_o is the mean pre-explosion density of the cloud = 8×10^{-20}kgm^{-3}.

Thus the age of the SNR, t = 10^5years. This is consistent with the age of W75N(B) calculated by Moore et al. (1988), and provides an explanation for their theory of coeval formation of the embedded stars. So it appears that a supernova remnant is **consistent** with all of the attributes of the dust loop in the W75 region.

References
Dame, T.M. & Thaddeus, P., 1985. Astrophys.J., **297**, 751.
Moore, T.J.T., Mountain, C.M., Yamashita, T. & Selby, M.J., 1988. M.N.R.A.S., **234**, 95.
Ward-Thompson, D., Robson, E.I., Gordon, M.A., Whittet, D.C.B., Duncan, W.D. & Walther, M., 1989 (Paper I). M.N.R.A.S., **241**, 119.
Ward-Thompson, D. & Robson, E.I., 1990a (Paper II). M.N.R.A.S.,**244**, 458.
Ward-Thompson, D. & Robson, E.I., 1990b(Paper III).M.N.R.A.S., In press.

An Extensive Study of Interstellar Matter in the IC 1396 Region using several Molecular Lines and Transitions of CO and Far-Infrared Maps from IRAS

H. Weikard[2,3], K. Sugitani[1,3], G. Duvert[2] and M. Miller[3]

[1] Department of Astrophysics, Nagoya University, Chikusa-ku, Nagoya 464, Japan

[2] Observatoire de Grenoble, B.P. 53 X, 38041 Grenoble CEDEX, France

[3] I. Physikalisches Institut der Universität zu Köln, Zülpicher Straße 77, 5000 Köln, FRG

IC 1396 is an H II region in Cepheus excited by the massive O6.5 star HD 206267. The distance is about 750 pc. The region exhibits a number of bright-rimmed molecular clouds in which outflows have been detected (Sugitani et al. 1989, Duvert et al. 1990).
We present preliminary results of a CO $(1 \to 0)$ survey of the entire region (Fig. 1) and a multi-line study (Fig. 2 as a first step) based on observations using three small millimeter wave telescopes (see Table 1) which are located at Nagoya, on Plateau de Bure (France) and on Gornergrat (Switzerland). The higher transition observations cover only the bright

Telescope	Line and transition	Positions observed	HPBW	Grid	Resolution in km/sec	r.m.s.-noise in K
Nagoya 4 m	^{12}CO $(1 \to 0)$	6050	2.8	2'	0.13	1.0
	^{13}CO $(1 \to 0)$	5000	2.9	2'	0.08 and 0.135	0.25
POM-2 2.5 m	^{12}CO $(2 \to 1)$	400	2.2	2'	0.20	0.45
	^{13}CO $(2 \to 1)$	400	2.3	2'	0.215	0.25
KOSMA 3 m	^{12}CO $(3 \to 2)$	730	1.5	1'	0.145	1.3

Table 1 : Observational parameters

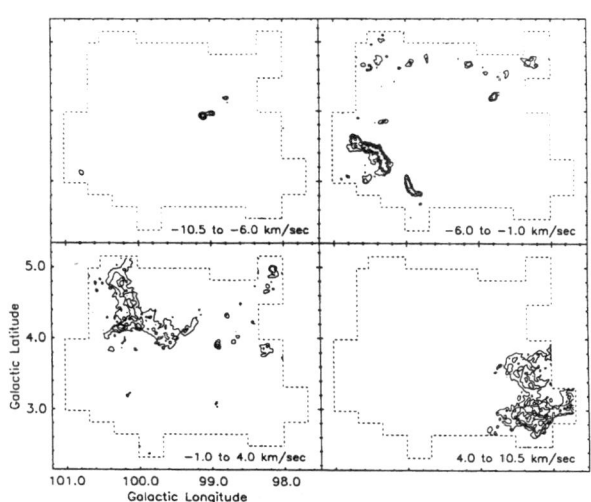

rims A, B and D to H (designations of Pottasch 1956, see Fig. 2 for some spectra taken on rim E).

Fig. 1 : Distribution and kinematic structure of the molecular material as observed in ^{12}CO $(1 \to 0)$ at Nagoya
Contour levels are drawn from 8 to 38 K·km/sec, the step is 5 K·km/sec. The most noticeable features are rim G (−4 to −2.5 km/sec), rim B (−5 km/sec) and the famous rim A (−8 km/sec). Rim E (0.5 km/sec) suffers from strong self-absorption (cf. Fig.2). The cloud which appears in the lower right map (8 km/sec) is probably in the foreground.
Note the absence of molecular emission in a region of more than one square degree around (99.4,3.3) and compare with Fig. 3!

514

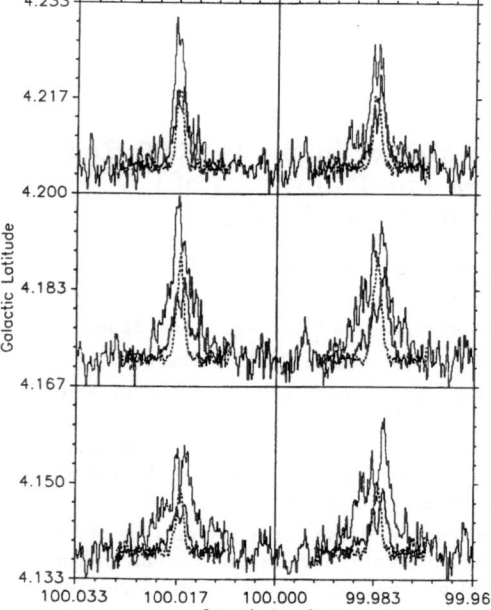

Galactic Latitude

Galactic Longitude

Fig. 2: Spectrum profile map of higher transition lines observed with POM-2 and KOSMA in rim E where an outflow had been detected

The range in LSR velocity is −24 to 24 km/sec, that in antenna temperature −1 to 6 K. The spacing is 2.'0.

In order to improve the signal-to-noise ratio and to make directly comparable the spectra originally taken with rather different beam sizes, 2 x 2 neighbouring $(2 \rightarrow 1)$ spectra and 3 x 3 neighbouring $(3 \rightarrow 2)$ spectra have been summed up, respectively. The effective beam size is then about 4.'3.

The ^{12}CO spectra $((2 \rightarrow 1)$ thick, $(3 \rightarrow 2)$ thin) show very strong self-absorption at the velocity of ^{13}CO $(2 \rightarrow 1)$ (dotted). Note the more intensive $(3 \rightarrow 2)$ emission! The IRAS source responsible for the outflow is at (99.983,4.170).

The bright rims are also very prominent on IRAS Sky Brightness Images in each of the four wavelength bands (cf. Fig. 3 for 100 μm).

An LVG formalism will be used to analyse the molecular data in order to get fundamental physical parameters like density, temperature and excitation conditions for the bright-rimmed dust clouds.

Besides, we want to apply a multicorrelation analysis to the millimeter and IRAS maps in all four wavelengths disclosing quantitative aspects.

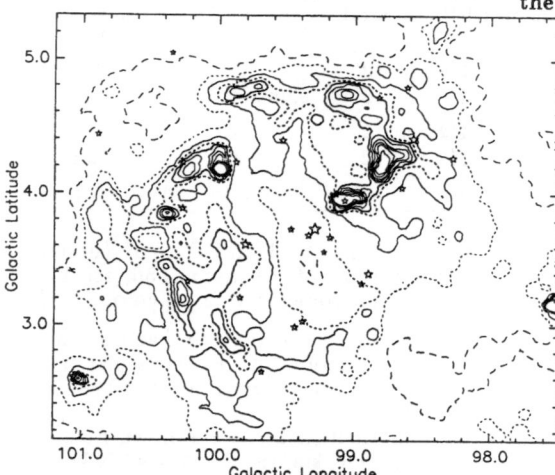

Galactic Latitude

Galactic Longitude

Fig. 3: 100 μm brightness (IRAS coadded map) of the same region as in Fig. 1

Contour levels are drawn from 70 (longdashed) to 250 MJy/sr, the step is 20 MJy/sr.

Stars indicate the positions of O and B stars: The brightest near the centre is HD 206267.

Acknowledgement: H.W. gratefully acknowledges the financial support from the European Economic Community under contract number ST2-0305.

References

Duvert, G., J. Cernicharo, R. Bachiller, and J. Gómez-González: Astron. Astrophys. **233**, 190 (1990)
Pottasch, S.: Bull. Astron. Inst. Netherlands **13**, 77 (1956)
Sugitani, K., Y. Fukui, A. Mizuno, and N. Ohashi: Astrophys. J. **342**, L87 (1989)

GRAVITATIONAL VIRIALIZATION OF MOLECULAR CLOUD FRAGMENTS

Anthony WHITWORTH
Department of Physics,
University of Wales College of Cardiff,
PO Box 913, Cardiff CF1 3TH,
Wales, UK.

ABSTRACT. We argue that if gravitational interactions play an important role in the amplification and transfer of turbulent energy between sub-clouds of different scales in a GMC, then the Second Law of Thermodynamics requires that the transfer occur from smaller to larger scales, and hence that the seed-energy be injected on relatively small scales.

1. Introduction

Scalo and Pumphrey (1982) have proposed that when an interstellar cloud contracts and fragments into sub-clouds, collisions between the sub-clouds are on average not very dissipative. A large fraction of the gravitational energy released by the contracting parent-cloud is channelled initially into randomized bulk kinetic energy of the sub-clouds. The parent-cloud is effectively virialized, in the sense that its subsequent contraction proceeds quasistatically on a time-scale which is much longer than a freefall time.

Scalo (1987) has termed this process gravitational virialization. He suggests that it may play an important role in the evolution of GMCs, as a means of generating turbulent kinetic energy and transferring this energy from one scale to another.

2. Assumptions

We assume — but do not unequivocally believe — (i) that Larson's relations ($R \propto M^\alpha$, $\alpha \sim 0.5$; $\Delta v \propto M^\beta$, $\beta \sim 0.25$) are real, rather than just an observational selection effect; (ii) that all clouds and sub-clouds subscribe to these relations; (iii) that all clouds and sub-clouds are virialized ($\alpha = 1 - 2\beta$); (iv) that all clouds and

515

sub-clouds belong to an homologous sequence; and (v) that there is an evolutionary connection between clouds and sub-clouds of different scales.

3. The Second Law of Thermodynamics

The 2nd Law of Thermodynamics requires evolution to proceed in that sense in which heat (here in the form of cooling radiation) flows out of the system (which is effectively at a temperature 10–100K) into the surroundings (which are effectively at a temperature 3–10K). Hence there must be a net release of gravitational potential energy, with half of the gravitational energy released going towards maintaining virialization (*i.e.* feeding the turbulent velocity dispersion), and the other half escaping as radiation.

4. Up → Down Gravitational Virialization

Consider a static, cold, spherical parent-cloud of mass M and radius R_o, which contracts and breaks up into N sub-clouds, each of mass $m = M/N$. If a fraction F of the gravitational potential energy released by the contracting parent-cloud is channelled into random bulk motions of the sub-clouds, whilst the rest is radiated away, the final virialized radius of the parent-cloud is

$$R = R_o \frac{(2F - 1)}{2F} \tag{1}$$

Gravitational virialization therefore requires $F > 1/2$.

Now consider the sub-clouds. Any gravitational potential energy they release in contracting to a radius $r_o = RN^{-1/3}$ has been used in virializing the parent-cloud. Consequently, if the sub-clouds are to be gravitationally virialized in their turn, they must contract still further to a radius

$$r = r_o \frac{(2f - 1)}{2f} = RN^{-1/3} \frac{(2f - 1)}{2f} \tag{2}$$

Here f is the fraction of the self gravitational potential energy of a sub-cloud which is channelled into random bulk motions of the sub-sub-clouds. Again gravitational virialization requires $f > 1/2$.

Larson's relations give $r = RN^{-\alpha}$, so equation (2) reduces to

$$N = \left\{ \frac{2f}{(2f - 1)} \right\}^{3/(3\alpha - 1)} \tag{3}$$

The following table gives values of N for representative combinations of f and α.

N	$f = 0.9$	$f = 0.8$	$f = 0.7$	$f = 0.6$
$\alpha = 0.4$	1.9×10^5	2.5×10^6	1.4×10^8	4.7×10^{11}
$\alpha = 0.5$	130	362	1838	46656
$\alpha = 0.6$	21	40	110	828

We conclude that N has to be very large (**unless** f is very high, *i.e.* gravitational virialization is nearly 100% efficient, **and** α is very much larger than inferred from observation). The parent-cloud must shatter into a myriad of much smaller sub-clouds. This is in stark contrast to the conventional picture of hierarchical fragmentation (*e.g.* Zinnecker 1984) in which at each level of the hierarchy a parent-cloud breaks into just a few sub-clouds, and the process then repeats itself in a more-or-less self-similar manner over a large spatial range.

5. Down → Up Gravitational Virialization

The alternative to the scenario outlined above is obtained if one turns hierarchical fragmentation on its head. Hierarchical fragmentation is in any case seriously flawed (*e.g.* Whitworth 1980), and without observational basis. If we turn it on its head, we have a scenario in which the dominant injection of turbulent energy is on small scales, and this energy is then transferred to larger scales and simultaneously amplified by gravitational virialization.

To establish that this will work, we consider an ensemble of N independent, virialized sub-clouds, each of mass m and radius r. If these sub-clouds combine to form a single collective-cloud with mass $M = Nm$ and radius $R = rN^\alpha$, the net gravitational energy released is positive (and hence the process is permitted by the 2nd Law of Thermodynamics) provided

$$N^{(1-\alpha)} > 1 - N^{(1-3\alpha)/3} \qquad (4)$$

This condition is satisfied for any positive integral N as long as $\alpha < 1$. In other words, the gravitational potential energy released by the sub-clouds falling together is more than enough to puff them up individually to the lower density required for the collective-cloud by Larson's relations.

6. Conclusions

We conclude that gravitational virialization is only likely to work if energy is transferred in the opposite sense to Kolmogorov turbulence, *i.e.* down→up, from smaller scales to larger ones. Although this goes against the theoretical concept of hierarchical fragmentation, there does not appear to be any unequivocal observational evidence to rule out the possibility that this is what occurs in GMCs. Other arguments supporting the down→up transfer and amplification of turbulent energy in GMCs will be assembled in a future paper.

References

Scalo, J.M., 1987, in *Interstellar Processes*
    (eds. D.J. Hollenbach, H.A. Thronson, Jnr.; Reidel, Dordrecht), 349
Scalo, J.M., Pumphrey, W.A., 1982, *Astrophys. J.* **258**, L29
Whitworth, A.P., 1980, in *Giant Molecular Clouds in the Galaxy*
    (eds. P.M. Solomon, M.G. Edmunds; Pergamon, Oxford), 285
Zinnecker, H., 1984, *Mon. Not. roy. astron. Soc.* **210**, 43

TURBULENT AMPLIFICATION OF INTERSTELLAR MAGNETIC FIELDS

Anthony WHITWORTH
Department of Physics,
University of Wales College of Cardiff,
PO Box 913, Cardiff CF1 3TH,
Wales, UK.

ABSTRACT. We show that the radiative cooling properties of the interstellar gas lead naturally to a scaling law for the magnetic field of the form $B \propto n^{1/2}$, *if* the magnetic field is amplified by a vigorous injection of turbulent energy due to expanding HII regions, stellar winds, supernova explosions, stellar jets, *etc.*

1. How is the interstellar magnetic field amplified?

In regions of the interstellar medium which are supersonically turbulent, the magnetic field is amplified until there is approximate equipartition of energy between the turbulence and the magnetic field. The more vigorous the input of turbulent energy, the stronger is the resulting magnetic field.

2. How is amplification limited?

This amplification process is limited by dissipation of the turbulence, either by ion-neutral friction or by radiation from shocks. As long as the processes injecting turbulent energy into the interstellar medium are coherent over time-scales longer than the ion-neutral coupling time-scale, radiation from shocks dominates the dissipation.

Once the field has been amplified to equipartition with the turbulence, disturbances propagate through the medium at speeds

$$v \sim v_A = (4\pi\rho)^{-1/2} B \tag{1}$$

where v_A is the Alfvén speed. v_A will therefore be the typical speed of dissipative shocks.

Because the radiative cooling efficiency of the interstellar gas increases steeply by several orders of magnitude above a critical temperature $T_C \sim 3000$–10000K, shocks with speeds

$$v > v_C \geq (16kT_C/3\overline{m})^{1/2} \tag{2}$$

(where $\overline{m}$ is the mean gas particle mass) are much more dissipative than those at lower speeds.

Turbulent amplification of the magnetic field is therefore relatively efficient as long as $v_A < v_C$, but much less efficient once $v_A > v_C$. Over a wide range of turbulent energy input rates the magnetic field will be amplified until v_A approaches v_C. The corresponding limit on the component of the magnetic field along the line of sight is

$$B_\| \sim (64\pi kT_C\rho/9\overline{m})^{1/2} \sim C\left(n/\mathrm{cm}^{-3}\right)^{1/2} \tag{3}$$

Here $n = \rho/(2.4 \times 10^{-24}\mathrm{gm})$ is the number density of hydrogen nuclei in all forms in a gas with population I composition.

For predominantly atomic gas with $\overline{m} \sim 2.1 \times 10^{-24}\mathrm{gm}$ and $T_C \sim 10^4$K (due to the onset of bound-bound and bound-free cooling by H°), $C_{\mathrm{H^\circ}} \sim 5.8\mu G$. For predominantly molecular gas with $\overline{m} \sim 3.5 \times 10^{-24}\mathrm{gm}$ and $T_C \sim 3000$K (due to the onset of dissociative cooling by H_2), $C_{\mathrm{H_2}} \sim 2.5\mu G$.

3. Comparison with observations

In the Figure, equation (3) is compared with observation. The solid diagonal line represents equation (3) with $C = C_{\mathrm{H_2}}$ as appropriate for predominantly molecular gas; whilst the broken diagonal line represents equation (3) with $C = C_{\mathrm{H^\circ}}$ as appropriate for predominantly atomic gas. The rest of the Figure is copied from Fiebig and Güsten (1989) — their Figure 4 — and displays the run of magnetic field strength (parallel to the line of sight) with density. There is a remarkably tight correlation, in the sense that observed values of $B_\|$ are concentrated just below the limit given by equation (3). This correlation provides strong support for the role of supersonic turbulence in amplifying the interstellar magnetic field.

The fact that very few values of $B_\|$ are measured well below the limit may be a selection effect, in the sense that larger magnetic fields are more readily detected. This is certainly the case for the H_2O maser observations of high density gas, as emphasized by Fiebig and Güsten (1989). However, it may also be an indication of how vigorously and frequently turbulent energy is injected into the interstellar medium on the scales observed.

4. Discussion

We note that the mechanism discussed here to explain the observed correlation of magnetic field and density is unrelated to that discussed by Mouschovias (1987). Mouschovias' discussion entails the gas being in gravitationally bound, magnetically supported configurations. As stressed by Fiebig and Güsten, these conditions do not obtain in the high density H_2O maser sources. They do not necessarily obtain in the lower density sources either.

These H_2O masers are believed to arise in strong MHD shocks (Kylafis and Norman 1987), which are precisely the conditions produced by vigorously stimulated turbulence.

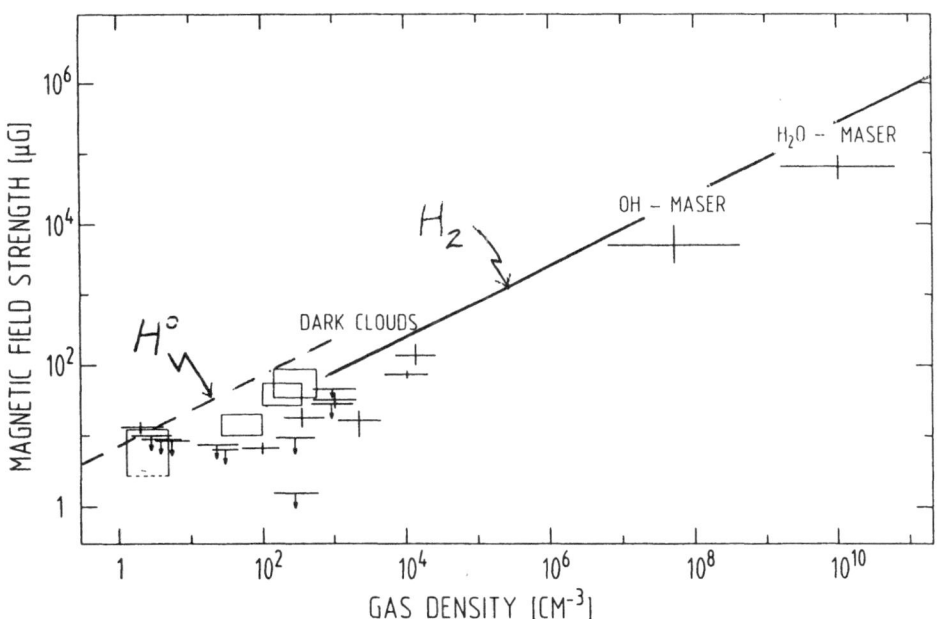

Figure. The variation of $B_\parallel$ with n. The diagonal lines represent equation (3) for predominantly molecular gas (full line) and predominantly atomic gas (broken line). The rest of the figure is copied from Fiebig and Güsten (1989) and represents a variety of observational estimates of $B_\parallel$ and n.

5. Conclusions

We conclude that the observed correlation of magnetic field strength and density may be evidence for a high level of interstellar turbulence in the regions observed; and for a high efficiency of magnetic field amplification resulting therefrom.

References

Fiebig, D., Güsten, R., 1989, *Astron. Astrophys.* **214**, 333

Kylafis, N.D., Norman, C., 1987, *Astrophys. J.* **323**, 346

Mouschovias, T.Ch., 1987, in *Physical Processes in Interstellar Clouds*
    (eds. G.E. Morfill, M. Scholer; Reidel, Dordrecht), 453

CLOUD-CLOUD COLLISIONS AND FRAGMENTATION

Anthony WHITWORTH and Helen PONGRACIC
Department of Physics,
University of Wales College of Cardiff,
PO Box 913, Cardiff CF1 3TH,
Wales, UK.

ABSTRACT. Supersonic head-on collisions between quiescent clouds produce flattened sheets of shocked gas. We derive the condition which the cooling law must satisfy if this sheet is to fragment into protostellar condensations (*i.e.* gravitationally unstable lumps). If this condition is not satisfied, colliding clouds are likely to be disrupted and dispersed. We show that under the conditions obtaining in GMCs, most cloud-cloud collisions probably do not result in fragmentation.

1. Virial equilibrium

Consider first a single quiescent cloud of mass M_0, dimension L_0, density of hydrogen nuclei in all forms n_0 ($\equiv n_{HI} + 2n_{H_2} + \cdots$) and sound speed a_0. Virial equilibrium requires

$$a_0 \sim (GM_0/L_0)^{1/2}. \tag{1}$$

If m is the mass associated with one hydrogen nucleus ($m \simeq 2.4 \times 10^{-24}$ gm for population I composition), then

$$M_0 \sim L_0^3 n_0 m; \tag{2}$$

$$a_0 \sim L_0 (Gn_0m)^{1/2}. \tag{3}$$

2. Thermal equilibrium

We shall assume that the cloud is optically thin to heating and cooling radiation, so that the heating rate per unit volume can be approximated by

$$\Gamma \sim \Gamma_r (n/n_r), \tag{4}$$

523

and — at least over a limited range — the cooling rate per unit volume can be approximated by

$$\Lambda \sim \Gamma_r \left(n/n_r\right)^2 \left(a/a_r\right)^\alpha .\tag{5}$$

n_r and a_r are simply reference values for the physical parameters n and a. The constant Γ_r is the same for both Γ and Λ because we want the reference state (n_r, a_r) to be a state of thermal equilibrium. $\Lambda \propto a^\alpha$ is roughly equivalent to $\Lambda \propto T^{\alpha/2}$. Typically $\alpha \sim 3$.

Equating equations (4) and (5) gives the thermal equilibrium condition:

$$(a/a_r) \sim (n/n_r)^{-1/\alpha} .\tag{6}$$

3. Reference values for physical parameters.

For the purposes of illustration we adopt $n_r = 100\,\mathrm{cm}^{-3}$ and $a_r = 0.5\,\mathrm{kms}^{-1}$. Equations (2) and (3) then give $L_r \sim 4$ pc and $M_r \sim 200 M_\odot$.

Combining equations (2), (3) and (6), we find that quiescent clouds (i.e. clouds in virial and thermal equilibrium) have,

$$(n_0/n_r) \sim (M_0/M_r)^{-2\alpha/(6+\alpha)} ;\tag{7}$$

$$(a_0/a_r) \sim (M_0/M_r)^{2/(6+\alpha)} ;\tag{8}$$

$$(L_0/L_r) \sim (M_0/M_r)^{(2+\alpha)/(6+\alpha)} .\tag{9}$$

In other words, a more massive cloud has to be hotter and more diffuse if it is to be in virial *and* thermal equilibrium.

Coincidentally (since we are here assuming only thermal pressure support), equations (7) to (9) with $\alpha \sim 3$ are compatible with Larson's relations, *viz.* $n \propto M^{-6/9}$, $a \propto M^{2/9}$ and $L \propto M^{5/9}$.

4. General cooling time-scale.

We shall adopt $\Gamma_r = 5 \times 10^{-27}$ erg cm^{-3} s^{-1}. This corresponds to a primary ionization rate of $\zeta \sim 10^{-17}$ s^{-1}. The cooling time-scale is then given by

$$t^{\mathrm{cool}} \sim \rho a^2/\Lambda \sim t_r^{\mathrm{cool}} \left(n/n_r\right)^{-1} \left(a/a_r\right)^{2-\alpha} ,\tag{10}$$

$$t_r^{\mathrm{cool}} \sim n_r m a_r^2/\Gamma_r \sim 4\,\mathrm{Myr} .\tag{11}$$

5. Collision and expansion time-scales.

Now consider two identical clouds involved in a head-on collision at relative speed $2v_0 = 2\mathcal{M}a_0$, where $\mathcal{M}$ is the Mach number. Assuming a strong shock ($\mathcal{M} >> 1$), we know that the density and sound-speed immediately following the shock are $n_i \sim 4n_0$ and $a_i \sim \mathcal{M}a_0 = v_0$. It follows that the collision time-scale and the time-scale on which the flattened sheet expands sideways in the absence of post-shock cooling are roughly equal:

$$t^{\text{coll}} \sim t^{\text{exp}} \sim L_0/\mathcal{M}a_0 \sim t_r^{\text{exp}}\mathcal{M}^{-1}\left(M_0/M_r\right)^{\alpha/(6+\alpha)}; \tag{12}$$

$$t_r^{\text{exp}} \sim L_r/a_r \sim 8\,\text{Myr}. \tag{13}$$

6. Post-shock cooling time-scale.

Since we expect $\alpha < 4$, and since the post-shock cooling regime will be approximately isobaric, cooling will be slowest at high temperatures and the cooling time-scale should be evaluated for the immediate post-shock density and sound-speed, *viz.*

$$t_i^{\text{cool}} \sim t_r^{\text{cool}}\left(n_i/n_r\right)^{-1}\left(a_i/a_r\right)^{(2-\alpha)} \sim t_r^{\text{cool}}\mathcal{M}^{(2-\alpha)}\left(M_0/M_r\right)^{4/(6+\alpha)}. \tag{14}$$

7. Fragmentation condition.

The flattened sheet can only fragment if it does not expand sideways significantly before it cools, *i.e.* if $t_i^{\text{cool}} << t^{\text{exp}}$ or

$$\mathcal{M}^{(\alpha-3)}\left(M_0/M_r\right)^{(\alpha-4)/(6+\alpha)} >> G^{1/2}\left(n_r m\right)^{3/2} a_r^2\, \Gamma_r^{-1} \tag{15}$$

Putting $\alpha = 3 + \epsilon$ ($\epsilon << 1$), and substituting for the reference parameters, this reduces to

$$\mathcal{M}^\epsilon\left(M_0/M_r\right)^{-1/9} >> 0.5. \tag{16}$$

Since the Mach number $\mathcal{M}$ is unlikely to exceed 10, this inequality can only be satisfied if the clouds are very small and dense $M_0 << M_r$.

8. Conclusion.

Unless the interstellar gas can cool much faster than we have assumed, the majority of cloud-cloud collisions result in disruption and dispersal of the clouds involved. Cloud coalescence is unlikely. Specifically, efficient fragmentation requires that the cooling law of equation (5) has $\alpha \geq 4$ and/or $\Gamma_r \geq 10^{-26}\text{ergcm}^{-3}\text{s}^{-1}$ (corresponding to $\zeta \geq 2. \times 10^{-17}\text{s}^{-1}$).

A BINARY STAR FORMATION MECHANISM
THROUGH THE FRAGMENTATION
OF PROLATE DENSE CORES
ROTATING END OVER END

HANS ZINNECKER

Institut für Astronomie und Astrophysik, Univ. Würzburg, Germany

Abstract. I propose and briefly elaborate on a major new mechanism for the formation of wide, low-mass binary stars : the fragmentation of a collapsing, initially elongated dense molecular core rotating end over end. This initial structure will develop into two independent gravitationally bound stellar condensations orbiting each other in a rather eccentric orbit.

1. Numerical calculations of binary star formation

The problems of various binary star formation processes have recently been critically reviewed by Pringle (1990, see also Zinnecker 1984). Here I discuss a new process, the fragmentation of slowly tumbling elongated fragments, first mentioned in my review on PMS binaries (Zinnecker 1989). 2D numerical collapse calculations of elongated fragments have been performed by Larson (1972), later by Bastien (1983) and very recently by Rouleau and Bastien (1990); the latter authors also suggest that fragmentation of elongated clouds is an important way of forming binary stars. However, none of these calculations considered or specified the rotational motion of the cylindrical fragments. It is the purpose of this contribution to point out the crucial role of "end over end" rotation in the fragmentation of elongated clouds into binary stars. With rotation of that kind taken into account, the fragmentation problem becomes truly 3D even if magnetic fields are ignored. Such 3D hydrodynamical collapse and fragmentation models of initially prolate, slowly tumbling fragments have never been calculated (except in the context of protostellar fission), yet it may be suspected that it is precisely these initial conditions which could lead to the duplicity of many young low-mass stars.

2. Observations of elongated fragments

The current idea occurred after some new observations of the shapes of dense cores in molecular clouds by Myers et al. (1990) became available. These observations show that most molecular cores, previously studied by Myers and collaborators, actually are not round but are elongated with an aspect ratio of 2 : 1 (perhaps 3 : 1 when deprojected). It is emphasized that elongated structures are seen in the half-power contour maps of at least two optically thin molecular tracers ($C^{18}O$ and CS) and also that these elongated structures cannot be due to rotational flattening of the cores along the rotation axis, because fast rotation (i.e. a substantial velocity gradient along the elongation of the cores) is simply not observed, as would be required if the cores were edge-on rotating disks. Therefore it is likely that the cores have a prolate, cigar-shaped structure. Such a geometry is also favored by the fact that many of these elongated cores are found as subcondensations embedded in larger filamentary structures of the parent cloud (Myers, these Proc.; Schneider and Elmegreen 1979). A nice example is TMC-1C (Fiebig and Güsten, priv. commun.) for which the observed position-velocity diagram suggests slow end-over-end rotation. In summary, typical parameters for the observed subcondensations are : length $\sim$ 0.2 pc, diameter $\sim$ 0.1 pc, density $\sim 10^4 - 10^5\,\mathrm{cm}^{-3}$, mass $\sim$ few $M_\odot$, angular velocity $\sim 3\cdot10^{-14}\,\mathrm{sec}^{-1}$, specific angular momentum = rot. vel. $\times$ length/6 $\sim 10^{21}\mathrm{cm}^2\,\mathrm{sec}^{-1}$.

3. Evolution of protobinary systems

Fig. 1 sketches prolate subcondensations in a filamentary cloud and the subsequent evolution of a subcondensation into a binary system. Even without detailed 3D calculations (still to be performed), some essential features of the evolution of a rotating, gravitationally unstable prolate clump may be anticipated : first the moderately prolate clump will collapse to a more prolate, thin bar which will subsequently break up into two (or perhaps in some cases three) fragments. Because of the non-axisymmetric bar-like configuration, specific angular momentum will be transported outward and the material trailing beyond the dumbbell-shaped, protobinary structure will exert some gravitational torque on the protobinary components, slowing down their orbital motion and shrinking their orbit.

Initially the components of the protobinary system will be on nearly radial orbits, since the components will fall towards each other due to their mutual gravitational attraction while centrifugal forces will initially be dynamically negligible at large separations (10000 AU). However, the initial angular momentum will prevent the components from merging into a single object, i.e. at smaller separations of the components (1000 AU) the centrifugal forces will grow. Meanwhile circumstellar disks of small extent (10-100 AU) may have formed around each component. When the components are getting closer, their disks are likely to interact either directly or tidally, and rapid outer disk disruption and inner disk accretion onto the star may be induced. Drag forces will make the orbit somewhat more circular but a certain non-zero eccentricity will survive after dissipation ceases. The end result is a gravitationally bound pair of T Tauri stars with component separation of order 100 - 1000 AU, with a rather eccentric orbit, and with little circumstellar disk material but possibly with remnant circumbinary debris of gas and dust (i.e. a pair of "naked" T Tauri stars, cf. Walter et al. 1988).

Observations indeed support one of the key predictions of this model : according to Duquennoy and Mayor (1989) the wide binaries in a complete sample of 210 F and G stars exhibit an eccentricity distribution of $f(e) = 2e$, i.e. most systems have substantial eccentricities, as predicted in the present scenario. Furthermore,

528

an example of a protobinary object formed from an elongated cloud might be IRAS 16293-2422 with an observed projected component separation of about 800 AU (Wootten 1989).

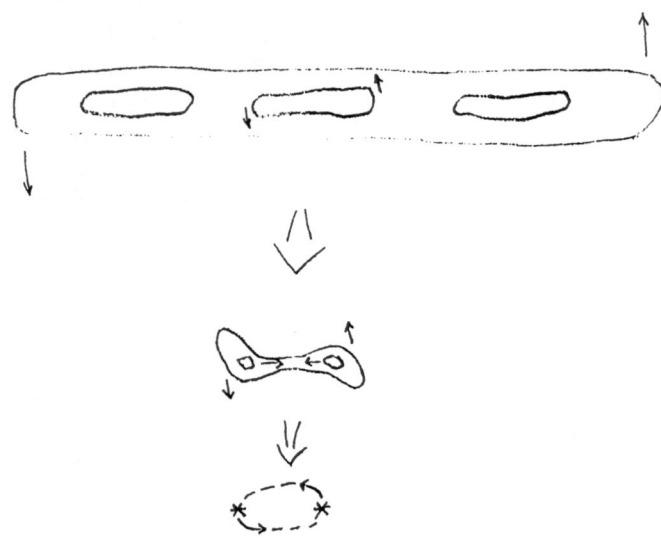

Fig. 1. Schematic evolution of a prolate subcondensation in a filament into a protobinary system. Here the angular momentum vector is perpendicular to the plane of the sky. The velocity gradient along the filament is $\sim 1 \, \text{km s}^{-1} \, \text{pc}^{-1}$. Arrows indicate the direction of motion. The initial length of the subcondensations is $\sim 0.2 \, \text{pc}$, the final major axis of the binary system is between 100 and 1000 AU.

References

Bastien, P. : 1983, *Astron. Astrophys.* **119**, 109
Duquennoy, A. and Mayor, M. : 1989, *Proc. XI European Regional Astronomy Meeting of the IAU*, ed. M. Vasquez, Cambridge Univ. Press
Larson, R.B. : 1972, *Mon. Not. Roy. Astron. Soc.* **156**, 437
Myers, P., Fuller, G., Goodman, A. and Benson, J. : 1990, submitted
Pringle, J. : 1990, in *The physics of star formation and early stellar evolution*, Nato ASI Crete (1990), eds. N. Kylafis and C. Lada (in press)
Rouleau, F. and Bastien, P. : 1990, *Astrophys. J.* **355**, 172
Schneider, S. and Elmegreen, B.G. : 1979, *Astrophys. J. Suppl.* **41**, 87
Walter et al. : 1988, *Astron. J.* **96**, 297
Wootten, A. : 1989, *Astrophys. J.* **337**, 858
Zinnecker, H. : 1984, *Astr. Sp. Sci.* **99**, 41
Zinnecker, H. : 1989, in *Formation of low-mass stars and pre-Main Sequence objects*, Proc. ESO Workshop, ed. B. Reipurth, p. 447

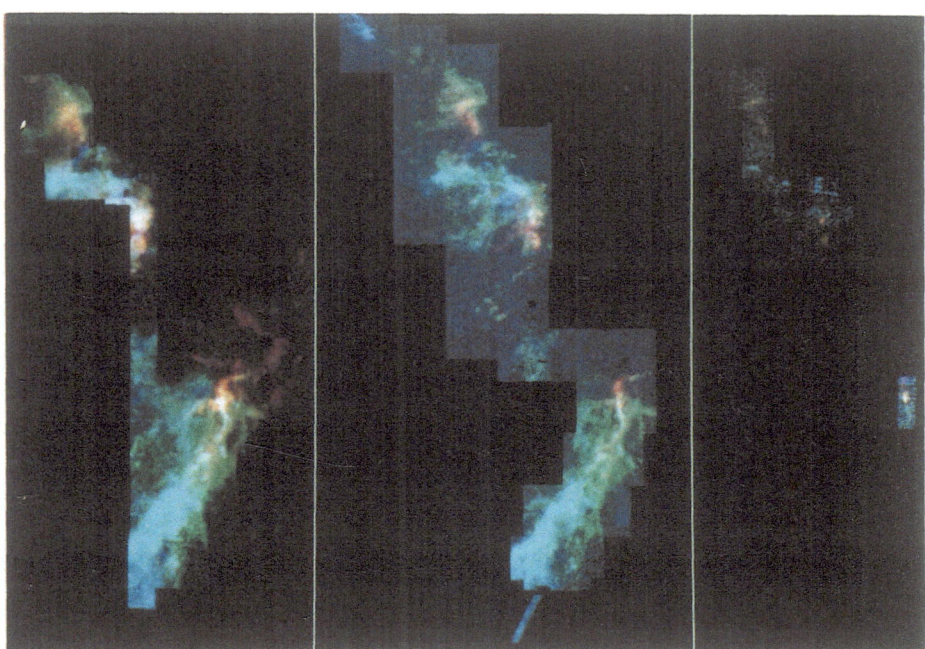

Figure 1: Three color images of the Orion region showing the major molecular clouds in ^{12}CO (left), ^{13}CO (middle), and CS J=2-1 (right). Each image shows the emission integrated from $V_{LSR} = -4$ km s^{-1} to $V_{LSR} = 16$ km s^{-1}, color coded with velocity so that blue corresponds to the velocity interval (-4,5), green (5,9) and red (9,16). The reflection nebulae NGC 2071 and NGC 2068 are located in the main concentration near the top of the figure, NGC 2023 and NGC 2024 is in the center, and Orion A (M42) is the region near the top of the comet-shaped L1641 region at the bottom.

530

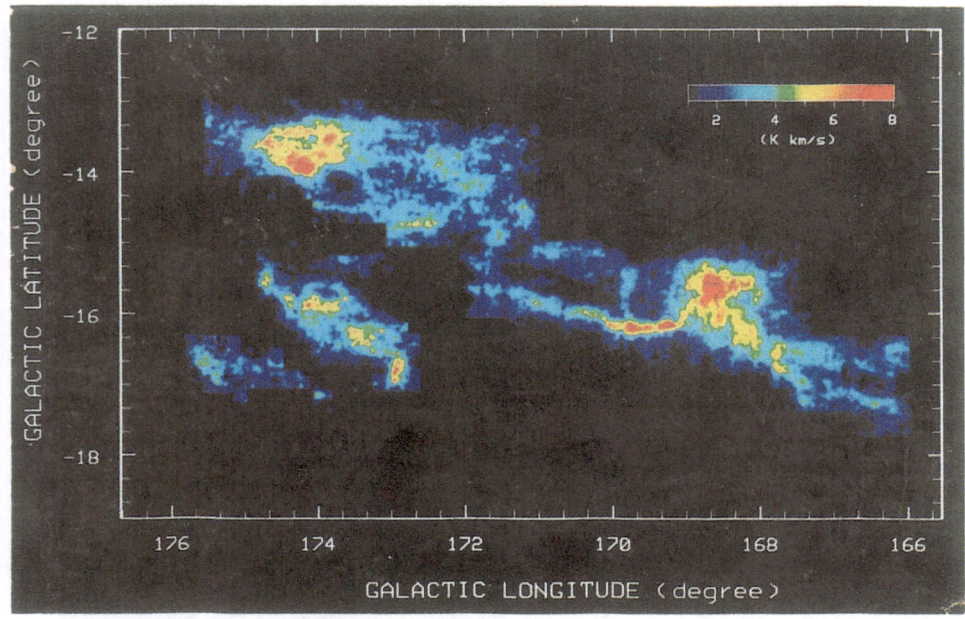

Figure 1   False-color image of total intensity in ^{13}CO (J=1-0) in the Taurus region.

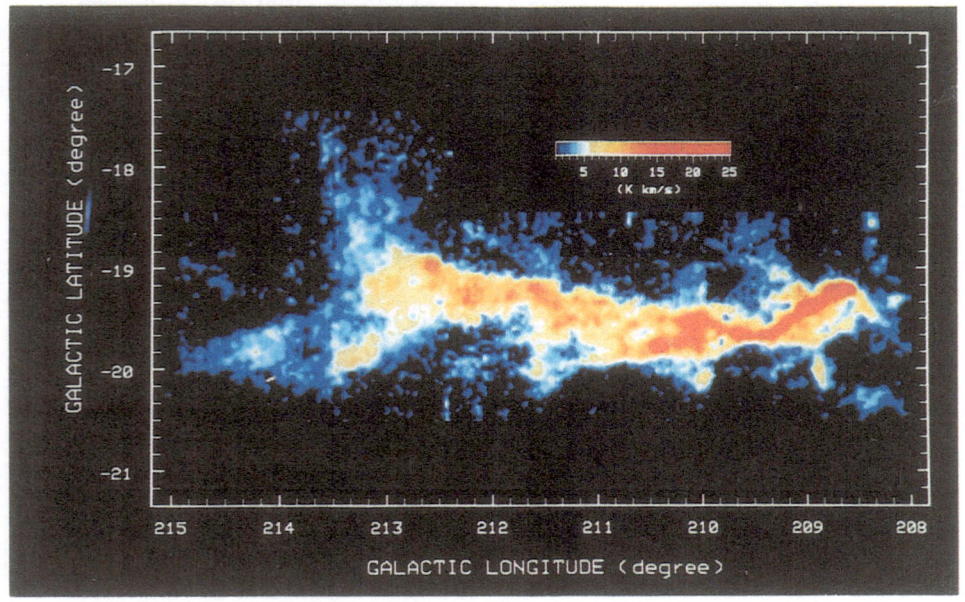

Figure 5   False-color image of the ^{13}CO (J=1-0) total intensity for the Orion south giant molecular cloud, involving the L1641 cloud.

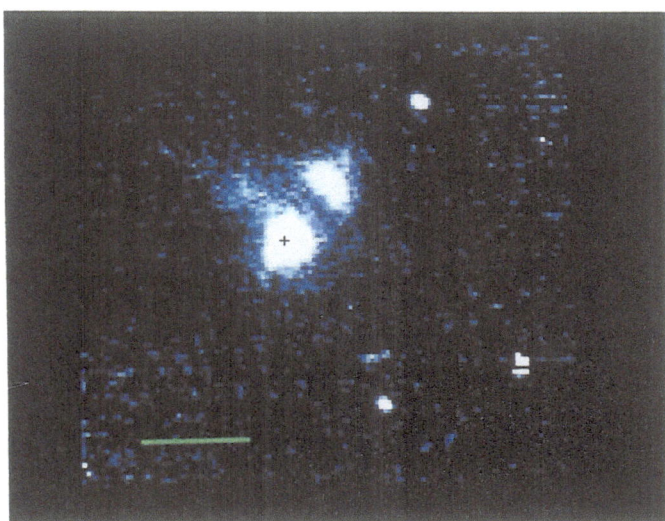

Fig. 1. A 2.2 μm image of the young embedded source TMR-1. The cross marks the infrared point source position. The reference line corresponds to 20″ (2800 AU). We interpret the spindle of emission extending to the NW to be a stellar wind cavity. Perpendicular to the direction of the spindle is a deep absorption feature. Two unrelated point sources lie to the NW and SW.

Fig. 2. A near infrared color composite image of the embedded source L1681B. Blue, green, and red correspond to 1.25, 1.65, and 2.2 μm, respectively. The reference line shows 20″ (3200 AU). There are three embedded sources separated by 3000 - 5000 AU. H_2O maser emission is detected coincident with the central and NE sources, thus demonstrating both sources are young. We interpret the nebulosity as scattered light that is tracing a poorly collimated stellar wind cavity.

532

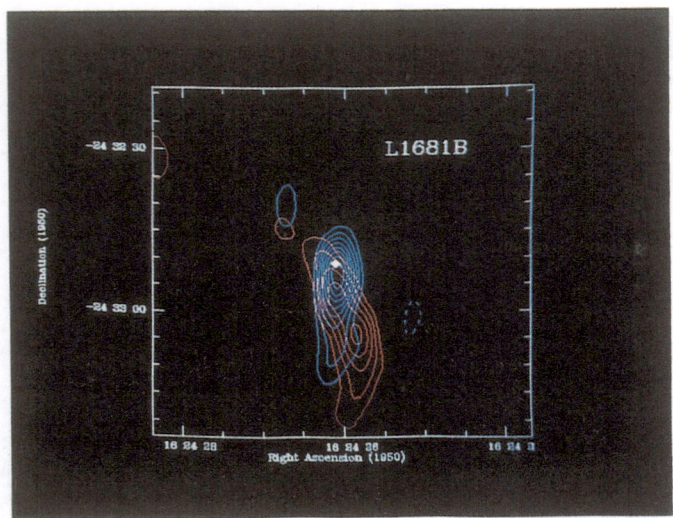

Fig. 3.   An interferometer map of the CO outflow seen towards the embedded source L1681B. The diamond marks the position of the infrared point source. Red-shifted and blue-shifted gas defines a stellar wind cavity extending south of the infrared source.

Index

Index

Index

Index

Many participants listening to official speeches by the Préfet, Edith Falgarone, and the city representative in the Salon of the Préfecture.

A friendly group at the Préfecture: Pat Thaddeus, Edith Falgarone and Phil Solomon above a picture of Alex Dalgarno and other astronomers applauding officials.

Top: Part of the organizing team ready to welcome participants: Alain Castets, Laurent Pagani and Claude Vianey-Liaud. Bottom: The Local Organizing Committee enjoying a "raclette" in Chamrousse: Elizabeth Palleau, Gisela Matoso, Claude Vianey-Liaud and Françoise Bouillet entertaining Laurent Pagani.

The youngest participants: Jean Terebey - van Buren accompanied by his father Dave (top). Too late for Elisabeth Keene - Phillips leaving the party with her parents.

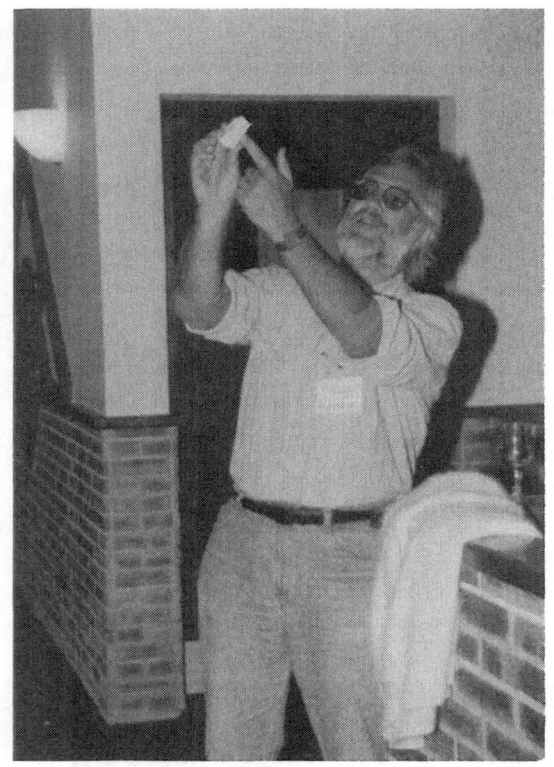

The many uses of a TAG (public transportation company in Grenoble) ticket by Don Cox. Do you remember what he said?

Who are they? Irina Val'tts (top), Vladimir Burdyuzha (forefront) and Gueorgui Rudnit-skii from Moscow.

At the beginning of the party at Chamrousse. René Laureijs and Patrick Boissé (right).

Scientific discussions days and nights. Night side: Bruce Elmegreen pointing his hands at Alexander Dudorov. Day time: Gilles Joncas and Don Cox looking at ... "H I channel maps".

Alain Omont talking over Yasuo Fukui and Al Wootten on the foreground. Top: Jean-Loup Puget with a flowered beard.

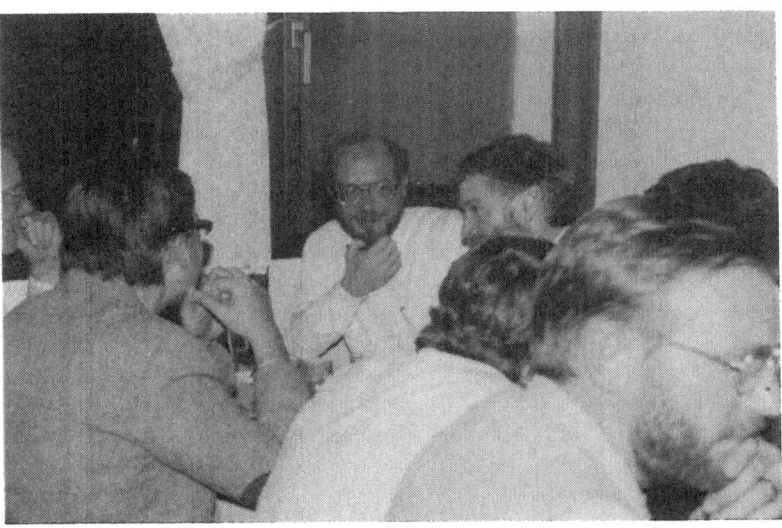

Ventaraman Radhakrishnan and Sachiko Okumura. Bottom, from left to right: Bruce Elmegreen, Richard Larson, John Black, M.R.W. Masheder, and Bruce Draine.

Happy participants: Phil Myers (top), Jacques Lépine, Lili Ephik, and Kazuyoshi Sunada.

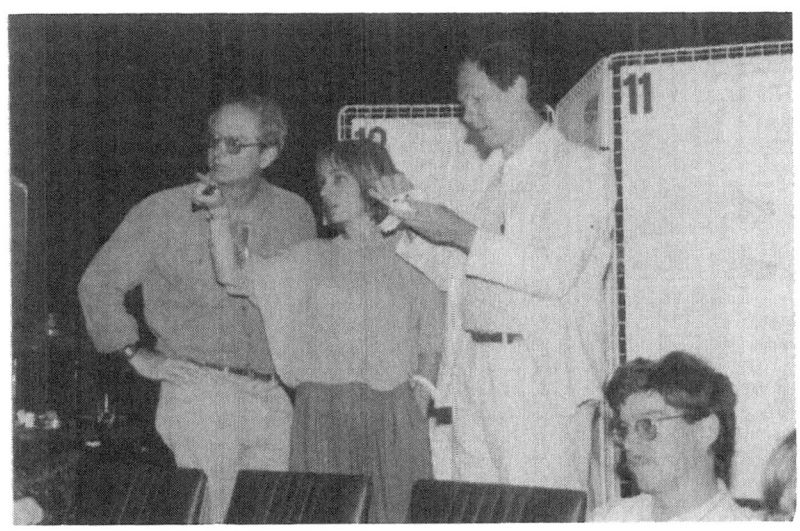

Carl Heiles, Edith Falgarone and John Bally exchanging views. Alan Glassgold talking to Pat Thaddeus.

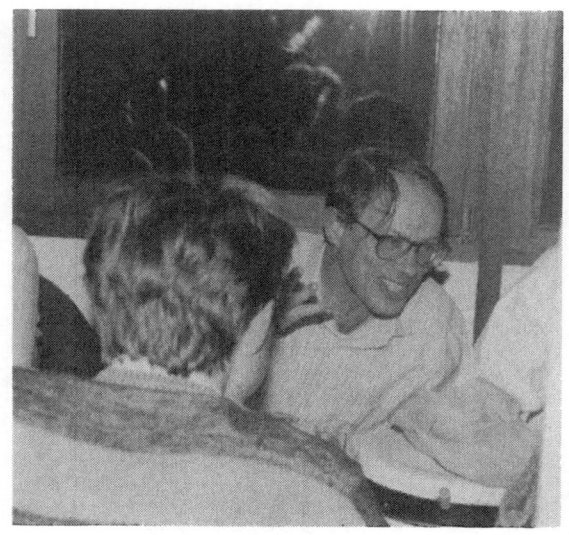

Further into the party: Malcolm Walmsley (top), Tom Troland with opened hand and Carl Heiles.